PROCESS SCALE BIOSEPARATIONS FOR THE BIOPHARMACEUTICAL INDUSTRY

BIOTECHNOLOGY AND BIOPROCESSING SERIES

1. Membrane Separations in Biotechnology, *edited by W. Courtney McGregor*
2. Commercial Production of Monoclonal Antibodies: A Guide for Scale-Up, *edited by Sally S. Seaver*
3. Handbook on Anaerobic Fermentations, *edited by Larry E. Erickson and Daniel Yee-Chak Fung*
4. Fermentation Process Development of Industrial Organisms, *edited by Justin O. Neway*
5. Yeast: Biotechnology and Biocatalysis, *edited by Hubert Verachtert and René De Mot*
6. Sensors in Bioprocess Control, *edited by John V. Twork and Alexander M. Yacynych*
7. Fundamentals of Protein Biotechnology, *edited by Stanley Stein*
8. Yeast Strain Selection, *edited by Chandra J. Panchal*
9. Separation Processes in Biotechnology, *edited by Juan A. Asenjo*
10. Large-Scale Mammalian Cell Culture Technology, *edited by Anthony S. Lubiniecki*
11. Extractive Bioconversions, *edited by Bo Mattiasson and Olle Holst*
12. Purification and Analysis of Recombinant Proteins, *edited by Ramnath Seetharam and Satish K. Sharma*
13. Drug Biotechnology Regulation: Scientific Basis and Practices, *edited by Yuan-yuan H. Chiu and John L. Gueriguian*
14. Protein Immobilization: Fundamentals and Applications, *edited by Richard F. Taylor*
15. Biosensor Principles and Applications, *edited by Loï'efc J. Blum and Pierre R. Coulet*
16. Industrial Application of Immobilized Biocatalysts, *edited by Atsuo Tanaka, Tetsuya Tosa, and Takeshi Kobayashi*
17. Insect Cell Culture Engineering, *edited by Mattheus F. A. Goosen, Andrew J. Daugulis, and Peter Faulkner*
18. Protein Purification Process Engineering, *edited by Roger G. Harrison*
19. Recombinant Microbes for Industrial and Agricultural Applications, *edited by Yoshikatsu Murooka and Tadayuki Imanaka*
20. Cell Adhesion: Fundamentals and Biotechnological Applications, *edited by Martin A. Hjortso and Joseph W. Roos*
21. Bioreactor System Design, *edited by Juan A. Asenjo and José C. Merchuk*
22. Gene Expression in Recombinant Microorganisms, *edited by Alan Smith*
23. Interfacial Phenomena and Bioproducts, *edited by John L. Brash and Peter W. Wojciechowski*
24. Metabolic Engineering, *edited by Sang Yup Lee and Eleftherios T. Papoutsakis*
25. Biopharmaceutical Process Validation, *edited by Gail Sofer and Dane W. Zabriskie*
26. Membrane Separations in Biotechnology: Second Edition, Revised and Expanded, *edited by William K. Wang*

27. Isolation and Purification of Proteins, *edited by Rajni Hatti-Kaul and Bo Mattiasson*
28. Biotransformation and Bioprocesses, *Mukesh Doble, Anil Kumar Kruthiventi, and Vilas Gajanan Gaikar*
29. Process Validation in Manufacturing of Biopharmaceuticals: Guidelines, Current Practices, and Industrial Case Studies, *edited by Anurag Singh Rathore and Gail Sofer*
30. Cell Culture Technology for Pharmaceutical and Cell-Based Therapies, *edited by Sadettin S. Ozturk and Wei-Shou Hu*
31. Process Scale Bioseparations for the Biopharmaceutical Industry, *edited by Abhinav A. Shukla, Mark R. Etzel, and Shishir Gadam*

PROCESS SCALE BIOSEPARATIONS FOR THE BIOPHARMACEUTICAL INDUSTRY

edited by

Abhinav A. Shukla
Amgen Inc.
Seattle, Washington, U.S.A.

Mark R. Etzel
University of Wisconsin-Madison
Madison, Wisconsin, U.S.A.

Shishir Gadam
Merck & Co., Inc.
West Point, Pennsylvania, U.S.A.

Taylor & Francis
Taylor & Francis Group
Boca Raton London New York

CRC is an imprint of the Taylor & Francis Group,
an informa business

Published in 2007 by
CRC Press
Taylor & Francis Group
6000 Broken Sound Parkway NW, Suite 300
Boca Raton, FL 33487-2742

Printed in the United States of America on acid-free paper
10 9 8 7 6 5 4 3 2 1

International Standard Book Number-10: 1-57444-517-0 (Hardcover)
International Standard Book Number-13: 978-1-57444-517-6 (Hardcover)
Library of Congress Card Number 2006002480

Library of Congress Cataloging-in-Publication Data

Process scale bioseparations for the biopharmaceutical industry /
Abhinav Shukla, Mark Etzel, and Shishir Gadam, editors.
p. ; cm. -- (Biotechnology and bioprocessing series ; 31)
Includes bibliographical references and index.
ISBN-13: 978-1-57444-517-6 (alk. paper)
ISBN-10: 1-57444-517-0 (alk. paper)
1. Proteins--Purification. 2. Cell culture. 3. Pharmaceutical biotechnology. 4. Chromatographic analysis. 5. Fermentation.
[DNLM: 1. Proteins--isolation & purification. 2. Cell Culture Techniques. 3. Chromatography--methods. 4. Fermentation. QU 55 P96406 2006]
I. Shukla, Abhinav. II. Etzel, Mark. III. Gadam, Shishir. IV. Series.

QP551.P693 2006
612.3'98--dc22 2006002480

informa
Taylor & Francis Group
is the Academic Division of Informa plc.

Visit the Taylor & Francis Web site at
http://www.taylorandfrancis.com

and the CRC Press Web site at
http://www.crcpress.com

Preface

Rapid growth in the biopharmaceutical industry in recent years — both in the value and variety of novel products —- has turned this once fledgling industry into an important player in the global economy. Commercial success of these products hinges on the successful development and implementation of robust, reliable, and economical production processes. Increasingly within the biopharmaceutical industry, bioprocess development is seen as a key source of competitive advantage. This trend will continue as the industry further matures. Bioseparations (also often called downstream processing) refers to the wide variety and combinations of production processes that are employed to recover and purify biomolecules from biological sources. Given the diversity of biomolecules and the complex nature of their biochemical properties it is no wonder that the field of bioseparations has evolved into a rich and varied one.

However, most of the developments in the bioseparations field are chronicled in a rather vast range of scientific papers, patents, and conference presentations that can pose a bewildering array for a newcomer to process development or for established scientists and engineers who are seeking to learn an unfamiliar technique. Biological systems are inherently complex and very often cannot be defined precisely by mathematical models. Due to these reasons, bioprocess development is often simultaneously an art and a science. Additionally, biopharmaceutical manufacturing is a highly regulated activity and hence several regulatory considerations often enter into choices made during process development. This book first lays a foundation of basic concepts and fundamental principles that are essential for understanding each topic, and then provides a set of rules of thumb that are based on hands-on industrial experience with actual large-scale processes, hindsight learning from scale-up problems, and regulatory issues that arise during development and licensure. The book follows a concise and practical approach and is replete with tables, flow charts, and schematics that provide a perspective on how process development is carried out in the biopharmaceutical industry. In each chapter, the authors attempt to bring together scientific principles, practical considerations, and empirical approaches that are closely intertwined in this field to give the reader a perspective on how purification process development is actually carried out in the biopharmaceutical industry. In this sense, the book is a departure from previous texts in this area that provide either a largely theoretical perspective into the field or are a compilation of review papers of scientific developments in any given topic.

The book contains three broad areas of focus. First is a focus on downstream unit operations, their fundamental principles, and considerations for process development. This includes chapters on unit operations that are widely accepted in the bioprocessing world as well as on methodologies that could find wider acceptance in the years to come. Next comes a focus on some highly essential ancillary aspects of downstream process development including viral validation and in-process analytical methods. The final chapters in the book deal with downstream process development for various classes of biomolecules and the strategies adopted for their process.

Chapter 1 presents a broad review of the principles of harvest clarification technologies (centrifugation, depth filtration, and tangential flow filtration) along with a case study of harvesting a therapeutic protein product from high cell density fermentation broth. The comprehensive literature review within this chapter should prove to be a valuable road-map for practitioners to navigate this vastly studied area of downstream processing. Chapter 2 presents theoretical and experimental frameworks and a real-life case study for development of expanded bed adsorption as an alternative to the more conventional techniques presented in Chapter 1. Given the increasing interest in this integrative technology in the last decade and its potential to reduce overall cost of goods, we believe this chapter will be a valuable resource for many readers. Chapter 3 presents another novel technology (High Gradient Magnetic Fishing) that can allow integration of harvest-clarification with chromatographic capture and purification. Magnetic adsorbents have a very powerful and unique "hook" that permits one to "fish" them from crude cell culture and fermentation broths simply through the application of a magnetic field. Because of its potential for rapid processing of large volumes of cell harvest, the future industrial prospects for High Gradient Magnetic Fishing look bright. And that's no fish story.

Chapter 4 spells out the fundamental principles of protein refolding and provides the reader with experimental strategies to develop and optimize a refolding process. A variety of useful points to consider during development of large-scale protein refolding operations are provided throughout the chapter.

Chapter 5 is on bulk protein crystallization — a technique that is generating increasing interest for (1) early stage recovery, (2) generating ultra-high purity product in the polishing stages of a downstream process, (3) improving product stability and shelf life, and (4) providing new dosage formats for protein therapeutics. The chapter sets the stage for new researchers by describing the basic principles and key equations needed for design of experiments, illustrating methods of data analysis, and providing case studies of industrial practice.

Chromatographic separations are ubiquitous in the manufacture of biopharmaceuticals. For this reason, the next three chapters provide a tutorial, some hands-on advice, and a gaze into the crystal ball for process chromatography. Chapter 6 focuses on the different modes of chromatography. All the major modes of process chromatography are discussed: affinity, ion exchange, hydrophobic interaction, reversed phase, hydroxyapatite, immobilized metal affinity, thiophilic interaction, mixed mode, and size exclusion. Rules of thumb and heuristics based on large-scale experience are provided in each section to serve as a practical guide for the reader.

Chapter 7 presents practical considerations and methodologies for screening and selecting chromatographic resins for industrial separation processes. Of special mention are tools such as high throughput screening, retentate chromatography, and cumulative yield–impurity plots that greatly facilitate the task of stationary phase selection.

Chapter 8 on *a priori* prediction of chromatographic separations from protein structure data describes a technique that might radically alter the current paradigm of process development. While the preceding two chapters concentrated on currently applied empirical methods for defining chromatographic unit operations, this chapter describes some major strides toward the "holy grail" of researchers in being able to predict chromatographic performance by simulation alone. The chapter describes methodologies for predicting chromatographic parameters of proteins and presents an overview of recent advances made in extending these predictive techniques beyond the small molecule realm for which they have been employed so far.

The next three chapters deal with membrane-based unit operations. Chapter 9 presents simple mathematical models for predicting breakthrough curves in membrane chromatography systems, and explains how to use these models to analyze laboratory and large-scale data. This analysis is extended to the prediction of viral clearance in membrane chromatographic systems. Chapter 10 focuses on the design and implementation of an ultrafiltration step in an industrial scale process. Ultrafiltration/diafiltration is ubiquitous in downstream processing for concentration, buffer exchange, and final formulation. Special attention is paid to addressing engineering constraints faced during scale-up of ultrafiltration systems. Chapter 11 —- on virus filtration process design and implementation — provides an overview of viral filter selection, process design and optimization (in both normal and tangential flow mode), and a detailed overview of the actual operating procedures for-full scale implementation.

Recognizing the economic potential of transgenic sources for biopharmaceutical production, the book includes a chapter on recovery of proteins from emerging transgenic sources. Chapter 12 provides a comprehensive educational overview of the field of protein production and recovery from transgenic sources and also discusses the challenges that need to be addressed for transgenic sourced products to become a reality in the future.

The next several chapters in the book deal with additional efforts aimed at ensuring purity, efficacy, and safety in the development of downstream processes for biopharmaceutical manufacturing. Key amongst these are in-process analytical methods, that form the eyes and ears of downstream process developers. Chapter 13 deals with various quality and efficacy attributes of biopharmaceuticals and provides practical guidance on analytical methods that can be employed to assess them. Chapter 14 provides an excellent tutorial on the elements that need to be in place in a downstream process to ensure safety from viral contamination. The chapter provides an overview of the various potential means of viral introduction into the process stream, design of appropriate virus clearance studies, considerations in selecting appropriate model viruses, and design of scale-down models. Chapter 15 describes the latest trends in viral clearance, a regulatory perspective of this area, and an in-depth comparison of the various established and upcoming methods of achieving viral clearance.

The final set of chapters in the book focus on various classes of biomolecules and provide insight into their process scale purification. Monoclonal antibodies have emerged as one of the most important classes of biopharmaceuticals today and their downstream processing aspects are covered in three chapters. Chapter 16 provides a comprehensive introduction to this class of therapeutics and provides detailed practical guidance for developing Protein A chromatography as the key purification step in antibody downstream processing. Chapter 17 describes the development of polishing chromatographic steps for monoclonal antibody downstream processing. In addition to providing useful practical advice, the chapter also provides several useful process templates. Purification of large biomolecules such as gene therapy vectors present significant challenges during scale-up when conventional chromatographic resins are used due to their low binding capacities. Chapter 18 provides an industrial case-study for an approved biopharmaceutical (Remicade®) dealing with post licensure process changes and strategies employed for their regulatory approval.

Chapter 19 describes the purification of a bacterial polysaccharide vaccine. Investigation of an unexpected problem during scale-up of ultrafiltration for this molecule led to a troubleshooting investigation that led to a better process understanding.

Chapter 20 further highlights the advantages of convective transport in membrane chromatography in overcoming capacity limitations in conventional beaded chromatographic resins for larger biomolecules. The chapter also provides a comprehensive literature review for the purification of gene therapy vectors —- another emerging class of biopharmaceuticals.

We hope *Process Scale Bioseparations for the Biopharmaceutical Industry* will be a valuable text for the growing numbers of scientific staff involved in process development in the biopharmaceutical and biotechnology industries, academia, and government laboratories. In some ways, this book was motivated by the shared feeling among the editors and authors that there was need for a comprehensive tutorial text combining fundamental principles and empirical guidelines originating from large-scale experience in the bioseparations arena. This makes us confident that this will also be a timely book for graduate students and senior level undergraduates who are preparing for a career in bioprocessing. We believe this book will find a worldwide audience in the rapidly growing biopharmaceutical sector.

Abhinav A. Shukla
Mark R. Etzel
Shishir Gadam

Editors

Abhinav A. Shukla is Principal Scientist in Purification Process Development at Amgen Inc. in Seattle, Washington. His group is responsible for the development, characterization, validation, and transfer of downstream processes for both early- and late-stage biopharmaceuticals. Dr. Shukla has been instrumental in setting up Amgen's platform strategy for the purification of monoclonal antibody therapeutics. He has also started several technology initiatives at Amgen, and authored numerous publications and presentations in bioseparations. Prior to joining Amgen in 2000, Dr. Shukla held a similar role in process development at ICOS Corporation, Bothell, Washington dealing with downstream processing of bacterial and mammalian cell-culture-derived products. He received his Ph.D. in chemical engineering from Rensselaer Polytechnic Institute, Troy, New York and his B.S. and M.S. in biochemical engineering and biotechnology from the Indian Institute of Technology.

Mark R. Etzel is Professor of Chemical and Biological Engineering at the University of Wisconsin-Madison. Dr. Etzel has seventeen years of teaching, research, and consulting experience in biological separation processes including membrane adsorption and filtration, freeze drying and spray drying, ion exchange and affinity chromatography, and protein crystallization. Professor Etzel received a B.S. at Purdue University, and a Ph.D. at the University of California at Berkeley, both in the department of chemical engineering. Dr. Etzel worked in industry for six years before returning to academia.

Shishir Gadam is part of the Bioprocess Research and Development division at Merck and Co., Inc., West Point, Pennsylvania, and is responsible for clinical manufacturing of bulk vaccine and therapeutic proteins using Merck's state-of-the-art multiproduct biologics pilot plant. During his last nine years of stay at Merck he has contributed in various areas within biologics including process development, scale up, technology transfer to manufacturing, GMP manufacturing for clinical use, and process validation. Prior to joining Merck in 1997, Dr. Gadam spent three years at Millipore Corporation, Bedford, Massachusetts, developing new and novel membrane purification technologies. Dr. Gadam obtained his Ph.D. in chemical engineering from Rensselaer Polytechnic Institute, Troy, New York.

Contributors

Mahesh K. Bhalgat
Amgen Inc.
West Greenwich, Rhode Island

Glen Bolton
Purification Development
Wyeth
Andover, Massachusetts

Curt M. Breneman
Department of Chemistry and Chemical Biology
Rensselaer Polytechnic Institute
Troy, New York

Kurt Brorson
Division of Monoclonal Antibodies
CDER/FDA
Bethesda, Maryland

Cynthia Cowgill
Regulatory Affairs
Chiron Corporation
Emeryville, California

Steven M. Cramer
Department of Chemical and Biological Engineering
Rensselaer Polytechnic Institute
Troy, New York

Christopher Daniels
Bioprocess Research and Development
Merck Research Laboratories
Merck & Co., Inc.
West Point, Pennsylvania

Niklas Ebner
Institute for Technical Chemistry
Forschungszentrum Karlsruhe
Eggenstein-Leopoldshafen, Germany

Mark R. Etzel
Department of Chemical and Biological Engineering
University of Wisconsin
Madison, Wisconsin

Matthias Franzreb
Institute for Technical Chemistry
Forschungszentrum Karlsruhe
Eggenstein-Leopoldshafen, Germany

Shishir Gadam
Bioprocess Research and Development
Merck Research Laboratories
Merck & Co., Inc.
West Point, Pennsylvania

Pete Gagnon
Bio-Rad Laboratories
Hercules, California

Marshall G. Gayton
Bioprocess Research and Development
Merck Research Laboratories
Merck & Co., Inc.
West Point, Pennsylvania

Sanchayita Ghose
Purification Process Development
Amgen Inc.
Seattle, Washington

Brian Gierl
University of Pittsburgh School of Medicine
Pittsburgh, Pennsylvania

Xuejun Sean Han
Purification Process Development
Amgen Inc.
Seattle, Washington

Timothy J. Hobley
Center for Microbial Biotechnology BioCentrum-DTU
Technical University of Denmark
Kgs. Lyngby, Denmark

Brian Hubbard
Purification Process Development
Amgen Inc.
Seattle, Washington

Drew N. Kelner
Amgen Inc.
Thousand Oaks, California

Mani Krishnan
Millipore Corporation
Bedford, Massachusetts

Amitava Kundu
Manufacturing Sciences and Engineering
PDL BioPharma, Inc.
Brooklyn Park, Minnesota

Robert Kutner
Gene Therapy Vector Core
Louisiana State University Health Sciences Center
New Orleans, Louisiana

Asif Ladiwala
Department of Chemical and Biological Engineering
Rensselaer Polytechnic Institute
Troy, New York

Ajay R. Lajmi
Pall Life Sciences
Pensacola, Florida

Timothy Laverty
Centocor Research and Development, Inc.,
Malvern, Pennsylvania

Ann L. Lee
Process Development
Genentech Inc.
San Francisco, California

John J. Lewnard
Millipore Corporation
Bedford, Massachusetts

Herb Lutz
Millipore Corporation
Billerica, Massachusetts

Thomas McNerney
Purification Process Development
Amgen Inc.
Seattle, Washington

Michele M. Myers
Global Biologics Supply Chain, LLC
Malvern, Pennsylvania

Asuman G. Ozturk
Pharmaceutical Development
Centocor, Inc.
Radnor, Pennsylvania

Mark Perreault
GTC Biotherapeutics
Framingham, Massachusetts

Michael W. Phillips
Millipore Corporation
Bedford, Massachusetts

Narahari S. Pujar
Bioprocess Research and Development
Merck Research Laboratories
Merck & Co., Inc.
West Point, Pennsylvania

Bala Raghunath
Millipore Corporation
Billerica, Massachusetts

R. Andrew Ramelmeier
Centocor Research and Development, Inc.
Global Biologics Supply Chain, LLC
Malvern, Pennsylvania

Anurag S. Rathore
Manufacturing Science and Technology
Amgen Inc.
Thousand Oaks, California

Karl Reindel
Manufacturing Sciences and Engineering
PDL BioPharma, Inc.
Brooklyn Park, Minnesota

Jakob Reiser
Gene Therapy Vector Core
Louisiana State University Health Sciences Center
New Orleans, Louisiana

William T. Riordan
Department of Chemical and Biological Engineering
University of Wisconsin
Madison, Wisconsin

John Rozembersky
Bioprocess Research and Development
Merck Research Laboratories
Merck & Co., Inc.
West Point, Pennsylvania

Elisabeth Russell
Manufacturing Science and Technology
Amgen Inc.
Thousand Oaks, California

David Serway
Chemical Technology and Engineering
Merck Manufacturing Division
Merck & Co., Inc.
West Point, Pennsylvania

Abhinav A. Shukla
Purification Process Development
Amgen Inc.
Seattle, Washington

Richard C. Siegel
Centocor Research and Development, Inc.
Global Biologics Supply Chain, LLC
Malvern, Pennsylvania

Martin Siemann-Herzberg
Institute of Biochemical Engineering
University of Stuttgart
Stuttgart, Germany

Hendrik I. Smit
Centocor B.V.
Leiden, The Netherlands

Alan Sonnenfeld
Project Management
Merck & Co., Inc.
Rahway, New Jersey

Richard St. John
Process Development
Chiron Corporation
Emeryville, California

Owen R.T. Thomas
Department of Chemical Engineering, School of Engineering
The University of Birmingham
Birmingham, United Kingdom

Jörg Thömmes
Department of BioProcess Development
Biogen Idec Inc.
San Diego, California

Kevin E. Van Cott
Department of Chemical and Biomolecular Engineering
Biological Process Development Facility
University of Nebraska-Lincoln
Lincoln, Nebraska

Paul J. Voronko
Centocor Research and Development Inc.
Global Biologics Supply Chain, LLC
Malvern, Pennsylvania

Alice Wang
Manufacturing Science and Technology
Amgen Inc.
Thousand Oaks, California

Peter W. Wojciechowski
Global Biologics Supply Chain, LLC
Malvern, Pennsylvania

P.K. Yegneswaran
Science and Technology
Merck Manufacturing Division
Merck & Co., Inc.
West Point, Pennsylvania

Yinges Yigzaw
Purification Process Development
Amgen, Inc.
Seattle, Washington

Chenming (Mike) Zhang
Department of Biological Systems Engineering
Virginia Polytechnic Institute and State University
Blacksburg, Virginia

Contents

1 Harvest of a Therapeutic Protein Product from High Cell Density Fermentation Broths: Principles and Case Study

Elisabeth Russell, Alice Wang, and Anurag S. Rathore

CONTENTS

1.1 INTRODUCTION

Harvest of biotechnology products from cell culture or fermentation process streams is often performed by a combination of several unit operations. The drivers for the process design include maximizing product recovery, scalability, robustness, and clarification of the process stream while operating in a physical and chemical environment where the product is stable.

The harvest approach is dependent on the mode of expression of the target protein. For the case of intracellular expression, the first step is cell concentration. This is followed by cell disruption to release the target protein into solution. After cell disruption, the cell debris is removed and the protein solution is further clarified using filtration. If the target protein is expressed extracellularly, then the first step in harvest is to remove the cells via centrifugation or microfiltration, followed by depth filtration, if further clarification is necessary. The density of the cell culture, shear sensitivity of the cells, and stability of the product are all important cell line-specific characteristics that influence the harvest process design.

This chapter will focus on harvest of yeast cells and the application of the above-mentioned unit operations for clarifying cell broth. Yeast cells can express proteins both intracellularly and extracellularly. Relative to other types of cells used in expression of protein products, yeast cells exhibit minimal sensitivity toward shear, which is an advantage as there is reduced risk of cell rupture and generation of cell debris. Yeast cells have a generation time of 4 to 5 h and can reach a cell density of 10^9 cells/ml and dry cell weight of as much as 100 g/l [1]. The high cell density of yeast fermentations creates challenges for designing an efficient and robust harvest process.

Centrifugation is used in harvest operations for various purposes: cell removal, cell recovery, cell debris removal, and recovery of precipitate. Most industrial applications use disc stack centrifuges (DSCs) to remove cells and cell debris [2,3]. These machines are preferred as they are scalable, perform continuous operation, and have capacity to handle a wide variety of feedstock. For the case of intracellular expression, the homogenization step (cell disruption) is followed by centrifugation to remove cell debris and recover protein in solution. Centrifugation can be challenging due to the broad range of particle sizes in the homogenate and particularly small, highly hydrated particles with very little density difference between the particles and the liquid. Further, breakage of cells leads to release of the cell components such as nucleic acids that can cause a significant increase in viscosity of the feed stream. Separation of sediments can be hampered by blockage of discharge ports from viscous solids. Low flow rates required for high density feed streams can increase the temperature of the feed, which may damage the protein [4]. In some cases, flocculating agents have been used to aid in product isolation. Mosqueira et al. [5] tested three different centrifuges to clarify disrupted yeast cell (baker's yeast) homogenates: two Westfalia DSCs (SAOOH 205 and SAMR 3036) and a 6P Sharples tubular bowl centrifuge. They found that for a suspension of 20% (w/v) they could remove 80 to 90% of the solids in the feed using any of the three centrifuges. The performance of one machine could be predicted from another and from data obtained in a laboratory centrifuge. Clarkson et al. [6] used a Westfalia SAOOH 205 DSC to remove >90% of the initial solids when feeding a yeast cell homogenate at 280 g/l (wet weight of whole cell suspension). For most yeast cell centrifugations, removal of >90% solids is difficult to achieve as the remaining 10% consists of very small particles that cannot be separated due to hindered settling effects and fundamental limitations imposed by the critical particle diameter. Bentham et al. [4] used a scroll decanter centrifuge (low centrifugal force) to recover protein from a flocculated yeast cell homogenate solution (18% wet solids). A 0.1 *M* borax solution (aides in flocculation of yeast cell wall fragments) was mixed equal parts with the homogenate to flocculate only the cell debris, allowing the target protein to remain in solution. They found that at the pilot plant scale they could achieve 93% product recovery and 85% solids removal, which was comparable to their laboratory scale centrifugation. Additionally, the supernatant clarity was comparable to that from a laboratory centrifuge. For processes using intracellular expression, protein precipitation followed by centrifugation has also been used for clarification and isolation of the target protein. One disadvantage to protein precipitation is that protein precipitates have been shown to be shear-sensitive. The feed zone in large-scale centrifuges generates substantial shear and can cause a reduction in protein recovery. Varga et al. [7] used a Westfalia SAOOH 205 DSC to recover an intracellular yeast protein (pe-ADH). The initial cell recovery centrifugation step used two passes on the centrifuge and both steps had >90% product recovery. Following cell debris removal the protein was precipitated and centrifuged twice to recover the product. Two subsequent centrifugations yielded 85 to 90% for the first precipitate recovery and >90% for the second precipitate recovery.

For the case of extracellular expression, it is significantly simpler to design the centrifugation step. This is due to the larger density difference between cells and liquid and a larger particle diameter. The challenge here is the limitation of the machine to handle high percent solids generated in yeast fermentations. To facilitate clarification, the fermentation is often diluted to a more acceptable percent solids level. The dilution, however, results in longer process times and lower product concentration. Mosqueira et al. [5] showed that there was an increase in viscosity and non-Newtonian behavior as the cell concentration in fermentation broth increased (10 to 40% w/v). For a 45% w/v concentration the viscosity was 0.006 Pa·sec with a density of 1100 kg/m^3; at 1% solids concentrate the viscosity was 0.001 Pa·sec with a density of 1000 kg/m^3. Increased viscosity significantly reduces solid–liquid separation efficiency in centrifugation. Varga et al. [7] found that >95% of particles larger than 1.6 μm could be removed from a 2% (v/v) whole cell baker's yeast suspension using a Westfalia SAOOH 205 DSC.

Combination of continuous centrifugation followed by depth filtration has been widely used for harvest of large-scale cell culture or fermentation processes in the biopharmaceutical industry [8–12]. Recently, Yavorsky and Mcgee [10] presented an approach toward selection and sizing of a depth filtration step for clarification of cell culture and fermentation broths. They also presented a discussion of the various strategies that can be adapted while designing a depth filtration step in order to achieve process compression, improved yield, lower operating costs, and reduced process footprint [8]. Use of filter aids to enhance the capacity of a filtration step has also been proposed in literature [13–18]. Heertjes and Zuideveld [13,14] performed depth filtration experiments with polystyrene particles as model impurities to characterize filter aids using effective particle diameter and pore diameter in the filter aid cake. They found that the type of precoat and the way in which it was formed were very important. They also observed that electrostatic repulsion played an important role and that interception and straining were the key mechanisms of capture in depth filtration. Other applications using filter aid assisted filtration for recovery of plasmid DNA [15] and of yeast cells [16] from fermentation broth have also been published. Reynolds et al. [17] presented a design for predicting changes in cake compressibility with time to allow for accurate estimation of flux profiles that are obtained in large-scale filtration steps that use filter aids. Literature from the vendors also provides useful information about the underlying principles that govern the performance of a filter-aid-enhanced depth filtration step [18].

Perhaps the most common approach used for harvesting product from fermentation broths, in particular high-density fermentations such as with yeast cells, is microfiltration. Both plate and frame and hollow fiber formats have been shown to be useful in these applications [11,19–29]. Recently published literature have reviewed the progress made in this area [19,20]. Bell and Davies [11] presented the several advantages that cross-flow filtration (CFF) offers for harvesting yeast fermentations over centrifugation and found that the performance of CFF depends on several factors including viscosity, concentration, membrane fouling due to media components, and influence of osmotic pressure. Patel et al. [21] have compared the different filter formats: pleated-sheet microfilter, tubular microfilter, and hollow fiber ultrafiltration (UF), in terms of flux and cell yields obtained with CFF of yeast cell suspensions. They found that the UF module had much lower fouling rate than with the pleated-sheet microfilter that had rapid plugging and significant cleaning issues. Bailey and Meagher [27] performed a similar comparison between the hollow fiber and plate and frame formats for microfiltration of recombinant *Escherichia coli* lysate and found both options to be comparable in performance under optimized conditions. Sheehan et al. [22] performed a comparison of the centrifugation vs. membrane-based separations of extracellular bacterial protease and found the membrane process to be twice as cost effective as the centrifuge and equivalent to a precoated filter, on the basis of unit cost of enzyme product recovered. Industrial studies demonstrating robust operation of tangential flow filtration (TFF) for harvest of mammalian cell culture [23] and CFF for harvest of recombinant yeast cell product [26] have also been reported. More fundamental studies investigating the various aspects of filtration processes such as membrane fouling, mathematical modeling, and critical flux determination have also been published [24,28,29].

More recently, operating at constant flux rather than at constant trans-membrane pressure (TMP) has been proposed for microfiltration applications [9,22,23,29]. It has been suggested that it is very important to operate below the critical flux, which is the maximum permeate flux that the system can sustain before the membrane becomes polarized. Once critical flux is reached, the cross-flow can no longer sweep solids from the surface as quickly as the permeate flow brings the solids to the membrane surface [29]. It has been observed that severe and often permanent fouling can occur when operating under constant TMP. Fouling due to very high initial permeate fluxes can be avoided by operating under constant flux [9]. Sheehan et al. [22] observed an average flux increase of 2.5× and protein transmission of 90% upon using permeate flow control for recovery of an extracellular protease. Harvest of mammalian cell culture using constant permeate flux at industrial scale has been shown to provide an average yield of 99% with the total cell number and viability maintained throughout the process [23]. Several efforts had been devoted to overcoming membrane-fouling limitations via improved fluid mechanics across the membrane. Lee et al. [30] applied rotating disk dynamic filtration to harvest yeast cells at laboratory scale and demonstrated dramatic improvement by introducing high shear rate, thus minimizing cake formation and fouling. A similar approach for performing shear-enhanced microfiltration of bacterial lysate was taken and increased flux rates were observed [31,32]. Other approaches to improve filtration performance such as use of Dean vortex microfiltration of *E. coli* inclusion bodies [33] and use of coiled hollow-fiber module for microfiltration of microbial suspensions [34] have also been reported in the literature. Periodic backpulsing of the permeate fluid to the feed for fouling reduction was reviewed by Davis [35]. Although these cutting edge technologies may provide significant benefits in terms of sustaining flux and reducing fouling, they have not gained wide industrial acceptance so far and the large-scale equipment that would be required to perform these in a manufacturing environment is not available.

Expanded bed adsorption (EBA) chromatography can also be used to capture target protein, enabling clarification and purification in a single step. Anspach et al. [36] provided a comprehensive review of EBA in protein primary recovery including the theoretical as well as practical aspects of this technology. The authors concluded that the general performance of an EBA is comparable to a packed bed column, but the optimal conditions are more restricted than a conventional packed bed. The influence of resin particle size and density as well as feedstock composition and viscosity in column performance were discussed. Hjorth [37] reviewed industrial application of EBA and found it to be widely used in different biological systems, including *E. coli* cell suspension, *E. coli* homogenate, *E. coli* periplasmic extract, yeast cell suspension, and cell culture. Lyddiatt [38] addressed current constraints and future development options for fluidized bed chromatography. Noda et al. [39] purified a human serum albumin from *Pichia pastoris* at production scale. The process was developed using a 5 cm expanded bed column and scale up to 100 cm column for production. It was claimed that the EBA process provided higher product purity and yield than the conventional approaches. Trinh et al. [40] recovered mouse endostatin from *P. pastoris* fermentation broth using EBA. The fermentation broth (39% v/v wet cells) was first adjusted to the desired pH, conductivity, and biomass concentration and then processed using a cation exchange EBA chromatography step (Streamline SP XL). They found that the expanded bed process has shorter run time, better process economics, and higher product-specific activity.

In this chapter, we review the theoretical principles that govern separation in the commonly used harvest unit operations of centrifugation, depth filtration, and microfiltration. We also present results from a case study involving recovery of a therapeutic protein from *P. pastoris* fermentation broth that will help us in comparison of the different harvest approaches.

1.2 THEORY

1.2.1 CENTRIFUGATION

Application of the solid–liquid separation theory can result in predictions of clarification performance [41]. This theory is useful for predicting operating conditions but does not account for nonideal separation conditions. Nonideal factors include shear damage and hindered settling. The solid–liquid separation theory below is described for a continuous DSC but can be modified for the various types of centrifuges that are used commercially.

1.2.1.1 Solid–Liquid Separation Theory

Figure 1.1 illustrates the flow pattern in a DSC. The feed stream enters through the top of the centrifuge and proceeds down through the stationary feed tube to the feed zone where it is accelerated to the bowl speed. Figure 1.1 shows a standard feed inlet where feed is introduced to the feed zone by a straight pipe. Over the years, various improvements have been made to the method of feed delivery to the feed zone including hydrohermetic, hermetic, and disc inlets. These designs attempt to reduce the gas–liquid mixing at the interface that can cause significant shear damage to the cells. At the base of the bowl the liquid reverses direction and flows up between a series of discs. While the liquid is between the discs the light fluid continues up, the heavy solids are collected on the underside of a disc and then move down. A close-up view of the separation of a particle from the liquid is shown in Figure 1.2.

The particle is forced to the underside of the disc by a net velocity vector, v_G, which has two components: one due to the centrifugal force acting on the particle, v_1, and the other one due to the force of fluid flow, v_2. The net force drives the particle up to the disc underside at which point v_1 becomes

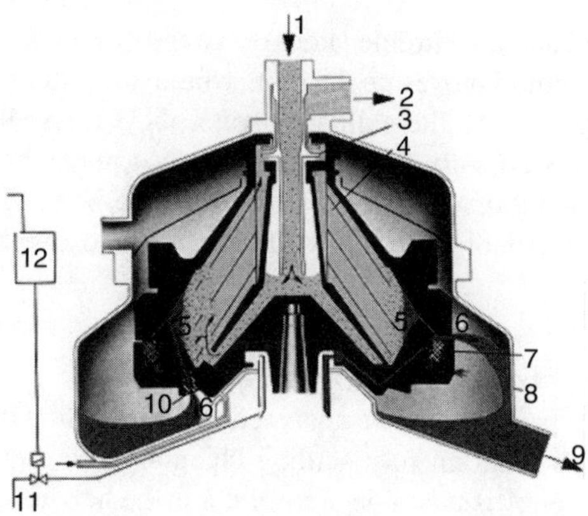

FIGURE 1.1 Diagram of the DSC. (1) Product feed; (2) clarified liquid discharge; (3) centripetal pump; (4) discs; (5) solids space; (6) solids discharge; (7) desludging mechanism; (8) concentrate catcher; (9) concentrate outlet; (10) nozzles; (11) operating water feed; (12) timing unit. (Figure reproduced courtesy of Westfalia Separator.)

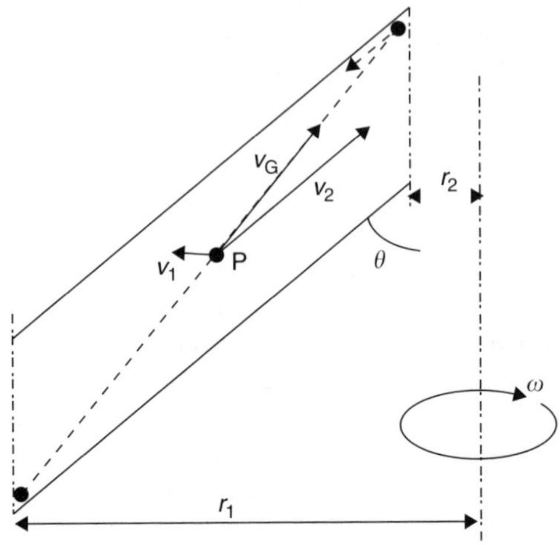

FIGURE 1.2 Solid–liquid separation between discs.

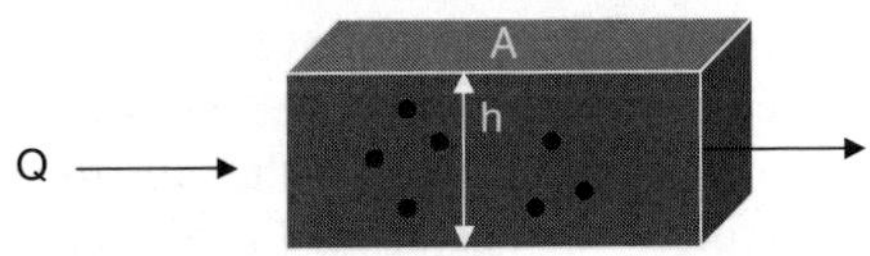

FIGURE 1.3 Separation in a settling tank.

the dominant force and the particle slides down the disc to the solids collection area. The clarified liquid moves up the discs where it is pumped out of the bowl by a centripetal pump. As the solids separate, they accumulate in the bowl. The DSC can be periodically desludged/discharged, meaning that the base of the bowl lowers and the solids are ejected from the bowl. The frequency of desludging is a function of the solids loading in the feed and the feed flow rate.

1.2.1.1.1 Driving Forces and Stokes' Equation

The solid–liquid separation can be analyzed using a settling chamber analogy. Figure 1.3 shows a diagram of a settling chamber where Q is the flow rate through the chamber, A is the table area, and h is the height.

In a settling chamber the time, t_S, it takes for a particle to settle is:

$$t_S = \frac{h}{v_S} \tag{1.1}$$

where ν_s is the settling velocity. The residence time in the chamber, t_R, is:

$$t_R = \frac{Ah}{Q} \tag{1.2}$$

For complete particle removal, one sets $t_R = t_s$, then solving for ν_s:

$$\nu_s = \frac{Q}{A} \tag{1.3}$$

In a DSC the equivalent area is termed the sigma factor, Σ. The settling velocity in the DSC is then:

$$\nu_s = \frac{Q}{\Sigma} \tag{1.4}$$

For a bottle centrifuge, a batch mode operation, the settling velocity is:

$$\nu_s = \frac{V}{t\Sigma} \tag{1.5}$$

where V is the volume of centrate and t is the time for centrifugation.

The settling velocity of a particle in a settling chamber is determined from Stokes' law by balancing the acting forces on the particle [1]:

$$F_D = 3\pi d\mu\nu \tag{1.6a}$$

$$F_B = \left[\frac{\pi d^3(\rho_s - \rho_L)}{6}\right] g \tag{1.6b}$$

where F_D is the drag force (Stokes' law), F_B is the buoyancy force, d is the diameter of the particle, ν is the velocity of the particle, ρ_s and ρ_L are the density of the solid and liquid, respectively, μ is the viscosity of the liquid, and g is the gravitational acceleration. Equating the forces and solving for the particle velocity gives:

$$\nu_s = \frac{d^2(\rho_S - \rho_L)}{18\mu} g \tag{1.7}$$

In a centrifuge the force separating the particle is the centrifugal force instead of gravity. By substituting centrifugal acceleration for the gravitational acceleration the settling velocity can be determined for a particle in a centrifuge.

$$a = r\omega^2 \tag{1.8a}$$

$$v_S = \frac{d^2(\rho_S - \rho_L)}{18\mu} r\omega^2 \tag{1.8b}$$

where a is the centrifugal acceleration, r is the distance the particle will settle through, and ω is the angular velocity. Combining Equation 1.4 and Equation 1.7 then solving for Q/Σ yields:

$$\frac{Q}{\Sigma} = \frac{d^2(\rho_S - \rho_L)}{18\mu} g \tag{1.9}$$

It is now possible to use Equation 1.9 to solve for the diameter of the smallest particle that can be separated, d_{min}:

$$d_{min} = \sqrt{\frac{18Q\mu}{\Sigma g(\rho_s - \rho_L)}} \tag{1.10}$$

Figure 1.4 shows the application of Equation 1.10 with the viscosity, μ, taken as 1.5 centipoises (cP) at 5°C. The Q/Σ range specified on the graph is a typical operating range for a pilot scale DSC.

In Figure 1.4, the x-axis is the settling velocity of a particle; in a disc stack the settling velocity is equivalent to Q/Σ. It is evident that a fivefold increase in the density difference will reduce the minimum particle size by 2.2 times.

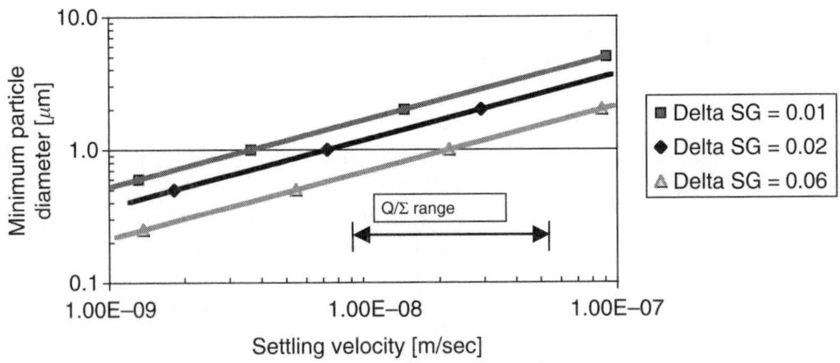

FIGURE 1.4 Plot showing prediction of theoretical flow rates for particles settling in a DSC. ■ Specific gravity difference of 0.01; ♦ specific gravity difference of 0.02; △ specific gravity difference of 0.06.

This emphasizes the difficulty of centrifuging feed material containing cell debris where the density difference is small. The density difference between a yeast cell and media is 0.125 g/cm^3 [7].

1.2.1.1.2 Sigma Factor

The sigma factor represents the equivalent settling area of a centrifuge. It is specific for each DSC and is based on the geometry of the centrifuge and the angular velocity. The sigma factor was first derived by Charles Ambler [42] and later modified by Maybury et al. [43] to account for acceleration and deceleration in batch centrifuges. For a DSC, sigma can be expressed as [42]:

$$\Sigma_{\mathrm{DSC}} = \frac{2\pi n\omega^2 (r_{\mathrm{o}}^3 - r_{\mathrm{i}}^3)}{3g \tan\theta} \tag{1.11}$$

where n is the number of discs, r_{o} and r_{i} are the outer and inner radii of the disc, and θ is the angle of the discs from the vertical axis. Equation 1.11 shows that the internal geometry of the centrifuge bowl, namely the number of discs, radii, and disc angle, determine the separation area available to separate solids and liquids during clarification.

For a laboratory centrifuge, the time it takes for acceleration and deceleration is significant and contributes to the settling of solids, and must therefore be accounted for. The sigma factor for a laboratory centrifuge is defined as [44]:

$$\Sigma_{\mathrm{Bottle}} = \frac{\omega^2 (3 - 2x - 2y)V}{\ln(2r_2/(r_1 + r_2))2g} \tag{1.12}$$

where x and y are the fraction of time for acceleration and deceleration, respectively, and r_1 and r_2 are the radii of the surface of the centrifuge liquid and the base of the centrifuge liquid, respectively. An important application of the sigma factor is for predicting equivalent recovery performance between centrifuges.

To remove the same size particle (i.e., same cut diameter) using two different centrifuges, it is necessary that:

$$\nu_{\mathrm{s}}^1 = \nu_{\mathrm{s}}^2 \tag{1.13}$$

Because the settling velocity is equivalent to Q/Σ for a continuous centrifuge, then the Q/Σ for the two different centrifuges should be equal. If each has a sigma factor, the cut-off particle diameter should be the same if:

$$\frac{Q_1}{\Sigma_1} = \frac{Q_2}{\Sigma_2} \tag{1.14}$$

Equation 1.14 could be used in a number of ways. If Σ_1 and Σ_2 are known, the ratio of flow rates, Q_1/Q_2, can be calculated to obtain the same separation performance on a given piece of equipment. To estimate the flow rate for a larger scale DSC from a lab bottle centrifuge, the Q/Σ ratios can be equated:

$$\frac{Q}{\Sigma_{\mathrm{DSC}}} = \frac{V}{t\Sigma_{\mathrm{Bottle}}} \tag{1.15}$$

Scaling up a centrifugation step while keeping Q/Σ ratio identical may still lead to underperformance at a large scale. This lower efficiency is generally attributed to shear damage or hindered settling in the continuous disc stack. The shear damages the cells and cell debris and generates submicron particles that cannot be separated by the centrifuge. The submicron particles stay in solution and exit in the centrate stream, and the higher particle load results in low clarification efficiency. Shear sensitivity is more of a concern for mammalian cell culture than yeast fermentations as the mammalian cells are more shear sensitive. A secondary problem associated with shear damage is the potential release of proteases that could affect the stability of the target protein.

1.2.1.2 Factors Affecting Solid–Liquid Separation

Many factors affect the outcome of a solid–liquid separation process. From Equation 1.4 and Equation 1.5, the settling velocity is defined by the sigma factor and flow rate for continuous centrifugation and by the volume per time for batch centrifugation. These three parameters can be varied to attain a desired separation performance. Variation of the separation comes from changing the sigma factor of the centrifuge, the density difference between the solid and liquid, RPM, and viscosity. Another parameter that has been shown to affect the solid–liquid separation is temperature, as it can alter both the viscosity and the stability of the product.

1.2.1.2.1 Sigma Factor

Equation 1.3 shows that an increased area corresponds to a lower particle settling velocity that in turn corresponds to a smaller particle diameter. For example, a bench scale bottle centrifuge has a sigma factor of about 55 m^2 and a pilot scale DSC has a sigma factor of about 1360 m^2. Keeping the flow rate and the properties of the liquid constant, the DSC is capable of removing particles five times smaller (square root of 1360/55 m^2) than the bottle centrifuge.

1.2.1.2.2 Density Difference

The density difference between the solid and liquid drives the settling of the particle. The greater its value, the faster a particle will settle out. The lines in

TABLE 1.1
Densities of Different Cell Culture Broths

Cell	Density (g/cm^3)
E. coli	1.09
Amoeba proteus	1.02
Saccharomyces pombe	1.09
Saccharomyces cerevisiae	1.11
Murine B cells	1.06
Chinese hamster ovary	1.06

Source: Kubitschek H. *Critical Reviews in Microbiology* 1987; 14: 73–97. With permission.

Figure 1.4 show the impact of the density difference. The densities for different cells are listed in Table 1.1.

From Table 1.1, it is clear that the density difference is smaller for mammalian cells than for yeast cells. Consequently, yeast cells are easier to separate. Equation 1.10 indicates that for systems with smaller density differences the d_{min} will be larger.

As seen in Table 1.2, for yeast cell broths, the density of the solution increases after homogenization. If the solution is precipitated the particle density increases significantly, aiding in the recovery of the precipitates.

1.2.1.2.3 Angular Velocity (RPM)

The RPM of a centrifuge is a specified operating variable. Bottle centrifuges have a wide range of settings from 500 to 4000 rpm and as high as 100,000 rpm for ultracentrifuges [3]. The sigma factor increases with the square of the RPM. Large-scale centrifuges are limited in RPM or G-force because of material stress limitation and safety concerns. Unlike the density difference, RPM is a controllable parameter and can have a strong impact on the separation performance.

1.2.1.2.4 Viscosity

The viscosity of the liquid affects the settling velocity of the particles. The settling velocity increases as the viscosity decreases. For a dilute aqueous stream, the viscosity decreases about 2% for every 1°C increase. Thus, increasing the temperature of the feed may lead to better clarification efficiency. For biological feed streams, this may need to be balanced with the concerns about the stability of the protein and the robustness of cells (see Section 2.1.2.6).

TABLE 1.2
Density Differences for Yeast Process Harvest Streams

Stage	Whole Cell Recovery Post-Fermentation	Cell Debris Removal Post-Homogenization	Solid Precipitate Removal	Solid Precipitate Recovery
Volumetric solid concentration (m^3/m^3)	0.02	0.25	0.03	0.05
Particle density (kg/m^3)	1130	1120	1200	1210
Suspending fluid density (kg/m^3)	1005	1020	1120	1170
Viscosity (mPa sec)	1.1	3.8	2.1	2.5
Mean particle diameter (μm)	3.8	2.9	0.3	1
Calculated d_c (μm)	0.8	1.3	0.7	1.1

Source: Varga EG, Titchener-Hooker NJ, and Dunnill P. *Biotechnology and Bioengineering* 2001; 74: 96–107. With permission.

1.2.1.2.5 Flow Rate and Residence Time

Ideally, low flow rates and long residence times lead to smaller particle removal. As seen in Equation 1.10, decreasing the flow rate will decrease d_{min}, since Q is proportional to the square of d_{min}. Unfortunately, with solids that are susceptible to shear damage, long residence times may be detrimental. This is especially true for shear sensitive protein precipitates. Therefore, there is an optimum flow rate that is low enough to separate small particles but high enough so that the residence time is sufficiently small and shear effects do not create more debris. For laboratory bottle centrifuges that are operated in batch mode, shear is not an issue and long residence times will not generate additional debris. This is because batch centrifugation does not have the initial shearing effects due to feed acceleration in the inlet, unlike continuous centrifugation. Consequently, batch centrifuges can be run as long as necessary for the desired removal of small particles. Maybury et al. [43] performed experiments with two centrifuges, a Beckman J2-MI lab centrifuge and a SAOOH-205 DSC. The authors used shear-sensitive ammonium sulfate protein precipitates (from baker's yeast cell homogenate) to compare the clarification efficiency of the two centrifuges. They found that at low flow rates there was increased deviation of the DSC clarification performance from that of the laboratory centrifuge. This indicates that the DSC is breaking up the protein precipitates during the long residence time, and confirms that low flow rates are detrimental to shear-sensitive material in continuous centrifugation.

1.2.1.2.6 Temperature

During large-scale continuous centrifugation, the temperature of the centrate will increase. This can be problematic if the increase in temperature leads to cell lysis and release of intracellular material. Additionally, the increase in temperature could denature the target protein, thereby rendering it inactive. To understand the temperature effects during centrifugation, Kempken et al. [44] studied the effects of temperature rise in a Westfalia CSA-1 high-speed DSC. Feeding a mammalian cell culture at different speeds and flow rates they found that there was a maximum temperature increase of 12°C for the centrate at 9000 rpm and 20 l/h. When the flow rate was 50 l/h the maximum temperature increase was only 5°C. This indicates that the longer the feed is in the bowl, the greater the temperature rise. Mosqueira [5] also observed an increase in temperature when operating the Westfalia SAMR 3036. While at very low flow rates ($Q/\Sigma = 3 \times 10^{-9}$ m/sec) a 16°C increase in temperature was observed, at normal flow rates the increase in temperature was only 2°C ($Q/\Sigma > 3 \times 10^{-8}$ m/sec). The temperature in the centrifuge can be better controlled using a cooling jacket.

1.2.1.3 Clarification Efficiency

The clarification efficiency of a centrifuge can be quantified by determining the relative amount of debris in the centrate and the filterability, or how well a centrate filters. The test results reveal how well the centrifuge removes small particles and whether particles are generated during the centrifugation process. A common approach to experimentally determine clarification efficiency is through turbidity measurements.

Measurements are taken of either the turbidity (in nephelometric turbidity units, NTU), a measure of the relative particle concentration in solution, or the optical density (at 600 or 670 nm) of the feed, centrate, and a very well-clarified feed sample. The well-clarified feed is a centrate sample (usually 4000 rpm using a bench centrifuge) that has been filtered through a 0.2 μm filter. The clarification efficiency is defined as the actual change in turbidity divided by the maximum change in turbidity [43], that is:

$$\text{Clarification efficiency} = \frac{\text{NTU}_{\text{feed}} - \text{NTU}_{\text{centrate}}}{\text{NTU}_{\text{feed}} - \text{NTU}_{\text{well-clarified}}} = 1 - \frac{\text{NTU}_{\text{centrate}} - \text{NTU}_{\text{well-clarified}}}{\text{NTU}_{\text{feed}} - \text{NTU}_{\text{well-clarified}}} \quad (1.16)$$

1.2.1.4 Definition of Key Parameters

Centrifuge operations can vary based on the equipment, application, and the feed stream. Centrifuges can be operated in batch mode, such as laboratory ultracentrifuge, basket and bottle centrifuges, or operated in a continuous mode, such as tubular bowl, decanter, and DSCs. The continuous DSC is most commonly used for large-scale biotechnology applications and will be the focus of the ensuing discussion.

1.2.1.4.1 Operational Parameters

Operational parameters, also known as process inputs, are those parameters that are directly controlled during operation. In the following, we briefly describe some of the key operational parameters that often impact centrifugation performance.

The percent solids in the feed stream and the shear sensitivity of the cells have a strong impact on the performance of a DSC. The lower the flow rate, the better is the solid–liquid separation (Equation 1.10). For high percent solid feed streams, the flow rate should be low to allow for adequate solid–liquid separation. The low flow rates lead to long residence times in the bowl, the longer the material is in the bowl the greater the exposure to shear. Therefore, shear sensitive material is run at the higher end of acceptable flow rates where the acceptable flow rate range is dictated by adequate clarification. Yeast fermentations have high percent solids (up to 60%) and are shear insensitive therefore the flow rates are low. Mammalian cell cultures have very low percent solids (1 to 6%). The flow rates for centrifugation of mammalian cell cultures are higher than for yeast fermentations because they are easier to clarify and because they are shear-sensitive.

Feed interval is the duration that feed is fed into the bowl. This time is calculated based on the solids volume capacity. There is a specified solids volume for each centrifuge and after this volume is filled solids are carried over into the centrate stream. A general practice is to fill the solids volume to 70 to 90% in order to avoid having excess solids in the centrate. There is a minimum feed interval for most centrifuges, so as to avoid long durations of frequent bowl discharges. Frequent bowl discharges may wear the internal parts of the centrifuge and cause mechanical problems. The equation for calculating the feed interval is:

$$\text{Feed interval [min]} = \frac{\text{solids volume [l]} \times \text{fraction of solids volume filled}}{(\text{percent solids in the feed}/100) \times (\text{feed flow rate [l/min]})} \tag{1.17}$$

Discharge type can be full, partial, or ratio. The discharge type determines how much bowl contents are ejected during a bowl discharge. A full discharge evacuates the entire contents of the bowl, including liquid and solids. A partial discharge opens the bowl for a fraction of a full discharge duration and allows only part of the bowl contents to be discharged. Ratio discharge mode uses both partial discharges and full discharges in some ratio.

Discharge time is the duration that the bowl is open during a discharge, is specified by the vendor for a full discharge, and is specified by the process for a partial discharge. This parameter is usually optimized during development so as to minimize liquid loss during solids ejection. Partial discharge times are less than full discharges in order to reduce the volume of contents ejected.

Discharge ratio is the ratio of partial discharges to full discharges for ratio mode operations. The ratio should be as high as possible to avoid excessive product loss during the full discharges (assuming no flushing between discharges).

Predischarge flush volume is the volume of displacement liquid (water or buffer) that is needed to push the liquid contents out of the bowl. The purpose of predischarge flushes is to recover any product containing liquid in the bowl before a discharge. The volume is determined either by measuring the turbidity or the protein concentration in the centrate during the flush and defining when the bowl is sufficiently clear of product (usually 2 to 4 bowl volumes). Postdischarge flush volume is the volume of displacement liquid that is fed into the bowl to displace the air in the bowl after a discharge. This flush is used for shear sensitive material. Introduction of shear sensitive material into an empty bowl can shear the cells and this can be avoided by initially filling the bowl with liquid. The flush volume is usually 1 to 2 bowl volume ensuring that the bowl is completely filled. Flushing is most commonly used before full shots.

Backpressure is the pressure applied on the centrate line downstream of the bowl. Backpressure is used to ensure that the bowl remains filled when there is flow to the bowl. Most new DSCs have a modified feed inlet that ensures that there is reduced air entrainment at the inlet and therefore reduced shear. For the hydrohermetic feed inlet and the disc feed inlet, enough backpressure is required to ensure a flooded inlet to properly reduce the shear. Too little backpressure can reduce the clarification efficiency. Too much backpressure will overflow the bowl and lead to loss of product. The backpressure is experimentally determined during operation or estimated from a vendor pump curve. For hermetic centrifuges backpressure is not required to flood the inlet but may be required to eliminate air entrainment at the outlet, which could affect the accuracy of in-line centrate turbidity measurement.

1.2.1.4.2 Performance Parameters

Performance parameters, also known as process outputs, are those parameters that illustrate the performance of the step. In the following, we briefly describe some of the key performance parameters that are often monitored to assess how well the centrifugation performed.

Target protein concentration determined for feed, centrate, sludge (solids that are discharged), and filtrate pools is often used to calculate product recovery. Mass balances on the liquid phase can be performed using weights of the feed, sludge, and centrate.

Particle size distribution (PSD) analysis can be useful to determine how well the centrifuge is clarifying the feed stream by indicating whether solids carried over into the supernatant consist of large yeast cells or smaller feed components. It is also used to determine if the size of the cells varies between harvests. Any changes in cell diameter would affect both the ease of sedimentation and the extent of dewatering that can be achieved in the centrifuge bowl. According to Equation 1.8b, the settling velocity increases with the square of the particle diameter and thus, larger particles are easier to separate.

Viscosity measurements allow for tracking of any changes in the fermentation that could potentially effect the centrifugation. Viscosity affects the solid–liquid separation process, in that a higher viscosity decreases the settling velocity of a particle. As the settling velocity decreases, it takes longer for a particle to settle and results in more particles remaining in the centrate. The relationship between the minimum diameter of the particle settled and the viscosity is: $\mu \propto d^2$. Doubling the viscosity therefore has a significant impact on the minimum diameter of a particle that can be settled. Equally important for centrifugation of concentrated suspensions is the influence of viscosity on dewatering of the large amounts of sediment deposited in the bowl, where an increase of the liquid viscosity causes a proportional increase of the pressure required to express liquid from the packed solids.

Sludge weights are also indicative of how the centrifuge is performing. For centrifuges that are performing properly the sludge weights should be constant throughout the run. For partial shot operations, the sludge weight should be close to the solids hold up volume of the bowl (corrected for density). For full shot operations, the sludge weights should be the weight of the bowl volume (corrected for density). Low sludge weights indicate that solids are accumulating in the bowl and that a full discharge may be necessary. If the sludge weights are too large then too much liquid is being ejected during the bowl discharge. Large sludge weights lead to significant product loss if the discharge time (duration the bowl is open) is not adjusted. Monitoring the sludge weight can be used for real time analysis of the centrifuge performance, as it provides important information to the dynamics of solids deposition in the bowl, and is especially valuable as an indicator of bowl clogging. In a high-solids centrifugation step, tracking the discharge masses is of crucial importance to the success of the separation. This is in contrast to a low-solids centrifugation where the precise timing and size of a discharge affects the overall process to a much lesser degree.

Percent solids measured on the feed, centrates, and sludges can be useful to track changes in the sludge concentration and to ensure high product recovery, by adjusting discharge times as required, or initiating full discharges to clear the bowl.

1.2.2 Filtration

1.2.2.1 Normal Flow Filtration

There are generally two operating modes for normal flow filtration (NFF) application. When particles are trapped within the pores or body of filter medium, it is called depth filtration or clarifying filtration. On the other hand, when solids are stopped at the surface of the medium and accumulated as a cake of increasing thickness, the separation is called cake filtration.

1.2.2.1.1 Depth Filtration

Depth filtration sometimes is also called deep-bed filtration. Mechanisms for separation that have been proposed include electrostatic attraction, van der Waal's forces, and physical adsorption. Deep-bed filtration has no positive cut-off of particle size, but rather removes a proportion of all sizes of particles. The effectiveness of deep-bed filtration depends on the flow rate, quantity, and size of the particles to be removed. Belfort [46] reviewed and summarized the theory of depth filtration. The capture of fine particles in depth filtration may be considered in two steps: attachment and transport. The efficiency of particle retention has been expressed through the so-called filter coefficient, λ, as

$$\frac{\partial C}{\partial z} = \lambda C \tag{1.18}$$

where C is the particle concentration in the bed and z is the depth of the bed. The filtration coefficient (λ) is related to the total collection efficiency η, which is determined by all the mechanisms of capture including Brownian diffusion, sedimentation, inertia, interception, and hydrodynamics. The transport capture mechanisms for particle through a depth filter are illustrated in Figure 1.5 [47].

$$\lambda = \frac{3(1-\varepsilon)}{2d_{\mathrm{g}}}\eta \tag{1.19}$$

$$\eta = \text{constant} \times N_{\mathrm{diff}}^{\beta} N_{\mathrm{sed}}^{\gamma} N_{\mathrm{iner}}^{\xi} N_{\mathrm{int}}^{\alpha} N_{\mathrm{hydro}}^{\delta} \tag{1.20a}$$

where N_{diff}, N_{sed}, N_{iner}, N_{int}, and N_{hydro} are dimensionless numbers for diffusion, sedimentation, inertia, interception, and hydrodynamic mechanisms. Upon replacing the dimensionless number with physical parameters and

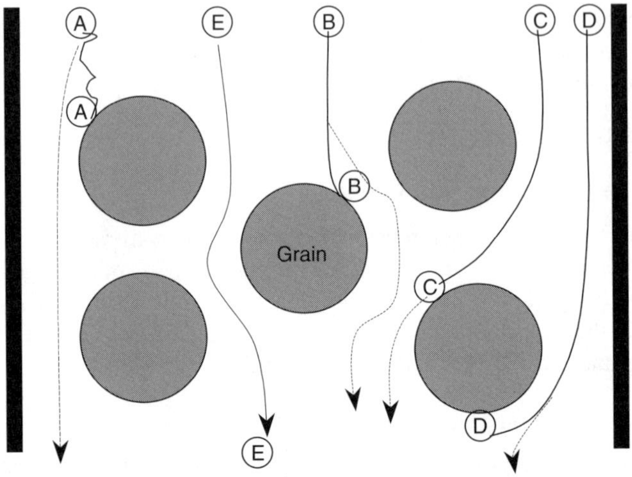

FIGURE 1.5 Transport capture mechanism for particles passing through a depth filter. Transport capture mechanisms: (A) Brownian diffusion; (B) sedimentation or inertia; (C) interception; (D) hydrodynamic; (E) escaped particles. (Adapted from Atkinson B and Mavituna F. *Biochemical Engineering and Biotechnology Handbook*. New York: Stockton Press, 1991, pp. 957–959.)

combining items together, we have:

$$\eta = \text{constant}\frac{d_{\mathrm{p}}^{\mathrm{g}}(v\rho_{\mathrm{s}})^{\xi}[\rho_{\mathrm{L}}f(\eta')]^{-\delta}(kT)^{\beta}(\rho_{\mathrm{p}}-\rho_{\mathrm{L}})^{\gamma}}{d_{\mathrm{g}}^{\mathrm{b}}\mu^{\mathrm{c}}v^{\mathrm{d}}} \tag{1.20b}$$

where $b, c, d, g, \alpha, \beta, \gamma, \delta$, and ξ are exponents, k is Boltzmann's constant, T is absolute temperature, μ is viscosity, d_{p} is particle diameter, d_{g} is grain diameter, ρ_{p} is particle density, ρ_{L} is liquid density, v is approach velocity, $f(\eta')$ is a function of the lateral position in pore with value of 0 to 1, and ε is bed porosity. This equation indicates that under normal circumstances, the total collection efficiency η increases with an increase in particle diameter, d_{p}, or a decrease in grain diameter, d_{g}, viscosity, μ, and velocity, v [47].

1.2.2.1.2 Cake Filtration

In cake filtration, the filtration rate is related to the driving force, pressure, fluid viscosity, and total resistance from cake and filter medium as follows [48]:

$$\frac{1}{A}\frac{\mathrm{d}V}{\mathrm{d}t} = \frac{\Delta P}{\mu[r_{\mathrm{c}}(W/A)+R_{\mathrm{m}}]} \tag{1.21}$$

where V is the filtrate volume, A is the area of filter surface, t is time, ΔP is the total pressure difference across the filter medium and the cake, μ is the viscosity of the filtrate, R_m is the resistance of the filter medium, and W is the mass of cake solids corresponding to V, and r_c is the specific cake resistance. W is related to V by mass balance:

$$W = C_p V \tag{1.22}$$

where C_p is particle concentration in the feed. Further, the average specific cake resistance, r_c, can be related to characteristics of cake formed and the pressure applied by the expression in the following manner [48]:

$$r_c = r_c' \Delta P^s \tag{1.23}$$

where r_c' is a constant determined by particle size forming the cake and porosity of cake layer and s is the compressibility of cake, varying from 0 for rigid cake to 1 for highly compressible cake. Equation 1.21 can be rewritten as:

$$\frac{d(V/A)}{dt} = \frac{\Delta P}{\mu(r_c' \Delta P^s C_p(V/A) + R_m)} \tag{1.24}$$

Filtration can be performed in three different modes: constant rate filtration which utilizes positive displacement pumps, constant pressure filtration, under which the constant pressure is maintained by compressed gas; and various pressure and various rate filtration when centrifugal pumps are used and the discharge rate decreased with increasing back pressure.

1.2.2.1.2.1 Constant Flow Filtration In this case, Flux = J, and Equation 1.24 can be integrated as follows:

$$\Delta P = \mu r_c' \Delta P^s C_P J^2 t + \mu R_m J \tag{1.25}$$

or

$$t = \frac{1}{\mu r_c' J^2 C_P} \Delta P^{1-s} - \frac{R_m}{r_c' J C_P} \Delta P^{-s} \tag{1.26}$$

Generally, R_m is negligible compared with the cake resistance, and so Equation 1.25 and Equation 1.26 can be reduced to

$$\Delta P = \left(\mu r_c' C_P J^2 t\right)^{1/(1-s)} = \left(\mu r_c' C_P J \frac{V}{A}\right)^{1/(1-s)} \tag{1.27}$$

or,

$$\frac{V}{A} = Jt = \frac{1}{\mu r'_{c} C_{P} J} \Delta P^{1-s} \tag{1.28}$$

Equation 1.28 indicates that filtration throughput (V/A) at constant flow rate condition is related to fluid properties such as filtrate viscosity (μ) and solids content (C_p); cake properties such as cake specific resistance (r_c) and cake compressibility (s); and process conditions including flow rate (J) and differential pressure (ΔP).

The impact of some of the above parameters on filtration performance is illustrated in Figure 1.6. Figure 1.6a shows that filtration capacity can be significantly improved by decreasing cake compressibility and Figure 1.6b and

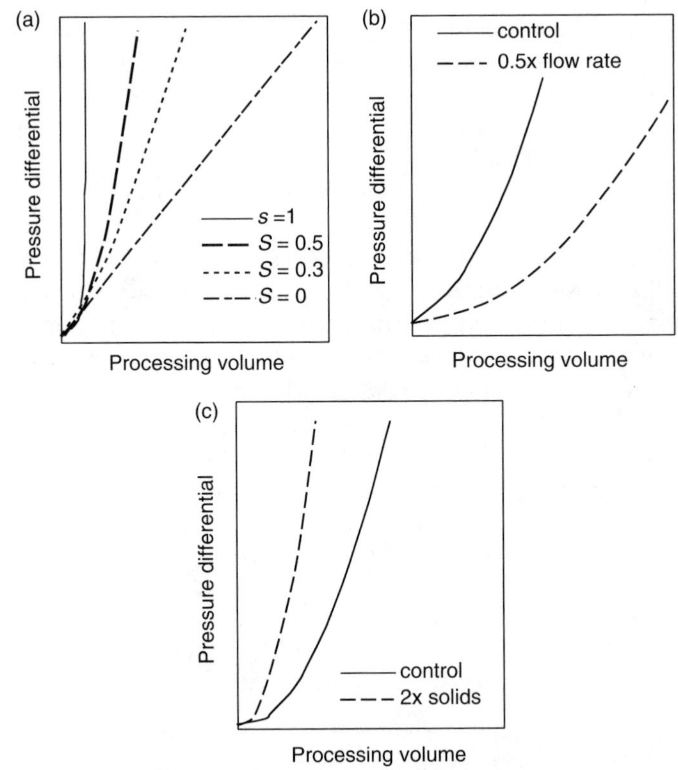

FIGURE 1.6 Effect of feed characteristics and process parameters on filtration profile under constant flow filtration mode. (a) Effect of cake compressibility (s), (b) effect of flow rate, and (c) effect of solids content in the feedstock.

Figure 1.6c show that capacity can also be improved by reducing flow rate or reducing solids content.

1.2.2.1.2.2 Constant Pressure Filtration For constant pressure (ΔP) filtration, Equation 1.24 can be integrated to give the following relationship among filtration time (t), throughput (V/A), pressure (ΔP), and other parameters (feed characteristics and membrane resistance).

$$\frac{t}{V/A} = \frac{\mu r_c' C_P}{2\Delta P^{1-s}}\frac{V}{A} + \frac{\mu R_m}{\Delta P} \tag{1.29}$$

According to Equation 1.29, a $t/(V/A)$ vs. V/A plot yields a straight line such that the slope and the intercept of the plot can be used to calculate membrane and cake resistances. When R_m is negligible compared to cake resistance, Equation 1.29 can be simplified as:

$$t = \frac{\mu r_c' C_P}{2\Delta P^{1-s}}\left(\frac{V}{A}\right)^2 \tag{1.30}$$

or

$$\frac{V}{A} = \sqrt{\frac{2\Delta P^{1-s}}{\mu r_c' C_p}t} \tag{1.31}$$

Equation 1.31 indicates that under constant pressure filtration mode throughput can be improved by increasing the driving force (differential pressure, ΔP) or processing time (t), or reducing cake specific resistance (r_c') or compressibility (s) using filter aids, or reducing fluid viscosity (μ) and solids content (C_p).

Figure 1.7 illustrates the filtration profile and the impact of compressibility (s), differential pressure (ΔP), and solids content (C_p) in the feedstock under constant pressure condition. Figure 1.7a shows that reducing compressibility (s) results in capacity improvement. Figure 1.7b shows that doubling differential pressure or processing time will not double filtration capacity due to nonlinear relationship between throughput and process parameters. A 2× dilution of feed material to reduce solids content can improve filtration throughput by 1.4× as indicated in Figure 1.7c, but the increased feed volume (2×) may offset the benefit of throughput improvement and result in longer processing time.

1.2.2.1.2.3 Variable Pressure and Flow Rate Filtration This case occurs when centrifugal pump is used and the discharge rate decreases with increasing backpressure. Integration of Equation 1.24 for this case can be complex. However, it can be solved by numerical method when the characteristic of the feed pump is known [48].

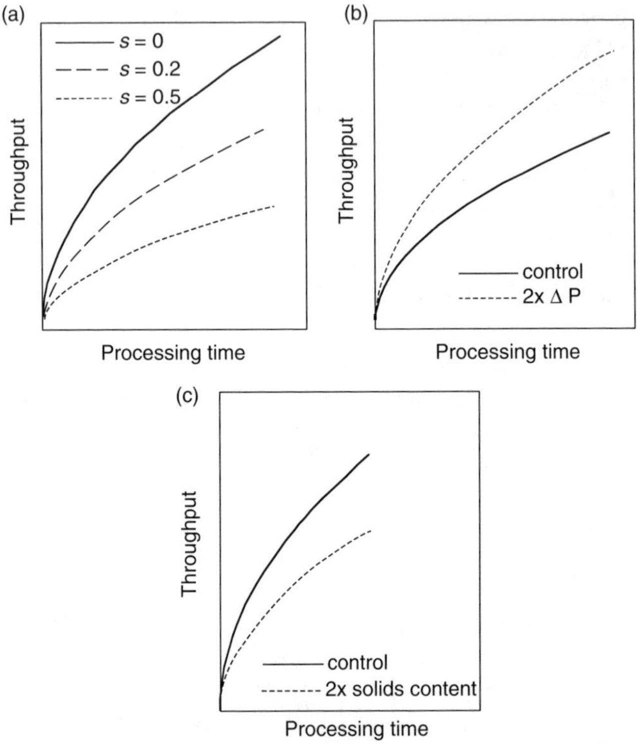

FIGURE 1.7 Effect of feed characteristics and process parameters on filtration profile under constant pressure filtration mode. (a) Effect of cake compressibility (s), (b) effect of differential pressure (ΔP), and (c) effect of solids content in the feedstock.

1.2.2.1.3 Filter Aid Assisted Depth Filtration

Various types of filter aids have been used for enhancing capacity of a filtration step [49–51]. These include those based on diatomaceous earth, perlite, and cellulose. Figure 1.8 shows images of the variety of diatoms that are found in some of the filter aids [51]. The mechanism of filter aid assisted filtration is illustrated in Figure 1.9 [18]. The filter aid is added to the feed material and when depth filtration is performed, the filter aid particles maintain separation of the otherwise impermeable solids in the feed and thus can cause significant improvement in permeability of the filter cake, yielding higher flux and filter capacity.

The cake filtration theory as described in Equation 1.24, Equation 1.28, and Equation 1.31 still applies to filter aid-assisted filtration. With filter aid addition, r_c' is the combined specific resistance from the mixture of filter aid and original solids in the feed; C_p is the total mass of filter aid and original

FIGURE 1.8 Variety of diatoms that are found in some of the commercially available filter aids. (From Hunt T. In *Encyclopedia of Bioprocess Technology: Fermentation, Biocatalysis and Bioseparation*. New York: John Wiley & Sons, 1999. With permission.)

solids in the suspension; and s is the combined compressibility of filter aid and original solids.

1.2.2.2 Tangential Flow Filtration

Tangential filtration, like NFF, is also a pressure-driven separation process. Fluid flows across membrane surface and only a small fraction of solvent and permeable materials penetrate through the membrane. The fluid circulation minimizes the formation of the filtered solids on the membrane, consequently maintaining flux without increasing pressure. The comparison of TFF vs. NFF is illustrated in Figure 1.10. TFF can be further divided into microfiltration (MF) and UF according to the pore size of the membrane.

Microfiltration is usually used in upstream recovery process to separate intact cells and some cell debris/lysates from the rest of the components in the feed stream. Either the retained cells or the clarified filtrate can be the product stream. Membrane pore size cutoffs used for this type of separation are typically in the range of 0.05 to 1.0 μm. Ultrafiltration/diafiltration (UF/DF) is one of the most widely used forms of TFF and is used to separate proteins from buffer components for buffer exchange, desalting, or concentration. Depending on the size and other physicochemical properties of the target protein, membrane NMWL (nominal molecular weight limit) in the range of 1 to 1000 kDa are commonly used.

The driving force through the membrane is determined by TMP which is defined as the difference between the average pressure on the retentate and permeate side:

$$\text{TMP} = \frac{P_{\text{feed}} + P_{\text{retentate}}}{2} - P_{\text{permeate}} \tag{1.32}$$

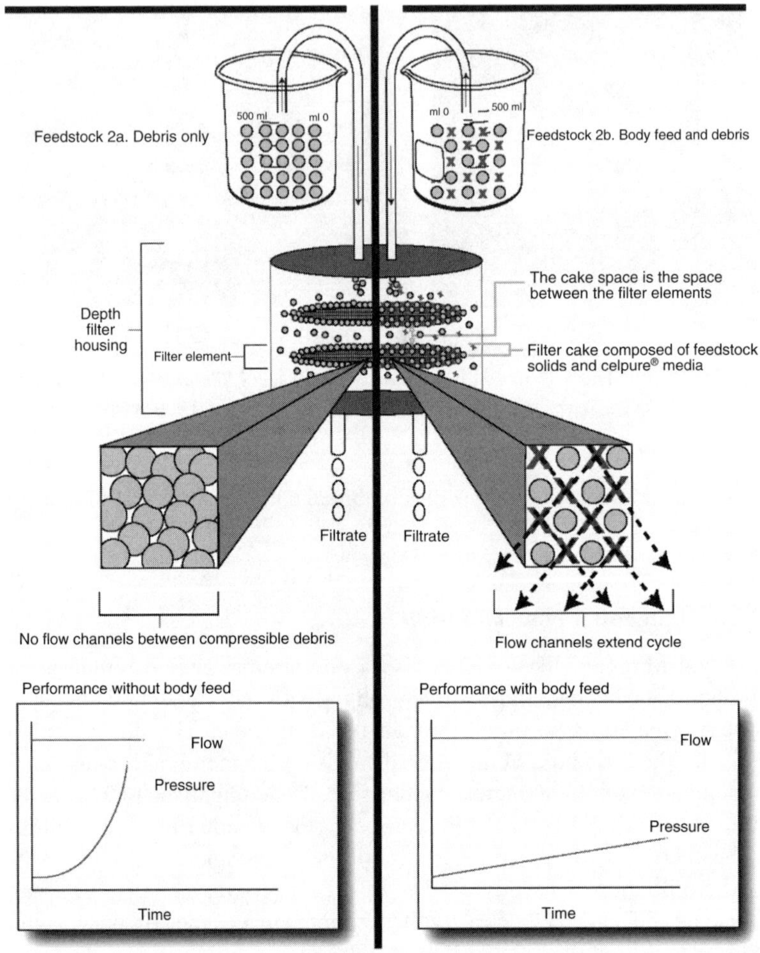

FIGURE 1.9 The mechanism of filter aid assisted depth filtration. (Adapted from Dynamic depth-filtration: Proof of principle. Technical note AMC06 v. 3.1, Advanced Minerals Corporation, 2004.)

where P_{feed} is the feed inlet pressure, $P_{\text{retentate}}$ is the retentate outlet pressure and, P_{permeate} is the permeate pressure. Normally in CFF process, flux initially increases with increasing TMP and then levels off as shown in Figure 1.11.

1.2.2.2.1 Pressure Controlled Model

Several models have been developed to predict flux as a function of process parameters and fluid characteristics. It is generally believed that there is

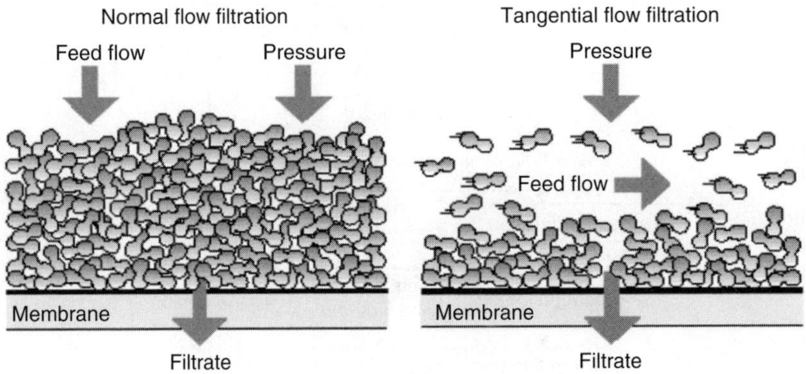

FIGURE 1.10 Normal flow filtration vs. TFF. (Courtesy of Millipore Corporation.)

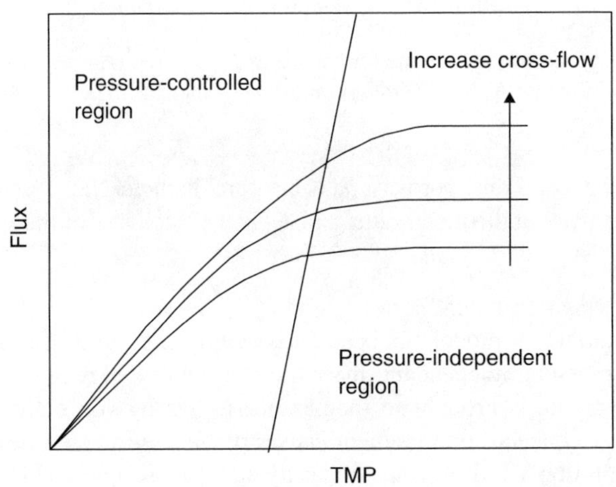

FIGURE 1.11 Effect of TMP on permeate flux, pressure-controlled and pressure-independent region.

no concentration polarization or membrane fouling in the pressure-controlled region as shown in Figure 1.12. Hagen–Poiseuille law is often used to model fluid flow through microporous membrane as follows [52]:

$$J = \frac{\text{TMP}}{\mu R_\text{m}} = \frac{\varepsilon d_\text{p}^2 \text{TMP}}{32\mu\delta_\text{m}} \tag{1.33}$$

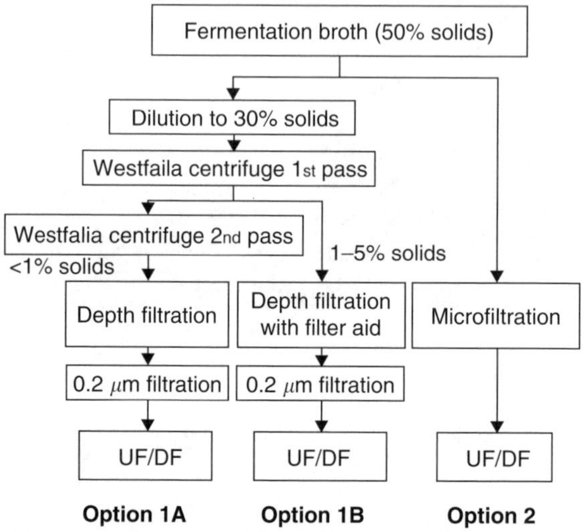

FIGURE 1.12 Case study outline for harvest of *P. pastoris*. (Adapted from Wang A, Lewus R, and Rathore A. *Biotechnology and Bioengineering*, in press, 2006.)

where ε is the membrane porosity, d_p is the pore diameter, μ is the viscosity of the fluid permeating the membrane, and δ_m is the thickness of the membrane.

1.2.2.2.2 Mass Transfer Model

The gel polarization model has been successfully used to describe permeate flux in the pressure independent, mass transfer controlled region [52]. In this model, the solute is brought to membrane surface by convective transport. The resulting concentration gradient causes the solute to be transported back into bulk solution via diffusion. At steady state, these two mechanisms will balance each other and flux (J) can be integrated over the boundary layer to give [52]

$$J = \frac{D}{\delta_g} \ln \frac{C_g}{C_b} \tag{1.34}$$

where D is diffusion coefficient, δ_g is the thickness of gel layer, C_g is the gel concentration, and C_b is the bulk concentration. The diffusion coefficient, bulk, and gel layer concentrations are determined by physicochemical properties of feed. Equation 1.34 indicates that in the pressure independent region, flux can be improved through reducing thickness of gel layer by increasing shear rate, increasing diffusion coefficient by increasing temperature, or reducing bulk protein concentration.

Although the gel polarization model has been widely used in protein UF, this model was found to underestimate the flux of cross-flow MF [53]. The calculated filtrate flux based on this model is one to two orders of magnitude lower than experimental observations. The unexpected flux behavior observed in particle MF was referred to as the flux paradox.

Belfort et al. [53] proposed inertial lift model that assumes that permeate flux bring particles deposit on the membrane and particle inertial lift force drag particles diffuse back to bulk stream. For fast laminar flow with thin fouling layers, the steady state flux predicted by the inertial lift theory is expressed as

$$J = 0.0036\rho_L r_p^3 \gamma_w^2 \mu^{-1} \tag{1.35}$$

where ρ_L and μ are fluid density and viscosity, γ_w is the wall shear rate, and r_p is the particle radius.

Zydney and Colton [54] modified the gel-polarization model by replacing Brownian diffusivity with shear-induced diffusivity and proposed that the filtrate flux during cross-flow MF in an open channel could be described as

$$J = 0.078\left(\frac{r_p^4}{L}\right)^{1/3} \gamma_w \ln\left(\frac{\varphi_w}{\varphi_b}\right) \qquad \varphi_w - \varphi_b \ll \varphi_w \tag{1.36}$$

$$J = 0.126\left(\frac{r_p^4}{L}\right)^{1/3} \gamma_w \left(\frac{\varphi_w}{\varphi_b}\right)^{1/3} \qquad \varphi_b \ll \varphi_w \tag{1.37}$$

where ϕ_w and ϕ_b are particle volume fraction at membrane surface and bulk solution, respectively. Further, ϕ_w is assumed to be close-packed particle concentration ranging from 0.6 for rigid particles and 0.8 to 0.9 for deformable particles. L is the channel length, γ_w is the wall shear rate, and r_p is the particle radius.

Belfort et al. [53] compared gel-polarization model, inertial lift model, and shear-induced diffusion model specifically as they apply to MF. They concluded that for open channel module operated at laminar flow (Reynolds number, Re < 2000), Brownian diffusion is the dominant mechanism for very small particles (diameter < 0.1 μm), shear-induced diffusion is dominant for particle size range from 1 to 10 μm, and inertial lift model becomes dominant for particle size higher than 100 μm.

1.2.2.2.3 Resistance-In-Series Model

This model assumes that the total filtration resistance is the sum of membrane intrinsic resistance (R_m), membrane pore fouling resistance (R_f), and cake resistance due to deposition of particulates on the membrane surface (R_c). Filtration flux is expressed by Darcy's law [52] in the following manner:

$$J = \frac{\text{TMP}}{\mu(R_m + R_f + R_c)} \tag{1.38}$$

Cheryan [52] believed that the fouling resistance (R_f) is related to physicochemical interactions between membrane and feed material, and is relatively unaffected by operating parameters. Cake resistance (R_c) is a function of cake thickness, permeability, and applied pressure.

1.2.2.3 Membrane Fouling

A major limitation of membrane separation is membrane fouling or flux decay over time during TFF operation. Three mechanisms have been proposed to explain the flux decline associated with particle deposition during membrane filtration: pore blockage, pore constriction, and cake filtration [55].

1.2.2.3.1 Pore Blockage Model

In this model, it is assumed that a portion of pores is completely blocked by the particles. The rate of pore blockage is related to the rate of particle convection to the membrane surface [55]:

$$\frac{dN}{dt} = -\alpha_{\text{block}} AJC_b \tag{1.39}$$

where A is membrane surface area, J is permeate flux, C_b is the bulk concentration, N is number of pores, and α_{block} is pore blockage efficiency. The cake formation is assumed to be negligible and membrane resistance can be calculated using Hagen–Poiseuille equation assuming uniform pore size distribution.

$$R_m = \frac{8\delta_m}{N\pi r_{\text{pore}}^4} \tag{1.40}$$

where δ_m is the membrane thickness and r_{pore} is the pore radius. Under constant TMP, Equation 1.39 can be integrated to yield

$$\frac{J}{J_0} = \exp\left(-\frac{\alpha_{\text{block}} AJ_0 C_b}{N_0} t\right) \tag{1.41}$$

where J_0 and N_0 are initial permeate flux and initial number of pores.

1.2.2.3.2 Pore Constriction Model

This model assumes the pore volume change is proportional to particle convection rate [55]:

$$\frac{\mathrm{d}(\pi r_{\text{pore}}^2 \delta_{\text{m}})}{\mathrm{d}t} = -\alpha_{\text{pore}} A J C_{\text{b}} \tag{1.42}$$

where α_{pore} is the pore constriction coefficient. Integration of Equation 1.42 yields

$$\frac{J}{J_0} = \left(1 + \frac{\alpha_{\text{pore}} A J_0 C_{\text{b}}}{\pi r_{\text{pore}}^2 \delta_{\text{m}}} t\right)^2 \tag{1.43}$$

1.2.2.3.3 Cake Filtration Model

This model assumes that the cake resistance (R_{c}) is proportional to cake mass (W) and cake specific resistance (r_{c}). Cake formation rate can be related to particle convection rate as follows [55]:

$$R_{\text{c}} = \left(\frac{r_{\text{c}}}{A}\right) W \tag{1.44}$$

$$\frac{\mathrm{d}W}{\mathrm{d}t} = AJC_{\text{b}} \tag{1.45}$$

Flux decline under cake filtration model can be described as

$$\frac{J}{J_0} = \left(1 + \frac{2r_{\text{c}} J_0 C_{\text{b}}}{R_{\text{m}}} t\right)^{-1/2} \tag{1.46}$$

1.2.2.4 Definition of Key Parameters

1.2.2.4.1 Operational Parameters

Inlet pressure for NFF and TMP for TFF applications are often considered key as they directly impact the flux or the throughput that is generated in the system. As TMP increases, the flux across the membrane typically increases such that the slope of the curve keeps decreasing with increasing TMP. These curves serve as a good indicator of the performance of a filtration step and are commonly used as a qualitative measurement.

Cross-flow rate in TFF applications can be key due to its direct impact on membrane fouling. It is defined as the volumetric flow rate of fluid through the retentate flow channel. It is also referred to as the recirculation rate.

Membrane loading is defined as the amount of product that is loaded on a filtration step divided by the area of the filter. This parameter is often used for scale-up and scale-down of filtration steps such that the scaling approach involves keeping membrane loading identical.

Temperature of a processing step also impacts filter performance as it affects the physical properties (such as viscosity) and the chemical properties (such as product stability). During concentration portion of a UF/DF step, an initial volume V_0 is concentrated to final retentate volume, V, and thus, volume concentration factor (VCF) is defined as

$$\text{VCF} = \frac{V_0}{V} \tag{1.47}$$

During diafiltration, a diavolume (DV) is a measure of the extent of washing that has been performed during a diafiltration step. If a constant-volume diafiltration is being performed, where the retentate volume (V) is held constant and diafiltration buffer (V_d) enters at the same rate that filtrate leaves, a DV is calculated as

$$\text{DV} = \frac{V_\text{d}}{V} \tag{1.48}$$

Other operational parameters that may impact step performance include filter design (channel size and shape) and material of construction. These are generally evaluated during screening of filters for an application.

1.2.2.4.2 Performance Parameters

Several parameters are commonly used as indicators of performance of a filtration step.

Percent recovery of NWP (normalized water permeability) is perhaps the most commonly used performance parameter for monitoring the integrity of a UF/DF membrane. This parameter measures the permeability of the membrane using water and allows for a comparison of the integrity of the membrane pre- and post-use. Percent recovery of NWP typically declines with number of uses since every time the membrane is used, product or other species in the feed material can bind to the pores of the membrane causing decay in the permeability. It is very common to use NWP criteria for determining the number of cycles a membrane should be used, for example, 75 to 125% of original NWP.

Integrity tests are commonly used to identify problems such as macroscopic holes in the membrane, cracks in the seals, or improperly seated modules, which can lead to product leakage and unsatisfactory clearance of impurities. A common way to do this is via an air diffusion test. When air is applied to the retentate side at a controlled pressure, it diffuses through water in the pores at a predictable rate. However, in the presence of any defects the air flows through at a significantly higher rate and, thus, fails the test value. Besides air diffusion, several other tests are also employed to evaluate membrane integrity. These include bubble point determination and pressure hold-decay test. It is recommended that the reader evaluate the applicability of these different tests to the application under consideration and then pick the appropriate integrity testing method.

Protein transmission factor is one of the most important concepts in CFF. Sometimes it is also called sieving coefficient (S). It is defined as the fraction of the target protein that passes through the membrane to the filtrate stream based on the measurable protein concentrations in the feed and filtrate streams. On the other hand, retention, also called rejection, is the fraction of a particular protein that is retained by the membrane.

$$S = \frac{C_{\text{permeate}}}{C_{\text{retentate}}} \tag{1.49}$$

Product yield is related to concentration factor, DV, and protein rejection coefficient. If product of interest is retained in the retentate such as an UF process for concentrating protein solution, product recoveries (Y) during concentration and diafiltration are

$$Y_{\text{concentration}} = (\text{VCF})^{-S} \tag{1.50}$$

and

$$Y_{\text{diafiltration}} = \mathrm{e}^{-S\times \text{DV}} \tag{1.51}$$

But if the product of interest is in the permeate, for example, a microfiltration process to remove extra cellular protein from whole cells and cell debris, then the product recoveries for concentration and diafiltration are:

$$Y_{\text{concentration}} = 1 - (\text{VCF})^{-S} \tag{1.52}$$

and

$$Y_{\text{diafiltration}} = 1 - \mathrm{e}^{-S\times \text{DV}} \tag{1.53}$$

1.3 CASE STUDY: HARVEST OF A THERAPEUTIC PROTEIN EXPRESSED IN *P.* PASTORIS

In this section we provide data from a comparison of the different approaches toward harvest of a target protein expressed in yeast cells. The fermentation broth had approximately 50% solids and the target protein was expressed extracellularly. Figure 1.12 illustrates the different approaches that were compared. Option 1A and 1B involved a combination of centrifugation and depth filtration. Option 2 involved using a MF step for clarification of the feed stream.

1.3.1 MATERIALS

High cell density *P. pastoris* fermentation broth was produced at 300 l scale and then diluted to 30 to 45% solids by the addition of purified water. The centrifuge used for the yeast centrifugation development work was a Westfalia CSA-8 DSC for which the machine specific values are given in Table 1.3. The sigma value was provided by the vendor but was confirmed using Equation 1.11. The centrifuge used for bench scale studies and measuring percent solids was a Beckman JC-HC bottle centrifuge with a temperature-controlled compartment.

After centrifugation, centrate was stored in the cold room (~4°C) prior to depth filtration. In order to account for the effect of feed properties, two batches of centrate with solids content of 0 and 0.7% were used for depth filtration study. For filter aid studies the feedstock was centrate from one pass on the Westfalia centrifuge, containing 7% solids. Eleven different filters from three manufacturers were carefully chosen and their properties are listed in Table 1.4. For centrate with high percent solids, the depth filter train was designed in

TABLE 1.3
Design Parameters for the Westfalia CSA-8 Centrifuge

Parameter	Value
Number of discs	90
Disc angle	55°
Bowl volume (l)	3
Solids volume (l)	1.5
Disc thickness (mm)	0.46
Disc spacer (mm)	0.32
Sigma (m^2)	11,366

stages by using an open filter (such as CUNO 10SP, Pall Supra 80P) ahead of a tighter grade. Some of the depth filters in Table 1.4, including the Millipore Millistak+A1HC, Millistak+B1HC, and the CUNO 90M08, 120 M08 combine sequential grades of media in one filter. Lab scale disposable filter disks were utilized in the early development stage and 16 in. (1.8 m^2) cartridges were used for pilot scale demonstration runs.

Filter aid Celpure 100, a high purity pharmaceutical grade filter aid, was purchased from Sigma-Aldrich.

Hollow fiber cartridge with pore size of 0.1 μm, lumen diameter of 1 mm, pass length of 30 cm, and surface area of 0.12 m^2 was obtained from GE Healthcare (Model# CFP-1-E-5A).

1.3.2 Methods

Table 1.5 lists the assays used to assess centrifuge and filtration performance. Target protein recovery was determined using anion exchange chromatography. Target protein concentration was determined for feed, centrate, sludge, and filtrate pools and used to calculated product recovery. PSD analysis was performed on feed and centrate samples. Viscosity was measured on some of the feed and centrate samples. Turbidity was measured using Hach®portable turbidity meter (Cole-Parmer Cat# EW-99511-00) in the unit of NTU. Absorbance at 600 nm was used to determine centrate clarity.

In addition to the assays presented in Table 1.5, several other measurements were recorded to assess centrifuge performance: sludge weight, percent solids, and product pool weights. Sludge weights were determined by either measuring the increase in the sludge pool weight or by pulling the sludge from each individual discharge and weighing them separately. The sludge weights were recorded for every discharge. Percent solids were measured on the feed, centrates, and sludges. The percent solids were measured by centrifuging ~1.5 ml of sample in an eppendorf tube at 14,000 rpm for 1 min in a benchtop microcentrifuge. The weight of the solid and liquid were measured, and the percent solids were calculated based on weight fractions. Mass balances were performed using weights of all the feeds, sludges, and centrates. All the collection tanks were tared so that accurate mass and volumes could be calculated for mass balance.

Prior to filtration, depth filters were first flushed with sufficient amount of buffer to thoroughly wet filter media and reduce the level of extractables. The centrate that was agitated and maintained cold (2 to 8°C) was then pumped through the filters at constant flow rate of 250 LMH until reaching differential pressure of 30 psi. Filtrate volume, filtrate turbidity, and differential pressure were recorded at different time intervals.

TABLE 1.4
Depth Filters Used in Filter Screening and Their Properties

Vendor	Filter	Pore Size (μm)	Description
Millipore	Millistak + A1HC	0.1–0.4 (DE65)/<0.1 (DE75)	Two layers of inorganic filter aid (DE) and 0.1 μm nominal cellulosic membrane (RW01)
Millipore	Millistak + B1HC	0.2–0.7 (DE50)/<0.1 (DE75)	Two layers of inorganic filter aid (DE) and 0.1 μm nominal cellulosic membrane (RW01)
CUNO	10SP	0.8–4	Single layer of pharmaceutical grade media Coarse pre-filter
CUNO	90SP	0.2–0.65	Single layer of pharmaceutical grade media
CUNO	30M03	0.8–4 (10SP)/ 0.6–2 (30SP)	Dual-zone construction, high contaminant holding capacity
CUNO	90M08	0.45–0.8 (60SP)/0.2–0.65 (90SP)	Dual-zone construction, high contaminant holding capacity
CUNO	120M08	0.45–0.8 (60SP)/0.1–0.45 (120SP)	Dual-zone construction, high contaminant holding capacity
Pall	SupraEK1P	0.2–4	P series depth filter, combination of cellulose fibers, DE and perlite, pyrogen removal capability.
Pall	EKSP	0.1–0.3	P series depth filter, combination of cellulose fibers, DE and perlite, pyrogen removal capability.
Pall	Supra 80P	1–3	P series depth filter, combination of cellulose fibers, DE and perlite, pyrogen removal capability.
Pall	K150	2.5–4	K series depth filter, combination of cellulose fibers, DE and perlite.

Source: Rathore AS, Wang A, Menon M, Riske F, Campbell J, Goodrich E, and Martin J. *Biopharm International* 2004;17:50–58. With permission.

TABLE 1.5
List of the Different Analytical Techniques That Were Used for Assessing Centrifugation and Filtration Performance

Assay Description	Purpose
Ion exchange chromatography	Target protein concentration determination
Malvern laser light diffraction	Particle size distribution
Rheometer	Viscosity
Turbiditimeter	Turbidity of centrates and filtrates
UV/VIS spectroscopy	Absorbance of centrates and filtrates at A600 nm
SDS-PAGE densitometry	Protein concentration

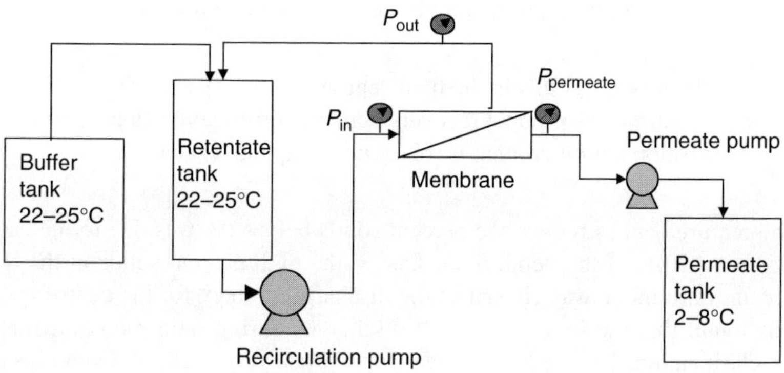

FIGURE 1.13 Microfiltration system.

For filter aid studies, the filter pad was first precoated with 3 mm Celpure media (0.1 g/cm^2 surface area). Then 0, 18, 30, and 54 g/l Celpure 100 (also called body feed) was added to the feedstock and were maintained in suspension by agitation. The body-fed feedstock was then filtered at a constant flow rate of 350 LMH until either the differential pressure reached 30 psi *g*, or the filter assembly no longer had any available volume in the headspace. Throughput and differential pressure were recorded at different time intervals.

The apparatus used to perform the microfiltration experiment is presented schematically in Figure 1.13. Two peristaltic pumps were used to circulate fermentation broth and control permeate flux. Normalized clean water permeability was tested prior to loading of feed material to the MF system. The targeted protein is stable in postproduction broth at room temperature. Therefore, during MF process, the retentate tank was maintained at room temperature

(22 to 25°C) to reduce fluid viscosity and maximize flux. The permeate pool was chilled to 2 to 8°C to minimize product degradation. For MF studies protein concentration in the feed and filtrate were analyzed by SDS-PAGE using 4 to 20% Tris-Glycine gel from Invitrogen (Cat# EC6025). The gels were stained by coomassie blue, imaged with a scanning laser densitometer (BioRad Model GS800), and the band intensity was quantified by Quantity One imaging software (Version 4.2.1).

1.3.3 Results and Discussion

1.3.3.1 Centrifugation

This unit operation was common between Options 1A and 1B and used a Westfalia CSA-8 DSC. There were three goals to the development work:

- Obtain <1% solids in the final centrate pool
- Obtain >85% product recovery across centrifugation step
- Develop a protocol that is robust and simple to execute

The requirement to reduce the percent solids below 1% was due to the capacity constraints of the depth filter. Due to the high percent solids in the feed stream, ratio mode was chosen as the discharge strategy for the centrifuge. It was found that the intermittent full discharges during ratio mode facilitated the clarification. The feed stream was found to be very sticky and would accumulate in the bowl between discharges and eventually cause solids carryover into the centrate stream. The intermittent full discharges allowed for removal of any accumulated solids. To develop the process to handle a dynamic clarification, the discharge time was allowed to vary during a run (manually changed as needed).

Seven runs were performed using 300 l fermentation batches to determine the optimal conditions for centrifugation and the final scheme utilized a combination of operating variables. Variation of the discharge ratio allowed for better control of solids accumulation in the bowl and the discharge time was adjusted throughout centrifugation to maintain a target sludge weight. The target sludge weight was based on mass balance for the desired 85% liquid recovery. The feed interval (see Equation 1.17 for calculation) was adjusted for each harvest based on the percent solids in the feed.

It was observed that an increased dilution to 30–35% solids vs. 40–45% greatly facilitated clarification and reduced the solids accumulation in the bowl. Further, as shown in Figure 1.14, with more diluted feed, only a single pass was required to achieve the targeted clarification.

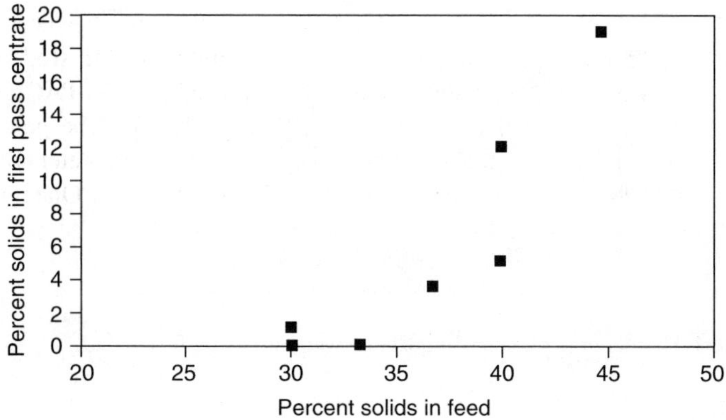

FIGURE 1.14 Plot showing effect of feed dilution on centrate clarification.

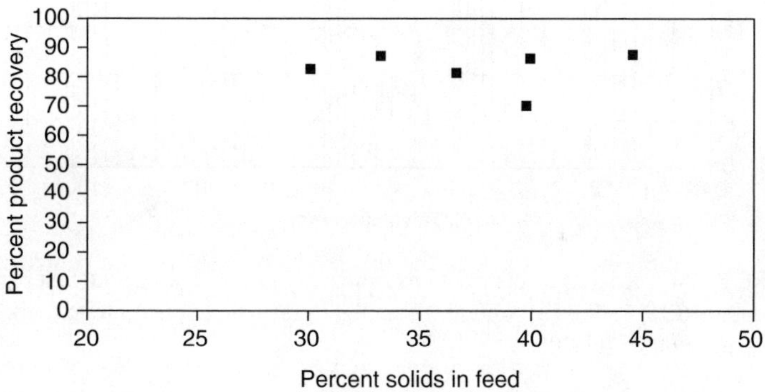

FIGURE 1.15 Plot showing effect of feed dilution on product recovery.

Data presented in Figure 1.15 suggests that the dilution did not have a strong impact on the product recovery. This is likely due to the fact that the sludge weights were controlled ensuring that there was no excess product loss during solids ejection.

Percent solids were also measured in the sludges during the centrifugation to help characterize the discharges and predict the product recovery. By measuring percent solids in the sludge during processing, the operating conditions could be optimized to maximize product recovery. It was found that after a full discharge, the percent solids in the sludge usually decreased for a few discharges and then started increasing as the bowl accumulated solids.

The weight of each discharge was measured to help monitor the efficiency of solids' removal by the centrifuge. If solids accumulate in the bowl and are

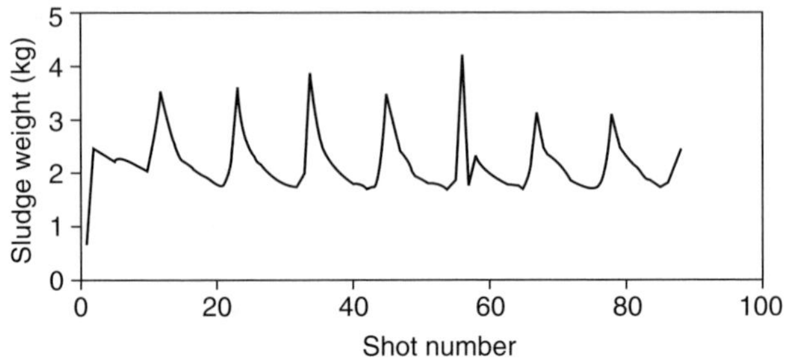

FIGURE 1.16 Sample plot showing sludge weights during first pass.

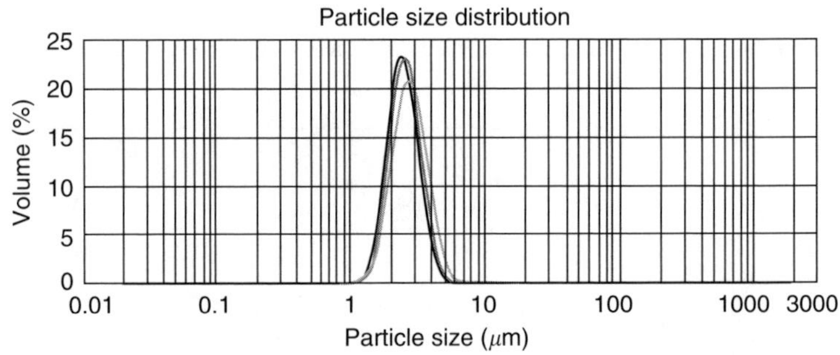

FIGURE 1.17 Particle size distributions from a poor harvest where the first pass centrate contained 19% solids by weight. Feed (light gray); first pass centrate (dark gray); second pass centrate (black).

not removed during a partial discharge, the available solids volume decreases, which leads to solids carryover in the centrate and a decrease in clarification efficiency. It is desirable to keep the sludge weight at a target so as to ensure that all the solids in the bowl have been ejected and that the percent solids in the sludge are within a desirable range. Figure 1.16 shows the sludge weight decreasing leading up to a full discharge.

Figure 1.17 shows particle size distribution from one of the development runs where the centrifuge did not perform optimally. A large fraction of particles were still present in the centrate, resulting in similar particle size distributions for the feed, 1st pass centrate, and 2nd pass centrate. As expected, the different samples contain different amounts of these particulates.

Figure 1.18 shows the particle size distribution from one of the development runs that yielded satisfactory performance. For this case, the particle size distribution of the centrate could be differentiated from the feed stream,

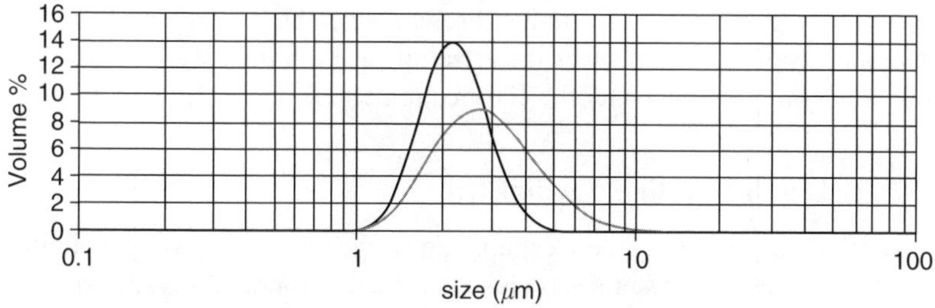

FIGURE 1.18 Particle size distributions from a successful harvest where the percent solids in the first pass was 1% by weight. Feed (gray); first pass centrate (black).

TABLE 1.6
Viscosity Results for Feed Streams Containing Different Percentage of Solids

Sample	Viscosity (cP)
Feed with ~45% solids	4.27
Feed with ~40% solids	4.5
First pass centrate with ~19% solids	1.88
First pass centrate with ~12% solids	1.7
Second pass centrate with ~15% solids	1.6

illustrating the usefulness of this analysis in development of a centrifugation step.

Viscosity was measured on a variety of feed and centrate samples and the results are shown in Table 1.6. However, for our application, a correlation could not be made between the feed viscosity and the clarification efficiency. This can be explained by using Equation 1.9 to calculate the minimum particle diameter that can be separated. Using $g = 9.8$ m/sec^2, $\Sigma = 10,350$ m^2, $\Delta\rho = 110$ kg/m^3, and viscosity of 4.5 cP, the minimum diameter of a particle that can be separated is 24.2×10^{-6} μm, while for a viscosity of 4.27 cP the minimum particle diameter is 23.0×10^{-6} μm. Both conditions ensure that yeast cells (diameter 2 to 3 μm) will be separated and the change in viscosity should not affect the clarification efficiency of the step.

The final centrifuge process was a ratio mode operation with intermittent full discharges. It was found that full discharges were necessary in order to remove accumulated solids from the bowl due to incomplete solids ejection during the partial discharges. Additionally, discharge time was varied throughout the process in order to maintain a target discharge weight. The key finding in the development of the process was the effect of feed dilution. We found that the greater the dilution, the better the clarification efficiency.

1.3.3.2 Depth Filtration (Option 1A)

This option involved using the Westfalia centrifuge followed by a depth filtration step. In this section, we summarize results from development, characterization, and scale-up of the depth filtration step. The objective was to develop a robust and scalable unit operation that could handle variations in the percentage of solids in the feed stream resulting from underperformance of the centrifuge.

Typical filtration profiles for feed with 0% solids and 0.7% solids are presented in Figure 1.19. For feed containing 0% solids, an appreciable turbidity breakthrough was observed in Figure 1.19a. For feed containing 0.7% solids, no turbidity breakthrough was observed, but rather pressure breakthrough was the primary limitation as shown in Figure 1.19b. Figure 1.19 indicates that two different capture mechanisms dominate depth filtration under different circumstances. When the solids level is low, fine particles bind to the filter media due to electrostatic or other physicochemical interactions. Once binding sites in the filter media are fully occupied, fine particles flow through in the filtrate, resulting in turbidity breakthrough. On the other hand, with high solids containing feed, pore, or flow channels within the filter media were progressively clogged with captured solids and the differential pressure across the filter gradually increases in response. In this case, mechanical sieving is the main capture mechanism. Since the characteristics (solids content, particle size distribution) of feedstock can vary with centrifugation performance, both mechanisms need to be considered while developing this application.

1.3.3.2.1 Filter Screening

Screening criteria considered were filter capacity, filtrate quality as determined by filtrate turbidity, target protein mass balance, as well as robustness toward different centrate characteristics, such as percentage of solids and feed turbidity. The throughput at 50% of maximum differential pressure for all the filters is summarized in Table 1.7. While most filters exhibited good capacity with feed containing 0% solids, filter capacity declined significantly when using feed containing 0.7% solids [12]. The only exceptions to this observation were the performance of the Millistak + A1HC and B1HC, which have open diatomaceous earth (DE) media, tighter DE media, as well as 0.1 μm cellulosic membrane. Both filters provided adequate capacity for both feeds.

As shown in Figure 1.19 with feed containing 0.7% solids, there was no turbidity increase and the filtrate pool turbidity was about 3 NTU for all filters tested. On the other hand, with feed containing 0% solids, turbidity

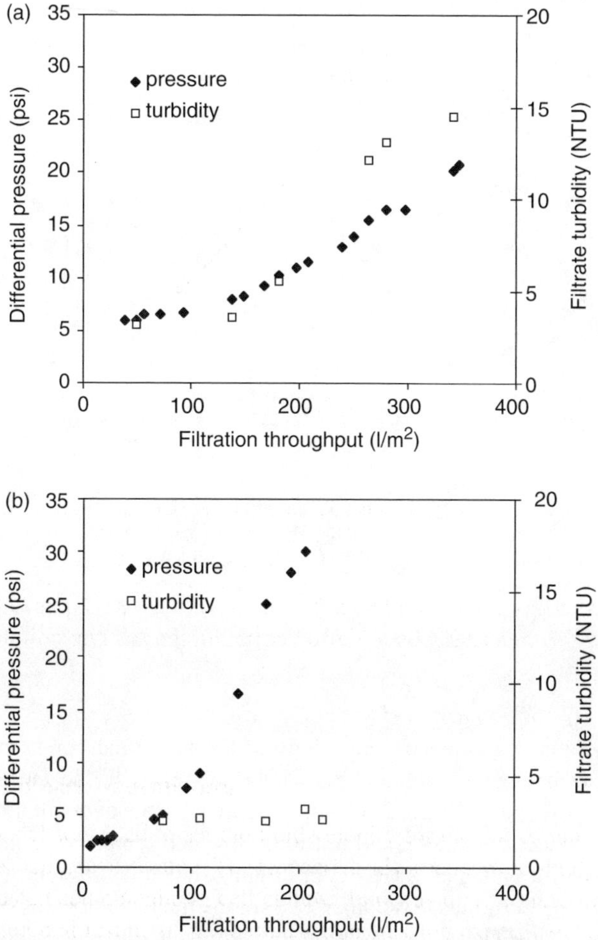

FIGURE 1.19 Filtration profile of A1HC depth filter. (a) Feedstock contains 0% solids, (b) feedstock contains 0.7% solids. Filtration was carried out at a flow rate of 250 LMH at 4°C. ♦ pressure; □ turbidity.

breakthrough was observed and the pool turbidity was 10 to 15 NTU for most filters. The recovery across the filters was measured by an ion exchange chromatography and was found to be between 80 and 95%.

Filter screening results indicated that pressure breakthrough is the primary limitation for this application. Upon further comparison of Millistak A1HC and B1HC, A1HC was selected as the most suitable filter due to significantly higher capacity with 0% solids and comparable capacity with 0.7% solids [12].

TABLE 1.7
Throughput Comparison (l/m^2 at 15 psi)

	Filter Train	Feed w/0% Solids	Feed w/0.7% Solids
CUNO	90SP	268	
	120M08	133	62.5
	10SP + 90SP		42.5
	30M08 + 120M08		71.4
	30M08 + 90M08		89.3
Millipore	A1HC	260	137
	B1HC	174	160
Pall	EKSP	40.4	
	SupraEK1P	161.5	69.2
	Supra80P + EK1P		54.8
	K150 + EK1P		115

Source: Rathore AS, Wang A, Menon M, Riske F, Campbell J, Goodrich E, and Martin J. *Biopharm International* 2004;17:50–58. With permission.

1.3.3.2.2 Optimization and Scale-Up

Several operating parameters were examined for their effects on filter performance. These parameters included percentage of solids in the feed, lot-to-lot variation in feed, batch-to-batch variation in filters, scale of depth filter (bench, pilot, and production scales), and filtration flow and temperature. Figure 1.20 shows the performance of A1HC depth filter using two different lots of feed material. While both feeds contained 0% solids, the turbidity was 58 and 129 NTU, respectively. Lot-to-lot variation in feed resulted in earlier pressure and turbidity breakthrough in one case, indicating the need for a safety factor during scale-up. Filter sizing strategy from lab scale to pilot scale is outlined in Table 1.8. Filter surface area is scaled-up linearly and other operational parameters, such as temperature and linear flow rate were maintained constant. Figure 1.21 presents data comparing the filter performance at lab scale and pilot scale. Comparable performance with respect to product recovery, filter capacity, and filtrate turbidity are observed between two scales.

Figure 1.22 shows that the filter capacity decreases significantly with increasing percentage of solids in the feed stream. If the performance of the centrifugation step varies, this would cause variations in the percentage of solids in the feed stream and result in a large variation in the required filter area for the depth filtration step.

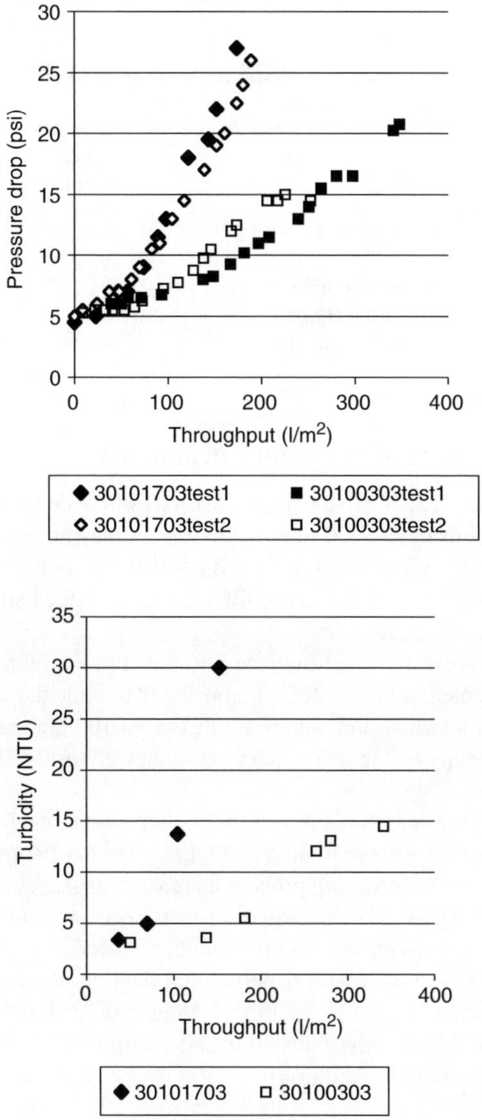

FIGURE 1.20 Impact of lot-to-lot feed variation on filtration performance. Load material turbidity is 129 NTU for lot 30101703 and 58 NTU for lot 30100303. (Adapted from Rathore AS, Wang A, Menon M, Riske F, Campbell J, Goodrich E, and Martin J. *Biopharm International* 2004;17:50–58.)

TABLE 1.8
Millistak+A1HC Filter Sizing for Lab and Pilot Scale

Parameters	Lab 1 l	Pilot 300 l
Max. pressure endpoint (psi)	30	30
Centrate volume (l)	0.5	250
Minimum required area (m^2)	0.0023	1.25
Process flux (LMH)	250	250
Process time (min)	30	30

1.3.3.3 Filter Aid Assisted Filtration (Option 1B)

As seen in Figure 1.22, due to the high sensitivity of the depth filter capacity on percentage of solids in the feed stream, Option 1A may not be optimal for feed stream containing >1% solids. Option 1B explores the potential of using filter aid to improve capacity of the depth filter for case of feed stream containing higher percentage of solids.

Differential pressure vs. normalized filtrate volume for different amount of filter aid is presented in Figure 1.23. It appears that while the depth filter clogs immediately without filter aid, addition of Celpure 100 dramatically improves the filtration capacity and higher concentration of Celpure 100 results in higher filtration capacity.

The underlying mechanism can be better understood upon applying Equation 1.28 on the data presented in Figure 1.6a and Figure 1.9. Although adding filter aids in the filtration process increases particle concentration (C_p), it significantly reduces cake specific cake resistance (r_c) and cake compressibility (s), thus improving filtration throughput. Modeling results based on Equation 1.28 fit very well with experimental data, as shown in Figure 1.23. Curve fitting results based on Equation 1.28 are summarized in Figure 1.24. It indicates that specific resistance (r_c) and compressibility (s) are reduced dramatically with filter aid addition.

1.3.3.4 Microfiltration (Option 2)

In this case study, we explored the opportunity of using microfiltration technology to harvest high cell density yeast fermentation without dilution and centrifugation. Because the fermentation broth already contains 50% solids,

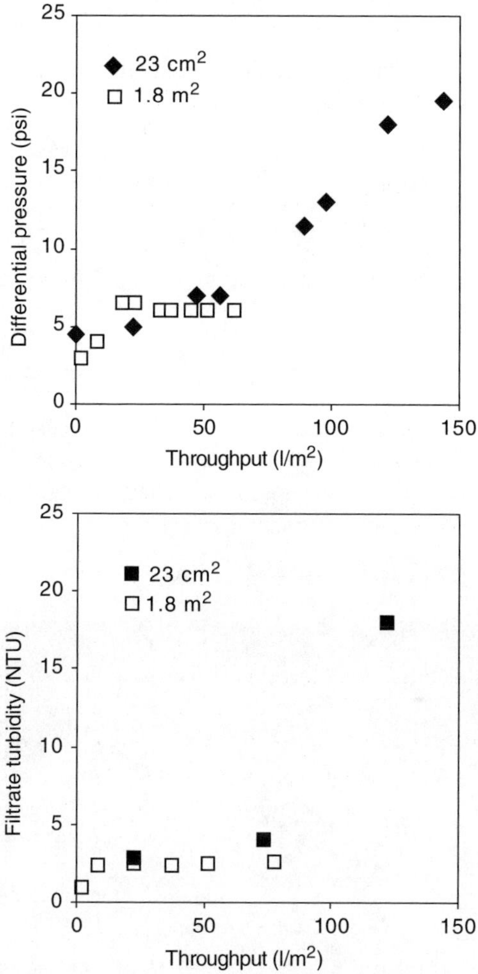

FIGURE 1.21 Comparison of Millstak A1HC performance at different scale. Filtration experiments were performed using same lot of feed material at flow rate of 250 LMH, 4°C. ♦ 23 cm^2 □ 1.8 m^2. (Adapted from Rathore AS, Wang A, Menon M, Riske F, Campbell J, Goodrich E, and Martin J. *Biopharm International* 2004; August.)

it was not further concentrated and two DVs were performed directly. Target protein is washed into permeate tank and yeast cells are retained in the retentate tank.

For cross-flow MF, the membrane permeability is so high that nearly all of the cross-flow is converted to filtrate with very little applied TMP. Often the

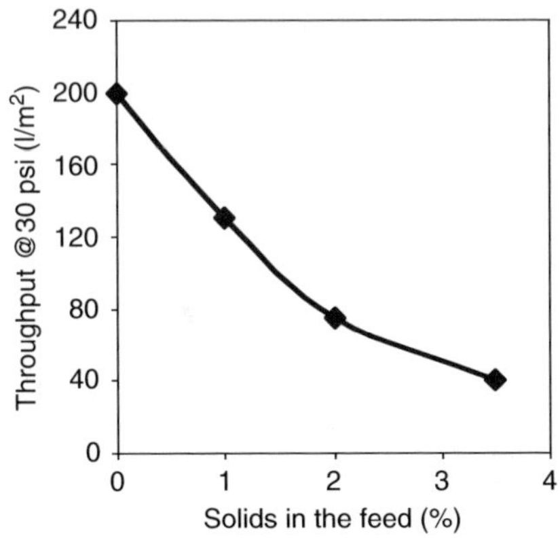

FIGURE 1.22 Graph showing a plot of filter capacity vs. percent solids. (Adapted from Wang A, Lewus R, and Rathore A. *Biotechnology and Bioengineering*, submitted.)

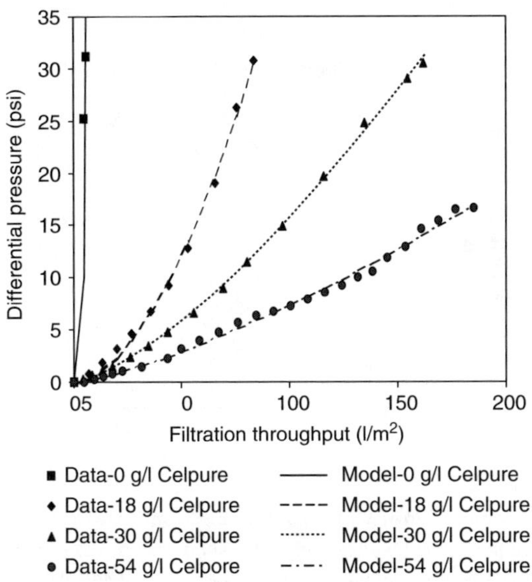

FIGURE 1.23 Effect of filter aid on filtration performance. Various amount of Celpure 100 (as specified in the figure) were added to a feed with 7% solids. Filtration was performed at constant flow rate of 350 LMH at 4°C. Markers are experimental data and lines are mathematical modeling based on Equation 1.28. ■ data- 0 g/l Celpure; ♦ data 18 g/l Celpure; ▲ data 30 g/l Celpure; • data 54 g/l Celpure. Model- 0 g/l Celpure (—); Model 18 g/l Celpure (- - -); Model- 30 g/l Celpure (. . .); Model- 54 g/l Celpure (-.-.-). (Adapted from Wang A, Lewus R, and Rathore A. *Biotechnology and Bioengineering*, submitted.)

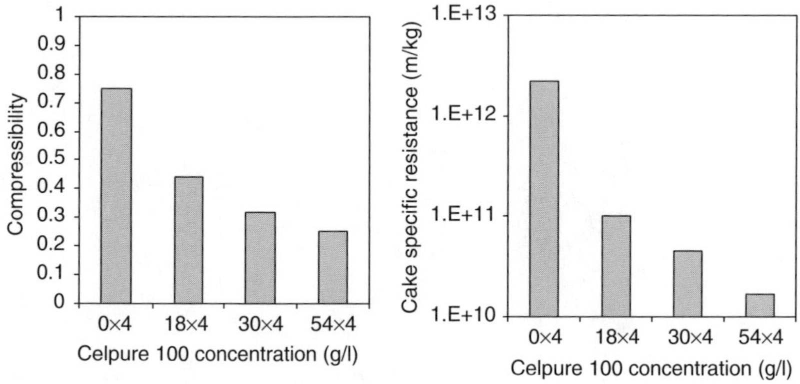

FIGURE 1.24 Summary of calculated compressibility and specific resistance under different Celpure 100 concentrations.

process starts with very high initial flux and follows by dramatic flux decay due to high wall concentrations and high membrane fouling. Operating under constant permeate flux instead of constant TMP is typically recommended for MF application to avoid excessive membrane fouling [9,22,23,29]. Under permeate control mode, flux is maintained constant, and the TMP is allowed to vary accordingly. Increase in TMP indicates higher resistance generated during microfiltration due to higher cake resistance (higher R_c), as illustrated in Equation 1.38.

Water permeability test results show that NWP for 0.1 μm membrane is 250 LMH/psi. Since viscosity of water at room temperature is 1 cP (0.001 Pa sec), membrane resistance R_m can be calculated using Equation 1.33 and is found to be 9.8×10^{10} m^{-1}.

Diafiltration of yeast fermentation broth was carried out at cross-flow of 33 l/min/m^2 using GE Healthcare 0.1 μm hollow fiber membrane. Permeate flux was controlled at 55 LMH. Two diafiltration volumes were performed at a membrane loading of 100 l fermentation broth/m^2 membrane area. TMP and permeate flux were plotted against process volume and are shown in Figure 1.25. It is seen that TMP is very stable at 2 to 3 psi during the entire diafiltration.

The viscosity of permeate was assumed to be 1 cP at room temperature since permeate is a very diluted protein solution with low salt concentration. Besides, the viscosity of Westfalia centrate with as much as 15% solids at 20°C was measured to be 1.6 cP (Section 1.3.3.1). Total resistance, cake, and fouling resistance can be calculated according to Equation 1.38 and are listed in Table 1.9. The results indicated that membrane resistance is

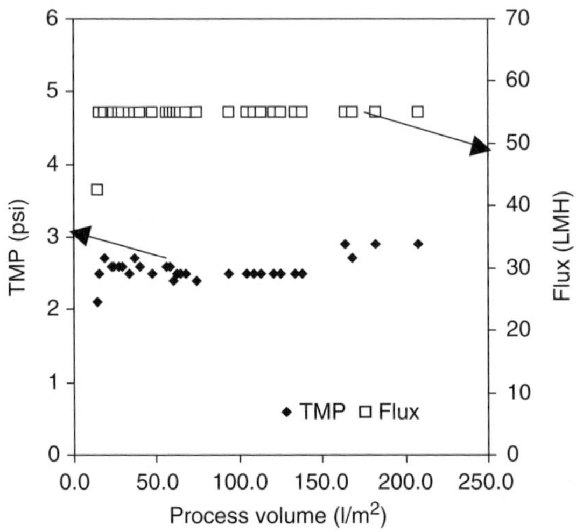

FIGURE 1.25 Harvest of yeast fermentation broth using hollow fiber cartridge. Two diafiltration volumes were performed at cross-flow of 33 l/min/m^2 and room temperature. Membrane loading is 100 l fermentation broth/m^2 membrane area. ◆ TMP; □ flux. (Adapted from Wang A, Lewus R, and Rathore A. *Biotechnology and Bioengineering*, in print, 2006.)

TABLE 1.9
Calculated Filtration Resistances during MF Using Hollow Fiber Cartridges

Parameters	Resistance (10^{11} l/m)
Total resistance	11
Cake and fouling resistance	10
Membrane resistance	0.98

negligible; the main resistance is from the cell cake and membrane fouling. We believe that MF of high cell density fermentation needs to be operated at conditions of high shear rate and low TMP. The high shear rate reduces the cell cake thickness and low TMP prevents overcompressing of the cell cake.

Target protein concentration in feed, permeate pools (every 0.5 DV), and final retentate were analyzed by SDS-PAGE and densitometry. It was confirmed that most of the target protein was washed out in permeate during diafiltration

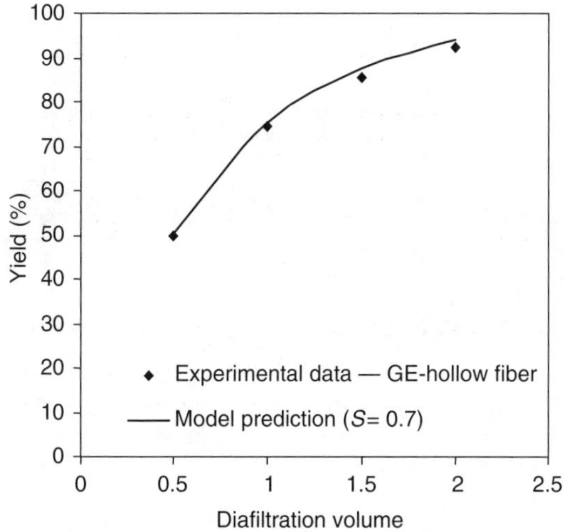

FIGURE 1.26 Comparison of product yield: experimental data using the GE hollow fiber (♦) vs. calculated values using model with $S = 0.7$ (solid line). (From Wang A, Lewus R, and Rathore A. *Biotechnology and Bioengineering*, in press, 2006. With permission.)

and only small amounts remain in the retentate at the end of diafiltration. Protein transmission factor (sieving) is calculated to be 0.7 by comparing permeate and retentate concentration. Based on SDS-PAGE results, protein mass in feed, each permeate pool, and final retentate were calculated. Mass balance for target protein is 100% and protein recovery at 1.5 and 2.0 DV are 85 and 92%, respectively. The product yield calculated based on Equation 1.53 using observed protein transmission factor also matches well with experimental observations, as shown in Figure 1.26.

The scale-up calculation is based on maintaining membrane loading constant. A membrane area of 30 m^2 is required for 3000 l scale harvest. The membrane configuration is presented in Figure 1.27. This would require three 30 cm path length cartridges (surface area of 2 m^2 each) in series and five banks in parallel with individually controlled permeate streams to manage frictional pressure drop, and to maintain same TMP for upstream and downstream cartridges. Processing time and pump requirement are presented in Table 1.10.

1.4 CONCLUSIONS

In this chapter we presented a discussion of the theoretical principles that govern separations for centrifugation, depth filtration, and MF. Theoretical models for

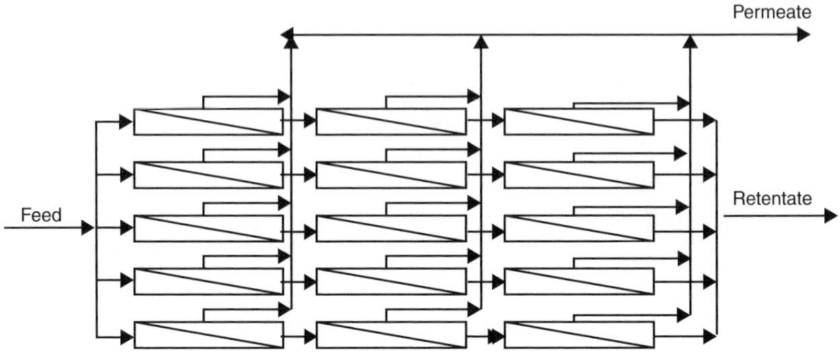

FIGURE 1.27 Proposed configuration for 30 m^2 hollow fiber cartridges. Permeate streams are individually controlled to manage frictional pressure drop, and to maintain same TMP for upstream and downstream cartridges.

TABLE 1.10
Scale-Up Parameters for 3000 l Harvest

Process Parameters	Values
Flux (LMH)	55
Retenate flow (l/min/m^2)	33
TMP (psi)	2–3
Loading (l/m^2)	100
Total membrane area (m^2)	30 (2 × 15)
Pool volume (l)	4500
Processing time (h)	2.7
Membrane configuration	3 in series, 5 banks
Pump requirement (LPM)	330

these steps have also been reviewed. Finally, a case study on clarification and isolation of a target protein expressed in *P. pastoris* is presented and performance of different harvest approaches consisting of these unit operations have been compared. Table 1.11 summarizes these results. It is evident that all three approaches offer feasible methods for harvest and offer comparable product recovery, clarification, and processing times. However, they differ significantly in the time required for development and optimization studies, scalability, cost of consumables, capital cost, and other attributes. Thus, it is recommended that these considerations be taken into account while choosing the optimal harvest approach for an application.

TABLE 1.11
Process Comparisons of Different Options

	Option 1A	Option 1B[a]	Option 2[b]
Process			
Membrane/filter area[b]	21.4	10	30
Processing time for filtration step	0.5	1	3
Number of processing	3	3	2
Total processing time for harvest	13	14	11
Harvest yield (%)	80	80	86
Filtrate pool turbidity	2–3	5–6	2–3
Economic Factors			
Capital	High (centrifuge required)	High (centrifuge required)	Low (no centrifuge)
Consumables	Low	Low	High (MF membrane cost)
Reuse validation	Low (no reuse required)	Low (no reuse required)	High (reuse required)
Manufacturability			
Ease of scale-up	Medium (centrifuge scale-up can be challenging)	Medium (centrifuge scale-up can be challenging)	High (linear scale-up is straightforward)

[a] Assuming feed w/3% solids and 20 g/l Celpure 100.
[b] Assuming 1.5 DV at flux of 50 LMH.

Source: Wang A, Lewus R, and Rathore A. *Biotechnology and Bioengineering*, in press, 2006. With permission.

ACKNOWLEDGMENTS

The authors would like to acknowledge Aurelie Edwards, Ph.D. (Tufts University, Medford, MA) for helpful discussions and guidance. The authors would also like to thank Rachael Lewus (University of Virginia, Department of Chemical Engineering); Sarah Hove, Kara Lounsbury (both from Millipore Corporation); Sharon Squires, Craig Robinson (both from GE Healthcare); Glenn Hiroyasu (Pall Corporation); Fred Hutchison (Sartorius Corporation); and finally, Lars Pampel, Ph.D., Aylin Vance, Stephanie Tozer, Matt Karpen, Tina Kim, Raj Krishan, Ph.D., and Steve Decker (all from Amgen Inc.) for their help with some of the experiments outlined in this chapter.

NOMENCLATURE

a	acceleration
A	area
C	concentration
d	diameter
D	diffusivity coefficient
DV	diavolume
F_B	buoyancy force
F_D	drag force
g	gravity
h	depth of settling tank
J	flux
k	Boltzmann's constant
L	channel length
n	number of discs
N	number of pores
P	pressure
Q	flow rate
r	radius
r_c	average specific cake resistance
R	resistance
s	compressibility
s_L	thickness of the liquid layer between the discs
S	sieving coefficient
t	time
t_s	settling time
T	temperature
TMP	transmembrane pressure
v	approach velocity
V	volume
VCF	volume concentration factor
W	mass of cake solids
x	fraction of centrifugation time accelerating
y	fraction of centrifugation time decelerating
Y	product recovery
z	bed depth
$b, c, d, g, \alpha, \beta, \gamma, \delta, \xi, r'_c$	constants
α_{block}	pore blockage coefficient
α_{pore}	pore constriction coefficient
γ_w	wall shear rate
δ_m	membrane thickness

δ_g gel layer thickness
ε bed porosity
η total collection efficiency
θ disc angle
λ filtration coefficient
μ viscosity
ν velocity
ν_s settling velocity
ν_g gravitational settling velocity
ρ density
Σ equivalent settling area in a centrifuge
ϕ volume fraction
ω angular velocity

REFERENCES

1. Stephenne J. Production in yeast versus mammalian cells of the first recombinant DNA human vaccine and its proved safety, efficacy and economy: Hepatitis B vaccine. In Mizrahi A and Wezel AL van, Eds., *Advances in Biotechnological Processes: Viral Vaccines*, Vol. 14. New York: Alan R. Liss, 1990.
2. Belter PA, Cussler EL, and Hu WS. *Bioseparations: Downstream Processing for Biotechnology*. New York: John Wiley & Sons, Inc., 1988.
3. Sharma BP. Cell separation systems. In Lydersen B, D'Elia N, and Nelson K, Eds., *Bioprocess Engineering: System, Equipment and Facilities*. New York: John Wiley & Sons, Inc., 1994, pp. 87–117.
4. Bentham AC, Bonnerjea J, Orsborn CB, Ward PN, and Hoare M. The separation of affinity flocculated yeast cell debris using pilot-plant scroll decanter centrifuge. *Biotechnology and Bioengineering* 1990; 36: 397–401.
5. Mosqueira FG, Higgins JJ, Dunnill P, and Lilly MD. Characteristics of mechanically disrupted baker's yeast in relation to its separation in industrial centrifuges. *Biotechnology and Bioengineering* 1981; 23: 335–343.
6. Clarkson AI, Bulmer M, and Titchener-Hooker NJ. Pilot-scale verification of a computer-based simulation for the centrifugal recovery of biological particles. *Bioprocess Engineering* 1996; 14: 81–89.
7. Varga EG, Titchener-Hooker NJ, and Dunnill P. Prediction of the pilot-scale recovery of a recombinant yeast enzyme using integrated models. *Biotechnology and Bioengineering* 2001; 74: 96–107.
8. Yavorsky D, Blanck R, Lambalot C, and Brunkow R. The clarification of bioreactor cell cultures for biopharmaceuticals. *Pharmaceutical Technology* 2003; 62–76.
9. van Reis R and Zydney A. Membrane separations in biotechnology. *Current Opinion in Biotechnology* 2001; 12: 208–211.
10. Yavorsky D and Mcgee S. Selection and sizing of clarification depth filters. *Genetic Engineering News* 2002; 22: 44–45.

11. Bell DJ and Davies RJ. Cell harvesting of oleaginous yeast by crossflow filtration. *Biotechnology and Bioengineering* 1987; 29: 1176–1178.
12. Rathore AS, Wang A, Menon M, Riske F, Campbell J, Goodrich E, and Martin J. Optimization, scale-up and validation issues in filtration of biopharmaceuticals — part I. *Biopharm International* 2004; 17(8): 50–58.
13. Heertjes PM and Zuideveld PL. Clarification of liquids using filter aids Part I. Mechanisms of filtration. *Powder Technology* 1978; 19: 17–30.
14. Heertjes PM and Zuideveld PL. Clarification of liquids using filter aids Part II. Depth filtration. *Powder Technology* 1978; 19: 31–43.
15. Theodossiou I, Collins IJ, Ward JM, Thomas ORT, and Dunnill P. The processing of a plasmid-based gene from *E. coli* primary recovery by filtration. *Bioprocess Engineering* 1997; 16: 175–183.
16. da Matta VM and Medronho RA. A new method for yeast recovery in batch ethanol fermentations: Filter aid filtration followed by separation of yeast from filter aid using hydrocyclones. *Bioseparation* 2000; 9: 43–53.
17. Reynolds T, Boychyn M, Sanderson T, Bulmer M, More J, and Hoare M. Scale-down of continuous filtration for rapid bioprocess design: Recovery and dewatering of protein precipitate suspensions. *Biotechnology and Bioengineering* 2003; 83: 454–464.
18. Dynamic depth-filtration: Proof of principle. Technical note AMC06 v. 3.1, Advanced Minerals Corporation, 2004.
19. Russotti G and Goklen KE. Crossflow membrane filtration of fermentation broth. In William K Wang, Ed., *Membrane Separations in Biotechnology*. New York: Marcel Dekker, 2001, pp. 85–160.
20. Schlegel VL and Meagher M. Effects of different membrane modular systems on the performance of crossflow filtration of *Pichia pastoris* suspensions. In William K Wang, Ed., *Membrane Separations in Biotechnology*. New York: Marcel Dekker, 2001, pp. 189–204.
21. Patel PN, Mehaia MA, and Cheryan M. Crossflow membrane filtration of yeast suspensions. *Journal of Biotechnology* 1987; 5: 1–16.
22. Sheehan JJ, Hamilton BK, and Levy PF. Pilot-scale membrane filtration process for the recovery of an extracellular bacterial protease. *ACS Symposium*, Series 419, 1988, pp. 130–155.
23. van Reis R, Leonard L, Hsu C, and Builder S. Industry scale harvest of protein from mammalian cell culture by tangential flow filtration. *Biotechnology and Bioengineering* 1991; 38: 413–422.
24. Redkar SG and Davis RH. Crossflow microfiltration of yeast suspensions in tubular filters. *Biotechnology Progress* 1993; 9: 625–634.
25. Tanaka T, Kamimura R, Itoh K, and Nakanishi K. Factors affecting the performance of crossflow filtration of yeast cell suspension. *Biotechnology and Bioengineering* 1992; 41: 617–624.
26. Russotti G, Osawa AE, Sitrin RD, Buckland BC, Adams WR, and Lee SS. Pilot plant harvest of recombinant yeast employing microfiltration: A case study. *Journal of Biotechnology* 1995; 42: 235–246.

27. Bailey SM and Meagher MM. Crossflow microfiltration of recombinant *Escherichia coli* lysates after high pressure homogenization. *Biotechnology and Bioengineering* 1997; 56: 304–310.
28. Jacob J, Prádanos P, Calvo JI, Hernández A, and Jonsson G. Fouling kinetics and associated dynamics of structural modifications. *Colloids and Surfaces A: Physicochemical and Engineering Aspects* 1998; 138: 173–183.
29. Kwon DY, Vigneswaran S, Fane AG, and Ben AR. Experimental determination of critical flux in crossflow microfiltration. *Separation Purification Technology* 2000; 19: 169–181.
30. Lee SS, Burt A, Russotti G, and Buckland B. Microfiltration of recombinant yeast cells using a rotating disk dynamic filtration system. *Biotechnology and Bioengineering* 1995; 48: 386–400.
31. Frenaander U and Jonsson A. Cell harvesting by cross-flow microfiltration using a shear-enhanced module. *Biotechnology and Bioengineering* 1996; 52: 397–403.
32. Parnham CD and Davis RH. Protein recovery from cell debris using rotary and tangential cross-flow microfiltration. *Biotechnology and Bioengineering* 1995; 47: 155–164.
33. Schutyser M, Rupp R, Wideman J, and Belfort G. Dean vortex membrane microfiltration and diafiltration of rBNDF *E. coli* inclusion bodies. *Biotechnology Progress* 2002; 18: 322–329.
34. Luque S, Mallubhotla H, Gehlert G, Kuriyel R, Dzengeleski S, Pearl S, and Belfort G. A new coiled hollow-fiber module design for enhanced microfiltration performance in biotechnology. *Biotechnology and Bioengineering* 1999; 65: 247–257.
35. Davis RH. Crossflow microfiltration with back pulsing. In William K Wang, Ed., *Membrane Separations in Biotechnology*. New York: Marcel Dekker, 2001, pp. 161–188.
36. Anspach FB, Curbelo D, Hartmann R, Garke G, and Deckwer W. Expanded bed chromatography in primary protein purification. *Journal of Chromatography A* 1999; 865: 129–144.
37. Hjorth R. Expanded-bed adsorption in industrial bioprocessing: Recent developments. *TIBTECH* 1997; 15: 230–235.
38. Lyddiatt A. Process chromatography: Current constraints and future options for the adsorptive recovery of bioproducts. *Current Opinion in Biotechnology* 2002; 13: 95–103.
39. Noda M, Sumi A, Ohmura T, and Yokoyama K. European patent application EP0699687A2.
40. Trinh L, Noronha S, Fannon M, and Shiloach J. Recovery of mouse endostatin produced by *Pichia pastoris* using expanded bed adsorption. *Bioseparation* 2000; 9: 223–230.
41. Russell E. Evaluation of disc stack centrifugation for the clarification of mammalian cell culture. M.S. thesis 2003, Tufts University.
42. Ambler C. The evaluation of centrifuge performance. *Chemical Engineering Process* 1952; 48: 150–158.

43. Maybury JP, Hoare M, and Dunnill P. The use of laboratory centrifugation studies to predict performance of industrial machines: Studies of shear-insensitive and shear-sensitive materials. *Biotechnology and Bioengineering* 2000; 67: 265–270.
44. Kempken R, Preissmann A, and Berthold W. Assessment of a disc stack centrifuge for the use in mammalian cell separation. *Biotechnology and Bioengineering* 1994; 46: 132–138.
45. Kubitschek H. Buoyant density variation during cell cycle in microorganisms. *Critical Reviews in Microbiology* 1987; 14: 73–97.
46. Belfort G. Membrane separation technology: An overview. In Bungay HR and Belfort G, Eds., *Advanced Biochemical Engineering*. Chapter 8, New York: Wiley, 1987, pp. 187–217.
47. Atkinson B and Mavituna F. *Biochemical Engineering and Biotechnology Handbook*. New York: Stockton Press, 1991, pp. 957–959.
48. Perry R. *Perry's Chemical Engineers' Handbook*, 6th ed. New York: McGraw-Hill, Inc., 1984, pp. 19–66.
49. Smith G. Filter aid filtration. In *Filtration in the Biopharmaceutical Industry*. New York: Marcel Dekker, Inc., 1998.
50. Gadam S, Leong A, Lee A, and Kosinski M. Clarification of yeast lysate using diatomaceous earth. Presented at the recovery of Biological Products 10, Cancun, Mexico, June 3–8, 2001.
51. Hunt, T. Filter aids. In *Encyclopedia of Bioprocess Technology: Fermentation, Biocatalysis and Bioseparation*. New York: John Wiley & Sons, 1999.
52. Cheryan M. *Performance and Engineering Models: Ultrafiltration and Microfiltration Handbook*. Lancaster: Technomic, Inc., 1998, p. 133.
53. Belfort G, Davis R, and Zyndey A. The behavior of suspensions and macromolecular solutions in crossflow microfiltration. *Journal of Membrane Science* 1996; 96: 1–58.
54. Zydney A and Colton C. A concentration polarization model for the filtrate flux in crossflow microfiltration of particulate suspensions. *Chemical Engineering Communication* 1987; 47: 275–281.
55. Zeman L and Zyndey A. *Microfiltration and Ultrafiltration, Principles and Applications*. New York: Marcel Dekker, Inc., 1996.
56. Wang A, Lewus R, and Rathore A. Comparison of different options for harvest of a therapeutic protein product from high cell density yeast fermentation broth. *Biotechnology and Bioengineering*, in print, 2006.

2 Expanded Bed Adsorption for Capture from Crude Solution

Alan Sonnenfeld and Jörg Thömmes

CONTENTS

2.1 INTRODUCTION

Recent advances in both eukaryotic and prokaryotic cell culture technology have led to substantial increases in the volumetric productivity of bioreactor processes. These increases are mainly due to the ability to grow commonly used expression systems such as bacteria, yeast, and even mammalian cells to very high density. The sheer amount of biomass accumulated in these processes poses significant capacity and throughput challenges to solid–liquid separation steps preceding primary recovery. Additionally, high performance bioreactor processes result in a cell population that might be significantly more stressed, rendering it more susceptible to damage by conventional solid–liquid separation methods. Finally, in times of limited manufacturing capacity, plant throughput becomes a major focus of process development, demanding streamlined recovery operations, where two or more unit operations are combined in innovative single steps.

Expanded bed adsorption (EBA) has been discussed as a potential tool to fulfill some of the needs discussed above. EBA attempts to integrate solid–liquid separation and the first adsorption chromatography step. By fluidizing suitable adsorbent particles in a liquid stream directed upwards, a stable fluidized bed of increased interstitial volume (expanded bed) is formed. The increased voidage of this adsorbent bed allows the introduction of a particle containing feedstock, for example, a crude bioreactor suspension or a cell homogenate, without the operational risk of adsorbent bed fouling or blocking. If the fluid phase conditions are chosen appropriately, the fluidized adsorbent will be able to capture the target molecule from the crude suspension, thus eliminating the need to clarify the bioreactor content by filtration or centrifugation prior to the first adsorption step. Figure 2.1 schematically depicts the EBA concept as well as the main process steps.

An integrated processing step might eliminate many of the problems frequently encountered during harvest of high productivity bioreactors, for example, problematic separation due to wide cell size distribution and ever increasing filter area are due to increased transport resistance of fouled filters/membranes. Consequently, the streamlined process may deliver increased yields and significantly reduce operation time.

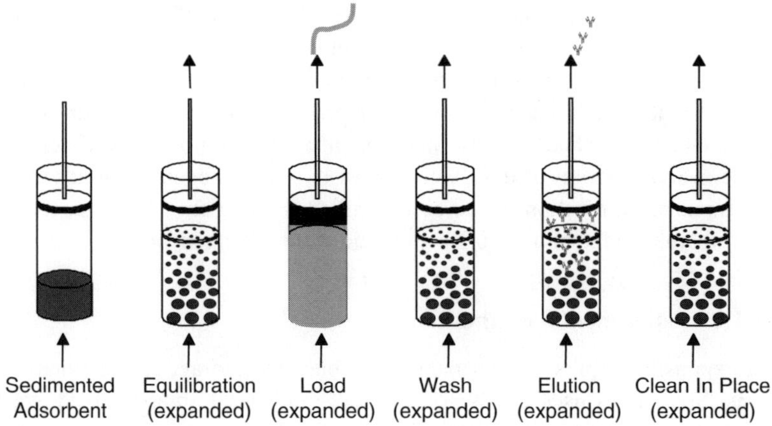

FIGURE 2.1 Pictorial representation of EBA process.

Adsorbing proteins from crude suspensions in an expanded bed necessitates a good understanding of how adsorbent particles are fluidized in the presence of biologic matter (cells or homogenate), particularly with regard to potential interactions of the expanded adsorbents with the bioparticles in the suspension. Since an adsorptive step is performed in the expanded bed, stable expansion without excessive back mixing is imperative. Furthermore, the adsorption process as such must be well understood too, in order to successfully integrate the two unit operations. Bringing the promise of EBA to the developer's bench requires a systematic methodology that can be quickly and efficiently used to understand the above-mentioned issues. The key of the platform method presented here is to screen biomass–adsorbent interactions upfront along with fully characterizing bed fluidization. Once interactions and dispersion effects are minimized, the developer leverages understanding of kinetics and chromatographic theory to guide small-scale experiments to optimize the adsorptive component of EBA. At the same time, buffer consumption is modeled and effectively minimized, enhancing the process economics. The excitement of efficient and effective small-scale development is transferred into intermediate scale-up where equipment issues are evaluated and addressed.

2.2 FUNDAMENTALS

2.2.1 FLUIDIZATION

Understanding and controlling the degree of expansion of the fluidized bed is essential in EBA processing. Since EBA is performed in columns of finite height, excessive expansion will result in elutriation of adsorbent or bed compaction at the top adaptor, both of which are highly undesirable events. By characterizing the bed fluidization, both of these events can be avoided. In addition, the performance of adsorption processes is governed among others by residence time, and thus by fluid velocity. Fluidization of particles is of course also a function of fluid velocity, so in order to control the adsorptive performance, fluidization has to be controlled. A predictive EBA model will integrate knowledge of adsorptive and fluidization components.

2.2.1.1 Experimental Methodology

Bed expansion results from a balance of buoyancy, drag, and gravitational forces. Since EBA uses a series of different fluids (equilibration, load, wash, elution etc.), in principle, all of them have to be characterized in terms of fluidization of the adsorbent particles used. It is often sufficient, however, to characterize the fluid with the highest density and viscosity, since this presents the worst case for overexpanding the bed.

The minimum and maximum fluid velocity necessary to establish and keep a stable expanded bed inside a column can be predicted. The minimal fluid velocity, U_{mf}, can be determined from Equation 2.1 [3].

$$U_{\mathrm{mf}} = 1.54 \times 10^{-2} \frac{\eta}{d_{\mathrm{p}}\rho_{\mathrm{p}}} \mathrm{Ga}^{0.66} \left(\frac{\rho_{\mathrm{p}} - \rho_{\mathrm{l}}}{\rho_{\mathrm{l}}} \right)^{0.7} \tag{2.1}$$

where Ga is the Gallileo number and can be calculated using Equation 2.2.

$$\mathrm{Ga} = \frac{\rho_{\mathrm{p}} g (\rho_{\mathrm{p}} - \rho_{\mathrm{l}}) d_{\mathrm{p}}^3}{\eta^2} \tag{2.2}$$

The maximum fluid velocity is estimated from the terminal settling velocity, U_{t}, above which the adsorbent particles will elutriate from the column. U_{t} is approximated by Stokes law, Equation 2.3, as being dependent on the particle diameter d_{p}, the density of the solid, ρ_{p}, and liquid, ρ_l, phase and the viscosity of the liquid phase η.

$$U_{\mathrm{t}} = \frac{(\rho_{\mathrm{p}} - \rho_{\mathrm{l}}) d_{\mathrm{p}}^2 g}{18\eta} \tag{2.3}$$

U_{t} and U_{mf} are workable estimates for fluid velocity boundary conditions. The force balance argument is an idealized case, however, because it neglects adhesion forces between particles and between particles and the column wall.

A simple correlation, known as the Richardson–Zaki (RZ) correlation, is available to predict the fluid velocity necessary to achieve a desired expanded bed height. Although the correlation was developed for monodispersed, spherical particles, the correlation satisfies the balance between accuracy and complexity for most EBA systems.

The RZ correlation describes the expansion of a bed of particles in a liquid flow by correlating the voidage of the bed ε with the fluid velocity, U, using two parameters: the terminal settling velocity of a single particle, U_t, and the expansion index, n, as shown in Equation 2.4 [2]. Values for n from 3 to 6 have been reported for expansion of adsorbent particles in biological feedstock [3].

$$U = U_t * \varepsilon^n \tag{2.4}$$

The bed voidage, ε is determined from Equation 2.5 [4] where H is the expanded bed height, H_0 is the sedimented bed height, and ε_0 is the void fraction of the sedimented bed.

$$\varepsilon = 1 - (1 - \varepsilon_0)\frac{H_0}{H} \tag{2.5}$$

When the log of both sides of the RZ equation is taken, a linear relationship results.

Experimentally, ε is determined by measuring H at multiple fluid velocities. Plotting U vs. ε on a double log plot, the y-intercept and the slope yields U_t and n. With the model defined, we can estimate the bed expansion for any fluid velocities.

2.2.1.2 Stability of Fluidized Beds

Although a stable bed height has been determined in the fluidization studies, the extent of mixing inside the bed has not been investigated. It is important to understand that efficient adsorption of proteins requires a flow pattern through the adsorbent bed that can be characterized as being close to plug flow. Therefore, back mixing in the expanded bed needs to be minimal. Theoretically, the specific weight and size polydispersity within the adsorbent bed minimizes the expanded bed mixing by creating a classified bed: more massive particles expand to a specific level, while lighter particles expand further. We confirm the stability of the bed with simple pulse response experiments and advanced mixing models. In literature on EBA, a stable fluidized bed where the absence of excessive back mixing leads to a plug flow type of fluid flow through the bed, is termed an expanded bed.

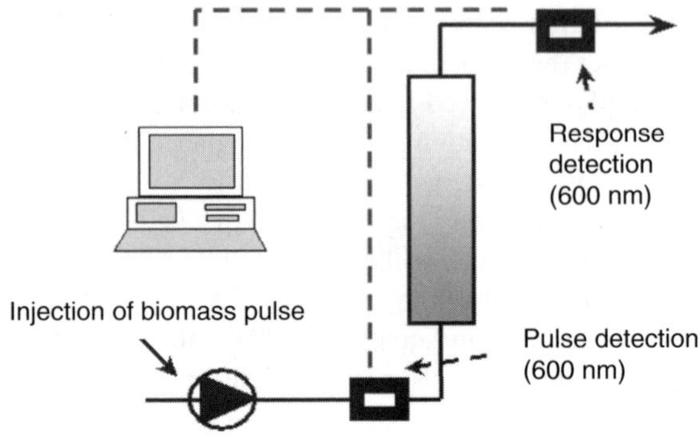

FIGURE 2.2 RTD experimental setup.

2.2.1.2.1 Residence Time Distribution Using Pulse Response Experiments

Before exposing the adsorbent to a crude, biomass containing suspension, the extent of axial mixing and the quality of fluidization due to buffer and equipment will be investigated using residence time distribution (RTD) pulse response method [5]. An illustrative example of the setup is shown in Figure 2.2.

From the instance the pulse is applied, time t_0, the exit concentration is monitored and recorded until the pulse response returns to baseline, time t_f. The bed's number of theoretical plates is calculated using moment's analysis, quantifying the extent of mixing in the expanded bed.

The number of theoretical plates (N) is calculated from moments by the equation shown in Equation 2.6. The higher the number of plates, the closer the fluid flow approaches plug flow.

$$\begin{aligned} \sigma_\Theta^2 &= \mu_2 - \mu_1^2 \\ N &= \frac{\mu_1^2}{\sigma_\Theta^2} \end{aligned} \tag{2.6}$$

where μ_1 and μ_2, the first and second moment, are calculated using Equation 2.7 and Equation 2.8.

$$\bar{t} = \mu_1 = \frac{\int_0^\infty Ct\mathrm{d}t}{\int_0^\infty C\mathrm{d}t} \cong \frac{\sum_i t_i C_i \Delta t_i}{\sum_i C_i \Delta t_i} \tag{2.7}$$

and

$$\mu_2 = \frac{\int_0^\infty C(t-\mu_1)^2 \mathrm{d}t}{\int_0^\infty C\mathrm{d}t} = \frac{\sum_i t_i^2 C_i \Delta t_i}{\sum_i C_i \Delta t_i} \tag{2.8}$$

Note that the first moment is also known as the mean residence time, or the mean time it takes for a particle or buffer to traverse the column.

2.2.1.2.2 Model to Describe the Phenomena Observed

To further investigate whether a stable bed has developed, a more quantitative analysis is performed on the result of the pulse response experiment. Villermaux and Van Swaij [6] originally introduced the PDE model to describe imperfect fluid flow through trickle bed reactors. Fernandez-Lahore et al. [7] employed this model as an advanced method of evaluating the quality of fluidization of expanded beds in real biological feedstock. In physical terms, the PDE model breaks the column into two sections: a dynamic stable zone where a perfectly classified (expanded) bed exists, and a stagnant zone where particles and feed material have aggregated and hinder proper fluidization. Figure 2.3 highlights the three key parameters for the model.

In the model shown in Equation 2.9, the key parameters are defined as:

- Fraction of the expanded bed which is stable (φ)
- Mass transfer between the stagnant and dynamic zones (N)
- Axial mixing in the stable zone defined as the Peclet number (Pe)

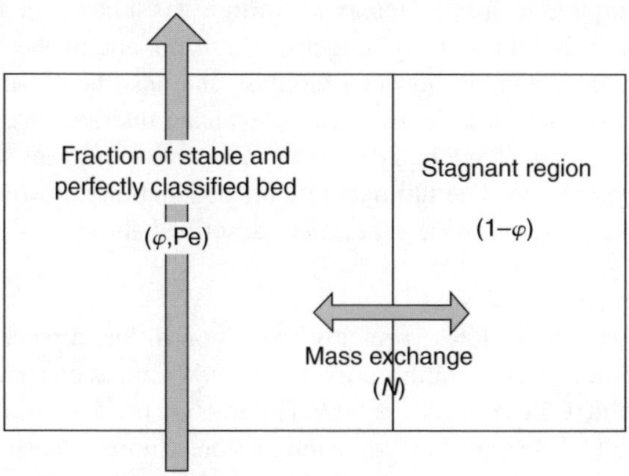

FIGURE 2.3 Visual explanation of PDE model.

$$C(s) = L[E(\theta)]$$

$$= \left\{ \sqrt{\text{Pe}} \exp\left(\frac{\text{Pe}}{2}\right) \cdot \exp\left[-\sqrt{\text{Pe}\varphi} \cdot \sqrt{s + \frac{N}{\varphi} + \frac{\text{Pe} \cdot \varphi}{4} - \frac{N^2 \cdot (1-\varphi)/\varphi}{s + (N/(1-\varphi))}} \right] \right\}$$

$$\times \left\{ \frac{\sqrt{\text{Pe}}}{2} + \sqrt{\varphi} \cdot \sqrt{s + \frac{N}{\varphi} + \frac{\text{Pe} \cdot \varphi}{4} - \frac{N^2(1-\varphi)/\varphi}{s + (N/(1-\varphi))}} \right\}^{-1} \quad (2.9)$$

The model describes the normalized pulse response data, the E curve, as a function of Θ with φ, N, and Pe as parameters. Unfortunately the model must be solved in the Laplacian domain. The analytical evaluation of the transformation back into the time domain is complex and is more easily solved numerically by software routines.

The model can be fitted to the experimental pulse response data, yielding values for φ, N, and Pe. An ideal bed is defined as having no channeling or aggregation ($\varphi = 1$) and a limited axial mixing (Pe > 40). Poor performance has been defined when $<80\%$ of the bed is properly fluidized, or when $\varphi = 0.8$.

Successful model analysis gives confidence that there is little interaction within the bed and that development should be moved forward.

2.2.1.3 Measuring Bed Stability in the Presence of Biomass

Stable bed expansion in buffer is a prerequisite for efficient EBA, but it does not guarantee stable fluidization in the crude suspension. In case biomass–adsorbent interactions occur, a stable expanded bed might not form and the protein adsorption efficiency can be compromised. If there are severe interactions, the interparticle space decreases and in a worst case, the agglomerates become so large that they can no longer be fluidized, and the bed collapses.

The microenvironment between biomass and adsorbent cannot be tested directly, but two methods are available to evaluate interactions. A quick preliminary pulse response method can be used as a very efficient screening tool for process conditions. The ultimate test for bed stability, however, is proper behavior in the pulse response experiment described above.

2.2.1.3.1 Screening Tests

The following method allows to quickly assess how different process conditions such as pH, conductivity, equilibration fluid, and biomass concentration in the cell culture fluid (CCF) impact a potential interaction of the expanded adsorbent with biomass. To determine the degree interaction, a biomass pulse is introduced to a stable expanded bed. UV absorbance at 405 nm is used to measure the amount of material flowing to and from the column. The area under the peak's

UV trace before and after passing through the column is used to calculate the transmission index shown in Equation 2.10.

$$P = \frac{A_{\mathrm{after}}}{A_{\mathrm{prior}}} \cdot 100 \quad [\%] \tag{2.10}$$

Stable bed expansion in the presence of biomass can be expected for a transmission index of at least 90%.

2.2.1.3.2 Residence Time Distribution

The RTD experiment and data analysis described for buffer fluidization is repeated with the CCF and a stable expanded bed is attained when the fraction of stable bed is >0.8 and the Peclet number is >40.

2.2.2 The Kinetics of Adsorbing Proteins in Fluidized Beds

In a simplified approach, protein breakthrough during adsorption to porous absorbents can be well described by a model presented by Hall et al. [8]. The model assumes irreversible equilibrium and limitation of sorption efficiency by fluid and particle side transport, represented by two number of transfer units, N_{f} and N_{p}, respectively.

2.2.2.1 Fluid Side Transport

In a dynamic system such as an expanded bed, the fluid side transport efficiency fluctuates depending on the level of bed expansion. As the bed expansion has been modeled with the RZ equation, the expansion can be easily incorporated into the external transport coefficient equation shown in Equation 2.11.

$$N_{\mathrm{f}} = \frac{3 \cdot k_{\mathrm{f}} \cdot L \cdot (1 - \varepsilon)}{r_{\mathrm{p}} \cdot U} \tag{2.11}$$

The only term in Equation 2.11 that cannot be directly evaluated is k_{f}, the transport coefficient.

The Nelson and Galloway [9] correlation has been shown in the literature to be a good estimate of k_{f} in expanded beds. Although more rigorous methods are available to evaluate k_{f}, this correlation considers velocity and voidage and is applicable for a wide range of Reynolds numbers, including the laminar flow regime, where EBA of proteins takes place.

2.2.2.2 Particle Side Transport

The particle side transport coefficient N_p is defined in Equation 2.12, where there is only one unknown, the apparent particle side diffusion coefficient D_e.

$$N_p \equiv \frac{15 \cdot D_e \cdot (1-\varepsilon) \cdot L}{U \cdot r_p^2} \tag{2.12}$$

Packed bed chromatography may be regarded as a limiting case where pore diffusion is the dominant resistance. For this case, the Hall model is reduced to Equation 2.13 where X is the fraction of the total concentration seen in the effluent (C/C_0) and T is the sorption efficiency, the ratio of the dynamic and equilibrium capacities.

$$X = 1 - \left[\frac{2.39 - N_p(T-1)}{3.59}\right]^2 \tag{2.13}$$

By generating a number of breakthrough curves in packed columns under varying residence times (at least four experiments are recommended), D_e can be determined by fitting Equation 2.13 to the breakthrough curve data.

2.3 THE FLUIDIZED BED ADSORPTION WORKING DIAGRAM

Employing the model discussed above, sorption efficiency can be correlated to expanded bed behavior. For various settled bed heights, a plot of sorption efficiency vs. fluid velocity is generated. Using the RZ correlation, the degree of bed expansion corresponding to the fluid velocity is overlaid onto the plot. The Hall model thus renders a working diagram.

2.3.1 INTEGRATING FLUID AND PARTICLE SIDE TRANSPORT

The full Hall model, including N_f, is presented in Equation 2.14. It should be noted that Φ is only approximated here, however, with negligible effect.

$$T = 1 + \left(\frac{1}{N_p} + \frac{1}{N_f}\right)\left(\frac{\Phi(X) + (N_p/N_f)\cdot(\ln(X+1))}{(N_p/N_f)+1}\right)$$

$$\Phi(X) \cong 2.39 - 3.59\sqrt{1-X} \tag{2.14}$$

$$T = \frac{V_{brk} C_0}{Q_{eq.} V_s}$$

In Equation 2.14, T is a loading parameter. It represents the ratio between the amounts of product loaded at a given load volume ($V_{\text{brk}} C_0$) and the available equilibrium capacity of the column ($Q_{\text{eq.}} V_S$). For example, at $T = 0.5$, enough feedstock has been loaded to saturate 50% of the entire equilibrium capacity of the column. Therefore, using the value of T at the termination of loading provides the user with a measurement of how efficiently a column was used. If the loading phase were to be ended at $T = 0.2$, for example, only 20% of the available capacity would be used, hence the process could be perceived as quite inefficient.

Using the methods to determine the fluid and particle side transport coefficients presented above, Equation 2.14 is used to generate a T vs. U plot as a function of settled bed height. The equipment-limited expansion, predicted by the RZ model, is overlaid onto the T vs. U plot to establish the operational zone.

2.4 BUFFER CONSUMPTION

Developing the wash step after product loading is crucial for EBA success. Reports in the literature have called for washing of nearly 20 settled bed volumes (SBV) to remove the cell containing suspension after loading, which could make the process potentially infeasible. A physical understanding of the system is necessary to minimize the amount of fluid needed to displace the cells.

2.4.1 DENSITY DISPLACEMENT

When a lower density fluid is introduced to the bottom of a column containing a higher density fluid, gross mixing occurs. The incoming fluid is not strong enough to evenly displace the buffer already in the column.

Fee and Liten [10] has developed a model to predict the number of SBV necessary to completely displace a higher density fluid using a two-tanks-in-series model. The model, shown in Equation 2.15 is based on two assumptions: there is little back mixing between the column and the head space, and the liquid entering the column is assumed to be non-compressible to allow a constant fluid velocity throughout the column. The equation is derived from a mass balance on the tank. Initially at time $t = 0$, the density of the headspace is the density of the feedstock, ρ_f. After the incoming fluid has been applied for an infinitely long time, $t =$ infinity, the headspace density is equal to the displacer solution, ρ_D.

$$\rho_{HS}(SBV) = \rho_F\left[e^{(-Vsb/Vhs)SBV} + \frac{e^{(-Vsb/Vhs)SBV}}{((V_{HS}/(X-1+\varepsilon_0)V_{SB})-1)} - \frac{e^{-SBV/(x-1+\varepsilon_0)}}{((V_{HS}/(X-1+\varepsilon_0)V_{SB})-1)}\right] + \rho_D\left[1 + \frac{e^{(-Vsb/Vhs)SBV}}{(((X-1+\varepsilon_0)V_{SB}/V_{HS})-1)} + \frac{e^{-SBV/(x-1+\varepsilon_0)}}{((V_{HS}/(X-1+\varepsilon_0)V_{SB})-1)}\right] \quad (2.15)$$

The model also takes into account, X, the fractional expansion of the bed, ε_0, the bed voidage, and V, the volume of the headspace (HS) and sedimented bed (SB).

When the incoming fluid is denser than the resident fluid, plug flow displacement rather than gross axial mixing occurs. Thus, a plug flow model has also been developed to predict the number of SBV needed to displace the feedstock fluid and reach the wash fluid density. The model is based on the difference in densities of the two fluids times a complex factor that includes the expansion, the axial mixing, Pe, and the volume of the sedimented bed, SBV.

$$\rho_{HS} = \rho_F + (\rho_{D1} - \rho_F)\cdot 0.5\cdot\left[1 - \mathrm{erf}\left[\frac{SBV/X}{\sqrt{(SBV/X)/Pe}}\right]\right] \quad (2.16)$$

A predictive model based on an if–then–else logic structure can be built from the two models. Applying the extended Fee model, a density profile that minimizes buffer consumption can be developed.

2.5 EQUIPMENT CHALLENGES

As a young, developing technique, the progress in modeling and developing an EBA process has outstripped the pace of equipment design, which is only just recently being effectively addressed.

2.5.1 Traditional Plate Distribution Systems

Traditional frit and distribution plate designs have been employed in early realizations of EBA hardware. Although not ideal, they should be tested, as in some cases the design has been successfully implemented in a robust process.

In our hands, however, experiences with a high cell density mammalian cell processes in such a traditional distribution system (STREAMLINE 25 and STREAMLINE 100, GE Healthcare, Uppsala, Sweden), however, have not been favorable. The cells tend to aggregate on the frit and clog the holes in the distribution plate causing an increase in backpressure and poor fluid distribution. The frit and plate were not effectively regenerated by the recommended clean in place (CIP) procedures. At small scale, solutions from low concentration sodium hydroxide to SDS (sodium dodecyl sulfate) mixtures were investigated with little success. Following each run, the column was emptied and the frit/plate manually cleaned.

At 10 cm diameter scale, severe pressure from fouled frits was seen. Pulsing and reverse flowing the feedstock had little effect on reducing the fouling.

Both the challenges of cleaning and the fouling at large scale motivated the investigation of nontraditional fluid distribution designs.

2.5.2 Alternate Distribution Systems

Improved designs such as an oscillating spider and check valve system have been discussed in the literature [11] and were also tested in our laboratories. In this design, fluid is distributed through motor driven oscillating inlet arms, without distributor plates or frits at the bottom of the column, as seen in Figure 2.4.

In this design, there is no opportunity to settle the bed subsequent to loading and washing for elution in a packed bed mode. All operations occur in an upward flow expanded bed regime. During tests in our laboratories, such a next generation STREAMLINE Spider column of 9.5 cm internal diameter significantly outperformed the traditional frit and plate design. No significant fouling was seen during loading and regeneration was much more complete, with only

FIGURE 2.4 Oscillating spider distribution system.

minor residual biomass observed upon disassembly, including a mechanical strainer upstream of the column completely eliminated visual biomass in the distribution system.

Drawbacks to the system were the start up procedures and the motor's positioning. Start up must be closely monitored as the moving parts can grind the resin if the bed is not fluidized, adding to the control complexity. In addition, if the drive shaft or motor require maintenance, the position directly underneath the column complicates the repair.

2.6 CASE STUDY

Following the fundamentals discussed above, a case study is presented here. The object of the development was to combine harvest and protein A capture step of a recombinant antibody from Chinese Hamster Ovary (CHO) cell culture suspension of 4% wet weight. Standard buffers and sequence of buffers were adapted from our traditional packed bed protein A adsorption chromatography process.

2.6.1 The Development Methodology

During fast paced development cycles, representative feedstock is at a premium. Implementing the modeling techniques presented above conserves this resource and speeds development by creating a platform approach.

2.6.1.1 Fluidization

Understanding the equipment and bed expansion characteristics is the first step in the development process. Expansion effects were modeled with the RZ equation for the densest/most viscous solution and the preload equilibration solutions by fluidizing the bed under at least five fluid velocities. With the RZ model, the expansion/fluid velocity prediction could be made quickly for subsequent development.

2.6.1.2 Bed Stability

Minimal biomass–absorbent interactions are imperative to operational success. The resin type and the feedstock condition play important roles in minimizing the interactions. From a process robustness and ease of processing standpoint, it would be ideal not to adjust the feedstock but load the suspension as is. Protein A adsorbents were investigated specifically for this reason. Screening multiple stationary phases and conditions with the pulse response/transmission index technique, we found the adsorbent and condition that met the >90% index criteria.

After identifying operating conditions, bed stability was analyzed by the pulse response technique. Using the RTD analysis procedure presented, the number of theoretical plates is determined. The PDE model was also be applied at this point to confirm the stability of fluidization in the column. In case these experiments reveal bed stability problems, a second round of conditions screening using the transmission test needs to follow.

2.6.1.3 Modeling Time

Having quickly identified and confirmed the adsorbent and biomass condition that experience minimal biomass–adsorbent interactions, the general working diagram and buffer consumption models should be developed.

2.6.1.3.1 Create Working Diagram

The Hall model described earlier is the basis for the working diagram. The external transport coefficient was modeled at various fluid velocities by incorporating the RZ model for the bed voidage term. For a fully representative model, a RZ model of the CCF should be experimentally determined.

The particle side transport piece of the Hall model requires the apparent particle side diffusion coefficient, D_e. At least four breakthrough curves in a packed bed column using the optimum conditions determined above were run under varying residence times. The simplified Hall model was then be fited to these curves and the resulting D_e averaged.

Incorporating both the fluid and particle side coefficients along with the RZ expansion model into the full Hall model created a working diagram. Overlaying the column expansion constraints from the RZ model, experimental development can be minimized, as the developer can see *a priori* relationship and constraint based on resin capacity, processing time, and equipment/resin costs.

2.6.1.3.2 Buffer Washout Strategies

Assuming that the wash and cleaning buffers from a packed bed platform process can be transferred to EBA, the initial density profile was established. Evaluating this profile with the extended Fee model, the feasibility of the current density profile for washing the cell containing feedstock out of the expanded bed was evaluated. If an excessive amount of buffer was predicted, the model was used to investigate the effects of novel buffer strategies, such as density enhancers.

While modeling the density strategy, it should be kept in mind that displacing a high-density fluid with a low-density fluid will cause gross mixing. The density profile strategy should be designed to ensure that the expanded bed is well established during product load and elution.

2.6.1.4 Perform Small-Scale Experiments

With the operation zone and buffer strategy modeled, a minimal number of experiments are needed to confirm operational parameters and generate material for downstream development.

2.6.1.4.1 Evaluate Process Results

Small-scale experimental results should be fully analyzed at this point to determine the direction of the development project. If results are promising, a team decision needs to be made whether to invest the resin, cell culture fluid, and time into scale-up activities. Although each development case is unique, several considerations to guide the decision are presented below.

2.6.1.4.2 Go/No-Go for Process Scale Up

- Is the protein recovery sufficiently high? (Remember that EBA combines two unit operation into one step, compare yield to the overall yield of a traditional harvest/capture process.)
- Do the analytics show equivalent or better product pool quality compared to a harvest/packed bed combination?
- How well does the CIP strategy work? If you can't clean it now, how are you going to clean it at large scale?
- Is stable bed expansion attained in a robust manner (e.g., with an expected variation of cell culture process parameters)?
- Is the expected amount of wash buffer consumed? If a high- to low-density transition is required, how long does it take to reestablish a stable bed?

2.6.2 Step-By-Step Example (Antibody Purified by Protein A Affinity EBA)

Examples from the experimental protocol described above for an EBA Protein A capture step for a recombinant antibody from a 4% wet weight CHO culture process are presented.

2.6.2.1 Fluidization

Richardson–Zaki fluidization experiments and analysis were performed on three resin candidates. The settled bed height and bed height at multiple fluid velocities was measured. Equation 2.5 was used to calculate these expansion heights to corresponding bed voidage values. A double log plot of fluid velocity vs. bed voidage yields the RZ constants U_t and n. Figure 2.5 is an example of the double log plot.

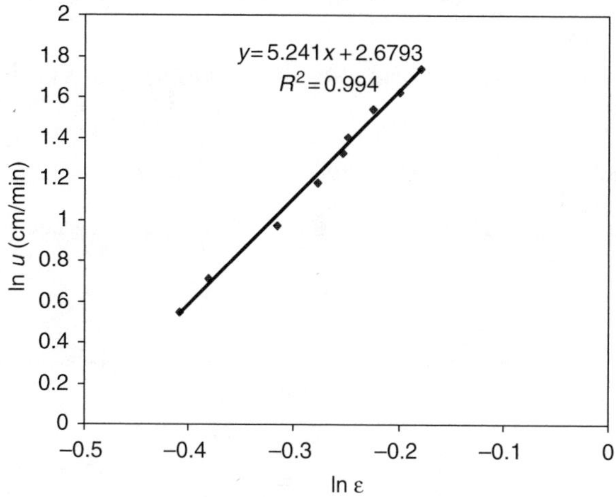

FIGURE 2.5 An example of experimental RZ data.

In this example $U_t = 14.6$ and $n = 5.2$. With the RZ models built for the densest solution and the equilibration buffers, bed expansion was now easily controlled.

2.6.2.2 Biomass Transmission

The RZ model was used to set the fluid velocity for the 3× expansion for the three resins being investigated. The respective adsorbents were stably expanded and unadjusted cell containing cell culture fluid was injected into the column. The experimental results of the biomass transmission experiments are shown in Table 2.1.

The STREAMLINE™ rProtein A and FastMabs AD demonstrated low enough interaction to warrant further development.

2.6.2.3 Stability of Expanded Beds

2.6.2.3.1 Without Biomass

The bed was expanded to three times the settled bed height based on the RZ model. A pulse tracer, 1 *M* NaCl, was injected into the column and the response was monitored and analyzed using moments analysis. Figure 2.6 presents the pulse response data for Streamline rProtein A resin.

TABLE 2.1
Percentage Transmission Results for Multiple EBA Resins

Vendor	GE Healthcare	Biosepra	Up-Front
Resin	Streamline rProtein A	CM Hyper Z	FastMabs AD
% Transmission	98	82	98

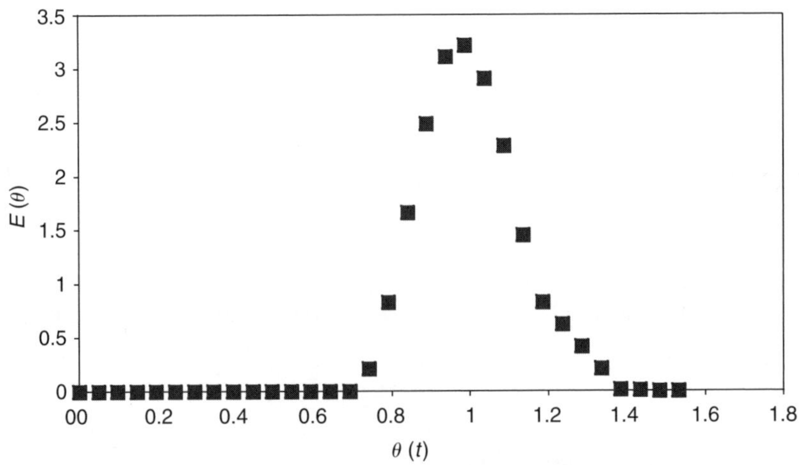

FIGURE 2.6 Buffer RTD experimental results.

Using the algorithm presented in Equation 2.6 to Equation 2.8, Equation 66 theoretical plates were calculated. Theoretical arguments for protein adsorption to porous affinity media estimate that approximately 30 plates should be sufficient to consider the system suitable for successful protein adsorption.

2.6.2.3.2 With Biomass

With the baseline fluidization fingerprint established, multiple RTD experiments in cell culture suspension were performed. Figure 2.7 presents four pulse-response experiments superimposed on the control fingerprint.

The CCF RTD moment analysis reported plate values of 58 to 76. Encouragingly, the control and CCF RTD number of theoretical plates were not significantly different (p value = 0.4, 95% confidence).

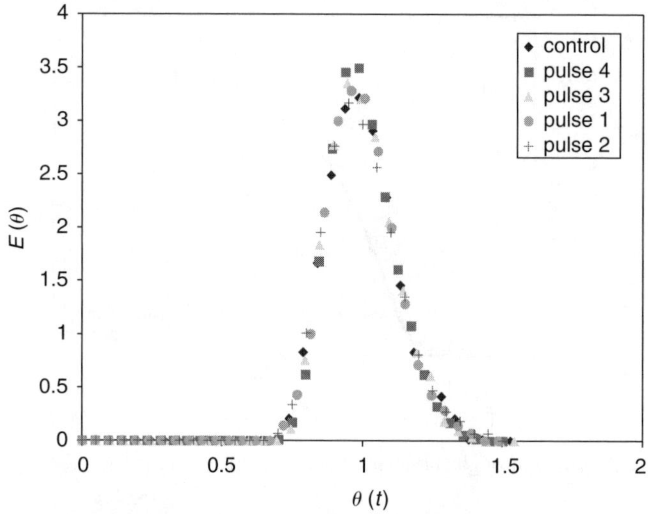

FIGURE 2.7 CCF RTD experiments overlaid onto buffer control data.

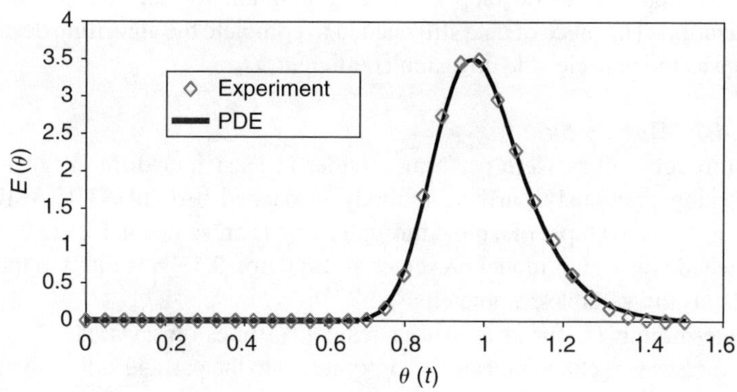

FIGURE 2.8 PDE model fit to experimental data.

Developing an enhanced quantitative picture, the PDE model was applied to the CCF pulse response data. Figure 2.8 demonstrates that the PDE model fits the pulse response experimental data well.

It has been shown in the literature that the fraction of well-fluidized bed should not be smaller than 90%. In our system, we achieved over 95% well-fluidized fraction. The PDE model confirms our RTD results, that there is minimal biomass–absorbent and therefore, further development of the Streamline rProtein A resin is possible.

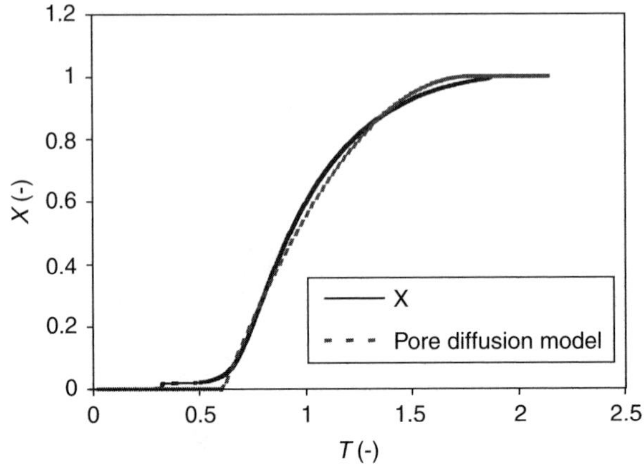

FIGURE 2.9 Breakthrough curve and fitted pore diffusion model.

2.6.2.4 Kinetics of Absorption

The next step was to develop a working diagram to enable rapid process optimization. The piece of data still needed to complete the algorithm described above was the particle side diffusion coefficient, D_e.

2.6.2.4.1 Particle Side

Breakthrough curves were performed under at least four different residence times using previously purified antibody in packed beds of STREAMLINE rProtein A. An example of a breakthrough curve is presented in Figure 2.9. The particle side only Hall model presented in Equation 2.13 was fitted to the data with D_e as the variable parameter.

Determining D_e for at least four residence times, an average D_e of 1×10^{-12} m^2/sec was calculated and incorporated into the particle side Hall model.

2.6.2.4.2 Load Optimization

Compiling the knowledge generated in the laboratory with the modeling theory presented earlier, a working diagram was created. The working diagram for our process, presented in Figure 2.10, illustrates the interactions between settled bed height, fluid velocity, and adsorbent capacity. The expansion markers reflect a 1-m equipment height constraint.

The diagram illustrates that for smaller settled bed heights, column performance is more sensitive to fluid velocities. Under faster fluid velocities, shorter processing times, a 15 cm bed will experience greater capacity decay than a 30 cm bed. Balancing resin costs with plant time, the model predicts optimal operations at 3× expansion, 30 cm settled bed height.

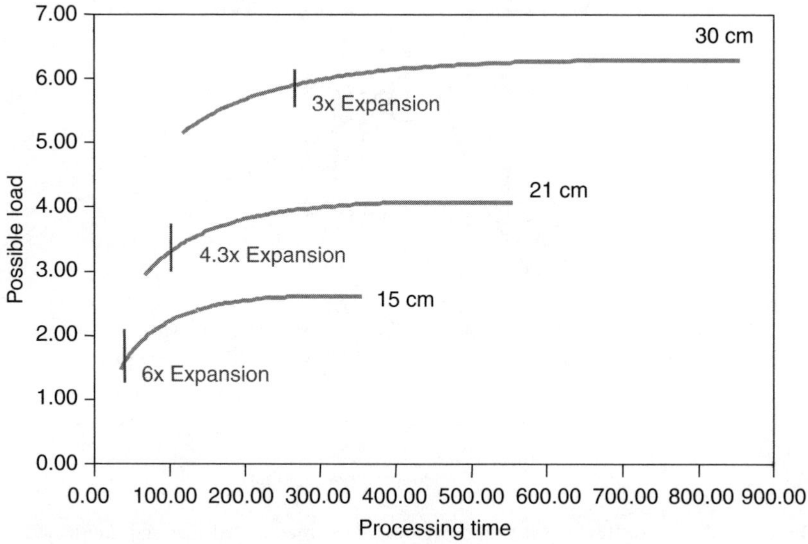

FIGURE 2.10 Working diagram with RZ fluidization model overlay.

Multiple small-scale experiments were performed under the working diagram's guidance. The process had an average yield of 90%, concentrated the product sixfold, and reduced host cell protein by 3.5 logs. Product quality in terms of monomer content and antibody integrity, quantified by CE-SDS and size exclusion chromatography, were comparable to packed bed chromatography. The eluate filterability was characterized by the V_{max} method and eluate turbidity and was also comparable to packed bed chromatography.

2.6.2.4.3 Wash Optimization

The extended Fee model was applied to our buffer wash strategy and is shown in Figure 2.11. The model predicted that the low-density buffer 3 would displace the high-density buffer 2 in approximately five column volumes. The model was tested over five experiments and shown to accurately predict the density profile.

The model also predicted that the cell culture fluid could be displaced within five column volumes, eliminating the need for density enhancers like glycerol or sucrose. With the buffer strategy confirmed, our fully modeled system was ready for go/no-go decision point.

2.7 CONCLUSIONS AND OUTLOOK

A process step that has the potential of reducing cost of goods sold may be worth investigating. EBA can deliver such savings by reducing process

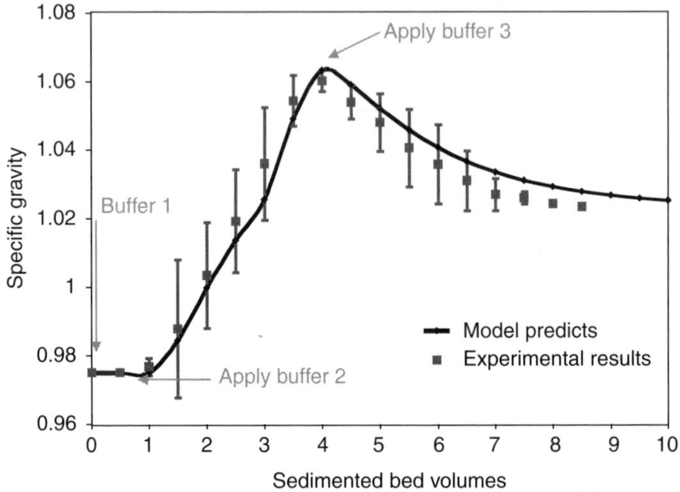

FIGURE 2.11 Buffer density profile modeled with extended Fee and confirmed in experimentation.

time, consumables, and process yield losses. To establish a robust process in the fast-paced development cycles typical of biopharmaceuticals, a platform methodology combined with robust equipment is essential.

This chapter should serve as a guide to achieving rapid development in minimal time and experimental costs. By investigating fluidization and interaction behavior up front, multiple resins can be screened with minimal feedstock. Once biomass–absorbent interactions have been quantified and are acceptable, minimal product is needed to model particle side and fluid side kinetics.

Combing the information learned to this point in a working diagram, a little adsorption theory can be leveraged to choose optimal conditions that will efficiently guide future development. Outside of biomass–absorbent interactions, the extended Fee model is fundamental to quickly confirming or developing low buffer consuming washout conditions.

Difference between cell type and culture conditions can have a significant impact on equipment performance. Based on our experience with low viability CHO cells, equipment cleaning and regeneration are the major issues facing EBA today. In our development with the frit and plate design, the distribution system frequently became blocked, resulting in unacceptable fluidization quality and even a total collapse of the expanded bed. Worse, the distribution system was not cleanable using CIP protocols compatible with a Protein A absorbent.

Limited experience with the second-generation spider distribution design has shown promise, particularly when a prestrainer is employed. The user is encouraged to pay close attention to the distribution system during initial development.

Making the go/no-go decision on future development is challenging, with the tools and art presented here, getting to that decision point should require minimal time and feedstock. We encourage you to investigate EBA with an open mind and lots of math.

REFERENCES

1. Anspach, F., Curbelo, D., Hatrmann, R., Garke, G., and Decwer, W. Expanded-bed chromatography in primary protein purification. *J. Chromatogr. A* 1999; 865:129–144.
2. Richardson, J.F. and Zaki, W.N. Sedimentation and fluidization: Part I. *Trans. Inst. Chem. Eng.* 1954; 32:35–52.
3. Thommes, J. Fluidized bed adsorption as a primary recovery step in protein purification. In Scheper, T. (Ed.), *Advances in Biochemical Engineering/ Biotechnology*, Springer, Vol. 58, 1997; pp. 185–230.
4. Thommes, J., Bader, A., Halfar, M., Karau, A., and Kula, M. Isolation of monoclonal antibodies from cell containing hybridoma broth using a protein A coated adsorbent in expanded beds. *J. Chromatogr. A* 1996; 153:111–122.
5. Fernandez-Lahore, H.M., Kleef, R., Kula, M.-R., and Thommes, J. The influences of complex biological feedstock on the fluidization and bed stability in the expanded bed adsorption. *Biotechnol. Bioeng.* 1999; 64:484–496.
6. Villermaux, J. and van Swaaij, W.P.M. Modele representativ de la distribution des temps de sejour dans un reacteur semi-infini a dispersion axiale avec zones stagnantes. Application a l'ecoulement ruisselant dans des colonnes d'anneaux raschig. *Chem. Eng. Sci.* 1969; 24:1097–1111.
7. Fernandez-Lahore, H.M., Geilenkirchen, S., Boldt, K., Nagel, A., Kula, M.-R., and Thommes, J. The influence of cell adsorbent interactions on protein adsorption in expanded beds. *J. Chromatogr. A* 2000; 873:195–208.
8. Hall, K., Eagleton, L., Acrivos, A., and Vermeulen, T. Pore- and solid-diffusion kinetics in fixed-bed adsorption under constant pattern conditions. *I&EC Fundam.* 1966; 5:212–222.
9. Nelson, P.A. and Galloway, T.R. Particle to fluid heat and mass transfer in multiparticle systems at low Reynolds numbers. *AIChE J.* 1975; 10:605–611.
10. Fee, C. and Liten, A. Buoyancy-induced mixing during wash and elution steps in expanded bed adsorption. *Bioseparation* 2001; 10:21–30.
11. Feuser, J., Barnfield Frej, K., Lundkvist, M., and Walter, J. EBA columns at technical scale. *EBA '02 Abstracts* 2002; pp. 30–31.

3 Product Recovery by High-Gradient Magnetic Fishing

Matthias Franzreb, Niklas Ebner,
Martin Siemann-Herzberg,
Timothy J. Hobley, and Owen R.T. Thomas

CONTENTS

Magnetic adsorbents possess a very powerful and unique handle that permits their selective manipulation within and from most (if not all) kinds of biological feedstock, simply through the application of a magnetic field. This ease of manipulation stands in stark contrast to all other adsorbents, and has been exploited in bioprocessing to develop a first capture step for proteins and other species from crude feedstocks known as high-gradient magnetic fishing (HGMF) [1–14].

3.1 BASIC CONCEPTS

The principle steps involved in using magnetic adsorbents are (i) binding of the protein of interest to the adsorbent; followed by (ii) separation of the loaded adsorbent; and subsequently (iii) washing and elution steps, including adsorbent cleaning if required. Adsorbents can be captured and manipulated on the lab bench by using test tubes and a simple bar magnet. However, for larger volumes in the lab, pilot plant, or at large scale, pumping through a magnetic separator captures the adsorbents most easily.

3.1.1 Batch Adsorption with Nonporous Magnetic Adsorbent Particles

If a small volume is being treated (e.g., up to 100 ml), then adsorbents that have been equilibrated in a binding buffer (e.g., to the correct pH and ionic strength) are mixed with the feedstock in a test-tube or flask. The mixture is allowed just a few minutes (5 min is generally ample) to come to equilibrium [4–6,13,14] given both the small particle size and essentially nonporous nature of the support. The adsorbents can then be collected by magnetic separation and the supernatant analyzed.

Sorption equilibria of biomolecules on magnetic adsorbent particles are, in common with conventional chromatographic media, usefully described by the simple Langmuir model [15]:

$$Q^* = \frac{Q_{\max} \cdot c^*}{K_d + c^*} \tag{3.1}$$

where Q^* denotes the equilibrium loading of the magnetic beads, c^* is the equilibrium concentration of the biomolecule remaining in the solution, and $Q_{\max}$

is the maximum binding capacity of the adsorbent. In this case, the equilibrium parameter K_d of the Langmuir model corresponds to the dissociation constant of the binary ligand–target molecule complex. Hence, K_d is a direct measure of the stability of this complex; the smaller the value of K_d, the more stable the complex. Another and perhaps a better illustration of the significance of the parameter of K_d is given by the verifiable fact that the theoretical loading of the adsorbent with the target molecule at $c^* = K_d$ equates to exactly half the maximum theoretical loading (i.e., to $\frac{1}{2}Q_{max}$). The Langmuir model assumes an energetically homogeneous adsorbent surface and monomolecular loading of it with the target molecule. In the present case of biomolecules binding to magnetic adsorbent particles, these assumptions are, in common with many other systems, rarely fulfilled. Nevertheless, this simple model frequently provides sound quantitative determinations of the equilibrium state. The apparent K_d for a magnetic adsorbent can be far below that for the free ligand in solution, which considerably expands the range of potentially useful ligands [1–4,6–9].

3.1.2 Case Study I: Simple Characterization of a Magnetic Adsorbent's Product Binding Behavior

The following illustrates the evaluation of an adsorbent for the recovery of added trypsin from crude cheese whey using magnetic supports derivatized with the serine protease inhibitor benzamidine [14]. In this case study, we wish to show simple systematic experiments that enable the following key questions to be answered:

1. Is the particular adsorbent chosen sufficiently good a binder to consider using in an HGMF process?
2. What amount of adsorbent will be required to quantitatively adsorb the product from a given volume of feedstock?
3. How much time is required for sufficient product sorption?

1. A first impression of the suitability of an adsorbent can, in most cases, be obtained from studies with monocomponent systems. Subsequently studies with the real feedstock should be undertaken at the earliest opportunity. The effectiveness of an adsorbent for use in batch adsorption-based separation processes such as HGMF, is critically dependent on the tightness of binding, which strictly speaking, is reflected by the initial slope of the isotherm (i.e., Q_{max}/K_d). In our experience an efficient magnetic adsorbent for HGMF will possess K_d values in the sub-micromolar range, a $Q_{max} \gg 100$ mg/g (preferably of the order of 200 to 300 mg/g) and a tightness of binding of $\gg$5 l/g.

Benzamidine-linked magnetic adsorbents were recovered from storage buffer using a bar magnet and equilibrated by resuspending in an equilibration buffer (100 mM Tris/HCl, 10 mM $CaCl_2$, pH 7.5). Aliquots containing 1 mg of adsorbent were added to a series of tubes. Subsequently, different amounts of trypsin (0 to 1.3 mg) prepared in equilibration buffer were added to each tube to give a final volume of 1 ml. After mixing at room temperature for 30 min the adsorbents were retrieved magnetically, and the liquid phases were removed and analyzed for residual trypsin content. The amount of trypsin bound to the adsorbents was determined by the difference and the data were plotted as shown in Figure 3.1a, and fitted to the simple Langmuir model [15]. Highly favorable adsorption behavior was found ($K_d = 1.2\ \mu M$, $Q_{max} = 225$ mg/g, initial slope = 8 l/g) and the adsorbent was deemed suitable for the next phase of experimentation.

2. The amount of adsorbent needed to remove added trypsin in crude whey was evaluated in a manner similar to that described above, but using the real test feedstock, and by varying the mass of adsorbent added to a given volume of it. Following equilibration, different quantities of benzamidine-linked magnetic adsorbents were aliquoted into each tube. The supports were magnetically separated and the liquid phases were removed prior to adding 1 ml of crude whey containing added trypsin at a concentration of 0.15 g/l. After 30 min of mixing, the adsorbent particles were magnetically retrieved from suspension and the liquid phases were analyzed for residual protein content and trypsin activity. Plotting the data as shown in Figure 3.1b gives an instant indication of the amount of adsorbent required to quantitatively bind the product (in this case ~5 g/l of feedstock).

3. The time required to reach equilibrium was determined by resuspending the magnetic adsorbents to a final concentration of 4.5 g/l, in 10 ml of whey previously supplemented with trypsin at 0.15 g/l, mixing with an overhead stirrer and collecting samples over a 10 min period. After various times the magnetic adsorbents were retrieved on a bar magnet (within 10 sec) and the liquid phases were analyzed for remaining trypsin activity. Inspection of Figure 3.1c confirms that binding was essentially complete with a 2 to 5 min time frame.

3.1.3 High-Gradient Magnetic Separation

The basic principle of high-gradient magnetic separation (HGMS) is simple and similar to that of deep-bed filtration [16–21]. Indeed HGMS can be described as a deep-bed filtration process in which a magnetic attraction force is added to the transport mechanism present in classical deep-bed filtration. A canister filled with a magnetizable separation matrix, usually composed of pads of stainless steel (400 series) wool or stacked layers of wire mesh, is introduced into an

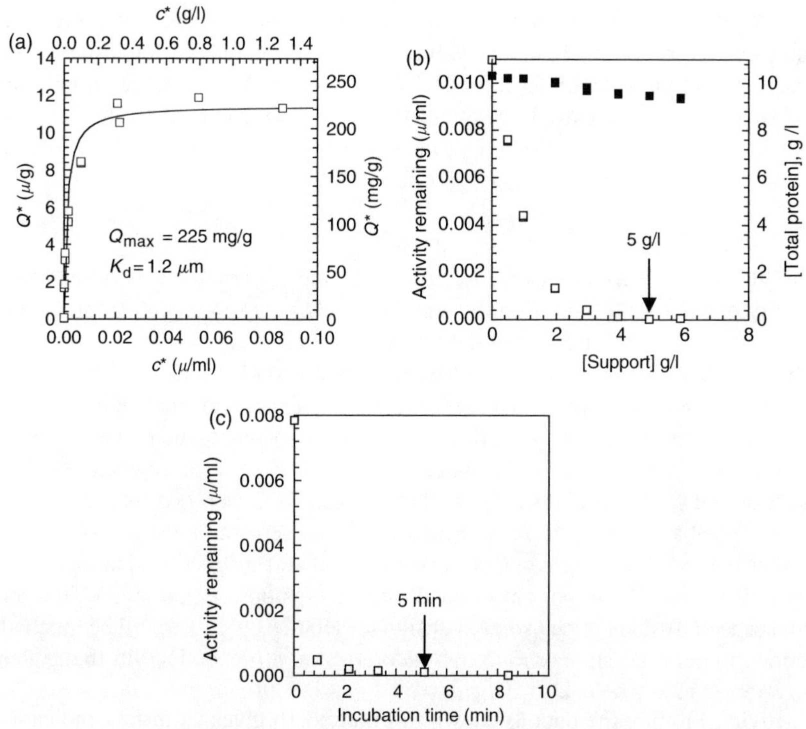

FIGURE 3.1 Trypsin binding characteristics of benzamidine-functionalized magnetic adsorbent particles of the type shown in Figure 3.7. (a) Equilibrium adsorption isotherm for trypsin. The line through the data represents the fit to the Langmuir model. (b) Effect of adsorbent concentration on the removal of added trypsin (□) and total protein (■) from crude whey. Trypsin was added to the whey feedstock at a final concentration of 0.15 g/l. (c) Time-course for the removal of added trypsin (0.15 g/l) from crude whey using benzamidine-linked magnetic adsorbents at a concentration of 4.5 g/l. (Adapted from Gomes CSG, Petersen TL, Hobley TJ, and Thomas ORT. In *Proceedings of the 7th World Congress of Chemical Engineering and 5th European Congress of Chemical Engineering*, Glasgow, July 10–14, 2005, ISBN 0 85295 494 8.)

external homogenous magnetic field. The filter matrix wires bundle the external magnetic field in their vicinity to generate distinct regions on their surfaces, which strongly attract paramagnetic, and especially ferromagnetic particles. Two general approaches are commonly employed to describe HGMS, namely (i) the macroscopic description of a whole filter based on its particle breakthrough behavior; and (ii) solving of a force balance for a microscopic system consisting of a magnetized ferromagnetic wire and a paramagnetic particle.

One of the most important parameters describing the efficiency of capture of magnetic particles by HGMS is the ratio (v_r) of the magnetic velocity (v_m) to the applied fluid velocity (v_0), which, for the capture of a magnetic particle on a single magnetized wire, is described by the following equation [16,17]:

$$v_r = \frac{v_m}{v_0} = \frac{2\mu_0(\chi_s - \chi_f)M_w H_0 b^2}{9\eta a v_0} \tag{3.2}$$

where μ_0 is the permeability of free space, χ_s and χ_f are the magnetic susceptibility of the support and liquid, respectively, M_w is the magnetization of the wire, H_0 is the field strength of the applied magnetic field, η is the viscosity of the liquid, a is the radius of the wire, and b the particle radius. Assuming the Stokes equation for hydrodynamic resistance to be valid, the magnetic velocity can be interpreted as the theoretical particle velocity due to the magnetic force in the immediate vicinity of the wire. At values of $v_r \gg 1$ the separation behavior of an HGMS will be similar to that of a classical deep-bed filter, that is, a relatively sharp loading front is formed within the separation matrix. At $v_r < 1$ rather extended loading fronts will be obtained and magnetic separation will be ineffective. The important implications of Equation 3.2 on magnetic filter and especially magnetic adsorbent design for HGMF processes will be touched upon in later sections. For thorough theoretical treatises on HGMS the reader is referred to key texts [16–20].

3.1.4 High-Gradient Magnetic Fishing

The integrated process consisting of coupling a batch-binding step to magnetic adsorbent handling (i.e., capture, washing, and elution) with a high-gradient magnetic filter has been termed HGMF [1]. It has already been applied for the capture of a wide range of proteins from different feedstocks [1–14]. Schematics of a typical HGMF process, and plan of a semitechnical HGMF pilot plant are presented in Figure 3.2 and Figure 3.3, respectively.

The typical approach for HGMF is to contact the feedstock and adsorbent in a stirred tank reactor for several minutes (G1, Figure 3.3) and then, following biomolecule sorption, pump the magnetic adsorbent particle/feedstock suspension through the filter canister (a velocity of ~25 m/h is typical) of the magnetic separator (MS) with the field switched on (Figure 3.2a). Alternatively, adsorption can be performed continuously en route to the magnetic separator, for example, by replacing the stirred batch adsorption tank with a pipe reactor [13]. This has the advantage of bringing the necessary contact time down to the order of 10 sec. In both cases the product-loaded magnetic adsorbents fed to the filter are retained within it, while all of the nonmagnetic components pass through unhindered. Shortly before adsorbent particle breakthrough, flow of

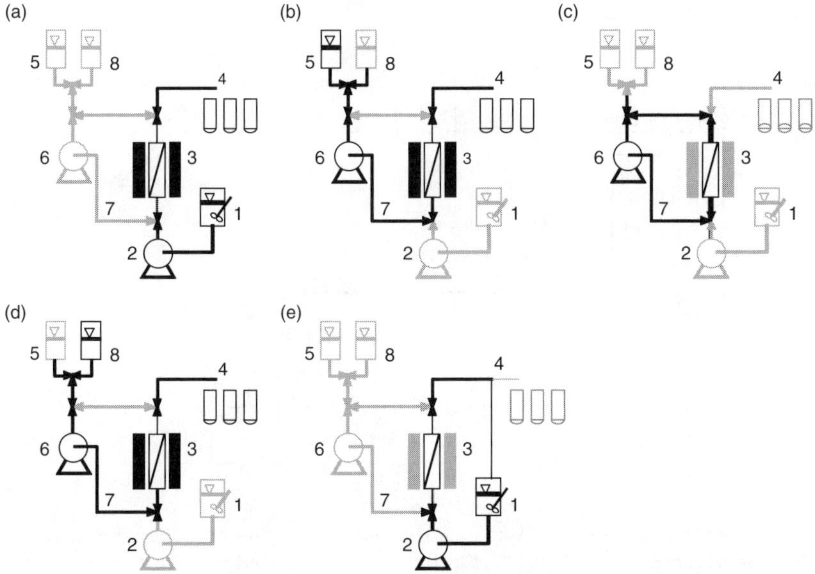

FIGURE 3.2 Stages in an HGMF process. (a) Batch adsorption and filter loading. (b) Filling of the loaded filter and recycle loop with wash buffer. (c) Redispersion of the adsorbents with the field off. (d) Filling of the loop and filter with elution buffer, with the field on. (e) Flushing adsorbent to the batch adsorption reactor for a semi continuous multicycle purification. (1) Batch adsorption reactor, (2) pump 1, (3) magnetic filter, (4) fraction collector, (5) buffer, (6) pump 2, (7) recycle loop, and (8) elution buffer.

the adsorbent/feedstock mixture to the separator is stopped, and, with the field still on, the system recycle loop is filled with a wash solution (Figure 3.2b) from the appropriate reservoir (G2, Figure 3.3). Subsequently, the recycle loop is closed, the magnetic field is switched off, and the adsorbent particles are flushed out of the filter into the recycle loop and circulated around the system (typically at ~80 m/h for several minutes). Following this, the washing solution is discharged and the now washed adsorbents are recovered within the filter by switching the magnetic field back on (Figure 3.2c). The same procedure is employed for subsequent elution steps (Figure 3.2d), which are performed with a buffer contained in reservoir G3. Cleaning, reequilibration or other process steps may then follow elution. In the case of immobilized metal affinity separations, for example, these might involve displacing the remaining eluant and re-equilibrating the adsorbent particles using the washing buffer (G2, Figure 3.3) followed by reloading immobilized chelating groups (e.g., iminodiacetate [IDA]) of the adsorbent with divalent metal ions, such as Cu^{2+} (G5, Figure 3.3). Finally, the revitalized adsorbent particles are recovered from

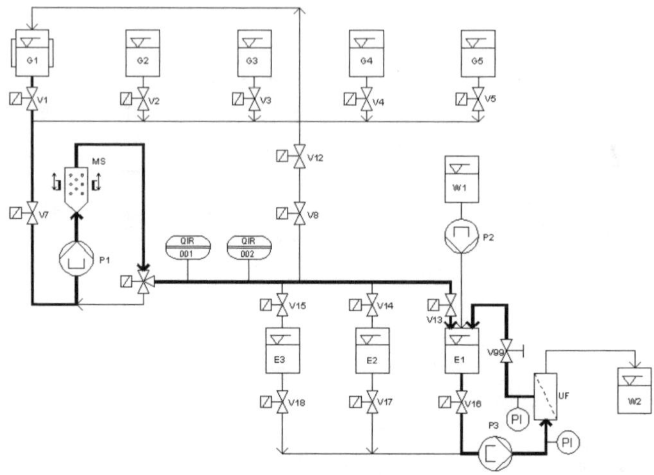

G1:	cooled sorption vessel	W1:	exchange buffer (UF)	QIR_1:	fluorescence detection
G2:	wash buffer	W2:	filtrate	QIR_2:	UV detection
G3:	elution buffer	E1:	eluate	UF:	cross-flow filtration
G4:	cooled feedstock	E2:	used Cu^{2+} buffer	PI:	UF pressure regulation
G5:	cu^{2+} conditioning buffer	E3:	waste	MS:	magnetic separator
Vx:	two way valves	Px:	peristaltic pumps		

FIGURE 3.3 The pilot-scale HGMF protein purification system (filtration volume 1.2 l) operated at the University of Stuttgart's Institute for Bioprocess Engineering.

the separation matrix and added back into the mixing vessel (G1) together with a fresh batch of crude bioprocess feedstock (G4, Figure 3.3) to begin a new cycle. All of the process steps during the purification cycle can be fully automated, and the pumping speeds, times, and valve settings are then simply controlled by means of a suitable graphic measurement and control software (e.g., Visual Designer 4.0, Texas Instruments). Figure 3.4a and Figure 3.4b, respectively, show photographs of the pilot plant HGMF system outlined above (Figure 3.3), and the magnetic separator it employs. Smaller, but nevertheless comparably automated test facilities are operated at the Forschungszentrum Karlsruhe (Germany), the Danish Technical University (Denmark), and the University of Birmingham (U.K.).

Single cycle HGMF processing of very large amounts of feedstock would require outsized magnetic filter canisters and separators. HGMF processing of large volumes is therefore best achieved by multicycling with smaller HGMF rigs. Multicycling in HGMF is made especially attractive given that the short sorption times and high fluid processing velocities typically employed translate

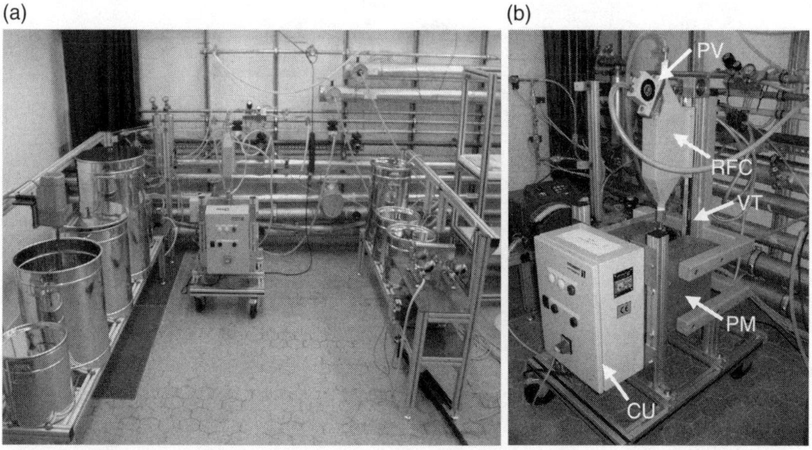

FIGURE 3.4 (a) Photograph of the University of Stuttgart's HGMF pilot plant pilot. (b) Close-up of the pilot-scale HGMS separator employed at Stuttgart. The system is operated via a control unit (CU) and features a permanent horse-shoe magnet block (PM) and a reciprocating 1.2 l magnetic filter (RCF) mounted on a vertical belt driven track (VT). Note the pneumatic vibrator (PV) attached to the top of the filter in (b) which is employed to enhance particle recovery from the filter during flushing.

into short overall cycle times and rapid turnaround. A flow sheet illustrating various options in multicycle HGMF processing is depicted in Figure 3.5. In this example, intensified multicycling begins directly after product elution from the adsorbents by flushing the adsorbent particles out of the magnetic filter back into the batch adsorption reactor using a fresh batch of crude bioprocess feedstock as the resuspending phase (Figure 3.2e). This allows for very fast cycling, but assumes that the trace amounts of eluant carried over into the batch adsorption reactor exert a negligible impact on product binding. In practice, the presence of trace levels of an eluant during the adsorption phase may often improve product purity by reducing the binding of species with weak affinity for the ligand. Multicycling and the effects of selective washing and elution steps are discussed further in Section 3.4.

3.1.5 Design of an HGMF Process

For the design of a basic HGMF process it is convenient to split development into a number of stages as represented in Figure 3.6 and reported in Reference 11. Process design should begin with identifying the most suitable ligand type and magnetic adsorbent, followed by optimization of binding, wash, and elution conditions. Next, the magnetic filter capture step, including filter

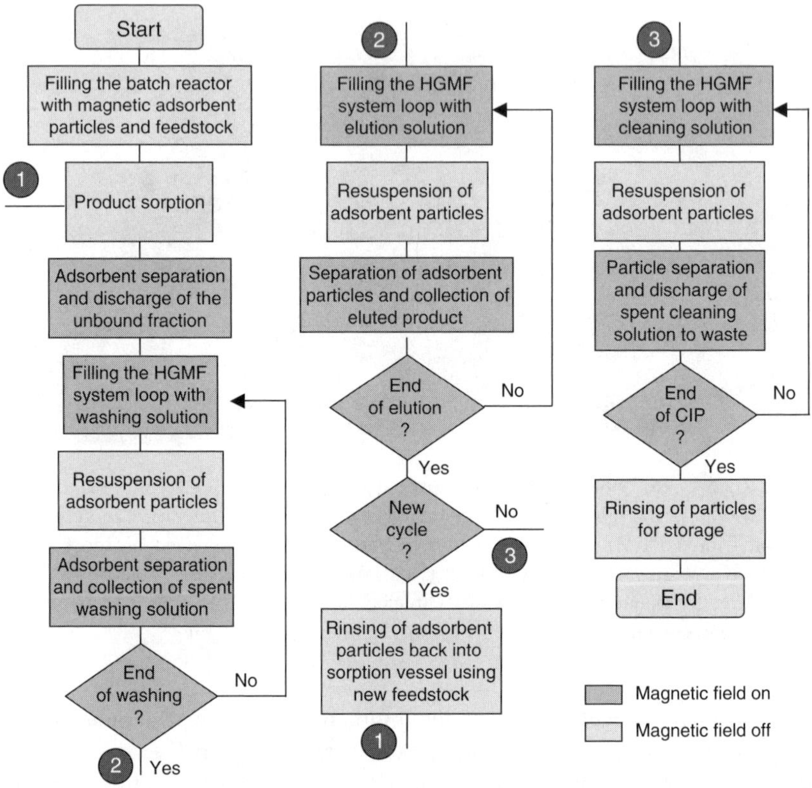

FIGURE 3.5 Flow sheet for the operation of a multicycle HGMF process. The numbers represent links to different parts of the flow sheet.

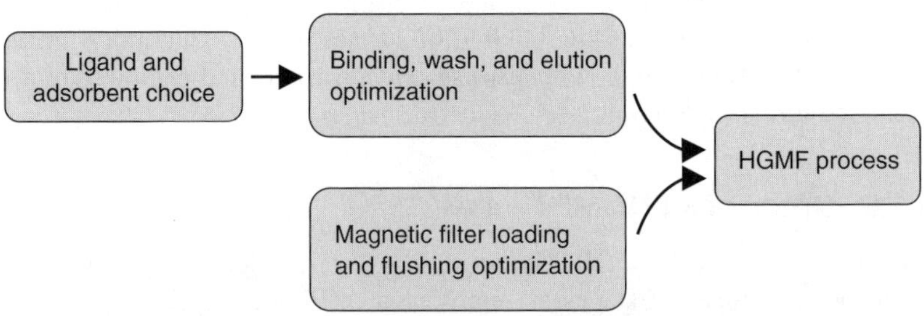

FIGURE 3.6 Steps involved in HGMF process design.

flushing during wash and elution should be considered. Subsequently all steps need to be combined and the process characterized and optimized under actual operating conditions. Aspects regarding the magnetic adsorbent are dealt with below (Section 3.2) and filter performance in Section 3.4.

3.2 SUITABLE ADSORBENTS FOR HGMF AND THEIR CONDITIONS FOR USE

3.2.1 MAGNETIC SUPPORT MATERIALS

A bewildering number of magnetic support materials of different designs are currently available commercially for use in small-scale biotech applications (plasmid isolation, cell sorting, routine diagnostics, etc.), and some of these are described in Table 3.1. Many of the features necessary for a successful magnetically responsive adsorbent are similar to those required for conventional chromatographic matrices, while others are quite unique. Very few of the materials listed in Table 3.1 meet the exacting requirements of an adsorbent material tailored for process-scale HGMF (summarized in Table 3.2) and for a detailed discourse on the subject the reader is referred to Thomas and Franzreb [21]. Most of the commercial magnetic supports listed in Table 3.1 exhibit that especially important property of superparamagnetism, that is, responsiveness to an applied magnetic field without any permanent magnetization. When a field is applied superparamagnetic particles are magnetized and agglomerate readily through interparticle forces to allow facile separation. However, unlike ferromagnetic materials, when the field is removed, an ideal superparamagnetic particle will retain no magnetic memory (i.e., no remanent magnetization or remanence). The absence of magnetic memory (or at least the possession of low remanence) in a magnetic adsorbent is crucial for large-scale use as it permits easy redispersion of the particles, efficient product elution from their surfaces, and allows their repeated use over many operating cycles.

Although magnetic supports are gaining popularity within the laboratory for routine use, their high cost and availability in limited quantities would appear to be major stumbling blocks to future large-scale use. This need not be the case as suitable magnetic supports can in fact be manufactured very cheaply and at vast scales. Indeed, some manufacturers (e.g., Merck KGaA, Chemagen) have already responded to the challenge of manufacturing magnetic support materials for process-scale HGMF [22]. Chemagen's MF-PVA adsorbent, for example (note, this is not the same as the M-PVA bead listed in Table 3.1), can be obtained in kilogram quantities on request, at a cost that approaches affordability at large scale (e.g., 20 Euro per gram). Alternatively, experimentalists may wish to make their own magnetic adsorbent for evaluation purposes. The high capacity

TABLE 3.1
Some Commercially Available Magnetic Adsorbent Particles

Manufacturer (Country)	Product	Description
Ademtech (France)	Monodisperse magnetic emulsions	Monodisperse magnetic nanobeads prepared by high shear fragmentation of a ferrofluid emulsion followed by droplet surface polymerization and functionalization (~80% w/w magnetic oxide; monosized, 0.20 and 0.30 μm; 0.1–1.0 μm possible; SA~100 $m^2\ g^{-1}$ for 0.2 μm bead)
Bangs Laboratories Inc. (USA)	Estapor®	Extensive range of microspheres based on the Estapor® "M" bead (see description under Merck Eurolab). Available with superparamagnetic crystal contents of 13, 24, 46, or 66% (w/w). Median sizes ranging from ~0.35 to 2.5 μm
	BioMag®	(see description under Polysciences)
Chemagen Biopolymer-Technologie AG (Germany)	M-PVA	Magnetite crystals encapsulated in crosslinked impervious polyvinyl alcohol bead (2 μm mean diameter).
Chemicell GmbH (Germany)	BeadMag	Magnetic crystals encapsulated in crosslinked starch (~1 μm)
	FluidMAG	Nonporous hydrophilic polymer coated magnetic crystals (0.05, 0.1, 0.25, and 0.5 μm)
	SiMAG	Uniform superparamagnetic-silica particle (80% w/w iron oxide) with highly porous surface (SA > 100 $m^2\ g^{-1}$; monosized, 0.25, 0.5, 0.75, and 1 μm)
Cortex Biochem (USA)	MagaPhase	Ultra-pure magnetite (33–60% w/w) encapsulated in polysaccharide or synthetic polymer beads. Various diameters available, for example, 1–10, 1–60 μm, and monosized 3.2 μm
CPG Inc. (USA)	MPG®	Porous borosilicate glass impregnated with magnetite crystals (SA ~60 $m^2\ g^{-1}$; ~5 μm)
Dynal Inc. (Norway)	Dynabeads®	Nonporous uniform monodisperse superparamagnetic beads (SA for 2.8, 4.5, and 5 μm beads quoted as 2–5, 4–8, and 1–4 $m^2\ g^{-1}$, respectively)
Immunicon (USA)	Ferrofluids	Nonporous protein-coated magnetite crystals (0.135 and 0.175 μm sizes)
Kisker GbR (Germany)	Magnetic polystyrene	Spherical nonporous polystyrene magnetite composite of relatively uniform size (e.g., ranging from 0.4–0.7 to 18–24 μm), prepared by layering polystyrene and magnetite onto polystyrene core particles
	Magnetic silica	Uniform nonporous silica/magnetite (80% w/w) composite particles (0.25, 0.5, 0.75, and 1 μm sizes)
	Magnetizable nanoparticles	Dextran based superparamagnetic nanoparticles (0.05, 0.1, 0.13, 0.25 μm sizes) with magnetite content of 90% (w/w)

Merck Eurolab (France) formerly Prolabo	Estapor® "M"	Superparamagnetic crystals (12, 20, 40, or 60% w/w) uniformly distributed in impervious polystyrene bead (various diameters available, e.g., 0.7–1.0, 0.7–1.3, 0.8–1.2, and 0.9–1.3 μm)
	Estapor "EM"	Superparamagnetic core particle (12, 20, 40, or 60% w/w) encased in impervious polystyrene (various diameters, e.g., 0.9–1.8 and 1.7–2.5 μm)
Merck KgaA (Germany)	SuperPara Magnetic Microspheres	Superparamagnetic microspheres based on the Classical Estapor® "M" bead (see above)
	MagPrep® Silica Particles	Irregularly shaped nonporous silica-coated magnetite (>95% w/w) particle ($\sim$1 μm, SA 16–22 m^2/g)
Micromod Partikeltechnologie GmbH (Germany)	Micromer®-M	Nonporous monodisperse supports prepared by encapsulation of magnetite within styrene–maleic acid copolymer matrix and subsequent coating with polysaccharide or silica (monosized, ranging from 2 to 12 μm)
Miltenyi Biotec (Germany)	MACS microbeads	Nonporous polysaccharide-coated magnetic iron oxide crystals ($\sim$0.05 μm)
Polysciences Inc. (USA)	BioMag	Irregular nonporous superparamagnetic silanized iron oxide (av. 1.8 μm)
	BioMag®Plus	Smaller and more uniform in size than BioMag® to deliver higher capacity and more predictable and consistent behavior during capture and magnetic separation steps
Promega GmbH (Germany)	MagneSil™	Magnetic core (55% w/w) coated in porous silica (45% w/w) shell (SA $\sim$27 m^2g^{-1}; 2–14 μm; av. 6.6 μm)
Qiagen GmbH (Germany)	BioMag	(see description under Polysciences)
Roche Diagnostics (Germany)	MGP	Magnetic core particle encased in substantially pore-free glass shell (no size information available from the manufacturer)
Seradyn Inc. (USA)	Sera-Mag™	Nonporous magnetic beads with highly textured cauliflower-like binding surface imparted by presence of a rough subsurface layer of magnetic crystals sandwiched between particle's core and exterior (monosized, ranging from 0.7 to 3 μm)
Scipac Ltd (U.K.)	Bioactivated Mag Particles	Porous cellulose iron-oxide (1–10 μm) and agarose iron-oxide particles (1–10 μm)
Sigma-Aldrich (USA)	Enzacryl FEO-(M)	Magnetite encapsulated in porous synthetic polymer matrix (40–70 μm)
Spherotech Inc. (USA)	SPHERO™	Nonporous magnetic sphere featuring layer of magnetite (typically 10–15% w/w) sandwiched between inner core particle and external coat, both constructed of polystyrene (monosized, ranging from 0.8 to 9 μm)
Whatman (U.K.)	Magarose	Magnetite encapsulated in porous cross-linked agarose bead (20–160 μm)

SA = Surface Area.

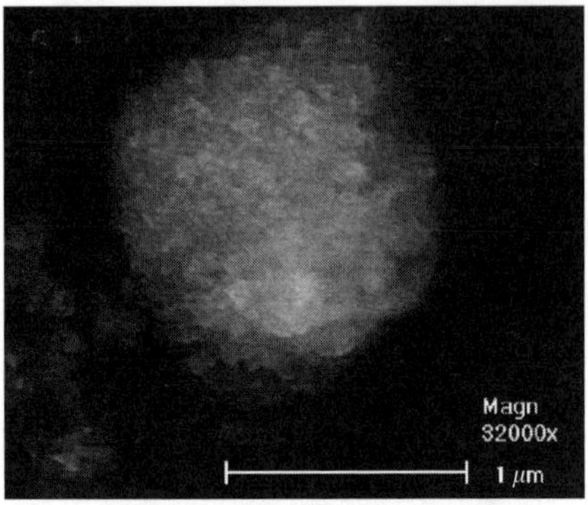

FIGURE 3.7 SEM of polyglutaraldehyde-grafted magnetic support particle described by Hubbuch and Thomas [2]. Note the adsorbent material's high degree of surface irregularity affords it product sorption capacities of up to 300 mg/g. (Courtesy of G. Beuchle, Forschungszentrum Karlsruhe GmbH, Germany.)

superparamagnetic adsorbent particle shown in Figure 3.7 can be recommended, as it possesses many of the ideal properties listed in Table 3.2, and is easy to produce in the laboratory in 10 to 100 g batches using relatively standard lab equipment [1–14], or in much larger quantities using pilot-scale apparatus.

3.2.2 Ligand Selection

Once a suitable superparamagnetic support base particle has been selected, the choice of an appropriate ligand will be critical. From knowledge of the protein of interest, the feedstock, and available literature, it may be possible to define an effective ligand. However, for a new protein, screening of adsorbents must be conducted. Due to the limited choice of appropriately derivatized magnetic supports at the present time, an approach based on conventional chromatographic screening with the clarified feedstock represents perhaps the best way to identify useful ligands [11]. Numerous chromatography manufacturers supply screening kits comprising of small prepacked columns or loose gels. If it is not subsequently possible to source a finished magnetic adsorbent with the required ligand, a base support should be obtained, and then functionalized inhouse. For an excellent source of functionalization techniques that can easily be adapted to coated magnetic adsorbents, the reader is referred to the text by Hermanson et al. [23]. Ligands of almost any type can be attached to magnetic adsorbents [1–14], but given the dirty process environments, harsh cleaning regimes, and demands for high sorption capacity in HGMF, small stable synthetic ligands are preferred to larger biological ones (Table 3.2).

TABLE 3.2
Idealized Requirements of a Magnetic Adsorbent for Process-Scale HGMF

Requirements	Reasons
I. Magnetic characteristics:	
(a) Superparamagnetic (or at least possessing low remanent magnetization, that is, magnetic memory)	Absence of magnetic memory affords easy redispersion at zero field, efficient product elution, and repeated use
(b) High M_s (i.e., >35 Am^2/kg) — usually obtained with a high magnetic iron oxide content (at least 30% w/w and preferably much higher, for example, 80–90%, especially for very small nanoparticles)	To ensure fast efficient separation through a high v_m value (see Equation 3.2)
II. Size, shape, density, uniformity, and stability:	
(a) Particle size between 0.5 (min) and 2 μm (max)	Relatively narrow optimum of sizes is defined by balance of surface area (SA) and v_m. A small diameter is required to afford sufficient SA, but too small a size leads to low v_m and therefore poor magnetic separation efficiency. A small particle size is also advantageous for use in agitated reactors from a mechanical durability perspective as a particle's susceptibility to attrition in agitated reactors is inversely related to its size
(b) Monosized	So that all adsorbent particles move with the same v_m in a magnetic field, thereby affording greater predictability during adsorbent particle capture
(c) Roughly spherical	To achieve high adsorbent packing densities within magnetic filters

(Continued)

TABLE 3.2
Continued

Requirements	Reasons
(d) Particle density between 2.5 and 4 g/cm^3	Both density and magnetic susceptibility are practically linked to how much of the support is made up by the magnetic component, which is much denser (e.g., magnetite has a density of 5.2 g/cm^3). Supports of 0.5–1 μm with magnetic core contents $\geq 50\%$ ($M_s > 35$ Am^2/kg; $\rho = 3–4$ g/cm^3) settle only very slowly at zero field
(e) Physico-chemically robust construction	To tolerate harsh chemicals during cleaning and regeneration and give long lifespan to the adsorbents. In this context measures to prevent the magnetic elements within the support from being corroded at a significant rate may prove necessary
III. Surface architecture and chemistry:	
(a) Nonporous, but with a highly folded surface (i.e., not smooth) and/or embellished with functionalized polymeric tentacles extending out from the surface	To afford better fouling resistance and easier cleaning cf. porous adsorbents. Further advantages of the nonporous design include improved ligand utilization, very fast adsorption/desorption kinetics, and possibly better resistance to mechanical attrition. Surface texturing can dramatically enhance product sorption capacity
(b) Target accessible SA of >50 m^2/g; preferably 100–150 m^2/g	To deliver sufficiently high target binding capacity, for example, for proteins >100 mg/g and preferably 200–300 mg/g
(c) Neutral, hydrophilic, and easy to derivatize binding surface	For low nonspecific binding and to enable ligands to be coupled at high densities
(d) Small cheap ligands	Generally preferred over biospecific varieties in HGMF, as they yield higher capacity adsorbents, and can tolerate harsh cleaning conditions
IV. Availability at low cost in kilogram to tonne quantities	To make HGMF a viable proposition

3.2.3 Case Study II: Use of Small-Scale Batch Experiments to Determine Conditions for Adsorbent Use in HGMF

Meyer and coworkers [11] recently demonstrated a general strategy for the development of a HGMF process (Figure 3.6) illustrated for the recovery of superoxide dismutase (SOD) from unconditioned cheese whey. They showed that optimal conditions for use of adsorbents in HGMF can, at all stages, be determined rapidly at the bench in small-scale (1 to 5 ml) batch experiments, before transferring directly to a HGMF process. The ligand employed on magnetic supports (of the type shown in Figure 3.7), that is, Cu^{2+}-IDA, was identified following screening of various candidate chromatographic matrices with the clarified feedstock (Figure 3.6 and Section 3.2.2). Using a procedure analogous to that described earlier in case study I (Section 3.1.2) and with the aid of various analytical techniques (enzyme assays, PAGE, and zymography), an adsorbent concentration required to recover all of the SOD from crude feedstock, the whey was determined to be 7 g/l.

For small-scale batch screening of washing and elution conditions suitable for HGMF the adsorbent concentration should be much higher than that employed during the binding step, given that during HGMF the adsorbents will be concentrated within the magnetic filter and recycle loop. In Meyer and coworkers' [11] case, the adsorbent concentration during wash and elution within the HGMF filter and recycle loop was predicted to be 20 to 30 g/l. Thus in their small-scale screening for improved wash conditions adsorbents were contacted with crude whey at a concentration of 7 g/l, magnetically separated and then washed with test buffers at an adsorbent concentration of 30 g/l. Elution optimization was performed systematically in a similar manner using adsorbent concentrations of 7 g/l for adsorption, and 20 to 30 g/l for washing and elution. The conditions defined for binding, washing, and elution in the lab-scale experiments were subsequently employed in an HGMF process, and it was found that these gave reasonably good predictions of HGMF process performance [11].

3.3 DESIGN AND SET-UP OF MAGNETIC SEPARATOR SYSTEMS

Due to the current absence of a market, no commercially available magnetic separator systems suitable for industrial downstream processing currently exist. That said, the physical principles, wide variety of available designs from parallel large-scale industries, and inherent advantages of magnetic separation techniques per se represent a sound basis for the imminent advancement of bespoke magnetic separation methods for industrial downstream processing.

Such designs will necessarily feature different materials of construction, as well as appropriate valves and seals, commensurate with containment, CIP, and SIP requirements.

For the separation of readily magnetizable particles from aqueous media the following types of separators are commonly employed at industrial scales of operation: chain-type magnetic separators, rod-type magnetic filters, wet-drum separators, and high-gradient magnetic separators. Of these the open-gradient design, and resultantly poor separation efficiency of chain and wet-drum type instruments for particles <5 μm, make such separators unsuitable for most biotech applications. Rod-type magnetic separators produce relatively weak magnetic fields (<0.2 T) that rapidly decline with increasing distance away from the rod. When operated in batch mode however, that is, so that the particle suspension remains within the separator for a sufficiently long time, rod-type separators can achieve separation of micron-sized magnetic particles. In more challenging situations, for example, those demanding continuous separation of particles smaller than ~2 μm high-gradient magnetic separators are the obvious instruments of choice, as these exert by far the highest magnetic forces upon particles compared to other types of magnetic separators (see Section 3.1.3). For HGMF processing, magnets producing magnetic flux densities of 0.3 to 0.6 T are normally sufficient for most adsorbent particle separation tasks. As implied earlier (Section 3.1.3), effective magnetic filters can be simply created by packing pads of wire wool, or rolled-up or stacked sheets of wire mesh into a nonmagnetic canister [1–4,6–11]. Even a seemingly tightly packed canister will usually have a voidage close to 90%. From Equation 3.2 it follows that thin wires of a highly magnetizable material (e.g., 400 series stainless steel) will deliver the highest values of v_{m}. In practice however, the need for adequate filter strength and long-life sets the lower limit for the wire diameter at ~100 μm. An improved filter design employing a cassette with an ordered array of meshes is illustrated in Figure 3.8a [5,12,14,24].

Cyclically operated HGMS systems are usually equipped with switchable permanent magnets or solenoids as the field source. The use of switchable permanent magnets (see e.g., Figure 3.8b) has the advantage of very low capital and operating costs [5,10,11,24], but is currently limited to systems with small to moderate matrix volumes (<~20 l) and magnetic field strengths (<0.5 T). Horseshoe like permanent magnet blocks can also be used, in which the filter is simply reciprocated in and out of the magnetic field (see earlier Figure 3.4a). Should higher matrix volumes or flux densities be needed, solenoid designs (see Figure 3.9) represent the only practical option. Depending on the size and maximum flux density desired, these magnetic separators are either simply cooled with air or fitted with a water or oil cooling system.

A brief, but by no means exhaustive, survey of manufacturers of high-gradient magnetic separators is given in Table 3.3. Solenoid designs with filter

(a)

(b)

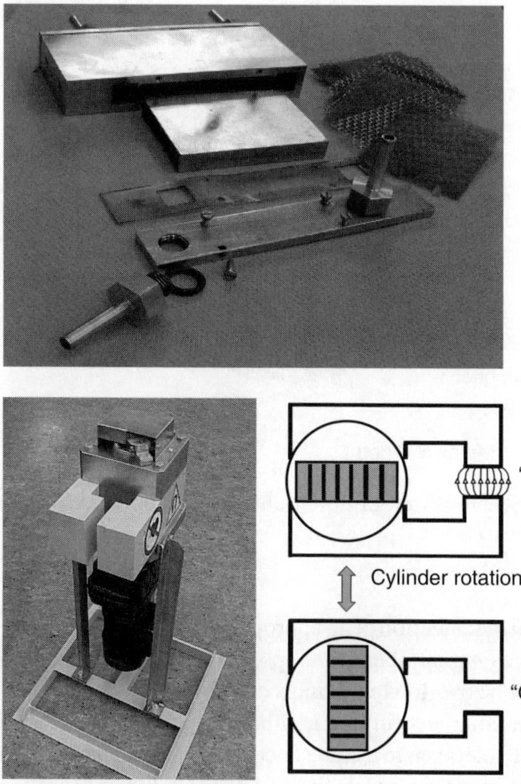

FIGURE 3.8 (a) Cassette type magnetic filter prototype of 40 ml volume with filter meshes, suitable for use in the separator shown in (b). (b) Mini-pilot scale (HGF-10, Steinert GmbH, Cologne, Germany) cyclically operated on–off permanent magnet. The magnet blocks are arranged within a cylindrical iron subyoke, which can be rotated along its central axis within the fixed iron yoke. The right-hand side shows the on–off switching principle. Magnets of this design are available in sizes able to accommodate filters of up to 20 l.

matrix diameters of up to 3 m corresponding to filter areas of more than 7 m^2 are employed in the cleaning of kaolin sludges. For bioproduct processing, such separators would be able to attain filtration rates of $\sim$175 m^3/h. However, when averaged over a complete HGMF cycle (i.e., to include washing and elution operations) the overall throughput of raw biosuspension is likely to be much lower than this value. Nevertheless, at target protein concentrations within the initial suspension of <1 g/l highly respectable overall throughputs of raw biosuspension of 50 to 100 m^3/h per magnetic separator should be feasible.

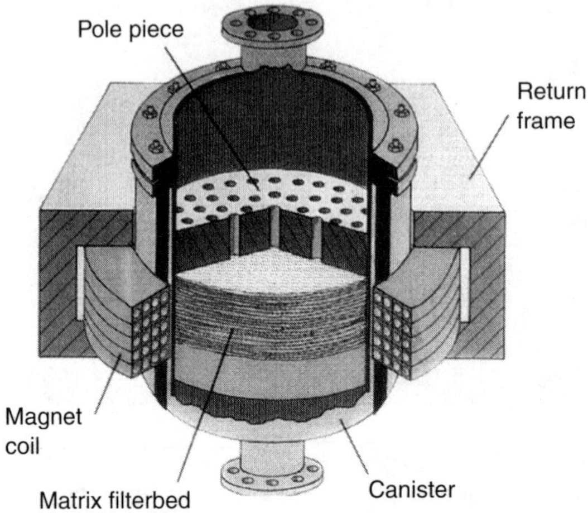

FIGURE 3.9 Sectional view of a cyclically operated solenoid type HGMS. (Courtesy of Metso Minerals.)

Apart from the selection of an appropriate magnetic separator, consideration of the type, size, and number of pumps required when setting up HGMF facilities should not be overlooked. Pumps employed for magnetic adsorbent-based protein purification must fulfill a number of special criteria, which include the following: high tolerance to the presence of small solid particles in the feed flow; the capacity to pump suspensions of elevated viscosity (up to $\sim$10 mPa sec); minimal back-mixing of the feed flow within the pump; easy cleaning and sterilization of the pump areas that come into contact with the biosuspension; and a wide working range vis à vis permissible throughputs. In relation to the last requirement, throughputs during adsorbent particle separation compared with resuspension should differ by a factor of roughly 5 or more. In our experience of operating HGMF at pilot scale, peristaltic pumps appear to satisfy all of the aforementioned tolerances, and are available with capacities ($>$20 m^3/h) sufficient for most potential applications.

3.4 PARAMETERS AFFECTING SYSTEM PERFORMANCE

3.4.1 INTRODUCTION

By far the most important parameter affecting the performance of a given HGMF process is the adsorbent particle's selectivity for the target product, or to put it another way, the equilibrium state developed within the adsorption vessel.

TABLE 3.3
Some Manufacturers of High-Gradient Magnetic Separators

Name	Address	Contact Details	Remarks
Eriez	2200 Asbury Road Eriez, PA 16506 USA	Phone: +1 814-835-6000 Fax: +1 814-838-4960 eriez@eriez.com www.eriez.com	All types of HGMS
Steinert Elektromagnetbau GmbH	Widdersdorfer Str. 329-331 D-50933 Cologne Germany	Phone: +49 221 49 84 0 Fax: +49 221 49 84102 sales@steinert.de www.steinert.de	HGMS based on switchable permanent magnets and solenoids
Slon Magnetic Separator Ltd.	36 Qingnian Road, Ganzhou Jiangxi Province 341000 China	Phone: +86-797-8186426 Fax: +86-797-8186436 slon@slon.com.cn www.slon.com.cn	Vertical ring and pulsating HGMS
Outokumpu Technology	Riihitontuntie 7 C, PO Box 86 02200 Espoo Finland	Phone: +358 9 4211 Fax: +358 9 3888 corporate.info@outokumpu.com www.outokumputechnology.com	Superconducting HGMS based on a reciprocating canister system
Master Magnets Ltd.	Burnt Meadow Road North Moons Moat Redditch, Worcs, B98 9PA UK	Phone: 01527 65858 Fax: 01527 65868 info@mastermagnets.co.uk www.mastermagnets.co.uk/	HGMS based on small- to medium-sized solenoids
Metso Minerals (formerly Sala)	Metso Minerals Oy P.O. Box 307, Lokomonkatu 3 FIN-33101 Tampere Finland	Phone: +358 20 484 100 Fax: +358 20 484 141 minerals.info@metso.com www.metsominerals.com	All types of HGMS

Although the relationships and laws that underpin our understanding of sorption processes involving multicomponent systems are well established, they have not yet been formulated, nor systematically applied, to the adsorption of biosubstances onto magnetic microparticle adsorbents. Accordingly, in the following sections we attempt to rectify this by describing the formal dependence of the parameters adsorption yield, purity, and yield factor upon the equilibrium state of such systems. Subsequently, an example illustration of the validity and usefulness of this framework for interpreting adsorption of competing proteins on magnetic adsorbents is demonstrated.

3.4.2 Simplified Yield Estimation

In the following sections assuming a simple model for adsorption, we derive useful general equations for estimation of yield and productivity in batch contacting systems. When adsorption is carried out in a closed, mixed vessel, the system mass balance can be represented as follows:

$$c_0 \cdot V_{\text{batch}} - c^* \cdot V_{\text{batch}} = Q^* \cdot m_{\text{p}} \tag{3.3}$$

where c_0 is the initial biomolecule concentration, V_{batch} is the biosuspension volume, and m_{p} is the particle mass in the batch reactor. By inserting Equation 3.1 into Equation 3.3 and solving for c^*, the biomolecule concentration in the liquid phase at equilibrium, the following expression is obtained:

$$c^* = \frac{1}{2}\left[c_0 - K_{\text{d}} - \frac{m_{\text{p}} \cdot Q_{\max}}{V_{\text{batch}}} + \sqrt{4 \cdot c_0 \cdot K_{\text{d}} + \left(\frac{m_{\text{p}} \cdot Q_{\max}}{V_{\text{batch}}} - c_0 + K_{\text{d}}\right)^2}\,\right] \tag{3.4}$$

To represent the achievable yield as a function of the adsorbent particle concentration used, we define a dimensionless capacity ratio, CR:

$$\text{CR} = \frac{m_{\text{p}} \cdot Q_{\max}}{c_0 \cdot V_{\text{batch}}} \tag{3.5}$$

CR describes the ratio between the maximum amount of biomolecule to be adsorbed by the amount of adsorbent particles supplied and the original amount of biomolecule available in the batch volume. With the aid of this parameter,

Equation 3.4 may be transformed to yield the following relationship for c^*/c_0:

$$\frac{c^*}{c_0} = \frac{1}{2}\left[1 - \frac{K_d}{c_0} - \mathrm{CR} + \sqrt{4 \cdot \frac{K_d}{c_0} + \left(\mathrm{CR} - 1 + \frac{K_d}{c_0}\right)^2}\right] \tag{3.6}$$

Hence, the biomolecule concentration remaining in the liquid phase at equilibrium is dependent on both the CR and the ratio between the dissociation constant and the initial concentration of the target molecule (i.e., K_d/c_0). The adsorbed product yield, Y, is given by:

$$Y = 1 - \frac{c^*}{c_0} \tag{3.7}$$

Figure 3.10 shows how the product yield varies as a function of CR and K_d/c_0, with the latter ratio being varied between 0.1 and 1.0.

As expected, product yield decreases with increasing values of K_d/c_0, given that higher values of this ratio are caused by either a smaller product concentration or a higher dissociation constant, that is, a reduced binding affinity. It is also evident that an increase in the CR value improves the product yield, and that to bind 80 to 90% of the initially available product typically requires high CR values (≥ 2 or even ≥ 3). This can therefore mean that the amount of particles practically required may actually be two or three times greater than that estimated from the simple assumption of $Q^* = Q_{max}$.

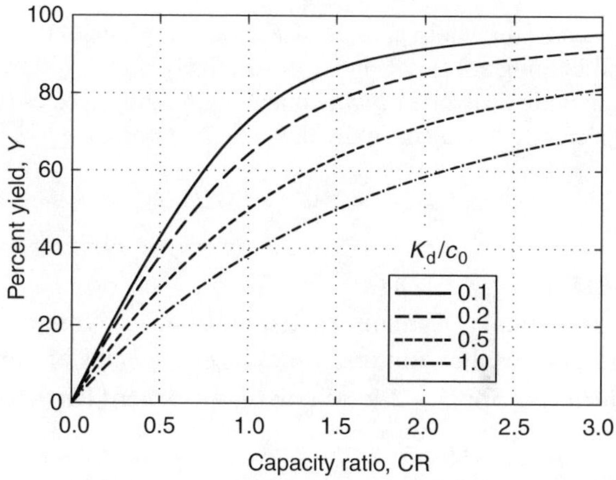

FIGURE 3.10 Product yield, Y, as a function of the capacity ratio, CR, at various values of K_d/c_0.

3.4.3 Multi-Component Systems

In solutions containing more than one adsorbing species, competition for occupancy of available sites on the adsorbent surface will undoubtedly occur. As a consequence, the achievable adsorbent loading of an individual adsorbing entity at a certain equilibrium concentration will be reduced compared with its adsorbent loading in the absence of competitive binding species (i.e., in a pure monocomponent binding system). In the multicomponent binding case, the loading isotherm for the target component will also be influenced by the equilibrium concentrations, $c_i^* i \neq 1$, of the other binding species. For reasons of simplicity, only one other binding substance shall be considered below, although the considerations we make may also be extended to cover additional competing binding species.

Starting from Langmuirian assumptions, Butler and Ockrent [25] developed a model to describe the adsorption of multicomponent mixtures. For a two-component system, their model described the loading of the individual components in the mixture as follows:

$$Q_1^* = \frac{Q_{\mathrm{max},1} \cdot (c_1^*/K_{\mathrm{d},1})}{1 + (c_1^*/K_{\mathrm{d},1}) + (c_2^*/K_{\mathrm{d},2})} \tag{3.8}$$

$$Q_2^* = \frac{Q_{\mathrm{max},2} \cdot (c_2^*/K_{\mathrm{d},2})}{1 + (c_1^*/K_{\mathrm{d},1}) + (c_2^*/K_{\mathrm{d},2})} \tag{3.9}$$

where $Q_{\mathrm{max},1}$ and $K_{\mathrm{d},1}$, and $Q_{\mathrm{max},2}$ and $K_{\mathrm{d},2}$ are the Langmuir constants for the individual binding species 1 and 2, respectively. In the following text we consider the hypothetical case of two competing binding species (target 1 and interfering species 2), having the individual Langmuir binding parameters cited in Table 3.4.

TABLE 3.4
Hypothetical Langmuir Equilibrium Parameters Describing the Binding of Target Species 1 and Interfering Species 2 to Magnetic Adsorbent Particles

$K_{\mathrm{d},1}$ (g/l)	$Q_{\mathrm{max},1}$ (mg/g)	$K_{\mathrm{d},2}$ (g/l)	$Q_{\mathrm{max},2}$ (mg/g)
10^{-2}	100	1	95

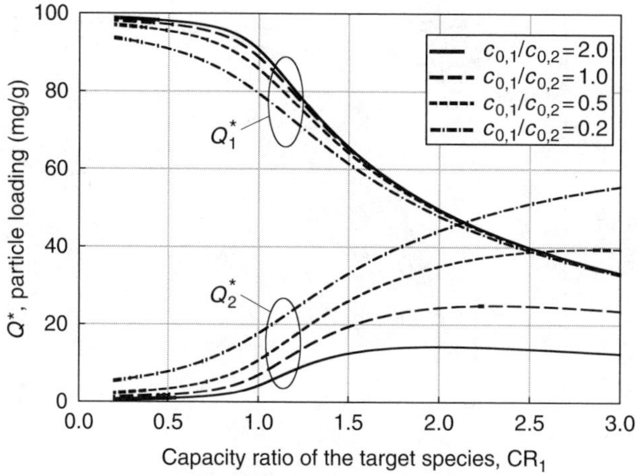

FIGURE 3.11 Equilibrium loads of competing binding species as a function of the capacity ratio of the more selectively binding species 1 (CR_1) at variable ratios of initial concentrations (with a fixed $c_{0,1}$ of 2.0 g/l).

Comparison of the dissociation constants $K_{d,1}$ and $K_{d,2}$ indicates a 100-fold higher affinity of the target species 1 for the magnetic adsorbent compared with the representative interfering substance 2. Differences in binding affinity of such magnitude are frequently observed in bioprocessing. The assumption of an initial concentration value for the target substance ($c_{0,1}$) of 2 g/l is considered reasonably representative of that observed with modern expression systems. Free and bound equilibrium concentrations of each species were obtained using Butler and Ockrent's [25] model (Equation 3.8 and Equation 3.9). Figure 3.11 shows the achievable magnetic adsorbent loadings for both the binding species plotted against the CR of the target binding species (CR_1). Figure 3.12 illustrates the influence of CR_1 on the yield and purity of the target species in the adsorbed state (i.e., it's purity while still immobilized on the adsorbent prior to elution, or to put it differently, the purity assuming 100% elution of both species).

Inspection of Figure 3.11 shows that at all initial concentration ratios ($c_{0,1}/c_{0,2}$) the equilibrium adsorbent loadings of the more selectively binding target species 1 decrease gradually between CR_1 values of about 0.7 to 0.9, to roughly converge after CR_1 values somewhat higher than 1, along a common downward curve. Despite complete adsorption of the target species at CR_1 values greater than 1 (Figure 3.12) the decrease in normalized adsorbent loading values for species 1 (Figure 3.11) reflects the increased total mass of magnetic adsorbent. With this elevation in adsorbent mass comes a concomitant rise in

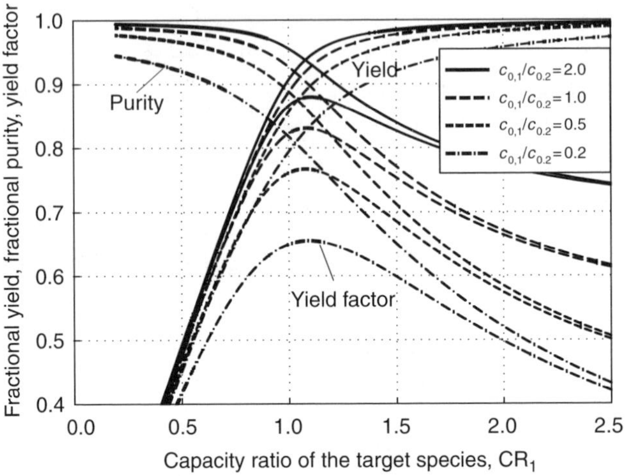

FIGURE 3.12 Purity, yield, and yield factors of target species 1 during competitive adsorption as functions of the capacity ratio (CR_1) of the target species employed. The initial target concentration ($c_{0,1}$) was 2.0 g/l and the ratio of the initial concentrations was varied.

the number of available adsorption sites, and therefore enhanced binding of the interfering component 2. At higher CR_1 values, and all $(c_{0,1}/c_{0,2})$ ratios the interfering component 2 exhibits apparent maxima of adsorbent equilibrium loading. With continued increase in CR_1 the increasing number of binding sites offered by the adsorbent can no longer be occupied and thus the loading values begin to drop. In contrast to the adsorbent loading values which have been normalized with respect to the particle mass, the absolute amounts of both bound species (i.e., 1 and 2) rise with increasing magnetic adsorbent concentration, that is, with increasing CR_1. At CR_1 values less than roughly 0.7, the purity of bound target species is high (in all cases above 90%; see Figure 3.12) given its much higher binding affinity. This high purity during adsorption is however, obtained at the expense of a low-target yield (Figure 3.11).

Beyond capacity ratios CR_1 of about 0.7 to 1, the purity of the target species begins to fall steeply (Figure 3.12), whereas high yields are only attainable when CR_1 is >1. It should come as no surprise therefore, that in an HGMF process, the magnetic adsorbent particle concentration employed plays a highly important role, with both over- and under-dosing leading to unsatisfactory results. A favorable operation point may be defined mathematically by the yield factor, that is, the product of the purity and yield of the target species [11,26]. As is evident from Figure 3.12, the maximum yield factor of the hypothetical system presented here is reached at a CR_1 value equal to 1.1, irrespective of the initial concentration ($c_{0,1}/c_{0,2}$) ratio employed. As a general rule of thumb, provided that the initial concentrations and affinities of the target species are not too low, a CR_1 value of 1 represents a good starting point for the optimization of magnetic adsorbent particle concentration for employment in an HGMF process.

3.4.4 Case Study III: Optimization of the Capacity Ratio Used

The applicability of the formulations derived in Section 3.4.3 for predicting competing adsorption of different species on magnetic adsorbent particles was tested experimentally. A detailed description of the model system (adsorbent, test species, conditions) employed for this illustration can be found elsewhere [12], and is only briefly described here. The magnetic adsorbents used (M-PVA from Chemagen Biopolymertechnologie AG, Baesweiler, Germany) were spherical 1 to 2 μm nonporous polyvinyl alcohol particles impregnated with superparamagnetic iron oxide crystals. The surface of the adsorbent was functionalized with immobilized metal affinity ligands charged with Cu^{2+} ions, and the model protein species tested were a hexahistidine-tagged green fluorescent protein (GFP) and a maltose binding protein — hexahistidine tagged streptavidin fusion protein — hereafter abbreviated as MalE. The Langmuir equilibrium binding parameters listed in Table 3.5 were determined for the binding of the individual proteins to the magnetic metal affinity adsorbent.

Although both test proteins carry polyhistidine tags it is clear (Table 3.5) that the preference of the magnetic adsorbent particles for GFP compared with MalE is even stronger than that considered for the hypothetical case described in Section 3.4.3. The most probable reason for the low-binding affinity of MalE fusion protein is an unfavorable steric arrangement of the hexahistidine tag, such that it is buried rather than surface exposed [12]. Figure 3.13 shows the

TABLE 3.5
Langmuir Parameters for the Binding of Hexahistidine Tagged GFP and the MalE Fusion Protein on Magnetic Cu^{2+} — Charged Immobilized Metal Affinity Adsorbent Particles

$K_{d,GFP}$ (g/l)	$Q_{max,GFP}$ (mg/g)	$K_{d,MalE}$ (g/l)	$Q_{max,MalE}$ (mg/g)
1.1×10^{-3}	99	0.5	94

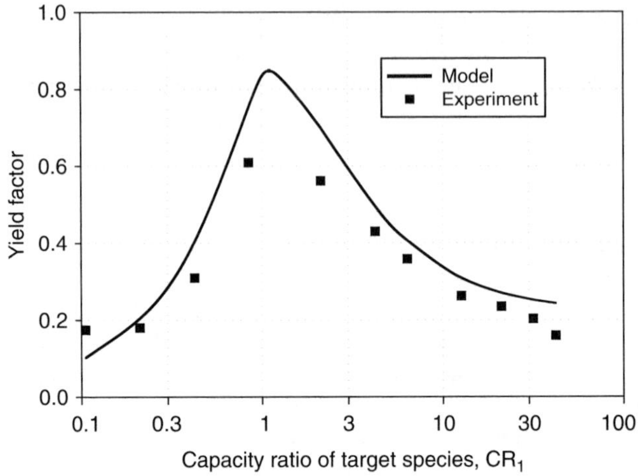

FIGURE 3.13 Yield factor (fractional yield × fractional purity) as a function of the capacity ratio for the target protein, GFP.

calculated and experimentally determined yield factors of the system, plotted against the logarithmic representation of capacity factors employed for the target protein substance, GFP. All of the experimentally obtained yield factors lie somewhat below the theoretically predicted values. This is not surprising, as the theoretical purity and yields described by the model refer to what is attainable during the adsorption step only, whereas the experimental yield factors were determined following washing and elution from the adsorbent. In the latter case some product loss during washing is inevitable and elution efficiencies of 100% are unlikely. The model nevertheless appears to describe the position of the optimal productivity factor reasonably well. It may be concluded that the adsorption model formulated and tested here enables useful initial predictions of purities, yields, and yield factors to be obtained. An additional strength of the model is that it can be used to narrow down the capacity ratios that should be entertained. Obvious benefits of this are significant reductions in the number of experiments required and time taken to identify optimal operating conditions.

3.4.5 Process Productivity

The productivity P of an HGMF system can be defined as:

$$P = \frac{m_{\text{prot}}}{t_{\text{cycle}} \cdot V_{\text{sep}}} \tag{3.10}$$

where m_{prot} is the mass of the isolated protein, V_{sep} is the volume of the magnetic separator used for particle separation, and t_{cycle} the time per process cycle. In principle, the mass of the isolated protein should be calculated by combining mass balances with Equation 3.8 and Equation 3.9 that account for competitive binding from other components within the feedstock. Should however, apparent Langmuir binding parameters for the target biomolecule be determined employing the very same feedstock that is intended for use in an HGMF process, then Equation 3.6 can be used to provide a first approximation of the mass of protein to be isolated.

For determining the amount of bioproduct produced by an HGMF process an elution efficiency of 100% is assumed, inferring that the total amount of bound protein (i.e., target and contaminants) will be recovered during elution. Consequently, the maximum amount of product produced per cycle will be given by:

$$m_{prot} = \frac{V_{batch} \cdot c_0}{2} \cdot \left[CR + 1 + \frac{K_d}{c_0} - \sqrt{4 \cdot \frac{K_d}{c_0} + \left(CR - 1 + \frac{K_d}{c_0} \right)^2} \right] \tag{3.11}$$

Here, the value of CR that is selected will depend on the purity and yield that are required, and if CR is fixed, the maximum batch volume to be processed per cycle can simply be determined from the maximum usable particle mass, m_p, which depends on the filtration capacity of the magnetic separator, and its volume. The filtration capacity, σ, is understood to be the particle mass retained within the magnetic separator per volume unit of separation matrix. As a general rule, this capacity is assumed to be about 10 to 20% below the maximum filtration capacity in order to ensure safe operation. With the separator volume V_{sep}, the maximum usable particle mass is given by:

$$m_p = \sigma \cdot V_{sep} \tag{3.12}$$

And the maximum batch volume to be processed will be:

$$V_{batch} = \frac{\sigma \cdot V_{sep} \cdot Q_{max}}{CR \cdot c_0} \tag{3.13}$$

In addition to the amount of product produced per cycle, determination of productivity requires estimation of the cycle time. The period for this considered below comprises: protein sorption, separation of product-loaded magnetic adsorbents from the bulk liquid phase (i.e., the feedstock), removal of impurities in two washing steps, and two stages of elution to recover the target species.

Periods for the possible conditioning of magnetic beads between cycles and for CIP are not taken into account. The washing and elution process steps include the following sequential substeps: filling of the system recycle loop with the buffer, resuspension, and circulation of the adsorbent particles; and reseparation of the adsorbents from the circulating flow.

As the size of the magnetic adsorbent particles is very small ($X_p \approx 1\ \mu$m), equilibrium adsorption is very fast and the time required for sorption $t_{sorption}$ (<2 min) is negligible compared to that of the overall process (t_{cycle}). The times needed for washing (t_{wash}) and elution ($t_{elution}$) are primarily determined by mixing during circulation and recapture of the adsorbent particles within the filter, rather than the time for desorption per se, which, like adsorption, is typically very short. As plant parameters, such as the volumes of the separator and recycle loop, the filtration capacity, and filtration rate will be constant, the washing and elution times that are obtained will, in the majority of cases, be constant even when different model systems are applied. In summary:

$$t_{sorption} \ll t_{cycle} \quad \text{and} \quad t_{wash} + t_{elution} = t_{fix} \tag{3.14}$$

The only cycle time variable is the time needed for the separation of the magnetic adsorbent particles from the feedstock, $t_{primary\ sep}$. This time may be derived from the expression:

$$t_{primary\ sep} = \frac{V_{batch}}{u_0 \cdot A_{sep}} \tag{3.15}$$

where u_0 is the filtration rate, A_{sep} is the cross-sectional area of the separator. The time required to complete a single processing cycle is given by:

$$t_{cycle} = t_{fix} + t_{primary\ sep} \tag{3.16}$$

For optimization of the overall process, the productivity can be represented by Equation 3.10 to Equation 3.16, once again as a function of the capacity ratio CR:

$$P = \frac{(V_{batch} \cdot c_0)/2 \cdot \left[\mathrm{CR}+1+(K_d/c_0)-\sqrt{4 \cdot (K_d/c_0)+(\mathrm{CR}-1+(K_d/c_0))^2}\right]}{\left(t_{fix} + (V_{batch}/(u_0 \cdot A_{sep}))\right) \cdot V_{sep}} \tag{3.17}$$

For calculating productivity using this equation, various process parameters must be defined. Typical representative values are presented in Table 3.6, and used below to illustrate HGMF process productivity.

TABLE 3.6
Process Parameters for the Calculation of Productivity

u_0 (m/h)	t_{fix} (min)	V_{sep} (l)	A_{sep} (cm^2)	σ (g/l)	K_d (g/l)	Q_{max} (mg/g)
25	15	1.0	60	100	0.05	100

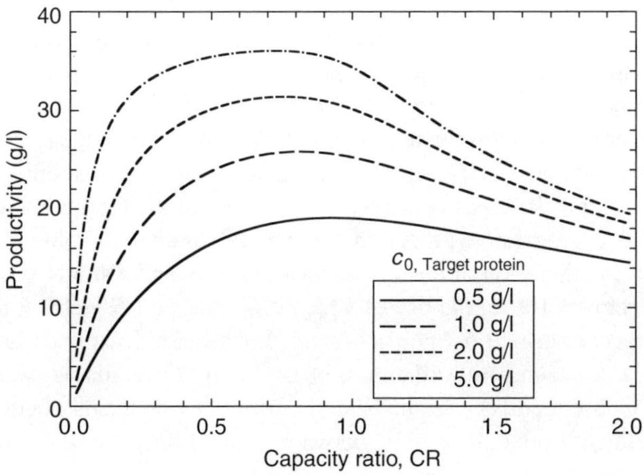

FIGURE 3.14 Productivity of an HGMF process as a function of the capacity ratio, at various initial concentrations of target species.

Figure 3.14 illustrates how the productivity of HGMF varies as a function of capacity ratio at various initial concentrations of the target species in the range of 0.5 to 5 g/l. At the lowest initial target protein concentration of 0.5 g/l and low CR values, productivity is as expected, low. Raising the amount of adsorbent particles employed results in increased productivity and a broad optimum is reached between capacity factors of 0.5 to 1. As the initial target species concentration is raised the productivity optimum tends to broaden and shifts toward lower CR values. At all initial target concentrations further increase in CR past the optimum range in each case results in reduced productivity. At CR values beyond the optimum further benefits in yield are only very small (see e.g., Figure 3.10); more importantly, however, the permissible batch volume per cycle is much reduced.

In addition to its influence on productivity, the filtration capacity of a given magnetic separator system can exert a strong impact on the degree of concentration of the target species attainable in the HGMF process. The volume of the loop used for the washing and elution operations and the magnetic adsorbent's binding capacity are also important determinants of concentrating power in HGMF. Accordingly the maximum product concentration attainable in the elution buffer can be found from the expression:

$$c_{e,1} = \frac{\sigma \cdot V_{sep} \cdot Q_{max}}{V_{loop}} \tag{3.18}$$

where V_{loop} denotes the total loop volume used by the circulating flow during the washing and elution processes (i.e., the loop and separator volumes combined). As is evident from Equation 3.18, the ratio of loop and separator volumes exerts a strong influence on the degree of product concentration that can be achieved. When designing a system for HGMF, one should aim for a V_{loop}/V_{sep} ratio as small as possible, that is, close to the theoretical minimum of 1. For the representative process parameter values cited in Table 3.6, assuming maximum utilization of the adsorbent's capacity and a single stage elution with efficiency of 100%, the use of V_{loop}/V_{sep} ratio of 1.5 yields a maximum product concentration of 6.7 g/l. In practice however, at least two elution steps will be necessary, and full utilization of the support's available capacity will not occur. Consequently, with the best of current magnetic adsorbents realistic values of eluted product tend to lie between 1 and 3 g/l.

3.4.6 Case Study IV: Influence of Washing and Elution Steps

The possibility of increasing the selectivity of affinity-based adsorptive separation processes through the use of mild elution conditions in a washing step prior to product elution is a well-established practice and is based on the principle that at low eluting agent concentrations bound impurities desorb in preference to the tighter binding target product. To experimentally investigate the potential benefits of this practice for the model system described earlier in Section 3.4.4, the washing buffer was supplemented with low concentrations (1 to 6 m*M*) of imidazole. The experiments (conducted at values of CR = 2.1, 4.2, and 6.3) served a particularly useful purpose, that is to identify the loss of yield of the target substance that must be accepted in order to deliver a certain level of purity.

Compared to the control case (lacking imidazole in the wash buffer) the incorporation of low levels of imidazole during washing increased the purity of GFP in the elution step in all test cases, albeit always at the expense of

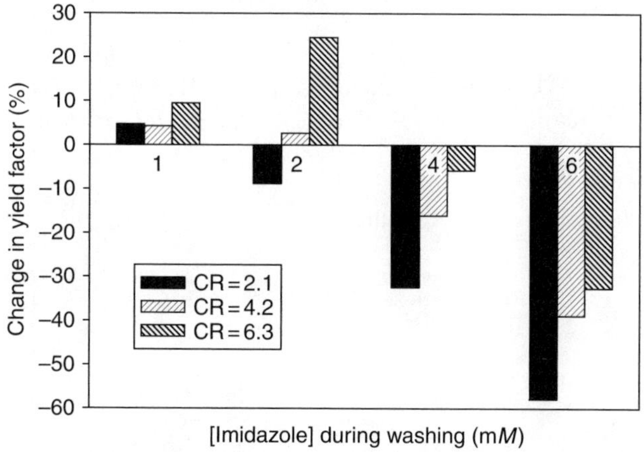

FIGURE 3.15 Influence of imidazole concentration during washing on the percentage change in yield factor. Experiments were conducted at CR values of 2.1, 4.2, and 6.3. For each test, the change (increase or decrease) in its productivity relative to the control case (i.e., lacking imidazole in the wash buffer) is expressed as a percentage of the productivity of the control.

loss in GFP yield. To ascertain whether on balance positive or negative effects predominate in each case (i.e., relative to the control), yield factors were again employed. Analysis of Figure 3.15 confirms that with an imidazole concentration of 1 m*M* the yield factor is slightly improved compared with the control at all CR values studied. At higher imidazole concentrations, however, this positive effect is rapidly reversed, and the losses in yield far outweigh the gains in purity. At any given imidazole concentration, the yield factor is observed to increase as the CR value (i.e., adsorbent concentration) is raised. The reason for this trend is wholly understandable. As the mass of adsorbent is increased a higher fraction of the surface will be occupied by impurities, thus the practical benefits of incorporating low levels of eluting agents during washing, prior to elution make greater sense in this instance.

3.4.7 Adsorbent Reuse

A vital condition for the future industrial application of HGMF processes will undoubtedly be the need to recycle the magnetic adsorbent particles over many process cycles. Clearly the reusability of a given adsorbent will be strongly dependent not only on its physical and chemical make-up, but also on the functionalization chemistry employed. The following example, involving magnetic

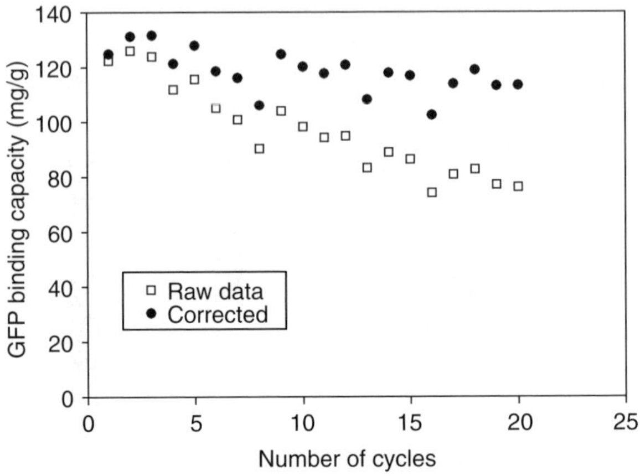

FIGURE 3.16 Variation of the GFP binding capacity with increasing adsorbent reuse.

polyvinyl-alcohol-based metal chelating adsorbents functionalized with Cu^{2+}, nevertheless gives a useful impression of what to expect.

The crude feedstock was a ball-milled recombinant *Escherichia coli* homogenate containing a polyhistidine tagged GFP [12]. In a given cycle the following steps were performed (i) loading of the magnetic particles with Cu^{2+} ions; (ii) washing; (iii) addition of the homogenate and adsorption; (iv) and (v) two washing operations; (vi) a single elution with imidazole; (vii) cleaning with EDTA; and finally (viii) and (ix) two further washing steps. In summary, for every cycle, the magnetic adsorbents were magnetically separated from 2 ml of suspension a total of 9 times using a handheld permanent magnet block.

Figure 3.16 shows how the GFP binding capacity varies over 20 cycles. The raw GFP binding capacities were calculated assuming a constant adsorbent particle concentration, whereas the corrected data take the actual particle loss occurring after each cycle into account. For the corrected data set, assuming exponential loss in GFP binding capacity with increasing number of cycles, the theoretical number of cycles that would elapse before the adsorbent's GFP binding capacity reached half of its original value, would be 97. Thus, the main factor limiting the possible number of cycles is not a reduction in the functionality of adsorbent particles, but rather adsorbent loss due to separation efficiencies below 100%. In the present example over 20 cycles, the adsorbent particles were separated magnetically 180 times, with each separation operation reaching an efficiency >99.7%. Nevertheless, over 20 cycles the accumulated adsorbent loss added up to nearly 33% of the initial mass of adsorbent employed. Assuming a constant percentage of adsorbent loss per cycle, it can be estimated that approximately half of the original adsorbent particles will remain after ~34 cycles of operation. This estimation holds for the separation of micron-sized magnetic adsorbent particles from small volumes using hand-held permanent magnets. Although high-gradient magnetic separators offer higher separation efficiencies, particle losses with larger magnetic separators can reach the same order. A possible solution to the problem of adsorbent particle loss is to collect and treat suspensions leaving the separator with a further polishing filtration which is performed by a dedicated high-gradient magnetic separator operated at reduced filtration rates (therefore delivering near 100% separation efficiency). Following a cleaning stage the recovered adsorbent particles can be returned to the main adsorbent pool.

3.4.8 Case Study V: Pilot Plant Efficiencies

Over the last five years or so, very large numbers of test runs — involving more than 20 different target molecules and nearly as many differing crude feedstocks (including crude cell homogenates and chemical lysates, whey, raw milk, and legume extracts, etc.) — have confirmed the suitability and attractive qualities (speed, robustness, efficiency) of HGMF for direct product recovery from tricky bioprocess liquors. The values of yield, purity, and process productivity achieved in each case are strongly dependent, not only on the selectivity of the chosen feedstock/magnetic adsorbent particle combination, but also on the operation parameters that are employed (of these the CR-value is especially important). Therefore, in order to gain a clearer picture of the inherent efficiency of these various HGMF pilot plants, the values in Table 3.7 have been normalized with respect to maximum theoretical performance possible in each of the test systems investigated. Accordingly, in the ideal case scenario for the HGMF the normalized adsorption and elution step efficiencies should both be 100%.

As is clear (Table 3.7), comparable performance is delivered regardless of the scale of the facilities. With the exception of the smallest unit the normalized sorption step efficiency reaches its theoretical value of 100% in all cases. In contrast the average normalized elution step efficiency only accounts for 75%. The most probable reason for this is that the sorption step is conducted in a stirred external sorption vessel, which guarantees good mixing between the magnetic adsorbents and the feedstock. For the elution step, in stark contrast, the adsorbents are flushed out of the separation matrix and mixed with the eluant by recirculating the resulting suspension within a closed loop. In such an elution procedure the first step of efficiently releasing the adsorbent particles from the filter matrix is especially crucial, and this will likely require further improvement en route to commercialization. This notwithstanding, the HGMF process is clearly matched to the task of delivering clarified partially purified products

TABLE 3.7
Normalized Process Efficiencies for HGMF Facilities of Different Scale

Investigated System (Species/Feedstock/Ligand)	Scale (Batch Size) (l)	% Normalized Sorption Step Efficiency	% Wash Step Loss	% Normalized Elution Step Efficiency	Ref.
$(His)_6$-tagged GFP/*E. coli* homogenate/Cu^{2+}-IDA	4.2	96	6	66	[12]
$(His)_6$-tagged GFP/*E. coli* homogenate/Cu^{2+}-IDA	2.2	108	5.6	79	[12]
Lactoferrin/ bovine whey/cation exchanger	2.2	103	2.2	67	[10]
Trypsin/bovine whey/benzamidine	0.06	105	0.6	74	[3]
Human papillomavirus coat protein L1/chemical *E. coli* extract/Cu^{2+}-IDA	0.015	87	3.1	85	[6]

in high yield from complex, dirty, difficult-to-handle bioprocess unclarified liquor.

3.5 CONCLUDING REMARKS

In this chapter, we have introduced necessary background, tools, and methodology to enable a downstream processor with no previous experience of HGMF to evaluate the technique's potential for the recovery of a target biomolecule of interest from a complex feedstock. We show that insightful predictions of HGMF process performance can be made with the aid of simple models and easily obtained data from experiments conducted at the bench.

High-gradient magnetic fishing is not yet a mature unit operation to be plucked from the shelf and immediately applied to the processing of a biotherapeutic product. Indeed, much work still needs to be done before complete HGMF packages (i.e., separator systems, magnetic filters, and adsorbents) become commercially available. In common with high-pressure homogenization, bead milling, and industrial centrifugation, HGMF too, has foreign origins with far less stringent requirements. Adaptation of these instruments to bioprocessing has principally involved gradual modifications to original designs and changing the materials employed in their construction to, for example, reduce potential shear-induced damage of fragile biological entities, meet various containment criteria, and afford easy CIP/SIP. HGMF is now undergoing similar development. Judged purely from a technical side, the future prospects for HGMF's adoption within the bioprocess industries look bright indeed, given its potential for very rapid processing of high volumes of crude bioprocess feedstocks.

REFERENCES

1. Hubbuch JJ, Matthiesen DB, Hobley TJ, and Thomas ORT. High gradient magnetic separation versus expanded bed adsorption: A first principle comparison. *Bioseparation* 2001; 10: 99–112.
2. Hubbuch JJ and Thomas ORT. High-gradient magnetic affinity separation of trypsin from porcine pancreatin. *Biotechnol Bioeng* 2002; 79: 301–313.
3. Hubbuch JJ. Development of adsorptive separation systems for recovery of proteins from crude bioprocess liquors. Ph.D. dissertation, Technical University of Denmark, Denmark, 2001.
4. Heebøll-Nielsen A. High-gradient magnetic fishing: Support functionalization and application for protein recovery from unclarified bioprocess liquors. Ph.D. dissertation, Technical University of Denmark, Denmark, 2002.

5. Hoffmann C. Einsatz magnetischer Separationsverfahren zur biotechnologischen Produktaufbereitung. Ph.D. dissertation, Forschungzentrum Karlsruhe, Karlsruhe, Germany, 2002.
6. Heebøll-Nielsen A, Choe WS, Middelberg APJ, and Thomas ORT. Efficient inclusion body processing using chemical extraction and high-gradient magnetic fishing. *Biotechnol Prog* 2003; 19: 887–898.
7. Heebøll-Nielsen A, Justesen SFL, Hobley TJ, and Thomas ORT. Superparamagnetic cation-exchange adsorbents for bioproduct recovery from crude process liquors by high-gradient magnetic fishing. *Sep Sci Technol* 2004; 39: 2891–2914.
8. Heebøll-Nielsen A, Justesen SFL, and Thomas ORT. Fractionation of whey proteins with high-capacity superparamagnetic ion-exchangers. *J Biotechnol* 2004; 113: 247–262.
9. Heebøll-Nielsen A, Dalkiær M, Hubbuch JJ, and Thomas ORT. Superparamagnetic adsorbents for high-gradient magnetic fishing of lectins out of legume extracts. *Biotechnol Bioeng* 2004; 87: 311–323.
10. Meyer A. Einsatz magnettechnologischer Trennverfahren zur Aufbereitung von Molkereiprodukten. Ph.D. dissertation, Forschungzentrum Karlsruhe, Karlsruhe, Germany, 2004.
11. Meyer A, Hansen DB, Gomes CSG, Hobley TJ, Thomas ORT, and Franzreb M. Demonstration of a strategy for product purification by high-gradient magnetic fishing: Recovery of superoxide dismutase from unconditioned whey. *Biotechnol Prog* 2005; 21: 244–254.
12. Ebner N. Einsatz von Magnettechnologie bei der Bioproduktaufarbeitung. Ph.D. dissertation, Forschungzentrum Karlsruhe, Karlsruhe, Germany, 2005.
13. Ferré H. Development of novel processes for protein refolding and primary recovery. Ph.D. dissertation, Technical University of Denmark, Denmark, 2005.
14. Gomes CSG, Petersen TL, Hobley TJ, and Thomas ORT. Controlling enzyme reactions in unclarified bioprocess liquors using high-gradient magnetic fishing (HGMF). In *Proceedings of the 7th World Congress of Chemical Engineering and 5th European Congress of Chemical Engineering*, Glasgow, July 10–14, 2005, ISBN 0 85295 494 8.
15. Langmuir I. The adsorption of gases on plane surfaces of glass, mica and platinum. *J Am Chem Soc* 1918; 44: 1361–1403.
16. Watson JHP. Magnetic filtration. *J Appl Phys* 1973; 44: 4209–4213.
17. Watson JHP. Theory of capture of particles in magnetic high-intensity filters. *IEEE Trans Magn* 1975; 11: 1597–1599.
18. Cummings DL, Himmelblau DA, and Oberteuffer JA. Capture of small paramagnetic particles by magnetic forces from low speed fluid flows. *AIChE J* 1976; 22: 569–575.
19. Gerber R and Birss RR. *High-Gradient Magnetic Separation*. Chichester, U.K.: John Wiley & Sons Ltd., 1983.
20. Svoboda J. Magnetic methods for the treatment of minerals. In: Fuerstenau DW, Ed. *Developments in Mineral Processing*, Vol. 8. Amsterdam: Elsevier Science Publication Co, 1987.

21. Thomas ORT and Franzreb M. Magnetic separations. In Kieran P, Cabral J, and Jungbauer A, Eds. *Bioseparation Processes*. Wiley, 2006: in press.
22. Holshuh K and Schwämmle A. Preparative purification of antibodies with protein A — an alternative to conventional chromatography. *J Magn Magn Mat* 2005; 203: 345–348.
23. Hermanson GT, Mallia AK, and Smith PK. *Immobilised Affinity Ligand Techniques*. London: Academic Press, 1992.
24. Hoffmann C, Franzreb M, and Höll WH. A novel high-gradient magnetic separator (HGMS) design for biotech applications. *IEEE T Appl Supercon* 2002; 12: 963–966.
25. Butler JAV and Ockrent C. Studies of electrocapillarity, Part 3. The surface tensions of solutions containing two surface-active solutes. *J Phys Chem* 1930; 34: 2841–2859.
26. Hearle DC, Aguilera-Soriano G, Wiksell E, and Titchener-Hooker NJ. Quantifying the fouling effects of a biological process stream on chromatographic supports. *I Chem E Res Event* 1994; 1: 174–176.

4 Protein Refolding and Scale Up

Cynthia Cowgill, Asuman G. Ozturk, and Richard St. John

CONTENTS

4.1 INTRODUCTION

Aggregation and protein misfolding are ubiquitous problems in production of protein therapeutics. Bacterial hosts are very efficient expression systems for production of recombinant protein products. However, overexpression of protein products in bacterial expression hosts generally yield large quantities of aggregated, inactive recombinant protein in the form of inclusion bodies. Other expression systems that utilize mammalian, fungal, or yeast cells may also produce either insoluble inclusion bodies, soluble aggregates, or improperly folded recombinant protein. Aggregation may also occur during processing due to filtration, agitation, or other processing steps. Aggregates are problematic because they are inactive and frequently cause immune reactions when injected into patients. Improperly folded protein, or misfolds, are usually considered impurities and must be purified from native protein, resulting in reduced process yields.

In order to produce an active product from these systems, the inclusion bodies, aggregates, or misfolds must be isolated and then refolded. Although isolation of inclusion bodies or aggregates with high process yields is fairly straightforward, refolding success is not guaranteed. When successful, significant process development is frequently required to obtain efficient refolding yields. In addition, soluble misfolded species of the protein product significantly complicate purification, as misfolds are often only subtly different in structure and chemical characteristics from the native protein. Finally, refolding processes generally use chemical denaturants that require expensive disposal fees. For these reasons, many companies have designed protein production processes utilizing alternative expression systems that excrete soluble, properly folded protein.

Bacterial expression systems and inclusion bodies, however, have several significant advantages over mammalian expression systems. Bacterial cells rapidly produce large quantities of recombinant protein with relatively inexpensive fermentation processes. Inclusion bodies generally contain >50% protein of interest. Consequently, when inclusion bodies are separated from host contaminants, the desired product is already partially purified. Bacterial systems are also fairly easy to transform, and therefore are excellent hosts for production of novel proteins with minimal development time. Finally, bacterial hosts do not support the growth of adventitious viruses capable of human infection. Thus, with bacterial expression of proteins, viral clearance steps are not required during purification. If an efficient refolding process can be developed, bacterial expression systems are the most simple, efficient, and cost-effective method for recombinant protein production.

In this chapter, we outline the basics of protein refolding and describe a series of experimental steps to guide a scientist in quickly developing and optimizing a refolding process. Special considerations during scale-up of a refold process are also presented. Finally, we explore exciting new refolding technologies that, to our knowledge, have yet to be fully implemented at industrial scale.

4.2 REFOLDING BASICS

Protein refolding is now a fairly well-established field of study. There are several reviews in the literature that cover the basic concepts and theoretical basis behind refolding experiments [1–4]. In a generic refolding process, inclusion bodies or aggregates are first solubilized in denaturing concentrations of urea or guanidine hydrochloride (GdnHCl). Refolding is induced by reducing chaotrope concentrations to levels that thermodynamically favor the native conformation. Detrimental side reactions, however, compete with proper folding, leading to formation of misfolded protein and aggregates. If a protein contains disulfide bonds, the disulfide bonds are broken during solubilization with the addition of a reducing agent. Formation of native disulfide bonds is then coupled with structural collapse of the protein. Disulfide bonds increase the complexity of the refolding reaction considerably, as formation of incorrect disulfide bonds causes nonnative covalent attachments that inhibit proper folding. Given this basic framework of a generic refold, the following section elaborates on the basic steps in refolding proteins from aggregates and inclusion bodies.

4.2.1 SOLUBILIZATION

Prior to refolding a protein from inclusion bodies, aggregates and inclusion bodies must first be solubilized and denatured. The most common denaturants used for solubilization are urea and GdnHCl. These chaotropes are small molecules that preferentially bind to proteins, disrupting the intramolecular and intermolecular forces responsible for protein structure [5,6]. In addition to urea and GdnHCl, denaturing surfactants such as Sarkosyl, at concentrations above the critical micelle concentration (CMC), are another common class of chemicals used for solubilization [7,8]. If disulfide bonds are present in the protein, intermolecular disulfide bonds may have formed in the inclusion bodies. As a result, proteins containing disulfides are treated with a reducing agent in the solubilization stage to break all existing disulfide bonds. The presence of a reducing agent in the solubilization buffer ensures that improper inter- and intramolecular disulfide bonds do not form until refolding is initiated. The resulting solution contains fully reduced, monomeric protein that is relatively flexible with minimal secondary or tertiary structure.

4.2.2 Refolding

Once the protein is fully solubilized, refolding is induced by transferring the protein into a solution that thermodynamically favors native structure. Most frequently, refolding is induced by reducing chaotrope concentrations to nondenaturing levels via dilution or dialysis. When denaturant concentrations are lowered sufficiently, intramolecular forces between amino acids drive collapse of protein conformation leading to formation of secondary and tertiary structure. As the protein collapses from an unstructured state, local structures form and populate as intermediates. Depending on the order in which these local structures form and assemble with each other to form tertiary structure, the energetic barriers to refolding vary. The variety of protein refolding energetic pathways can be graphically represented as energetic landscape funnels (Figure 4.1) [9]. In Figure 4.1, the upper rim of the funnel represents denatured protein. The native energy state (denoted as N in Figure 4.1) is generally thought to be the thermodynamically the most stable protein conformation, but energetic barriers along the folding pathway can trap folding intermediates as stable misfolds or precursors to aggregation. In some cases there can be multiple conformations with similar energetic states, leading to a fairly stable mixture of conformational misfolds. Chemical additives can be added to a refold buffer to modulate the energy landscape of a folding protein, minimizing energetic barriers limiting the refolding reaction, or favoring one conformation over another.

4.2.3 Disulfide Bonds

Coupled with the thermodynamically driven collapse of three-dimensional protein structure, disulfide bonds must also be properly formed to yield native protein. Although the statistical probability of forming correct disulfide bonds

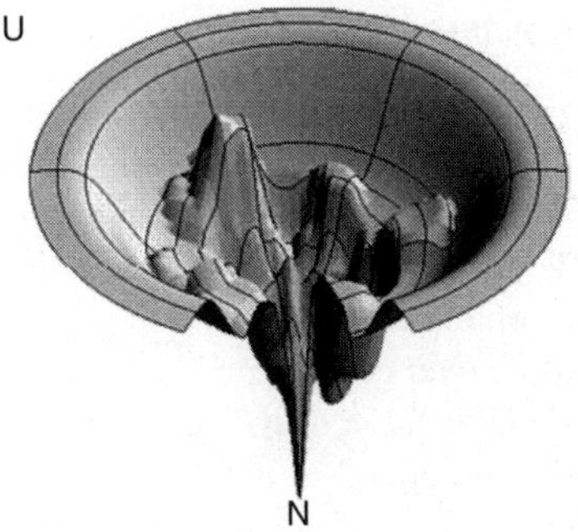

FIGURE 4.1 Protein folding pathways can be viewed as rugged energetic landscapes with kinetic traps and energy barriers between unfolded protein and native structure (N). A given protein may fold using multiple pathways to the native form, resulting in a mixture of native protein and kinetically trapped intermediates (misfolds). (Figure reproduced with kind permission from Ken Dill.)

during refolding is discouragingly small [10], native disulfide bonds can usually be formed with reasonable efficiency. Consequently, we infer that the collapse of protein structure orients cysteine residues in proximity to their native disulfide couples. Kinetically trapped intermediates, however, may induce formation of nonnative disulfide bridges, or a disulfide (native or nonnative) may induce a kinetic trap. Alternatively, a disulfide linkage may be required to overcome an energetic barrier in the refold process. Hence, the coordination between structural collapse and disulfide formation can be critical to success [11].

Formation of disulfide bonds from reduced cysteine residues is an oxidation reaction, and as a result, an oxidizing agent is required to drive disulfide bond formation. Because cysteine residues have a pKa in the range of pH 8 to 9 [12], a pH > 8 is usually required for timely disulfide formation. The most abundant oxidant available is oxygen, which is cheap and easily utilized. Residual levels of catalyzing metals commonly present in buffer components will facilitate oxidation of cysteines. A metal chelator such as ethylenediaminetetraacetic acid (EDTA) can be added to complex with catalyzing metals, but metal chelators may or may not inactivate the catalyzing properties of the metal ions [13–15]. The drawback of air oxidation lies in the irreversibility of disulfide bond formation during air oxidation. Nonnative disulfide bonds covalently lock the protein in nonnative conformations. Because nonnative disulfides are frequently formed in a refolding reaction, oxido-shuffling agents, such as reduced/oxidized glutathione or cysteine/cystine, are often introduced to reversibly break and form disulfide bonds during refolding. With oxido-shuffling agents present, a nonnative disulfide can be reversibly broken, allowing the protein to continue along the refolding pathway to native conformation.

4.2.4 Refolding Additives

When attempting to improve refolding yields, protein stability and solubility are integrally connected to success. If the native protein is not soluble in the refold buffer chosen, no refolding will take place. Once a refold buffer promoting solubility is determined, understanding formulations that improve stability and solubility of the native protein can aid in finding good refolding conditions that promote proper folding over competing side reactions. Scientific literature reports a wide range of chemical additives that have been shown to improve refolding yields of specific proteins. No single additive has proven to generically increase protein refolding yields, and consequently, additives must be screened and optimized for each protein individually. Table 4.1 offers a list of additives that appear most frequently in the literature as refolding enhancing additives.

4.3 DESIGNING A REFOLD PROCESS

There are many approaches to designing a refold process, and many examples in the literature that discuss specific techniques that work for individual proteins. In this chapter, we attempt to provide an approach to refolding that will help efficiently develop a successful refolding process for most proteins. Figure 4.2 outlines a decision tree for developing a refold process. This section will follow Figure 4.2, elaborating on critical decisions during design of a refold process.

4.3.1 Decide Whether to Refold

Before initiating an effort focused on refolding a protein, a company should ask the strategic question whether to invest in refolding a protein from inclusion bodies or to invest in expressing the protein in an alternative host that will produce properly folded protein. The specific experience of the cell culture development team may direct the project in a particular direction. Also, a fully glycosylated product expressed in mammalian cells will usually exhibit slower pharmacokinetic clearance in the clinic. Depending on the clinical strategy for the product, glycosylation may be desired. In some cases, glycosylation may be required for activity of the product. On the other hand, the market demand or expected sale price of the drug may favor bacterial expression and pursuing a refold process.

4.3.2 Do You Have Purified Protein?

If you plan to pursue a refolding process, access to purified protein can be very useful in directing the development. If purified protein is available, we suggest constructing an unfolding curve in both urea and GdnHCl. Several

TABLE 4.1
List of Additives Frequently Reported to Increase Protein Refolding Yields

Additives	Recommended Concentration
Amino Acids	
Arginine	0.25 *M*
Proline	1 *M*
Preferentially Excluded Solutes	
Sucrose or trehalose	0.5 *M*
PEG	0.05%
Glycerol	0.5 *M*
Chaotropes	
Urea	2 *M*
GdnHCl	1 *M*
Surfactants	
Tween	0.1%
CTAB	0.2%
Sarkosyl	0.4%
Triton X-100	10 m*M*
Dodecyl maltoside	5 m*M*
Short Chain Alcohols	
Ethanol	10%
n-pentanol	5 m*M*
cyclohexanol	5 m*M*
Salts	
NaCl	0.5 *M*
$MgCl_2$ or $CaCl_2$	5 m*M*
Ammonium sulfate	0.5 *M*
Oxido-Shuffling Agents	
Reduced and oxidized glutathione	5–15 m*M*
cysteine/cystine	5–15 m*M*

nice practical guides to constructing and interpreting unfolding curves can be found in the literature [16,17]. Unfolding curves are generally constructed by incubating purified protein in varying concentrations of chaotrope. Tertiary and secondary structural conformation of the protein is then measured in each solution, normally by a spectroscopic technique such as circular dichroism (CD), derivative UV spectroscopy, or fluorescence spectroscopy. In general, proteins undergo a synergistic unfolding, resulting in a significant change in the

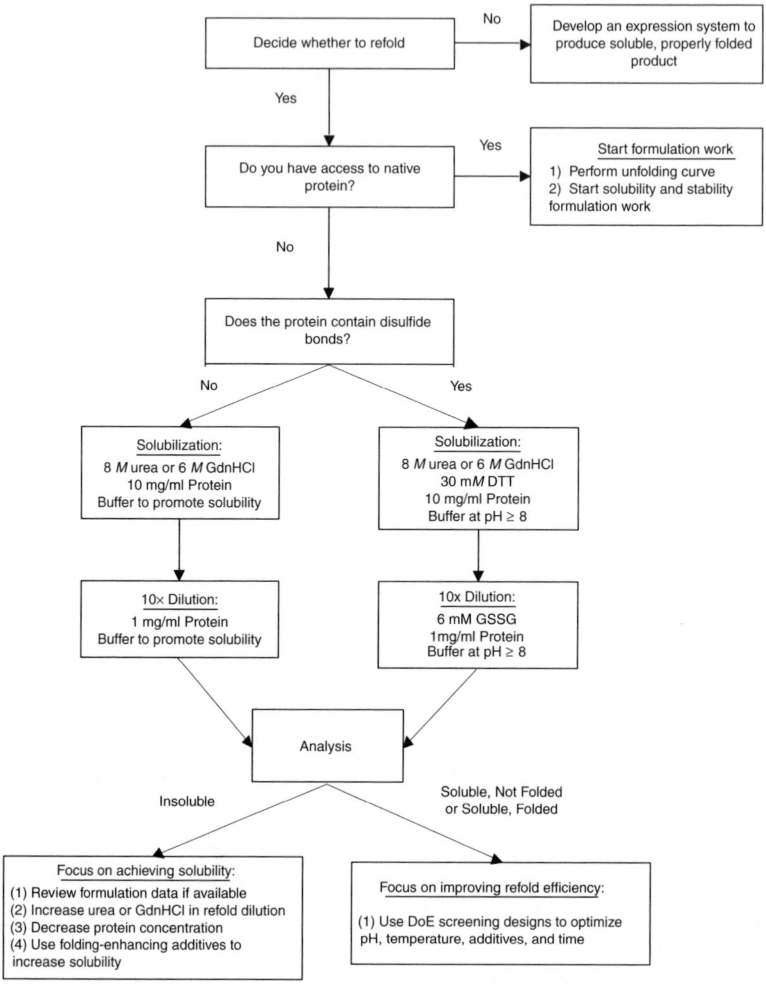

FIGURE 4.2 Protein refolding development decision tree. GdnHCl — Guanidine Hydrochloride, DTT — Dithiothreitol, GSSG — Oxidized Glutathione, DoE — Design of Experiments.

features of a protein's spectroscopic analysis. Shifts in spectroscopic features are plotted against chaotrope concentration and translated into an unfolding curve, assuming that unfolding is a two-state, equilibrium transition. A typical unfolding curve is shown in Figure 4.3.

An unfolding curve will first instruct the user on the concentration of urea or GdnHCl required to fully unfold the native protein. In Figure 4.3, for example,

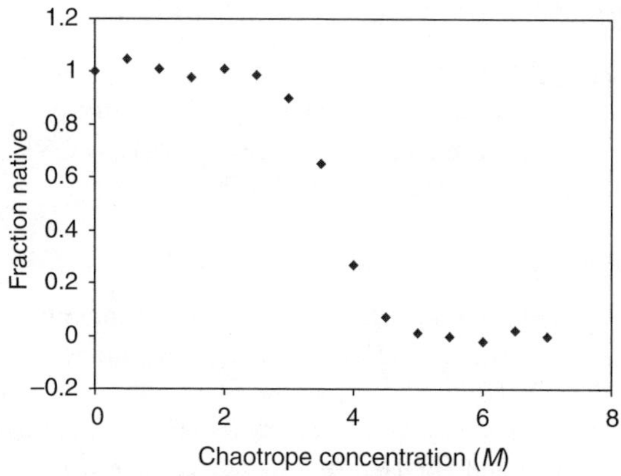

FIGURE 4.3 A typical unfolding curve for a generic protein. At low chaotrope concentration, the fraction of protein present in the native conformation is near unity. Between 3 and 5 *M* chaotrope, the protein exhibits unfolding transition, and above 5 *M* chaotrope, the protein is considered unfolded. For refolding, an unfolding curve can be used to understand the minimum chaotrope concentration required for dissolution, and to determine the maximum chaotrope concentration where a protein will refold.

a chaotrope concentration of >5 *M* would be required for complete unfolding of the native protein. A similar chaotrope concentration to the unfolding concentration will also be required for complete solubilization from inclusion bodies. Second, unfolding curves can provide information on the insolubility of the protein at intermediate chaotrope concentrations. During the transition between native and unfolded protein, intermediate chaotrope concentrations tend to stabilize highly aggregation prone folding intermediates. Careful observation of individual samples from the unfolding curve experiment may identify chaotrope concentrations where the protein product is insoluble. Because the protein must transition through intermediate chaotrope concentrations from solubilization to refolding, information on the insolubility of the protein at intermediate chaotrope concentrations helps understand the potential for aggregation during dilution refolding. Finally, unfolding curves identify chaotrope concentrations where native structure is thermodynamically favored (<3 *M* in Figure 4.3). Since low concentrations of chaotrope frequently improve refolding yields, it is helpful to understand the range of chaotrope concentrations where native structure is favored. In addition to unfolding curves, we recommend initiating, as early as possible, formulation studies to identify solution conditions that favor solubility of the native protein. Excipients identified as good stabilizers of native structure may be very useful as additives to improve native refolding yields.

4.3.3 Does the Protein Contain Disulfide Bonds?

The presence of disulfide bonds adds a degree of complexity to the development of a refold process. As mentioned earlier, disulfides must be reduced during the solubilization step and oxidized during the refold. Hence, we offer separate advice for proteins with and without disulfide bonds.

4.3.3.1 Refolding Proteins without Disulfide Bonds

When refolding proteins without disulfide bonds, first solubilize the protein in 8 *M* urea or 6 *M* GdnHCl, 10 mg/ml protein (we recommend screening both urea and GdnHCl separately at first, as the two chaotropes sometimes yield significantly different results). If an unfolding curve was performed, choose concentrations of urea and GdnHCl that induce complete unfolding of native protein based on the unfolding curve. The solubilization should be buffered to a pH that promotes solubility (use formulation data if available). Following solubilization (30 to 60 min is usually sufficient), we recommend a 10-fold dilution into refold buffer designed to support protein solubility (if formulation data are available, utilize solubility data at this step). The resulting refold solution then contains 1 mg/ml protein and 0.8 *M* urea or 0.6 *M* GdnHCl. Approximately 4 to 6 h of refold incubation is usually sufficient for complete refolding, but longer incubation times are sometimes required and an analytical assay should be used for confirmation.

4.3.3.2 Refolding Proteins Containing Disulfide Bonds

For proteins containing disulfide bonds, solubilize the protein in 8 *M* urea or 6 *M* GdnHCl, 30 m*M* dithiothreitol (DTT), 10 mg/ml protein. For complete reduction of disulfide bonds, the solution pH should be >8. If an unfolding curve was performed, let the unfolding curve dictate the urea or GdnHCl concentrations required for complete unfolding. Initiate refolding by diluting 10-fold into refold buffer (pH ≥ 8) containing oxidized glutathione (GSSG, final concentration in the diluted refold of 6 m*M*). Because of its low redox potential, DTT will reduce other thiol groups quantitatively [18]. Thus, the remaining reduced DTT that was not oxidized during reduction of the inclusion bodies will reduce approximately 3 m*M* GSSG, resulting in a reduced glutathione to oxidized glutathione (GSH : GSSG) ratio of approximately 6 : 3, or 2 : 1. Recent reviews on refolding recommend refolding with a reduced to oxidized ratio of between 10 : 1 and 1 : 1, with a total concentration between 5 and 15 m*M* [2,3]. It should be noted that, because DTT is such a strong reducing agent, DTT does not act as an oxido-shuffling agent.

Refolding reactions containing disulfide bonds usually require incubation between 1 and 24 h. For initial screening, we recommend allowing 24 h for complete refolding, but completion of the reaction should be examined with an analytical assay. When analyzing time points, care must be taken to assure that the refold reaction rate is adequately slowed to allow time for analysis. For refolds involving disulfide bonds, the activity of protein and oxido-shuffling sulfhydryls can be essentially stopped by lowering the pH to 5 to 6, or by blocking sulfhydryls with an effective blocking agent, such as iodoacetamide, that binds to free sulfhydryls [19,20]. pH adjustment or sulfhydryl blocking should be approached cautiously, as they may introduce aggregation or other changes in impurity profiles in the refold samples.

4.3.4 Analysis of the Refold

Analytical methods for determining success of a refold should be chosen depending upon the refolding result. Insoluble aggregates may be detected by light scattering techniques (UV absorbance at ~320 nm or fluorescent scattering). Insoluble product may also be spun down by centrifugation, and then resolubilized and analyzed for content by sodium dodecyl sulfate poly acrylamide gel electrophoresis (SDS PAGE). Also, after centrifugation, the supernatant can be analyzed for soluble aggregates, misfolds, and native protein. Soluble aggregates can be measured by size exclusion high performance liquid chromatography (HPLC), dynamic light scattering, field flow fractionation, or analytical ultracentrifugation. Reducing and nonreducing SDS PAGE can be helpful in determining whether aggregates are covalently attached. To detect native protein from monomeric misfolds, activity assays indicate functionality and are therefore the most representative of proper folding, but they can be low-throughput and can be inhibited by refold additives. Reversed phase HPLC, ion exchange HPLC, and nonreducing SDS PAGE analysis can also potentially separate native protein from misfolds. In general, separating misfolds from properly folded protein may be difficult, and orthogonal techniques should be used whenever possible.

When screening refold conditions, HPLC assays are very useful for detection of proper folding, as they allow for relatively rapid, automated analysis of samples. As discussed above, different HPLC methods can offer excellent separation of different conformational species. Several limitations of HPLC analysis for refolds should be kept in mind, however. Misfolds or chemically modified protein may co-elute with properly folded protein, and thus HPLC analysis should be interpreted cautiously. To help understand the elution characteristics of an HPLC technique, fractions can be taken of individual peaks and analyzed by orthogonal techniques (such as different HPLC assays, mass spectroscopy, or activity assays). As solution conditions are altered to improve refolding, some conditions or additives may interfere with HPLC analysis. For example, high salt concentrations are likely to interfere with ion exchange HPLC, and surfactants will likely interfere with reversed phase HPLC analysis.

If using HPLC as the primary mode of analysis, periodic verification against an activity assay will help ensure the validity of the HPLC result.

4.3.5 Refolding Results and Strategies for Improving Yields

Upon completion of this first refold screening, three main results may be observed: insoluble product, soluble but misfolded product, and soluble product containing measurable native protein. Strategies for designing subsequent experiments for improving this refold have different focuses depending on whether the first refold screening experiment yielded soluble or insoluble product.

4.3.5.1 Insoluble Product

If the initial refolding experiment resulted in formation of insoluble aggregates, experimental effort should be directed toward achieving solubility during refolding. If formulation data are available, solution conditions that increase solubility of native protein is likely to help improve the solubility of the protein during refolding. Increasing the concentration of urea or GdnHCl in the refold buffer usually improves the solubility of the product during refolding. Aggregation is a second to third order reaction with protein concentration, where folding displays first order kinetics [21,22]. As a result, lower protein concentrations in the refold will help to suppress aggregation during refolding. This can be achieved by increasing the magnitude of dilution or by solubilizing at reduced protein concentrations. Finally, if formulation data are not available, a screening design may be used to quickly evaluate additives (Table 4.1), combinations of additives, and pH values that will help to improve solubility [23]. Statistical design software and design of experiments (DOE) techniques (discussed below) will aid in rapid analysis of these variables.

4.3.5.2 Soluble Product

If the initial refold resulted in soluble product, experimental efforts should then be focused on improving the percentage of properly folded protein. Based on the initial refold conditions, build a screening design to study the combinatorial effects of pH, temperature, redox ratios, redox concentration, and chemical additives on refolding efficiency. If refold yields are low, consider exploring longer refold incubation times. On rare occasions, some refolding reactions may take as long as 2 to 3 days to reach completion. Again, a DOE approach can greatly enhance the development and optimization of a refold process.

4.3.5.3 Use of Design of Experiments

Using statistical screening designs can greatly enhance the efficiency of refold process development. There are several software packages such as Design-Expert® by Stat-Ease, Inc. or ECHIP® DOE software by ECHIP, Inc., which make statistical design and analysis of experiments easily accessible to the bench scientist with minimal statistical training. These software packages are well developed and user friendly, with excellent manuals and online support.

The basic approach using statistical screening usually entails a sequential approach. Initially, a design exploring the high and low range of a series of variables (a screening design) is chosen to evaluate the main effects of each additive on refold yield and possibly secondary interactions due to synergistic effects between the additives. These screening models are very powerful for understanding the general trends within selected concentrations. Screening models only examine or define the conditions at the extremes or limits. They do not capture the conditions in between (the curvature), so they should be used to eliminate clearly negative effects and to determine regions to target for optimization. Once the number of variables have been minimized and the region of interest identified, a design that incorporates curvature (a surface response design) will enable the rapid optimization of refolding conditions to maximize yield.

Screening of refold conditions can be executed in deep 96-well plates for rapid evaluation [24]. Robotic equipment can ease the preparation and enhance the reproducibility of these plates. An obvious limitation of these designs is the need for a rapid and reliable method for analysis. If the analytical method used is not robust, these screening methods could easily misrepresent the data. Careful design of the experimental methods becomes more critical with increasingly sparse designs.

4.3.6 Premade Kits

One strategy for getting started quickly with refolding a protein from inclusion bodies utilizes premade protein refolding kits. At least half a dozen premade protein refolding screening kits are currently available. Protein refolding kits are designed with all of the necessary solution components, simply requiring the addition of inclusion body to the refolding matrix. The majority of these kits utilize guanidine–HCl or urea as a denaturant, GSH/GSSG as disulfide-shuffling agents, and mixtures of detergents, sugars, amino acids, and salts as additives. In general, the available protein refolding kits simply package basic refolding conditions into a convenient, easy to use format, but do not offer innovative, proprietary refolding technology. Several notable kits include the FoldIt Screen by Hampton Research, the Pro-Matrix Protein Refolding Kit by Pierce, and the Protein Refolding Kit by Novagen. The Refolding CA Kit by

Takara may be the exception, utilizing a proprietary cycloamylose molecule in combination with surfactants to enhance protein refolding. In most cases, refolding kits claim to be statistically designed as a sparse matrix, allowing the user to implement statistical analysis of the refolding results.

4.3.7 Other Refolding Methods

In addition to dilution refolding, there are several other methods for transitioning a protein from denaturing conditions to conditions that favor refolding. Alternative methods for refolding a protein may induce different structural transitions or inhibit side reactions that compete with proper folding. Hence, though the basic theory of these different refold methods is the same, alternative physical techniques may improve refolding yields.

4.3.7.1 Staged Dilution

If refolding yields are improved by executing the refold at lower protein concentrations, a staged dilution method may help to improve refolding efficiency at increased protein concentrations [25,26]. The staged dilution method pulses concentrated, solubilized protein into a refold buffer in discrete pulses. Each pulse is given some time to refold or partially fold before the next pulse is added to the refold mixture. As a result, the effective concentration of partially folded, aggregation-prone protein remains low, reducing the rate of aggregation.

4.3.7.2 Diafiltration

Dilution refolding and the subsequent rapid, uncontrolled collapse of protein structure may induce protein aggregation or trap the protein in a kinetically trapped folding intermediate or misfolds. In some cases, a more gradual transition from denaturing chaotrope concentrations to nondenaturing chaotrope concentrations may be beneficial to refolding yields. A practical method for achieving this gradual chaotrope reduction utilizes ultrafiltration (UF) cassettes to gradually exchange a protein from one buffer composition to another. In this scenario, the UF cassette is chosen to retain the protein, while allowing buffer components to pass freely through the membrane. As denaturing solution is removed from the protein solution, the volume is replaced with refolding buffer. With modern UF systems, the speed of buffer exchange can be easily scaled up and carefully controlled in a manufacturing environment.

4.3.7.3 Solid Phase Refolding

Chromatographic refolding methods have been the subject of increased interest in protein refolding [27,28]. The major chromatographic techniques for refolding are ion exchange (IEX) and size exclusion chromatography (SEC). In the case of ion exchange, solubilized, denatured protein is bound to the resin, exchanged into a refold buffer, and then eluted from the column. By immobilizing the protein on the solid chromatographic surface, protein–protein interactions are theoretically eliminated, allowing the protein to refold into a monomeric species. While bound to the solid support, however, a protein may no longer have the structural flexibility to properly fold. In addition, ligand density may be of critical importance, as high ligand density will allow protein–protein interactions while immobilized on the column. SEC simply acts as a chromatographic buffer exchange device, where protein is loaded onto the column in a denaturing buffer, and eluted with a refolding buffer. Although SEC may be effective at small scale, it does not scale well and is not recommended for large-scale manufacturing. Other chromatographic methods such as HIC, immobilized metal, and immobilized catalyst have also been used for protein refolding.

4.4 SCALING UP YOUR REFOLD REACTION

So now you have a basic refold reaction and want to scale it up, let us say, to 10,000 l. Here we will discuss some general scale-up considerations, and focus on the dilution of soluble nonnative protein into mix tanks. Refold reactions occurring on column matrices or during ultrafiltration/diafiltration will not be discussed here, but use scale-up parameters appropriate to those unit operations in the absence of a refold reaction.

The decision to scale-up a protein refold reaction is best made during the bench scale development phase. Some important issues and questions that need to be answered before scale-up begins are presented in Figure 4.4. These include the acquisition of an appropriate mix tank, facility and personnel requirements for special additives, reactivity of process reagents toward proteins and equipment, commercial availability and expense of raw materials, waste management, component addition, mixing speed, and oxygen transport.

A reaction is chosen for scale-up taking into account the answers to the questions in Figure 4.4. Figure 4.5 outlines a simple decision tree for the scale-up activities based on modeling all proposed activities at the bench scale.

4.4.1 Components of the Refold Reaction

4.4.1.1 Components That Lead to Protein Microheterogeneities

There are a number of chemicals routinely used in refold reactions that react with proteins. If the overall time and temperature of the reaction remain the same upon scale-up (which is definitely the goal), the extent of side-reactions causing protein microheterogeneities should remain the same. However, verification of the purity of the protein refolded at large scale and its similarity to the

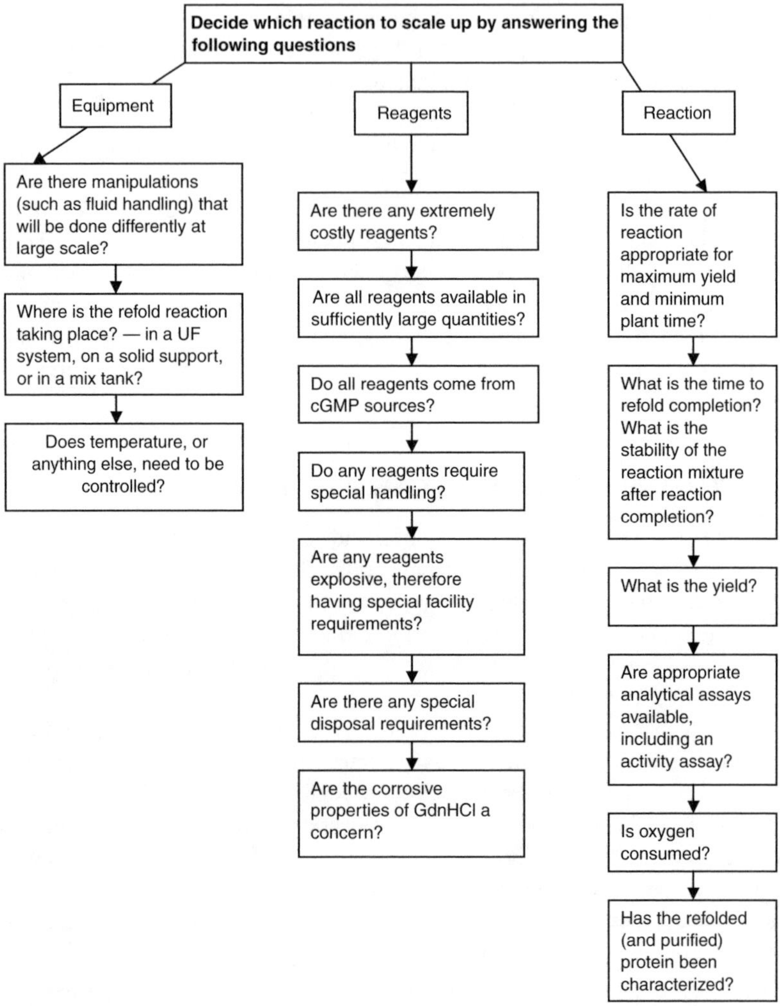

FIGURE 4.4 Some initial questions to answer before scaling up a protein refolding reaction.

small-scale product is a must. The most common side-reactions are oxidation of methionines, carbamylation and deamidation of amines, and redox reagent adducts [29]. Also to be considered is metal–ion catalyzed air oxidation of several amino acids [14]. All of these alterations have a potential for decreasing the activity of the protein, if the affected amino acid is in or near the active site.

Methionine can be oxidized to form methionine sulfoxide in the presence of oxygen and metal ion catalysts found contaminating the reagents. If there

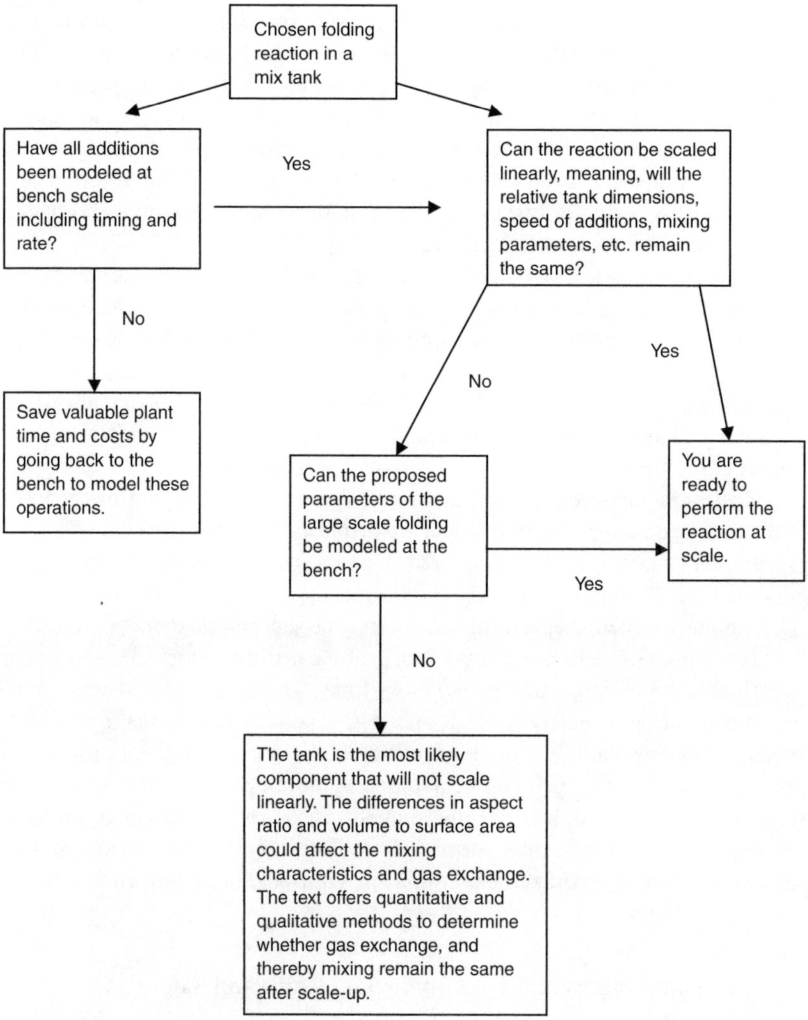

FIGURE 4.5 Protein refolding scale-up decision tree.

are particularly reactive methionines in the target protein, this reaction can be curtailed by folding in an oxygen-free environment, such as buffer sparged with nitrogen or helium.

Carbamylation of amines occurs when proteins are denatured and folded in urea. Urea breaks down to form cyanate ion (HNCO) that reacts with amino groups to form the stable carbamylated product. At equilibrium the cyanate concentration in 8 *M* urea (the concentration typically used for solubilization) is approximately 0.02 M. The carbamylation side-reaction can be controlled by limiting the time the protein spends in urea and by using lower urea concentrations. Cyanate ion concentration can also be controlled by proper preparation and timely use of the urea. There are ultra-pure urea products on the market or the urea can be recrystallized or acid treated. However, the best large-scale solution to preparing cyanate ion-free urea is to deionize it in solution directly before use, using in-line packed bed cylinders containing mixed bed (both cationic and anionic) ion exchange resins.

Deamidation is another side-reaction that occurs on the free amino group of the N-terminus and the epsilon amino group of lysine. The reaction occurs in the same pH range of 8 to 10 that is optimal for oxidation and disulfide exchange of cysteines during protein refolding. The rate of deamidation decreases with decreasing pH (which also slows down the desired reactions of oxidation and disulfide exchange). The development goal is to limit the total time the protein spends at high pH.

Most reductants (except dithiothreitol) form thiol adducts in the process of disulfide exchange. These adducts are displaced by a large excess of thiols. For this reason, the recommended redox couples for refolding reactions (see Section 4.2.3.2) consist of an excess of reductant.

Proteins in reactions performed in air may be susceptible to metal–ion catalyzed oxidation of methionine, cysteine, proline, histidine, arginine, and lysine as well as fragmentation and tyrosyl cross-links. The mechanism is believed to be via generation of Fe(II) or Cu(I) and H_2O_2, and the reaction is highly site-specific. This mechanism is not reported in the literature as a major problem in protein folding, however it can be avoided by developing refold reactions for sensitive proteins in an air-free environment. If the oxygen-sensitive proteins contain disulfide bonds, their formation would require using a redox couple, such as reduced and oxidized glutathione in a nitrogen or helium environment.

4.4.1.2 Components and Environmental Health and Safety Concerns

Refold reactions in large volumes typically contain huge quantities of denaturant guanidine or urea, a smelly reductant, expensive redox reagents, and perhaps exotic or explosive additives. Each of these components must be thrown away. It is important to work with your Environmental Health and Safety (EH&S) officer to determine whether there are restrictions to dumping any of the reagents (particularly in large volumes) into the city sewer. If the spent volume must be trucked away, how much will it cost? If there is a component that is expensive to dispose of, should it be replaced with another chemical, even if the rate or yield of the reaction is lowered?

Some chemicals are not safe in large quantities for manufacturing personnel. For example, dithiothreitol and other reductants produce fumes, as do organic solvents. A simple open addition on the bench, or preferably in the fume hood, must become a closed transfer from one tank to another at large scale. Organic solvents, such as ethanol, are sometimes used as folding enhancers. There may be restrictions on the volume of the solvent that can be in the plant at one time. Furthermore, facility modifications may be necessary to make the suite explosion proof. Again, early discussions with EH&S are essential.

4.4.1.3 Guanidine and Process Equipment

Guanidine–HCl and other ionic compounds in high concentrations are corrosive to stainless steel. The typical concentration for guanidine solubilization is 6 *M*. Tanks and lines can be protected by a schedule of electropolishing and passivation, and by limiting the time the chemical is in contact with them. Other materials, such as the nickel alloys, have greater corrosion resistance, but are more expensive to purchase. If the protein folds equally well in urea, a switch to urea (remembering the carbamylation issues discussed above) will resolve the equipment problem.

4.4.1.4 Commercial Availability and Expense of Raw Materials

All chemicals used in large-scale reactions destined for commercial production, must come from vendors capable of producing under current Good Manufacturing Practices (cGMP). Exotic folding enhancers may not be available commercially or only from a small single-source vendor with no cGMP capability. For these cases, alternatives should be considered. The expense of raw materials is another factor to consider for a refold reaction that will be scaled up. If a screen of additives reveals two that perform similarly, cost may become the deciding factor in choosing the reagent.

4.4.2 Addition and Mixing of Components

During the early development of a refold reaction, reagent addition and denatured protein dilution are both essentially instantaneous, such as pipetting 1 ml of protein denatured in 6 *M* guanidine–HCl into 100 ml of buffer in a beaker containing a magnetic stir bar. In the case of a 10,000 l refold reaction, 1000 l of denatured protein solution will be pumped, following cGMP [30], rather than dumped, into the refold tank that contains buffer. This step requires extra time and should be modeled on the bench. Alternatively, the two components can be mixed in-line on the way to the refold tank. However, this approach to instantaneous addition requires an extra process tank.

Another feature of the small-scale beaker system is that the mixing is quite efficient. The vessel is short and squat in comparison to many manufacturing mix tanks and contains a relatively large mixing device (the stir bar). This device can be adjusted to stir efficiently without causing protein shearing and foaming. We will discuss mixing further, in the context of oxygen mass transfer. Here we want to emphasize that providing proper mixing at the large scale, in a way that does not damage the protein being folded, is an important parameter that should be modeled at small scale.

4.4.3 Disulfide Bonding and Oxygen Mass Transfer

When the protein to be refolded contains disulfide bonds, the refold reaction must facilitate disulfide bond formation, often including thiol/disulfide exchange. The latter requires a redox couple [31] such as reduced and oxidized glutathione, and the former can utilize a redox couple or, in some cases, oxygen alone catalyzed by metal ions, such as Cu^{++} [32]. Both mechanisms may occur in a given refold reaction when a redox couple, metal ions, and oxygen are present. Metal ions commonly exist as contaminants in buffer components. Oxygen comes from air dissolved in solution and is replenished by mixing.

There are several approaches to the development and scale-up of refold reactions for proteins containing disulfide bonds. Most commonly a redox couple is added to the diluted reduced protein in the presence of air. For scale-up it is useful to determine whether the reaction requires oxygen. One approach is to perform a known reaction in a vessel sparged with helium or nitrogen and measure the rate and yield of protein refolding. A simple reaction vessel design, not involving a commercial tank is shown in Figure 4.6. If the yield and rate of native protein formation are equivalent to the reaction performed in air, then the reaction is independent of oxygen and can be scaled-up with little risk of major scalability issues. If the reaction is inhibited or the rate slows down, it requires oxygen.

Without these studies the bench scientist may develop a redox system, only to find upon scale-up that the reaction rate is slower, or that the reaction does not go to completion. When this happens it becomes necessary to examine tank mixing parameters and buffer making practices. For example, pumping hot WFI (water for injection) into a closed process tank and letting it cool before adding stock components may result in an oxygen depleted buffer with resulting refold kinetics that are different from the bench reaction.

Some proteins, such as our model protein and lysozyme [13], will refold efficiently to the native conformation by either mechanism — air oxidation or redox coupling. Disulfide bonding utilizing oxygen occurs efficiently (and cost effectively) after the protein is denatured in a buffer containing a reductant that does not participate in disulfide exchange, such as DTT [31]. In this special case,

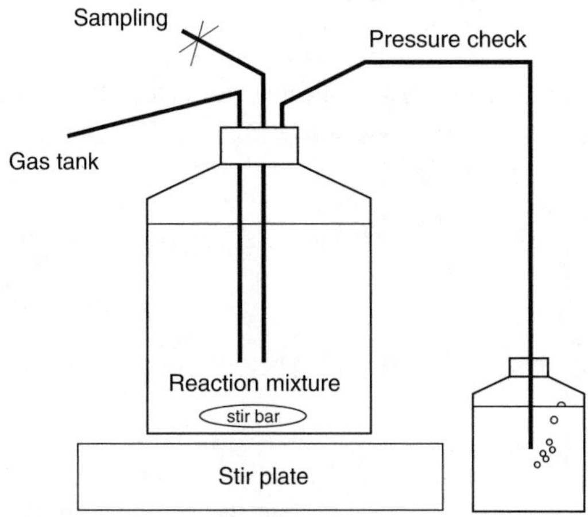

FIGURE 4.6 Bench-scale reaction vessel for reactions requiring control of dissolved gases. The vessel is used for experiments requiring gas compositions different from air. Air can be excluded from the protein refold reaction by attaching the line to a helium or nitrogen tank. The effect of different oxygen concentrations can also be studied by connecting a gas tank containing the desired percent oxygen. There is a sampling line so that the extent of refolding can be monitored.

the oxygen consumption rate could be approximated by either measuring the rate of SH (sulfhydryl) decay with time (using the DTNB [dithionitrobenzoic acid] assay [33]) and calculating 4 moles SH per mole O_2 [34] or by measuring the rate of rising reduction potential (mvolts) with time using a potentiometer (Figure 4.7). A more direct approach is to determine the oxygen concentration and utilization rate by using a dissolved oxygen (DO) probe. This approach will be discussed later.

Protein reduction and refolding can also be initiated using SH reagents, such as glutathione or cysteine. Partial air oxidation of the reductant results in the formation of a redox couple and the mechanism of disulfide bond formation is likely a mixture of air oxidation and disulfide exchange. The potentiometric time course of a redox refolding reaction performed in air is shown in Figure 4.7. The time course is distinctly different from air oxidation of SHs, since it is the result of multiple oxidation mechanisms.

Another approach involves protein originating from inclusion (refractile) bodies. This protein is typically in the reduced form [35]. Protein refolding can be initiated by adding oxidized glutathione or cystine. Reaction with the reduced protein will result in the generation of a redox couple and likely a mixed oxidative/redox reaction. Since large-scale recovery procedures often result in air oxidation of some SHs in the inclusion bodies, reduction with DTT may be required.

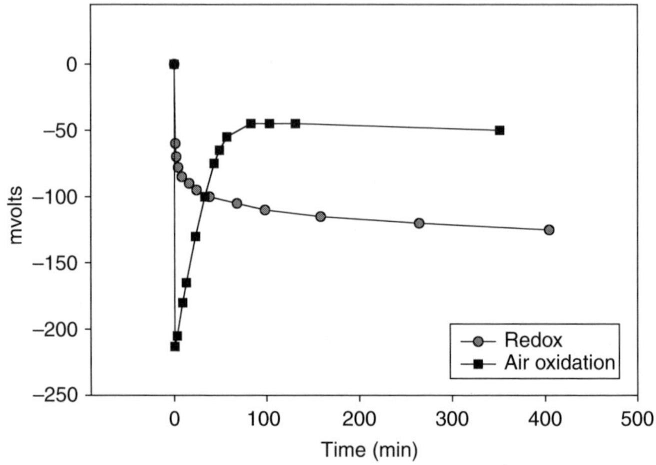

FIGURE 4.7 Potentiometric measurement of the refold of a model protein containing disulfides using oxygen in one reaction and a redox reagent in the other reaction. The reaction conditions for the reaction ulilizing oxygen and no redox reagents are found in Figure 4.8. The redox reaction contained, in addition, 1 m*M* 2-hydroxyethyl disulfide. Air was not excluded from the redox reaction. Reduction potential was measured using an Orion pH/potentiometer.

The most direct approach to determine if oxygen is a reactant in the refold reaction is to insert an oxygen probe into an open reaction vessel at the bench. If the oxygen concentration does not change with time, the reaction may be independent of oxygen or it may use oxygen at such a slow rate that exchange upon mixing compensates for its utilization. To distinguish between the two, the reaction could be performed in nitrogen or helium environment as described above. If the oxygen concentration in the reaction changes with time, oxygen is utilized and an approximate consumption rate could be calculated from the slope of the initial fall in concentration (Figure 4.8). The refold reaction can then either be redeveloped in a nitrogen or helium environment to be oxygen independent (solely a redox reaction), or the oxygen requirement can be studied and the dissolved oxygen concentration controlled. This latter case will be discussed below.

Oxygen required for the refold reaction is provided to the liquid phase by mass transfer from the gas phase (Figure 4.9). The efficiency of oxygen mass transfer is controlled by agitation, sparging, and pressurization [36,37]. The oxygen mass transfer rate must be sufficient to allow the refold reaction to go to completion in the desired time, while avoiding aggregation, denaturation, and foaming that can be caused by agitation and sparging. At a small scale, during refolding development, oxygen transfer is not usually a rate-limiting factor.

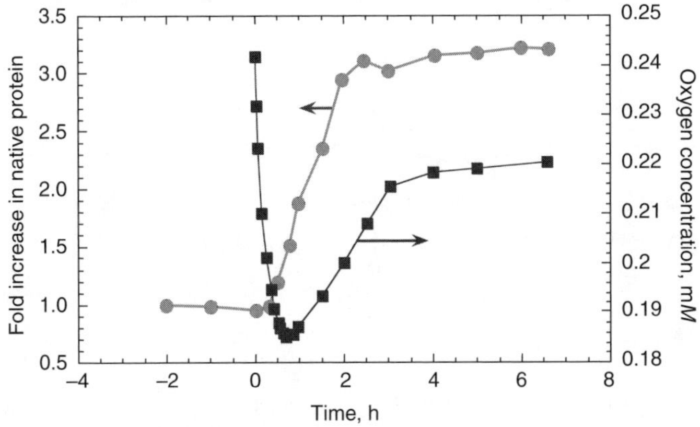

FIGURE 4.8 Oxygen consumption during folding of model protein containing disulfides. Protein was diluted in refold buffer containing urea and 2 μM Cu^{++}. Folding was initiated by the addition of 0.2 mM DTT. Oxygen concentration was measured using a dissolved oxygen probe (Ingold polarographic 25 mm probe and Cole Parmer dissolved oxygen meter). The probe was calibrated to 100 and 0% by sparging water with air and nitrogen before installation. Samples were taken to analyze for native protein formation using a reversed phase HPLC assay. Notice that the buffer is supersaturated with oxygen at the beginning of the reaction. This is a result of oxygen entrainment during the buffer making activities. (Oxygen saturation under this refold condition is 0.22 mM.) Oxygen depletion is the most severe during the first 45 min. The protein is fully folded by 2 h even though the buffer does not reach full oxygen saturation until 4 h.

In fact, oxygen consumption may be a significant component of a redox reaction that is not initially recognized by the developer until scale-up. Figure 4.8 shows the relationship between native protein formation and oxygen concentration for a model protein.

If oxygen is utilized in the refold reaction, it is necessary to determine the amount and type of metal–ion catalyst (such as Cu^{++}) that gives the desired yield and reaction time. For proteins, the range is usually 1 to 10 μM [38]. Buffers alone often provide 0.5 to 1 μM metal ions [32,39], therefore, the range of adequate metal–ion concentration should be determined and a value in the middle of the range chosen to assure that the process step will be robust. Too high a concentration may result in a fast reaction rate and protein aggregation (Figure 4.10) [40].

Oxygen mass balance in the presence of a reaction is expressed by the following equation [41]:

$$\frac{\mathrm{d}(C_{\mathrm{O}_2}^{\mathrm{L}})}{\mathrm{d}t} = k_{\mathrm{L}}a(H^*P_{\mathrm{O}_2}^{\mathrm{G}} - C_{\mathrm{O}_2}^{\mathrm{L}}) - Q \tag{4.1}$$

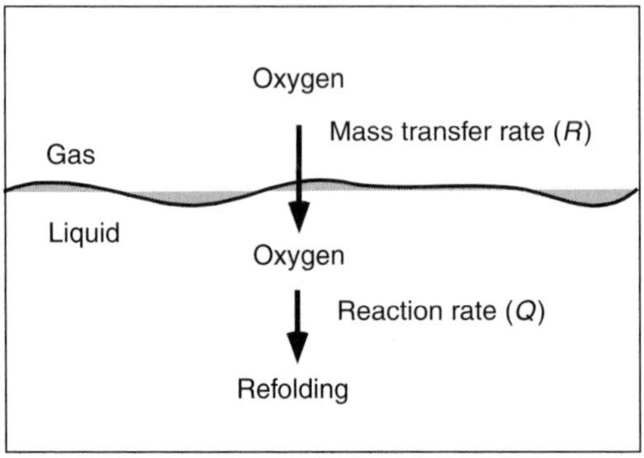

FIGURE 4.9 Mass transfer rate must supply sufficient oxygen for refold reactions that utilize oxygen.

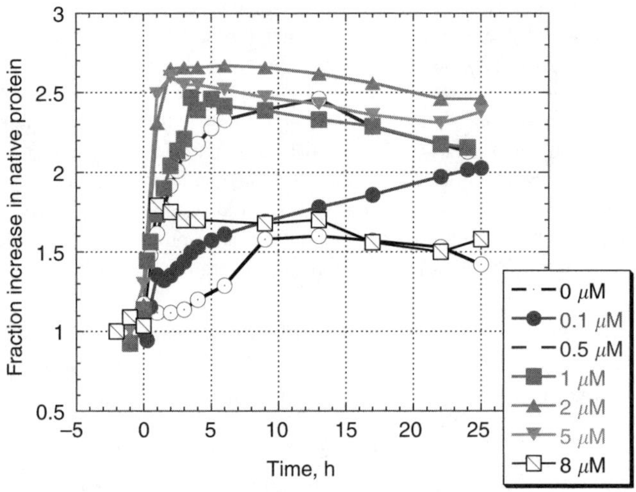

FIGURE 4.10 Copper ion concentration effects the rate and yield of protein refolding when disulfide oxidation occurs using molecular oxygen. The refold reaction was performed as described in Figure 4.8, varying only the Cu^{++} concentration. Some folding takes place in the absence of added Cu^{++}, because the buffer contains contaminating metal ions.

where:

$d(C^{L}_{O_2})/dt$ = the rate of change of the O_2 concentration in the liquid (mM/h)

$R = k_{L}a(H^{*}P^{G}_{O_2} - C^{L}_{O_2})$ = O_2 mass transfer rate to the liquid (mM/h)

Q = O_2 consumption rate (mM/h) during the refold reaction. It is assumed to be a constant or independent of oxygen concentration in the range studied.

Further:

$H^*P^{G}_{O_2} = C^*$, Dissolved oxygen concentration at saturation (mM)
C^{L} = Concentration of oxygen in the liquid phase (mM)
k_L = Oxygen mass transfer coefficient (cm/h)
a = Interfacial area per volume (1/cm)
H = PHenry's constant (1 mM/atm at 31°C)
k_La = Volumetric oxygen mass transfer coefficient (1/h)

When the differential equation is rearranged and integrated

$$C_L = m_1 - m_2 e^{-m_3 t} \tag{4.2}$$

where

$$C^* - \frac{Q}{k_L a} = m_1 \tag{4.3}$$

$$C^* - C^0 - \frac{Q}{k_L a} = m_2 \tag{4.4}$$

$$k_L a = m_3 \tag{4.5}$$

Equation 4.2 provides a useful way to describe oxygen supply and demand in refold reactions for several reasons. Only one experiment is necessary to determine both the oxygen consumption rate (Q) and the k_La needed to maintain the oxygen concentration of the solution. The equation can be used when Q is positive, as is the case during a reaction, fermentation, or cell culture. It can also be used when Q is set to 0, as is the case when k_La is determined in the absence of reaction. This method is used traditionally in the literature and k_La is determined by measuring the change in oxygen concentration in a tank during pressurization, depressurization, or sparging [42].

A model small-scale refold reaction is performed using the optimized conditions intended for large-scale. Oxygen concentration in the liquid (C_L) per time is measured throughout the reaction using a DO probe (Figure 4.11). The oxygen consumption rate (Q) and volumetric mass transfer coefficient (k_La)

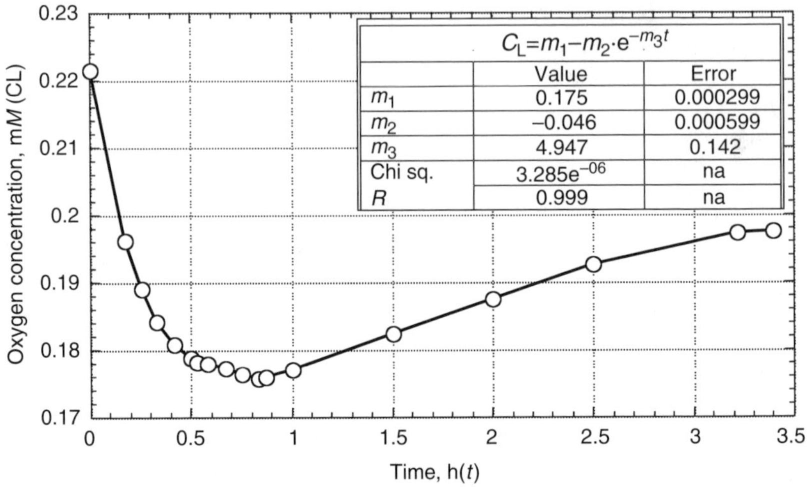

FIGURE 4.11 Fitting oxygen consumption data obtained during model protein refold into the mathematical model in Equation 4.1. The reaction took place in a 100 ml beaker containing a stir bar using the same conditions described in Figure 4.1. Oxygen concentration (C_L) was plotted against time (t), and from the fitted data m_1, m_2, and m_3 were obtained. $m_3 = k_L a = 4.95\,\text{h}^{-1}$ and $m_1 = C^* - Q/k_L a$, $Q = 0.17$ mM/h.

are determined by fitting the oxygen concentration data from the initial decay until the concentration stabilizes. The data is analyzed, using the mathematical model in Equation 4.2, determining the coefficients m_1, m_2, and m_3, and solving Equation 4.3 to Equation 4.5. Fitting data into the model is most readily accomplished using one of the commercially available computer graphing packages, such as Sigma Plot or Kaleidagraph. During each experiment, the degree of surface agitation is noted.

Since the rate of oxygen mass transfer from air to the liquid phase must be sufficient to complete the refold reaction in the desired time, the rate of the oxygen mass transfer (R in mM/h) must be equal to or greater than the oxygen consumption rate of the reaction (Q in mM/h). In other words:

$$R = k_L a(H^* P^{G}_{O_2} - C^{L}) \geq Q \tag{4.6}$$

The minimum required $k_L a$ is determined by setting the following criteria:

$$k_L a \geq \frac{Q}{(H^* P^{G}_{O_2} - C^{L})} \tag{4.7}$$

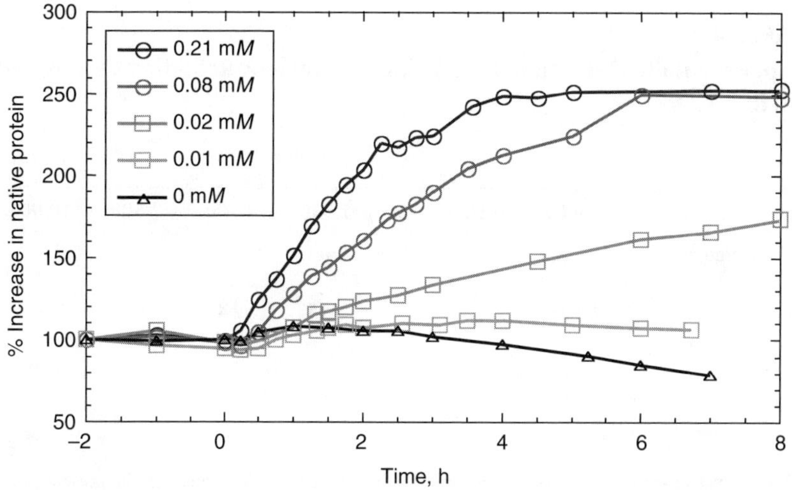

FIGURE 4.12 Dependence of refolding rate and yield on oxygen concentration. The model protein was refolded as described in Figure 4.8 using the reaction vessel pictured in Figure 4.5. A gas tank containing the specified oxygen concentration was attached for each experiment.

Once the oxygen consumption rate (Q) is known (from small-scale experiments), we can experimentally determine the oxygen concentration in the liquid (C_L) that supports this rate, and calculate the k_La required to give the (C_L) value. Controlling oxygen concentration in this way can be a useful tool to control the overall folding reaction rate, especially to slow it down to let other, noncovalent, folding interactions occur. An example of the effect of oxygen concentration on protein refolding rate and yield for a model protein is shown in Figure 4.12. The concentration at which oxygen becomes rate limiting is called the critical concentration. For the model protein the critical oxygen concentration is 0.08 m*M*.

4.4.4 Mixing and Type of Tank

Now that we know the k_La necessary to give us the desired yield and reaction rate, we need to find a proper tank and mixing regime. A desirable vessel is a fermentation tank with pressurization and sparging capability, as well as with multiple controls such as pH, temperature, and dissolved oxygen. An adequate and more common situation in the plant is that refolding takes place in a buffer tank with a simple single-blade impeller.

TABLE 4.2
Experimentally Determined k_La Values for Various Refolding Vessels and Mixing Styles

Mixing Regime	Volume, l 0.05	0.10	10	1,500	4,600	7,500	10,000
Swirling/no vortex	3	5	2		0.5–0.7 $N = 4$		
Gentle waves/ Small vortex	5			1.25	2.4	1.6	7.9
Turbulent			13.8		21	13.2	18
Excessive turbulence/ deep vortex	41					50	

The k_La values (in h^{-1}) for the vessels up through 4600 l were determined in the presence of a refolding reaction, as described in this chapter. The k_La values for the two larger vessels were determined by two methods. In the first, the closed tank, containing water or aqueous buffer, was sparged with air until 15 psig were reached, then the pressure was released, and the oxygen concentration was measured with time as outgassing occurred. In the second, the tank was pressurized to 25 psig, then the pressure was released and the dissolved oxygen was measured with time.

Source: Methods discussed in Reference 43.

Scale-up and manufacturing qualification runs are typically performed on a very tight plant schedule with little or no time for experiments at-scale. The k_La of a large tank can be estimated by observing the behavior of the liquid at the surface during mixing. Table 4.2 shows the relationship between surface agitation and k_La for vessels ranging from 50 ml to 10,000 l [43]. Going from least to greatest surface agitation, the surface categories are (1) swirling/no vortex, (2) gentle waves/small vortex, (3) turbulent, and (4) excessively turbulent/deep vortex. The first two categories represent the least disturbance to the surface and result in k_La values in the 1 to 10 h^{-1} range. The third category, a turbulent surface, results in k_La values in the 10 to 20 h^{-1} range. Finally, a very deep vortex and other extreme turbulence increase the k_La value up to the 40 to 50 h^{-1} range. The k_La limit for a simple tank and impeller is, approximately, 50 h^{-1}. This is well within the range needed for the oxidation of thiols in the concentrations used for most protein refolding reactions, and manyfold below the demand of bacterial fermentation (Table 4.3) [15,44–48].

The mixing regime in a simple tank can be adjusted by using the following empirical relationships. Aunins et al. [49] observed, in 500 ml tanks, that for a given impeller speed, whatever the k_La is when the impeller is greater than

TABLE 4.3
Oxygen Consumption Rates of Different Systems

Reaction	Conditions	Q (mM Oxygen/ h)	k_La (h^{-1})	Reference
Hybridoma cells	1×10^6 cells/ml	0.05–1.0	0.25–5	[44,45]
Small scale model protein oxygen-dependent refolding	0.2 mM DTT, 2 μM Cu^{++} 50 ml, 100 ml	0.32, 0.43/mM SH	0.8, 5	[15]
Large scale model protein oxygen-dependent refolding, $N = 4$	0.2 mM DTT, 2 μM Cu^{++} 4600 l	0.23/mM SH	0.66±0.09	[15]
Glutathione oxidation	0.1 mM GSH, 5 μM Cu^{++}	0.7/mM SH	NA	[46]
Glutathione oxidation	0.1 mM GSH, 50 μM Cu^{++}	2.28/mM SH	NA	[46]
E. coli fermentation	NA	>100	50–1000	[47,48]

a quarter of the way down from the surface, it will be approximately twice as large when the impeller is placed in the top quarter. Table 4.4 shows that this relationship also applies to 10,000 l tanks when the topmost of its three impellers is in the top quarter of the liquid. This table, furthermore, shows that when the impeller is in the top quarter, the k_La increases approximately 10-fold for each doubling of the impeller speed. This means that once the k_La has been approximated by observing the liquid surface behavior and the position of the impeller (or of the topmost impeller), k_La can be predictably altered by adjusting the height and speed of the impeller.

Two aeration conditions shown in Table 4.4 are not appropriate for protein folding. One is placing the impeller at the surface and the other is running the impeller at speeds that cause excessive turbulence with bubbling. Both these conditions raise the k_La dramatically, but may lead to protein foaming and denaturation.

4.4.5 Summary

Scale-up of a chemical reaction, such as protein refolding, should be preceded by scaled-down models of the process at the bench. In the model, all manipulations, including reagent transfers, are performed using the methods and the

TABLE 4.4
When the Mixing Regime Increases the Surface Area, It Increases the $k_L a$

	500 ml Tank	10,000 l Tank						
Impeller Distance From Liquid Surface	$k_L a$	Volume of Liquid in Tank (l)	$k_L a$ at 60 rpm (h^{-1})	Mixing Regime	$k_L a$ at 120 rpm (h^{-1})	Mixing Regime	$k_L a$ at 180 rpm (h^{-1})	Mixing Regime
>1/4 of liquid height	$1 \times k_L a$	10,000			7.9	Gentle waves	18	Turbulent
<1/4 of liquid height	$2 \times k_L a$	7,500	1.6	Gentle waves	13.1	Turbulent	50	Excessive turbulence with bubbling
0	$4.5 \times k_L a$							

The $k_L a$ value can be increased approximately two-fold by changing the impeller placement from the bottom or middle to the top quarter of the liquid volume. It can then be increased approximately 10-fold by doubling the impeller speed.

time frames that will be used at the large scale. This approach will assure that not only will the reaction rate and yield remain the same, but also the protein quality. This latter is important in drug development, especially when the scale-up occurs after initial clinical studies have been completed. The determination of comparability of the product made before and after a manufacturing change is usually required for regulatory submission.

Beyond the reaction itself, there are issues of personnel safety, availability and cost of reagents, waste removal, acceptable cGMP procedures, and appropriate and available equipment. Resolution of these issues may affect which reaction is scaled-up and exactly how it is performed. These decisions will need to be incorporated into the model reaction.

4.5 EMERGING REFOLDING TECHNOLOGIES

4.5.1 High Pressure Refolding

Although there are several developing refolding strategies helping to improve refolding efficiency, refolding proteins from aggregates using high hydrostatic pressure is the only emerging technology that offers an alternative mechanism of protein refolding to conventional methods. Although hydrostatic pressure has been used for decades to study protein unfolding, the application of pressure for refolding aggregates was first published in 1999 [50]. Since 1999, high pressure (1 to 3 kbar) has been shown to be a versatile protein refolding method by refolding noncovalent aggregates of both monomeric and multimeric industrial proteins, covalent aggregates, and inclusion bodies [51–53]. In addition, high pressure refolding has not exhibited the same protein concentration dependence as conventional chemical refolding. Proteins have been refolded from aggregates at >20 mg/ml without decrease in yields. An exciting new capability of high pressure refolding is realized in the ability to refold soluble aggregates in highly purified product or bulk drug substance. Because high pressure refolding frequently does not require chemical additives, noncovalent soluble aggregates can be refolded to native structure, hence reducing the immunogenicity profile of chronic medications and potentially eliminating the need for size exclusion chromatography or other methods for reducing aggregates. A new company has emerged (BaroFold, Inc.) with the vision of bringing high pressure refolding into industrial applications.

4.5.1.1 Mechanism of High Pressure Refolding

High pressure unfolding of proteins is thermodynamically driven by a decrease in system volume (volume of protein + volume of surrounding aqueous solution) associated with protein unfolding. As a protein unfolds under pressure, imperfectly packed void volumes in the interior of the protein structure are filled with water, and system volume decreases. In addition, organization of water molecules around charges or on hydrophobic surfaces results in a dense water layer, and a reduction in system volume. Thus, disrupting intraprotein ionic and hydrophobic contacts decreases the system volume. The combined volume decrease from void volumes and disruption of intraprotein contacts during unfolding is small, and therefore high pressures are required for unfolding.

Utilization of high pressure for protein refolding is based upon the observation that multimeric proteins generally unfold between 1 and 3 kbar, while monomeric proteins generally unfold between 4 and 8 kbar. If we treat aggregates as disorganized multimers, we can assume that there is a pressure window between 2 and 4 kbar where monomers are stable, but multimers dissolve. As a result, an aggregated protein solution can be held at a particular pressure where aggregates dissolve and monomers refold without kinetic competition with aggregation side reactions. Once refolded, the system can be depressurized without fear of aggregation. If refolding multimers or particularly unstable monomers, a refolding or reassociation of the native multimer may occur at lower pressures, and a slow depressurization or hold at intermediate pressure may be required to induce proper refolding.

4.5.1.2 Scale-Up of High Pressure Refolding

Unlike temperature and mass transfer, which are traditionally challenging variables to scale-up, pressure is transmitted through aqueous solutions nearly instantaneously. The only remaining challenges, then, lie in finding the appropriate equipment and dealing with safety concerns. Hydrostatic pressures up to 5 kbar are routinely used in the food industry for sterilization. Pressure vessels can be ordered as stock items up to roughly 1000 l. Smaller vessels can also be connected together to increase the effective volume of the pressure vessel. Safety concerns are not much different from common manufacturing equipment. Large-scale homogonizers frequently operate at 1 to 1.5 kbar. Water compressibility at 2 kbar is only 7%. Thus, a small equipment leak will cause a rapid drop in pressure. We have been assured by BaroFold Inc., that no special building requirements are required for safe installation of a pressure refolding device in manufacturing. Proper engineering and appropriate safety procedures are, of course, still required.

REFERENCES

1. Clark, E.D.B., Refolding of recombinant proteins. *Curr. Opin. Biotechnol.*, 1998; **9**: 157–163.

2. Clark, E.D., Protein refolding for industrial processes. *Curr. Opin. Biotechnol.*, 2001; **12**: 202–207.
3. Middelberg, A.R., Preparative protein refolding. *Trends Biotechnol.*, 2002; **20**: 437–443.
4. Rudolph, R. and H. Lilie, *In vitro* folding of inclusion body proteins. *FASEB J.*, 1996; **10**: 49–56.
5. Lee, J.C. and S.N. Timasheff, Partial specific volumes and interactions with solvent components of proteins in guanidine hydrochloride. *Biochemistry*, 1974; **13**: 257–265.
6. Prakash, V., C. Loucheux, S. Scheufele, M.J. Gorunoff, and S.N. Timasheff, Interactions of proteins with solvent components in 8 M urea. *Arch. Biochem. Biophys.*, 1981; **210**: 455–464.
7. Burgess, R.R., Purification of overproduced *Escherichia coli* RNA polymerase gamma factors by solubilizing inclusion bodies and refolding from sarkosyl. *Meth. Enzymol.*, 1996; **273**: 145–149.
8. Puri, N.K., E. Crivelli, M. Cardamone, R. Fiddes, J. Bertolini, B. Ninham, and M.R. Brandon, Solubilization of growth hormone and other recombinant proteins from *Escherichia coli* inclusion bodies by using a cationic surfactant. *Biochem. J.*, 1992; **285**: 871–879.
9. Dill, K.A. and H.S. Chan, From Levinthal to pathways to funnels. *Nat. Struct. Biol.*, 1997; **4**: 10–19.
10. Ghelis, C., *Protein Folding*. Academic Press: New York. 1982, pp. 262–263.
11. Welker, E., W.J. Wedemeyer, M. Narayan, and H.A. Scheraga, Coupling of conformational folding and disulfide-bond reactions in oxidative folding of proteins. *Biochemistry*, 2001; **40**: 9059–9064.
12. Shaked, Z., R.P. Szajewski, and G.M. Whitesides, Rates of thiol–disulfide interchange reactions involving proteins and kinetic measurements of thiol pKa values. *Biochemistry*, 1980; **19**: 4156–4166.
13. Perraudin, J.P., T.E. Torchia, and D.B. Wetlaufer, Multiple parameter kinetic studies of the oxidative folding of reduced lysozyme. *J. Biol. Chem.*, 1983; **258**: 11834–11839.
14. Stadtman, E.R., Metal ion-catalyzed oxidation of proteins: Biochemical mechanism and biological consequences. *Free Radical Biol. Med.*, 1990; **9**: 315–325.
15. Ozturk, A. and C. Cowgill, *Personal observation.*
16. Pace, C.N., Determination and analysis of urea and guanidine hydrochloride denaturation curves. *Meth. Enzymol.*, 1986; **131**: 266–280.
17. Shirley, B.A., Urea and guanidine hydrochloride denaturation curves, in *Protein Stability and Folding*, B.A. Shirley, Ed. Humana Press: Totowa, NJ. 1995, pp. 177–190.
18. Cleland, W.W., Dithiothreitol, a new protective reagent for SH groups. *Biochemistry*, 1963; **3**: 480–482.
19. Pace, C.N. and T. Creighton, The disulphide folding pathway of ribonuclease T1. *J. Mol. Biol.*, 1986; **188**: 477–486.
20. Happersberger, H.P., C. Cowgill, and M.O. Glocker, Structural characterization of monomeric folding intermediates of recombinant human macrophage-colony

stimulating factor β (rhM-CSFβ) by chemical trapping, chromatographic separation and mass spectrometric peptide mapping. *J. Chromatogr. B*, 2002; **782**: 393–404.

21. Zettlmeissl, G., R. Rudolph, and R. Jaenicke, Reconstitution of lactic dehydrogenase. Noncovalent aggregation vs. reactivation. 1. Physical properties and kinetics of aggregation. *Biochemistry*, 1979; **18**: 5567–5571.
22. Kiefhaber, T., R. Rudolph, H.H. Kohler, and J. Buchner, Protein aggregation *in vivo*: A quantitative model of the kinetic competition between folding and aggregation. *Bio/technology*, 1991; **9**: 825–829.
23. Clark, E.D., E. Schwarz, and R. Rudolph, Inhibition of aggregation side reactions during *in vitro* protein folding, in *Amyloid, Prions, and Other Protein Aggregates*, R. Wetzel, Ed., vol. 309. Academic Press, Inc.: San Diego. 1999, pp. 217–236.
24. Vincentelli, R., S. Canaan, V. Campanacci, C. Valencia, D. Maurin, F. Frassinetti, L. Scappucini-Calvo, Y. Bourne, C. Cambillau, and C. Bignon, High-throughput automated refolding screening of inclusion bodies. *Protein Sci.*, 2004; **13**: 2782–2792.
25. Fischer, B., B. Perry, I. Sumner, and P. Goodenough, A novel sequential procedure to enhance the renaturation of recombinant protein from *Escherichia coli* inclusion bodies. *Protein Eng.*, 1992; **5**: 593–596.
26. Vallejo, L.F. and U. Rinas, Optimized procedure for renaturation of recombinant human bone morphogenetic protein-2 at high protein concentration. *Biotechnol. Bioeng.*, 2004; **85**: 601–609.
27. Jungbauer, A., W. Kaar, and R. Schlegl, Folding and refolding of proteins in chromatographic beds. *Curr. Opin. Biotechnol.*, 2004; **15**: 487–494.
28. Li, M., Z.G. Su, and J.C. Janson, *In vitro* protein refolding by chromatographic procedures. *Protein Exp. Purif.*, 2004; **33**: 1–10.
29. Means, G.E. and R.E. Feeney, *Chemical Modification of Proteins*. Holden-Day, Inc.: San Francisco. 1971.
30. *Code of Federal Regulations, Title 21, Sections 110, 210, 820 found on the Food and Drug Administration's website, www.fda.gov*
31. Gilbert, H.F., The formation of native disulfide bonds, in *Mechanisms of Protein Folding*, R.H. Pain, Ed. IRL Press at Oxford University Press: New York. 1994, pp. 104–136.
32. Ahmed, A.K., S.W. Schaffer, and D.B. Wetlaufer, Nonenzymic reactivation of reduced bovine pancreatic ribonuclease by air oxidation and by glutathione oxidoreduction buffers. *J. Biol. Chem.*, 1975; **250**: 8477–8482.
33. Ellman, G.L., Tissue sulfhydryl groups. *Arch. Biochem. Biophys.*, 1959; **82**: 70–77.
34. Zwart, J., J.H.M.C. Van Wolput, J.C.J.M. Van Der Cammen, and D.C. Koningsberger, Accumulation and reactions of H_2O_2 during the copper ion catalysed autoxidation of cysteine in alkaline medium. *J. Mol. Catal.*, 1981; **11**: 69–82.
35. Mitraki, A., C. Haase-Pettingell, and J. King, Mechanisms of inclusion body formation in protein refolding, in *Protein Refolding*, G. Georgiou

and E.D. Clark, Eds. ACS Symposium Series: Washington, D.C. 1991, pp. 35–49.

36. Blanch, H.W. and D.S. Clark, *Biochemical Engineering*. Marcel Dekker: New York. 1997.
37. Aunins, J.G. and H.J. Henzler, Aeration in cell culture bioreactors, in *Biotechnology: A Multi-Volume Comprehensive Treatise*, G. Stephanopoulos, Ed. Weinheim: New York. 1993, pp. 219–281.
38. Chang, J.Y. and J.R. Schwartz, Single-step solubilization and folding of IGF-1 aggregates from *Escherichia coli*, in *Protein Folding: In Vivo and In Vitro*, J.L. Cleland, Ed. ACS Symposium Series: Washington, D.C. 1993, pp. 178–188.
39. Cowgill, C., *Personal observation.*
40. Takage, T. and T. Isemure, Accelerating effect of copper ion on the reactivation of reduced taka-amylase: A through catalysis of the oxidation of sulfhydryl groups. *J. Biochem.*, 1964; **56**: 344–350.
41. Stanbury, P.F., A. Whitaker, and S.J. Hall, *Principles of Fermentation Technology*. Butterworth/Heinemann Press: Oxford, UK. 1995.
42. Shuler, M.L. and F. Kargi, Bioprocess engineering basic concepts, in *Bioprocess Engineering: Basic Concepts*. Prentice Hall: Englewood Cliffs, NJ. 1992, p. 280.
43. Linek, V., T. Moucha, M. Dousova, and J. Sinkule, Measurement of $k_L a$ by dynamic pressure method in pilot-plant fermentor. *Biotechnol. Bioeng.*, 1994; **43**: 477–482.
44. Spier, R.E. and B. Griffiths, An examination of the data and concepts germane to the oxygenation of cultured animal cells. *Dev. Biol. Stand.*, 1983; **55**: 81–92.
45. Johnson, M., G. Andre, C. Chavarie, and J. Archambault, Oxygen transfer rates in a mammalian cell culture bioreactor equipped with a cell-lift impeller. *Biotechnol. Bioeng.*, 1990; **35**: 43–49.
46. Kachur, A.V., C.J. Koch, and J.E. Biaglow, Mechanism of copper-catalyzed oxidation of glutathione. *Free Rad. Res.*, 1998; **28**: 259–269.
47. Atkinson, B. and F. Mavituna, *Biochemical Engineering and Biotechnology Handbook*. Macmillan: London. 1983, pp. 773, 784.
48. Calik, P., P. Yilgor, P. Ayhan, and A.S. Demir, Oxygen transfer effects on recombinant benzaldehyde lyase production. *Chem. Eng. Sci.*, 2004; **59**: 5075–5083.
49. Aunins, J.G., B.A.J. Woodson, T.K. Hale, and D.I.C. Wang, Effects of paddle impeller geometry on power input and mass transfer in small-scale animal cell culture vessels. *Biotechnol. Bioeng.*, 1989; **34**: 1127–1132.
50. St. John, R.J., J.F. Carpenter, and T.W. Randolph, High pressure fosters protein refolding from aggregates at high concentrations. *Proc. Natl Acad. Sci. USA*, 1999; **96**: 13029–13033.
51. Webb, J.N., S.D. Webb, J.L. Cleland, J.F. Carpenter, and T.W. Randolph, Partial molar volume, surface area, and hydration changes for equilibrium unfolding and formation of aggregation transition state: High-pressure and cosolute studies on recombinant human IFN-gamma. *Proc. Natl Acad. Sci. USA*, 2001; **98**: 7259–7264.

52. St John, R.J., J.F. Carpenter, and T.W. Randolph, High-pressure refolding of disulfide-cross-linked lysozyme aggregates: Thermodynamics and optimization. *Biotechnol. Prog.*, 2002; **18**: 565–571.
53. Seefeldt, M.B., J. Ouyang, W.A. Froland, J.F. Carpenter, and T.W. Randolph, High-pressure refolding of bikunin: Efficacy and thermodynamics. *Protein Sci.*, 2004; **13**: 2639–2650.

5 Bulk Protein Crystallization—Principles and Methods

Mark R. Etzel

CONTENTS

5.1 INTRODUCTION

Crystallization is a powerful method of purifying proteins. For many decades, industry has used crystallization to purify insulin for treatment of diabetes, and proteases for use in laundry detergent. However, central focus has been on crystallization for the production of diffraction-quality single crystals for protein structure determination by crystallography. The demands on crystals are quite different for structure determination vs. purification during manufacture. Bulk protein crystallization has played a cameo role in the epic of protein crystallography research. Yet, much can be learned from protein crystallography and applied to bulk protein crystallization. High throughput screening techniques, statistical design-of-experiments approaches, and thermodynamic theories for protein solubility, all developed for protein crystallography research are also beneficial to the development of bulk protein crystallization processes. The time has come for rapid growth in the use of bulk crystallization for proteins. Purification is only one method of implementation. Formulation is another. In formulation, the increased stability, higher concentration doses, controlled release dosages, and potential for administration of doses by subcutaneous injection rather than intravenous infusion are all driving increased interest in bulk protein crystallization.

The purpose of this chapter is to set the stage for researchers new to the field of bulk protein crystallization. The fundamental principles, common experimental methods, and classic industrial examples are described with the beginner in mind. It is hoped that this chapter will facilitate increased usage of crystallization in the manufacture of proteins for biopharmaceutical applications.

5.2 PRINCIPLES OF PROTEIN CRYSTALLIZATION

Proteins crystallize from supersaturated solutions. This makes knowledge of protein solubility and the protein phase diagram essential to understanding the protein crystallization process. As stated by Price [1]: "Trying to develop or troubleshoot a solution crystallization process without knowledge of the solubility curve and metastable zone width is akin to hiking in the wilderness without a map or compass." Nevertheless, although hundreds of proteins have been crystallized, phase diagrams have only been determined in a few cases [2].

Phase diagrams can be divided into five regions (Figure 5.1) (1) soluble region, (2) metastable region, (3) secondary nucleation region, (4) primary nucleation region, and (5) precipitation region. In the soluble region, crystallization does not occur, because the protein concentration falls below the solubility curve. The solution is not supersaturated. In the metastable region, an existing seed crystal can grow, but new crystals cannot form. In the secondary nucleation region, new crystals form from parent crystals by either breakage, attrition from collisions, or shedding of ordered surface layers by fluid shear. In the primary nucleation region, new crystals form spontaneously from a previously crystal-free solution. In the precipitation region, proteins aggregate to form amorphous precipitates. This region is to be avoided in a crystallization process. Finding conditions that maximize the metastable region is an important goal in the development of a protein crystallization process.

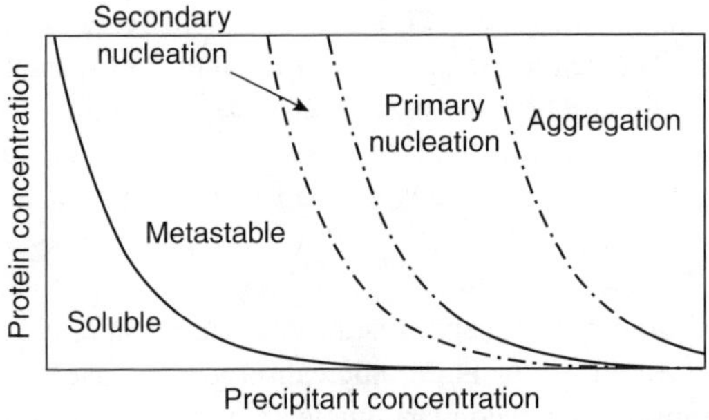

FIGURE 5.1 Phase diagram for protein crystallization.

Addition of a precipitant, and adjustment of the pH, ionic strength, and temperature of the solution alter protein solubility. Several methods to measure and calculate protein solubility are available. For example, Pusey and Munson [3] developed a miniature column technique for the rapid determination of protein solubility diagrams. They used this technique to measure the solubility of lysozyme as a function of pH, temperature, and sodium chloride concentration. Judge et al. [4] measured the solubility of ovalbumin at 30°C as a function of ammonium sulfate concentration and pH using stirred crystal slurries sealed in glass bottles.

Theories to calculate protein solubility are emerging. For example, the Gibbs–Duhem equation was used to analyze lysozyme solubility data [5]. Other workers used a solubility product to represent exchange of protein between the solution and crystal phases, and the UNIQUAC model to calculate solution activity coefficients [6]. In a landmark publication, George and Wilson [7] proposed the second virial coefficient, B_{22} (mmol-ml/g^2), as predictor of conditions for protein crystallization, and defined a "crystallization slot" of about $-0.8 \leq B_{22} \leq -0.1$ for promotion of protein crystallization. Positive values of B_{22} corresponded to providing a "good solvent" for the protein thus preventing crystallization, and more negative values of B_{22} corresponded to protein aggregation rather than crystallization. Protein solubility and B_{22} are not independent parameters: protein solubility increases nonlinearly with increasing B_{22}, but the relationship is in no way simple [8]. Other than the traditional method of static laser light scattering for measurement of B_{22}, self-interaction chromatography is a new method of measurement [9–11]. This new method was extended to lower protein surface coverage on the stationary phase of chromatographic supports, allowing lower mobile phase concentrations to be used, and to account for multibody interactions [12].

Supersaturation is the driving force for crystallization. Controlling the driving force for crystallization is important, because increasing the driving force both increases the rate of growth of the crystals, and decreases crystal size and purity. The rate of nucleation (B^0) is proportional to the supersaturation driving force raised to a power [13]:

$$B^0 = \frac{dN}{dt} = k_n \left(\frac{c - s}{s} \right)^a \tag{5.1}$$

where N is the number of crystals, c is the concentration of protein in solution, s is the protein solubility, k_n is the nucleation growth rate constant, and a is the nucleation order. For secondary nucleation, k_n increases with increasing concentration of the crystals in the solution.

The rate of crystal growth is also proportional to supersaturation [14]:

$$\frac{dl}{dt} = k_g \left(\frac{c - s}{s} \right)^g \tag{5.2}$$

where l is the characteristic length of the crystal, k_g is the linear growth rate constant, and g is the growth order.

To change the crystal-size distribution requires changing the relative nucleation and growth rates. This may be accomplished by adjusting the supersaturation over time. Generally, nucleation rates increase more rapidly than growth rates with increasing supersaturation ($a > g$). The growth order for primary nucleation is typically much higher than for secondary nucleation or crystal growth. In addition, the degree of supersaturation required for primary nucleation is typically much higher than for secondary nucleation, which in turn is much higher than the threshold for the metastable region. For example, the primary nucleation threshold for ovalbumin is at a supersaturation ratio ($S = c/s$) of $S_{primary} > 50$, and the corresponding value for secondary nucleation is $S_{secondary} > 20$ [15]. The metastable region falls in the range $1 < S_{metastable} < 20$, wherein crystals already present grow without the formation of any new crystals by primary or secondary nucleation. Therefore, operation at low supersaturation favors formation of large crystals, whereas high degrees of supersaturation, and operation close to the upper limit of the metastable region, favors formation of many nuclei, producing many small crystals.

Laser light scattering and digital microscopy are often used to measure the crystal-size distribution (CSD). Laser light scattering measures the volume equivalent sphere diameter. Changes in crystal shape cannot be observed. Digital microscopy can be used to measure a characteristic length of the crystal and the shape factors such as the ratio of the crystal dimensions. To convert the characteristic dimension to the volume equivalent sphere size requires the use of shape factors. Analysis of the CSD using distribution functions such as the normal distribution function or the log-normal distribution function can often provide an empirical representation of the data [16]. The mean and coefficient of variation of the distribution are perhaps the most valuable measures of the properties of the CSD.

Few data exist for the CSD of protein crystallization experiments. Ovalbumin crystals grown in the metastable region exhibit negligible secondary nucleation, and the coefficient of variation is constant as growth proceeds [15]. This would be the case if the crystal number was constant during crystallization and there was no dispersion in the crystal growth rate, that is, identical crystal faces grew at a single rate. In addition, Judge found that the crystal growth rate was not size-dependent, that is, k_g in Equation 5.2 was constant, and did not depend on the crystal size. It remains to be shown whether these observations are valid for other protein crystallization systems.

5.3 PRACTICAL CONSIDERATIONS IN DEVELOPING A PROTEIN CRYSTALLIZATION PROCESS

Implementation of a crystallization step into a purification process of a recombinant protein is a strategic issue. Placement further downstream provides a more pure and controlled feed solution to the crystallization process, and a more predictable outcome. However, placement earlier in the process provides the most advantages, because crystallization both concentrates and purifies the feed solution. Both implementation approaches have been taken. Lipase crystals [17] and subtilisin crystals [18] have been produced from a fermentation broth that was simply clarified by centrifugation and concentrated by ultrafiltration and diafiltration prior to the crystallization process.

Implementation of a crystallization step further downstream in the process has been described for production of aprotinin [19]. Figure 5.2 contains a photomicrograph of the aprotinin crystals. They mention that the stability of protein crystals is advantageous, because (1) crystals can be stored for years without

FIGURE 5.2 Crystals of aprotinin. (From Peters J, Minuth T, and Schroder W. *Protein Expr. Purif.* 2005; 39:43–53. With permission.)

significant product degradation, and (2) a time delay between manufacture of the protein therapeutic and formulation (fill and finish) can be tolerated. Often manufacture and formulation are performed at separate locations. There are several advantages of protein crystals for the delivery of protein therapeutics (1) high-concentration doses at low viscosity, (2) controlled release formulations, and (3) administration by subcutaneous injection rather than intravenous infusion [20–21]. Monoclonal antibody (mAb) products such as Herceptin, Remicade, and Rituxan have been crystallized for this purpose [22]. Thus, protein crystallization can be implemented either closer to cell culture and fermentation or closer to purification and polishing, depending on the objective.

Traditionally, the focus of protein crystallization was first and foremost to produce diffraction quality single crystals for x-ray crystallography. There are very different demands on making crystals for diffraction studies vs. protein purification (Table 5.1). In crystallography, microcrystallization techniques such as the hanging-drop and sitting-drop vapor diffusion techniques use spontaneous nucleation and growth to make large, single, highly ordered crystals from pure solutions. Increasingly this is done in 96-well or larger microplates using high-throughput screening (HTS) kits (e.g., JBScreen HTS Classic I, Jena Bioscience, Jena, Germany and Crystal Screen HT, Hampton Research, Aliso Viejo, California, USA). Wells contain a wide variety of precipitants that have been successful in past protein crystallization studies (Table 5.2). Many of these precipitants are not practical for process-scale separation processes. For example, high concentrations of polyethylene glycol (4 to 30%), ammonium sulfate (0.2 to 3.5 *M*), lithium sulfate (1.5 *M*), or 2-methyl-2,4-pentanediol

TABLE 5.1
Success Criteria for Crystallography vs. Bioseparation Processes

Criterion	Crystals for Crystallography	Crystals for Processes
Precipitants	Free choice	Nontoxic, nonhazardous
Precipitant costs	No issue	Important
Process compatibility	Not important	Essential
Crystal size	Large is best (150–500 μm)	Small okay (10–20 μm)
Crystal lattice resolution	High resolution	Not important
Crystallization yield	Not important	Very important
Growth kinetics	Often slow (days to months)	Fast (hours to days)
Redissolution	Not necessary	Necessary
Scalability of conditions	Not important	Very important
Protein available for screening	Critical restriction	No restriction

Source: Adapted from Peters J, Minuth T, and Schroder W. *Protein Expr. Purif.* 2005; 39:43–53.

TABLE 5.2
Contents of a Typical Crystallography Screening Kit

Buffer Salts	Precipitant Salts		Glycols and Alcohols	Polyethyleneglycol (PEG)
Bicine	NaCl	Ammonium acetate	Glycerol	PEG 400
Sodium citrate	$MgCl_2$	Calcium acetate	Ethylene glycol	PEG 550
Sodium acetate	$CaCl_2$	Magnesium acetate	2-Methyl-2,4-pentanediol	PEG 1000
HEPES	KCl	Zinc acetate	Methanol	PEG 1500
Tris	$NiCl_2$	Ammonium formate	Ethanol	PEG 2000
MES	LiCl	Sodium formate	2-Propanol	PEG 3000
Imidazole	$(NH_4)_2SO_4$	Ammonium phosphate	tert-Butanol	PEG 4000
CHES	Li_2SO_4	Sodium tartrate		PEG 5000
	$MgSO_4$	Potassium tartrate		PEG 6000
	$ZnSO_4$			PEG 8000
				PEG 10000
				PEG 20000

(10 to 70%) cause problems with wastewater treatment. New developments in HTS kits are the use of data mining and automation [23–25].

When a statistical design-of-experiments (DoE) approach using only process-compatible precipitants was compared to the screening kit approach, the hit rate was ten times higher for the DoE approach [19]. Thus, the initial screening for suitable crystallization conditions can be implemented using HTS, but should keep in mind the process compatibility of the precipitants.

The other factors to consider are protein concentration, precipitant concentration, pH, and temperature. Protein concentration determines the degree of supersaturation as discussed above. Higher protein concentration increases the growth rate, which shortens the crystallization time. But, if protein concentration is too high, then too much nucleation or perhaps aggregation can occur. The same situation occurs for precipitant concentration, because that also determines the degree of supersaturation by lowering the protein solubility. Therefore, a certain window of operation for protein concentration and precipitant concentration should be determined. Regarding pH, screening kits typically examine from pH 4.6 to 9.0, but protein stability should define the pH range examined. On the subject of temperature, most proteins are crystallized at either 4°C or ambient temperature (~22°C). As shown in Figure 5.3, increasing the precipitant concentration and decreasing the temperature greatly lowers

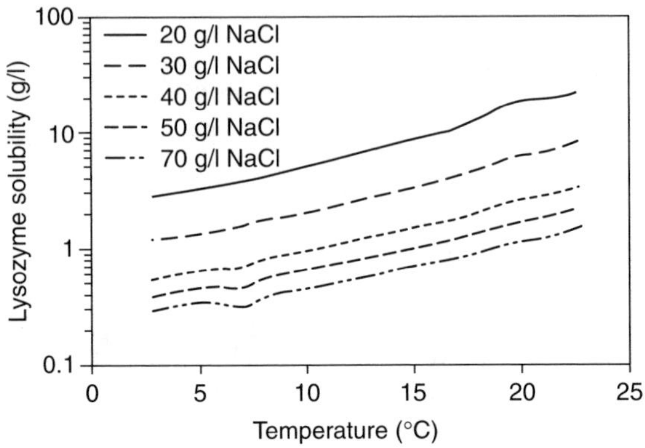

FIGURE 5.3 Lysozyme solubility vs. temperature at different sodium chloride concentrations. (From Forsythe EL, Judge RA, and Pusey ML. *J. Chem. Eng. Data* 1999; 44:637–640. With permission.)

lysozyme solubility [26]. Decreasing temperature alone from 22.6 to 4.3°C decreased solubility by 81% at 4% salt, and increasing salt concentration alone from 2 to 7%, decreased solubility by 93% at 22.6°C. Solubility decreases by 67-fold when the temperature is decreased and salt concentration is increased. Thus, the impact of precipitant concentration and temperature on supersaturation can be dramatic, and offer powerful tools for supersaturation control during crystallization [27].

Once screening experiments have been used to identify suitable crystallization conditions, and the solubility curve has been measured, then batch crystallization experiments may be attempted. Detailed procedures for isothermal batch protein crystallization have been published for lysozyme [14], ovalbumin [28], and lipase [17] among other proteins. Using lysozyme as an example, the experimental procedure is described below [14]:

1. Filtered solutions of protein (50 g/l lysozyme in 0.1 *M* sodium acetate, pH 4.8) and precipitant (80 g/l NaCl in 0.1 *M* sodium acetate) are prepared along with a seed crystal solution of 20 g/l lysozyme crystals in precipitant solution that is sheared to cause crystal breakage.
2. Protein solution is placed into a temperature-controlled vessel and a roughly equal volume of precipitant solution is added slowly while providing gentle agitation.
3. Seed crystal slurry is added and small samples taken periodically by syringe for protein concentration determination. The sample is

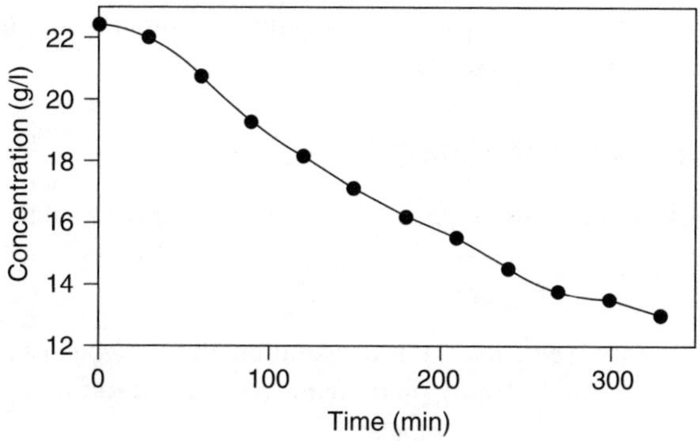

FIGURE 5.4 Desupersaturation curve for lysozyme crystallization at pH 4.8 and 22°C.

> filtered (0.22 μm syringe filter) to remove crystals and directly discharged into a tared volumetric flask. The sample is diluted to volume and absorbance at 280 nm measured to calculate protein concentration. Samples are taken from the vessel 24 h later for solubility determination.

The data from this experiment were used to construct the desupersaturation curve (Figure 5.4). If conditions are chosen where crystal growth falls in the metastable region, then primary and secondary nucleation can be neglected. In this case, crystal growth occurs on existing crystals only and the crystal growth rate is given by Equation 5.2. Two approaches can be used to determine the growth rate constant k_g (1) observation of the characteristic crystal length vs. time, and (2) mass balance calculations. Observation of the crystal length vs. time requires expensive instrumentation such as laser light scattering and digital microscopy, whereas the mass balance approach does not. Both methods will be described below.

5.3.1 Data Analysis Method 1

Dynamic laser light scattering measures the volume equivalent sphere diameter (L), which can be converted to the characteristic crystal length using the shape factor:

$$l = \sqrt[3]{\frac{\pi}{6\phi_V}}L \tag{5.3}$$

where ϕ_V is the volume shape factor ($\phi_{V,\text{cube}} = 1$, $\phi_{V,\text{sphere}} = \pi/6$). Digital microscopy measures the characteristic crystal length directly. By plotting

$\ln[\mathrm{d}l/\mathrm{d}t]$ vs. $\ln[(c - s)/s]$, the slope gives the rate order g and the intercept gives the linear growth rate constant k_g [28].

5.3.2 Data Analysis Method 2

Mass balances on the solution phase can be used to calculate the crystal mass at each time:

$$M = M_S + V(c_0 - c) \tag{5.4}$$

where M_s is the initial seed mass, V is the solution mass of water (which remains constant during crystallization), and c_0 is the initial protein concentration. Defining a dimensionless crystal mass:

$$x = \frac{M - M_S}{M_S} \tag{5.5}$$

and integration of the growth rate equation yields [14]:

$$\int_0^x \frac{\mathrm{d}x}{(x+1)^{2/3}(x_\infty - x)^g} = k_{\mathrm{rel}} \left(\frac{M_s}{sV}\right)^g t \tag{5.6}$$

where x_∞ is the dimensionless crystal mass at equilibrium, and k_{rel} is the relative mass deposition rate constant, which can be converted to the linear growth rate constant $[k_g = (l_s/3)k_{\mathrm{rel}}]$.

To use Equation 5.6, the desupersaturation curve is measured, and concentrations converted to crystal mass using Equation 5.4, and then to dimensionless crystal mass using Equation 5.5. The integral solution for the LHS of Equation 5.6 is then calculated using the values for the dimensionless crystal mass and plotted vs. time (Figure 5.5). An example calculation is shown in Table 5.3. Using this method, the LHS of Equation 5.6 is computed using straightforward measurements of mass and absorbance to obtain the values of M_s, V, c_0, c, and s, and the RHS term is determined from the slope of the plot. The rate order (g) can be found from plotting the LHS of Equation 5.6 for different values of g and choosing the value giving the best fit [29]. For lysozyme, $g = 2$ gave the best fit (data not shown), and matched values found in the literature [13].

Use of k_{rel} obviates the need to characterize the seed crystal size prior to data analysis. This is because only the slope, s, V, and M_s are required to determine k_{rel}. Thus, the need for expensive instrumentation for measurement of the crystal-size distribution by laser light scattering or digital microscopy is avoided by using k_{rel} for data analysis, and only instruments such as an analytical balance for solution preparation and a plate reader or spectrophotometer for measurement of the desupersaturation curve are required. Yet k_{rel} can be

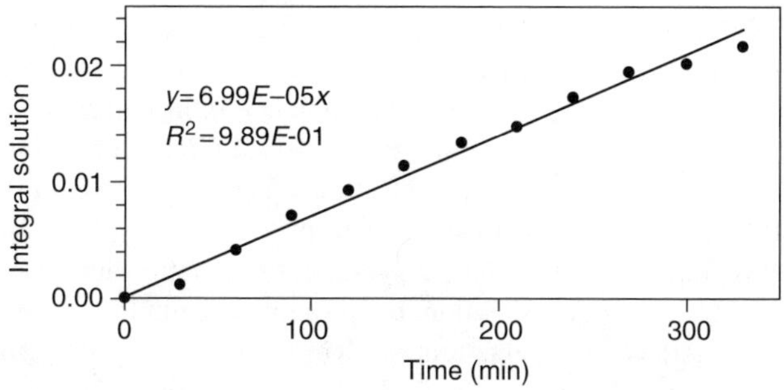

FIGURE 5.5 Plot of the LHS of Equation 5.6 vs. time for lysozyme crystallization.

TABLE 5.3
Example Calculation for Lysozyme Crystallization (s = 5.19 g/l, M_∞ = 273 mg, x_∞ = 17.4, V = 15 ml)

Time (min)	c (g/l)	M (mg)	x	Integral Solution
0	22.4	14.8	0.000	0.00000
30	22.0	20.8	0.404	0.00121
60	20.8	39.6	1.67	0.00418
90	19.3	61.8	3.17	0.00713
120	18.2	78.2	4.27	0.00923
150	17.1	94.3	5.36	0.0113
180	16.2	108	6.28	0.0132
210	15.5	118	6.99	0.0148
240	14.5	133	8.00	0.0173
270	13.8	145	8.77	0.0194
300	13.5	149	9.03	0.0202
330	13.0	156	9.51	0.0217

$k_{\text{rel}} = \text{slope} \times (sV/M_s)^g = 6.99 \times 10^{-5}\ \text{min}^{-1} \times (5.19\ \text{mg/ml} \times 15\ \text{ml}/14.8\ \text{mg})^2 = 1.93 \times 10^{-3}\text{min}^{-1}$
$k_g = (l_s/3)k_{\text{rel}} \approx (10\ \mu\text{m}/3) \times 1.93 \times 10^{-3}\ \text{min}^{-1} = 0.4\ \mu\text{m/h}$

used directly to assess the effects of changes in experimental conditions such as pH, initial concentration, temperature, and salt concentration, because k_{rel} in Equation 5.6 remains constant between experiments when seeds are taken from the same sample.

Scale up of the crystallization process can be facilitated by use of Equation 5.6 to examine the impact of changes in initial protein concentration, solution volume, solubility, precipitant concentration, seed crystal mass, seed crystal size, pH, temperature, and time. Maintaining complete mixing during scale up is another key factor because incomplete mixing can create pockets of nonuniform precipitant concentration or temperature that trigger locally elevated supersaturation and unwanted nucleation, and result in variations in crystal size, crystal shape, or batch-to-batch variations.

It is advantageous to maintain supersaturation control during crystallization [1]. During batch crystallization, the protein concentration of the solution falls thereby decreasing supersaturation. Thus, the driving force for crystallization is highest at time zero, and falls to zero as crystallization proceeds. This can lead to uncontrolled crystallization where nucleation and growth rates are fast initially, producing many fine, rapidly growing crystals, and crystal growth rates slow down to near zero at the end, decreasing productivity. Simultaneous decrease of protein solubility is required to maintain a constant supersaturation. Reducing the solubility over time has the added advantage of increasing yield [$\sim 1 - (s/c_0)$]. Most commercial crystallizers for sugars use programmed cooling to reduce solubility over time. This is called the cooling curve. Crystallizers for small organic molecules such as pharmaceuticals often use programmed precipitant addition to reduce solubility over time. Supersaturation control for protein crystallization is rare. A temperature-control strategy for growth of lysozyme crystals has been proposed, but only for production of diffraction-quality single crystals for x-ray crystallography [17]. Supersaturation control by addition of precipitant may be easier to manage for protein crystallization than temperature control, because protein solubility typically varies with precipitant concentration in a more predictable fashion than it does with temperature.

One other approach to supersaturation control is to add fresh protein to the crystallizer as the supernatant concentration falls as a result of crystal formation. This approach maintains supersaturation by holding the solution concentration and the solubility constant rather than decreasing solubility as the solution concentration falls. In both cases, the supersaturation ratio ($S = c/s$) is controlled to a set point value. An approach approximating this strategy has been described wherein, after subtilisin crystallization for 25 h, fresh feed solution and concentrated precipitant solution were added continuously to a batch crystallizer [30].

5.4 APPLICATIONS

5.4.1 Rubisco

Ribulose 1,5-biphosphate carboxylase/oxygenase (rubisco) is perhaps the most abundant protein in the world comprising 50 to 60% of total soluble protein in leaf extract [31]. Pursuit of leaf proteins for human consumption stems from the worldwide need for alternative sources of protein. Leaf protein extract by itself is not palatable, but the purification and concentration resulting from crystallization provides a route to human consumption of rubisco. An improved method was developed to crystallize rubisco from leaf extract without desalting or use of expensive buffers (Figure 5.6) [32]. Leaves (400 g) were homogenized in a Waring blender after addition of 2% sodium metabisulfite solution (200 ml): an inexpensive reducing agent that is generally recognized as safe (GRAS) for food applications. The homogenate was filtered using cheesecloth, adjusted to pH 5.6, and centrifuged to form a pellet consisting of residual plant insoluble material, and a clear brown supernatant containing the rubisco. After refrigeration for 24 to 48 h, crystals of rubisco that were formed were removed from the supernatant by centrifugation, leaving the supernatant (aka fraction-2 protein)

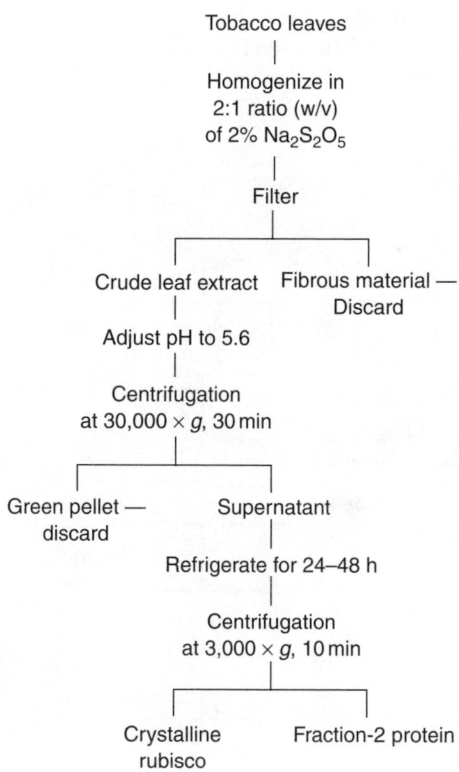

FIGURE 5.6 Flow diagram for recovery of rubisco from plant leaves using crystallization. (Reproduced from Kwanyuen P and Allina SM. *J. Proteome Res.* 2002, 1, 471–473. Copyright 2002 *Am. Chem. Soc.* With permission.)

quantitatively free of rubisco. In conclusion, this simple procedure enabled the purification and concentration of a single protein from a crude solution using crystallization.

5.4.2 Subtilisin

Laundry detergent contains proteases such as subtilisin to enhance cleaning action. Genencor International manufactures subtilisin and holds patents on methods for its crystallization [18,30]. A schematic diagram of the process is shown in Figure 5.7. Fermentation of *Bacillus subtilis* followed by cell separation yields a clear solution containing subtilisin. This solution is concentrated by ultrafiltration to 45 to 52 g/l subtilisin, adjusted to pH 4.8 to 5.4, and 15 to 40 g/l of either sodium chloride or sodium sulfate added as the precipitant. Seed crystals are added and the solution is held at 22 to 30°C for about 5 to 24 h to allow for crystallization. Raising the temperature increases the rate of crystal growth, shortening the time for crystallization from days to as little as 5 h. Nevertheless,

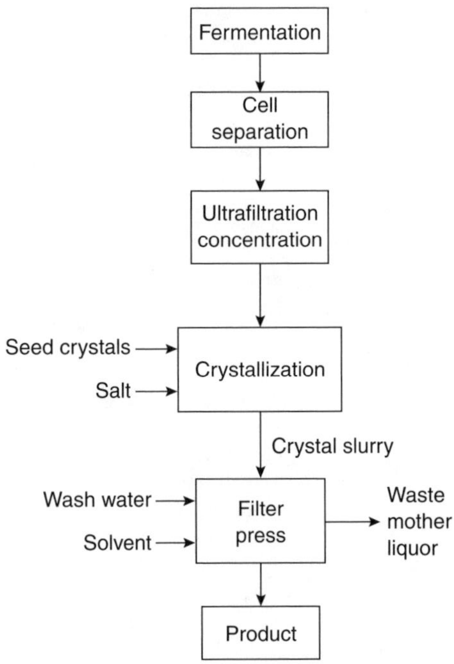

FIGURE 5.7 Flow diagram for recovery of subtilisin from fermentation broth using crystallization. (From Becker T and Lawlis VB. United States Patent 5,041,377, August 20, 1991.)

for subtilisin and rubisco (above) the protein crystallization process can take many hours, but in both cases can be accomplished using crude solutions and inexpensive nontoxic precipitants.

5.4.3 Aprotinin

Blood loss during cardiopulmonary bypass surgery can be cut in half using the recombinant protease inhibitor aprotinin. Implementation of a crystallization step into the manufacturing process for aprotinin has been described [19]. A flow diagram of the process is shown in Figure 5.8. The crystallization step was conducted at pH 4.8 and 20 to 25°C using a feed solution containing 5.3 g/l aprotinin and 50 m*M* NaCl, and took 72 h. Yield was 85%. Increasing pH

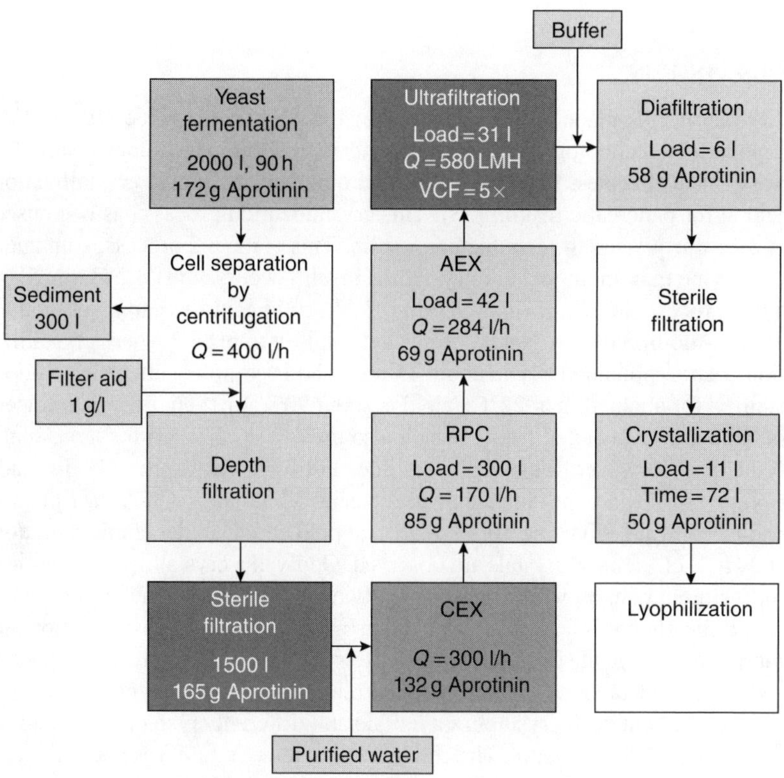

FIGURE 5.8 Flow diagram for aprotinin production. (Abbreviations: Q = flow rate; CEX = cation exchange column; RPC = reversed phase column; AEX = anion exchange column; LMH = l/m^2-h; VCF = volume concentration factor.)

increased crystal quality, but decreased yield. Increasing NaCl concentration decreased yield and crystal quality, but a small amount of NaCl was required to obtain any crystals. Decreasing or increasing the aprotinin concentration reduced yield and crystal quality. Thus, the conditions found represented an optimum. The crystallization step can be performed as a final step to improve product stability during formulation and storage, or on the RPC pool. Crystallization from the RPC pool was successfully scaled-up and may be used as a hold step prior to completion of the remaining purification steps. This work illustrated the process-scale implementation of a protein crystallization step for an active pharmaceutical ingredient. In addition, it showed the advantages of screening for crystallization conditions using a design of experiments approach with only ingredients that are compatible with process-scale manufacture of pharmaceutical ingredients, rather than screening using kits developed for crystallography.

5.4.4 Insulin

In 1982, Eli Lilly made history by launching the world's first successful product of modern biotechnology for human healthcare: recombinant human insulin for treatment of diabetes. In 1969, Lilly filed a patent on a novel crystallization method for pancreatic insulin [33]. This crystallization process has been used for over thirty years to manufacture insulin. This is the 8.2 process, so named because the maximum yield of crystalline insulin occurs at pH 8.2 (Table 5.4). In this process, insulin is dissolved in 0.5 *M* acetic acid to yield a solution at pH 3.6. Addition of 1 *M* NaOH brings the solution to pH 8.2 where crystallization occurs spontaneously in about 15 min, and is complete after the solution is stirred for about 18 h at 22°C. Yield is about 90%. In 1996, Lilly introduced fast-acting insulin called lispro, which also uses crystallization for large-scale production. However, the 8.2 process does not work for lispro [34]. Instead, a solution of 20 g/l lispro in 0.75 *M* acetic acid, 37.5 m*M* NaCl, 23 μM phenol is adjusted to pH 9.0 using 10% NaOH and held at 5°C under gentle agitation for 24 h. Well-defined crystals are observed. Oddly, the crystallization does not work without phenol, which was originally added to prevent bacterial growth. Even if the phenol is added, after pH adjustment and before crystallization, an amorphous precipitate forms along with a few crystals. If other preservatives are used such as meta-cresol, resorcinol, and methyl paraben, then crystallization is successful, but each crystal has a unique and different crystal habit, none of which match that for phenol. The crystallization temperature was not critical as tested in the range of 4 to 26°C. However, an optimum pH of 9.0 was observed with no crystals forming at slightly higher (pH 10.0) and lower (pH 8.2) values. These two examples are classic illustrations of the successful implementation of protein crystallization at process-scale in the biopharmaceutical industry.

TABLE 5.4
Effect of pH on Crystallization Time and Yield for Insulin Recovery from Pancreas Extract Using the 8.2 Process

pH	Time Required for Maximum Crystallization	Insulin Per Pound of Pancreas (mg)
7.0	2 days	11.3
7.4	1 day	32.8
7.8	75 min	56.5
8.2	5 min	46.9
8.6	5–10 min	43.7
9.0	1 day	41.7
9.4	3 days	32.5
9.8	4–5 days	46.8
10.2	—	—

Source: From Jackson RL. U.S. Patent 3,719,655, March 6, 1973.

5.5 THE FUTURE

Bulk protein crystallization is emerging as a new and upcoming unit operation in the manufacture of biopharmaceutical products. It is not without its problems. To date, proteins that have been successful in bulk crystallization have been stable and easy to crystallize. Expanding the method to more applications will require finding conditions where proteins crystallize reproducibly using pharmaceutical-grade buffers and precipitants. Proteins that are easily denatured by pH variation, changes in temperature, addition of precipitants, and agitation are more difficult to crystallize in bulk. Furthermore, bulk protein crystallization typically takes longer than the crystallization of sugars, amino acids, small molecule organics drugs, and other products. Time scales of hours to days are required for proteins, compared to minutes to hours for small molecules. During this time, proteins may be subject to attack by proteases or microbes if inhibitors are not added. Furthermore, protein crystals themselves are fragile; agitation only sufficient to suspend crystals in solution should be used. Despite these caveats, success stories are rapidly emerging. New protein crystallization techniques and theories spawned by the use of protein crystallography in the structural genomics revolution has built the base for rapid progress in bulk protein crystallization. Successful implementation in the manufacture of recombinant human insulin, subtilisin, and other protein therapeutics has also paved the way for rapid progress in the future. The recent success in the crystallization of monoclonal antibodies for subcutaneous delivery may place bulk protein crystallization at center stage in the biopharmaceutical industry and lead to a larger role in protein purification operations in the future.

REFERENCES

1. Price CJ. Take some solid steps to improve crystallization. *Chem. Eng. Prog.* 1997; 93:34–43.
2. Klyushnichenko V. Protein crystallization: From HTS to kilogram-scale. *Curr. Opin. Drug Discov. Devel.* 2003; 6:848–854.
3. Pusey ML and Munson S. Micro-apparatus for rapid determinations of protein solubilities. *J. Crystal Growth* 1991; 113:385–389.
4. Judge RA, Johns MR, and White ET. Solubility of ovalbumin in ammonium sulfate solutions. *J. Chem. Eng. Data* 1996; 41:422–424.
5. Knezic D, Zaccaro J, and Myerson AS. Thermodynamic properties of supersaturated protein solutions. *Crystal Growth Des.* 2004; 4:199–208.
6. Agena SM, Pusey ML, and Bogle IDL. Protein solubility modeling. *Biotechnol. Bioeng.* 1999; 64:144–150.
7. George A and Wilson WW. Predicting protein crystallization from a dilute-solution property. *Acta Crystallogr.* 1994; 50:361–365.
8. Haas C, Drenth J, and Wilson WW. Relation between the solubility of proteins in aqueous solutions and the second virial coefficient of the solution. *J. Phys. Chem.* 1999; 103:2808–2811.
9. Henry CS, Payne RW, Valente JJ, Wilson WW, and Manning MC. Self-interaction chromatography for screening. *Genet. Eng. News* 2005; 25:62–63.
10. Garcia CD, Hadley DJ, Wilson WW, and Henry CS. Measuring protein interactions by microchip self-interaction chromatography. *Biotechnol. Prog.* 2003; 19:1006–1010.
11. Tessier PM and Lenhoff AM. Measurements of protein self-association as a guide to crystallization. *Curr. Opin. Biotechnol.* 2003; 14:512–516.
12. Teske CA, Blanch HW, and Prausnitz JN. Measurement of lysozyme–lysozyme interactions with quantitative affinity chromatography. *J. Phys. Chem.* 2004; 108:7437–7444.
13. Saikumar MV, Glatz CE, and Larson MA. Lysozyme crystal growth and nucleation kinetics. *J. Crystal Growth* 1998; 187:277–288.
14. Carbone MN and Etzel MR. Seeded isothermal batch crystallization of lysozyme. *Biotechnol. Bioeng.* 2006; 93:1221–1224.
15. Judge RA. Investigating the bulk crystallization of proteins. 1995. Ph.D. thesis, University of Queensland, Australia.
16. Randolph AD and Larson MA. *Theory of Particulate Processes: Analysis and Techniques of Continuous Crystallization*, 2nd ed. 1988. Academic Press: New York.

17. Jacobsen C, Garside J, and Hoare M. Nucleation and growth of microbial lipase crystals from clarified concentrated fermentation broths. *Biotechnol. Bioeng.* 1998; 57:666–675.
18. Becker T and Lawlis VB. Subtilisin crystallization process. U.S. Patent 5,041,377, August 20, 1991.
19. Peters J, Minuth T, and Schroder W. Implementation of a crystallization step into the purification process of a recombinant protein. *Protein Expr. Purif.* 2005; 39:43–53.
20. Basu SK, Govardhan CP, Jung CW, and Margolin AL. Protein crystals for the delivery of biopharmaceuticals. *Expert Opin. Biol. Ther.* 2004; 4: 301–317.
21. Pechenov S, Shenoy B, Yang MX, Basu SK, and Margolin AL. Injectable controlled release formulations incorporating protein crystals. *J. Control. Release* 2004; 96:149–158.
22. Yang MX, Shenoy B, Disttler M, Patel R, McGrath M, Pechenov S, and Margolin AL. Crystalline monoclonal antibodies for subcutaneous delivery. *Proc. Natl Acad. Sci. USA* 2003; 100:6934–6939.
23. Page R and Stevens RC. Crystallization data mining in structural genomics: Using positive and negative results to optimize protein crystallization screens. *Methods* 2004; 34:373–389.
24. Mayo CJ, Diprose JM, Walter TS, Berry IM, Wilson J, Owens RJ, Jones EY, Harlos K, Stuart DI, and Esnouf RM. Benefits of automated crystallization plate tracking, imaging, and analysis. *Structure* 2005; 13:175–182.
25. Walter TS, Diprose JM, Mayo CJ, Siebold C, Pickford MG, Carter L, Sutton GC, Berrow NS, Brown J, Berry IM, Stewart-Jones GBE, Grimes JM, Stammers DK, Esnouf RM, Jones EY, Owens RJ, Stuart DI, and Harlos K. A procedure for setting up high-throughput nanolitre crystallization experiments. Crystallization workflow for initial screening, automated storage, imaging and optimization. *Acta Crystallogr.* 2005; 61:651–657.
26. Forsythe EL, Judge RA, and Pusey ML. Tetragonal chicken egg white lysozyme solubility in sodium chloride solutions. *J. Chem. Eng. Data* 1999; 44: 637–640.
27. Schall CA, Riley JS, Li E, Arnold E, and Wiencek JM. Application of temperature control strategies to the growth of hen egg-white lysozyme crystals. *J. Crystal Growth* 1996; 165:299–307.
28. Judge RA, Johns MR, and White ET. Protein purification by bulk crystallization: The recovery of ovalbumin. *Biotechnol. Bioeng.* 1995; 48:316–323.
29. Carbone MN, Judge RA, and Etzel MR. Evaluation of a model for seeded isothermal batch protein crystallization. *Biotechnol. Bioeng.* 2005; 91:84–90.
30. Gros EH and Cunefare JL. Crystalline protease and method for producing same. U.S. Patent 6,207,437, March 27, 2001.
31. Johal S, Bourque DP, Smith WW, Suh SW, and Eisenberg D. Crystallization and characterization of ribulose 1,5-biphosphate carboxylase/oxygenase from eight plant species. *J. Biol. Chem.* 1980; 255:8873–8880.

32. Kwanyuen P, Allina SM, Weissinger AK, and Wilson RF. A new form of crystalline rubisco and the conversion to its common dodecahedral form. *J. Proteome Res.* 2002; 1:471–473.
33. Jackson RL. Process for the crystallization of the ammonium and alkali metal salts in insulin. U.S. Patent 3,719,655, March 6, 1973.
34. Baker JC and Roberts BM. Preparation of stable insulin analog crystals. U.S. Patent 5,597,893, January 28, 1997.

6 Modes of Preparative Chromatography

Abhinav A. Shukla and Yinges Yigzaw

CONTENTS

6.1 INTRODUCTION

Chromatographic stationary phases are offered in a staggering array of chemical diversity. One of the key means of organizing stationary phases has been on the basis of their functional groups that enable classification into broad classes of operating modes. While much has been written about each of these operational modes in the literature, this chapter aims at providing a readily accessible, single source of basic information that a new practitioner in bioseparations would find useful. The chapter provides concise information about the interaction mechanism in each mode of chromatography and practical considerations that are important to bear in mind on each of them. Not included are topics such as mass transport and kinetics considerations, the stationary phase morphology or large-scale column packing and operation. Since these considerations are not necessarily specific to the mode of interaction, they are not considered here. The classification of chromatographic modes used in this chapter is shown in Figure 6.1.

6.2 LINEAR AND NONLINEAR RETENTION IN CHROMATOGRAPHY

In linear chromatography the equilibrium concentrations of a component in the stationary and mobile phase are proportional, that is, the adsorption isotherms are straight lines. Retention is characterized by an absence of intersolute competition for binding sites on the column. Such conditions are common under very low loadings (sometimes called analytical loadings) on the column.

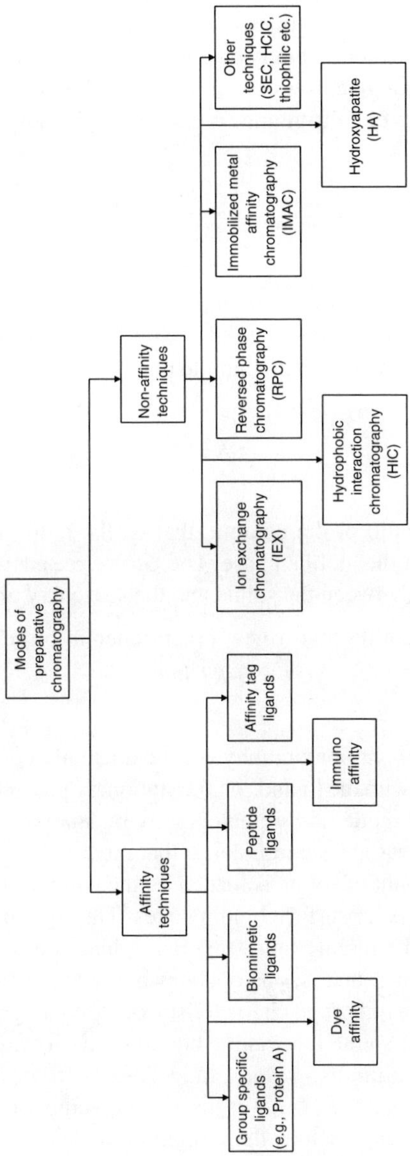

FIGURE 6.1 Classification of modes of preparative chromatography.

The retention of solutes under linear conditions is characterized by the dimensionless retention time (a.k.a isocratic retention factor). As the name suggests, this is a measure of the solute's retention on the column under isocratic conditions (i.e., constant inlet mobile-phase conditions).

k' can be measured experimentally by injecting a small amount of the solute onto the column under a given set of mobile-phase conditions and measuring the time it takes for it to elute from the column. k' is given by:

$$k' = \frac{t_r - t_0}{t_0} \tag{6.1}$$

where t_0 is the retention time for an unretained solute on the same column.

The retention factor is related to the equilibrium constant K for the distribution of the solute between the mobile phase and the stationary phase.

$$k' = K\phi \tag{6.2}$$

where ϕ is the phase ratio of the column, that is, the ratio of the stationary-phase volume to that of the mobile phase. The Gibbs free energy related to the strength of association between the solute and the stationary phase is given by:

$$\Delta G^0 = -RT \ln K \tag{6.3}$$

Thus linear retention in chromatography is fundamentally connected with the strength of association with the ligands on the stationary phase. Further insights can be gained for each mode of chromatography by studying the relationship of ΔG^0 to operational variables particular to that mode.

When a larger amount of solute is loaded on the column, solute molecules compete with each other for surface binding sites. The equilibrium concentrations of the solute in the mobile- and stationary-phase are no longer directly proportional to each other, that is, one operates in the nonlinear portion of the adsorption isotherm. An important characteristic of chromatographic stationary phases is the amount of solute that can be bound to the column under a given set of mobile-phase conditions. Several models exist for describing binding capacity under these conditions. Most of these are specific for a given mode of chromatography, but some, such as the Langmuir isotherm, can be generally applied across all modes of chromatography.

The Langmuir isotherm relates the concentration of solute that binds on the stationary phase (Q expressed in units of mass of protein bound per unit volume of the stationary phase) to the mobile-phase concentration of the same solute.

$$Q = \frac{Q_{\max} C}{1 + KC} \tag{6.4}$$

where K is the equilibrium constant for solute binding on the stationary phase, C is the mobile-phase solute concentration, and $Q_{\max}$ is the maximum concentration of solute that can be bound at equilibrium under the mobile-phase conditions the isotherm was measured in. The Langmuir model involves several assumptions [1] including monolayer coverage of the stationary-phase surface with the solute.

The Langmuir isotherm is the simplest model with wide applications for describing nonlinear adsorption onto chromatographic supports. It can capture the generally observed shape of protein adsorption isotherms very well. However, this model does not allow the extrapolation of measurements made under one set of mobile-phase conditions to other conditions. This predictive ability is often required and has led to the development of specific adsorption models for several modes of chromatography.

6.3 AFFINITY CHROMATOGRAPHY

Affinity chromatography refers to the use of an immobilized ligand that interacts specifically at a well-defined site on the desired biomolecule. This section deals solely with biospecific affinity chromatography, for example, the interaction of the natural ligand-binding site on a biomolecule with its ligand. Sometimes, chromatography techniques that involve fairly strong interactions with certain amino acid residues on the protein surface are also classified under affinity chromatography (e.g., immobilized metal ion affinity chromatography). Here, these techniques are considered separately since they do not involve binding to a single, well-defined site but instead to residues that are distributed over the protein surface.

Given the natural diversity of biomolecules and their ligands, it is no wonder that an almost infinite diversity of affinity chromatographic techniques exist. Most of these however, are employed for biomolecule isolation in small quantities. This chapter only considers techniques that have been employed for large-scale protein separations. The importance of the coupling chemistry employed, the base bead chemistry, and transport characteristics have been reviewed elsewhere [2,3] and will not be detailed here.

The energetics of biomolecule interactions with immobilized affinity ligands are often quite strong. This leads to a low K (Equation 6.2 and Equation 6.3) in the range of $\sim 10^{-3}$ to $10^{-9} M$ giving ΔG^0 for the association to be $\sim$10 to 30 kJ/mol. Such strengths of interaction cannot be achieved in nonaffinity modes of chromatography without simultaneous binding of the solute to multiple sites on the stationary phase.

6.3.1 Protein A Affinity Chromatography and Other Group Specific Natural Ligands

Protein A affinity chromatography is arguably the most widely utilized affinity technique for large-scale protein separations. This technique is discussed in detail in a separate chapter later in this book.

Synthetic protease inhibitors such as *m*- and *p*-aminobenzamidine which bind to the catalytic sites of trypsin family proteases have been used as affinity ligands for process-scale purification. These are employed in a batch or flow-through mode and can improve product stability by eliminating trypsin family proteases. *p*-amino benzamidine (pABA) affinity resins are available from GE Healthcare (formerly Amersham Pharmacia Biotech, Uppsala, Sweden) on a 90μ 4% cross-linked agarose base bead.

Heparin affinity columns have been employed for the purification of antithrombin III from blood plasma [4]. Commercial resins include Heparin Sepharose 6FF in a 90μ 6% cross-linked agarose bead (GE Healthcare) and Affigel Heparin (Biorad Laboratories, Hercules, CA). Lysine affinity columns have been employed for t-PA (tissue plasminogen activator) purification from both recombinant and plasma sources [5]. Lysine Sepharose 4B (90μ, 4% cross-linked agarose base bead) is available from GE Healthcare.

6.3.2 Dye Ligand Chromatography

Several dyes belonging to the triazine class have emerged as useful ligands for protein separations [2,6,7]. The most popular among them is Cibacron Blue F3G-A, which is thought to mimic the nucleotide binding sites of enzymes. This dye has found application in plasma protein fractionation and the recovery of albumin from Cohn Fraction IV [6,7]. This ligand is available commercially as Blue Sepharose 6FF™ on a 90μ cross-linked agarose bead from GE Healthcare. Procion Blue and Red triazine derivatives have been compared in the purification of α1-trypsinase inhibitor [8]. Procion Red H-E3B has been shown to possess specificity toward NADP-binding proteins. The reader is referred to Scopes [3] for a full history of the development of dyes as affinity ligands.

Dye ligands are often fairly cost-effective since these ligands can be synthesized chemically. A major disadvantage is the risk of ligand leaching and the unknown toxicity profile of new dye chemistries. The mechanism of binding interaction is also a matter of debate and may not be affinity-based with all but a few proteins. Instead, it is likely that a combination of ionic and hydrophobic interactions provide these resins with a mixed-mode functionality. Elution from

Cibacron Blue F3G-A is often effected by increasing salt concentration or decreasing pH, which might point to electrostatics being the dominant mechanism for many biomolecules. In addition, the presence of several phenyl rings provide an opportunity for hydrophobic interactions to occur. The use of commercially available dye ligand resins is largely restricted to cases where combinations of conventional nonaffinity resins do not provide adequate purification. A range of dye chemistries (including red, orange, yellow, green, and blue dyes) are available as a mimetic kit for screening from Sigma-Aldrich (St. Louis, MO).

6.3.3 Immunoaffinity Chromatography

Immunoaffinity chromatography utilizes the extremely high specificity and strength with which antibodies bind to their antigens. Typically, the immobilized antibody serves as a ligand for the purification of the antigen. While the high selectivity and the broad range of products that can be purified make immunoaffinity chromatography one of the most powerful chromatographic techniques, it is often difficult to elute the bound protein without harsh or denaturing conditions. Elution has been effected by low pH, high pH, organic solvents, or high concentrations of denaturing agents such as urea or guanidine hydrochloride [9]. There have been advances in screening for antibodies that release the product under mild, nondenaturing conditions [10]. However, since this technique requires the production of another protein (i.e., the antibody) at significant cost and effort, it is not generally practiced beyond the protein chemistry laboratory.

6.3.4 Biomimetic Ligands

Small molecule ligands that mimic the interaction of biomolecules with their natural ligands are termed as biomimetic ligands. These ligands have their origin in dye ligand chromatography [7]. Triazine serves as a stable and easily functionalized structure to enable the synthesis of large libraries of dye compounds. This has enabled the synthesis of large combinatorial libraries of compounds that can be screened for structures that can act as affinity ligands. In addition to combinatorial library screening, computer-aided design has also been employed to create designer dyes that can bind specifically to biomolecules [11–13]. Among other applications, this technique has been employed to generate high affinity ligands for alcohol dehydrogenase [11] and to mimic the binding pocket of Protein A to the Fc region of antibodies [14]. Clearly, this is a versatile technique that can be employed to create custom ligands for purifying virtually any biomolecule.

However, only a few of these novel structures are available commercially. The first biomimetic stationary phases to enter the market are MAbSorbent A1P and A2P (ProMetic Biosciences, Cambridge, U.K.) that are aimed at replacing Protein A affinity chromatography with small molecule ligands for monoclonal antibody purification. ProMetic also offers custom ligand generation services for designing small molecules as ligands for other biomolecules. These two resins, at least initially, have met with limited commercial success owing to poorer selectivity as compared to Protein A chromatography. This notwithstanding, biomimetic ligands may yet prove to be successful custom ligands for preparative purification of biomolecules.

6.3.5 Peptides as Ligands

Peptides comprised of the binding domains can serve as cheaper, smaller, and more stable substitutes for the natural ligands themselves in affinity chromatography [15]. Peptide sequences can be obtained in two ways — by identifying and engineering the binding domain to obtain a minimalistic sequence or in cases where a natural ligand might not exist to the biomolecule of interest, by screening a combinatorial library of peptide sequences. This latter technology has broadened the applicability of peptides as affinity ligands. Phage display techniques have allowed the rapid creation and screening of large libraries of peptides [16]. Amongst other companies, Dyax (Cambridge, MA) carries out contract screening work for identifying peptide ligands that can be employed in affinity separations [17]. Affibody AB (Bromma, Sweden) develops highly specific proteins that mimic antibodies for binding to a specific target. Immobilization techniques for peptide affinity tags have been reviewed elsewhere [18].

In addition to peptide ligands, specific oligonucleotide sequences called aptamers have also been employed as affinity ligands [19]. Once again, these can be readily screened using high throughput screening techniques and can be selected such that mild elution conditions (e.g., increase in salt concentration) are sufficient to elute the target protein.

6.3.6 Affinity Tag Ligands

A wide diversity of affinity tag procedures have been reported for proteins that do not exhibit strong affinity to a natural ligand. While most affinity tags are employed for protein isolation for characterization at the laboratory scale, some tags have made it into therapeutic proteins as well. Possibly, the most prominent example of affinity tags is that of fusion with the Fc portion of an antibody imparting strong and specific affinity toward Protein A ligands. At least two Fc fusion proteins have been approved for human therapeutic use (Enbrel from Amgen and Amevive from Biogen-Idec). Other examples of popular tags include a poly-His tag which imparts strong affinity to immobilized metal affinity columns (IMAC); β-galactosidase which binds to *p*-aminophenyl β-thiogalactoside, and maltose binding protein binding to amylose. Other tags that are in common use include the cellulose binding domain and a poly(arginine) tag that imparts a positive charge making cation exchange purification possible [20]. Affinity tails for protein recovery have been comprehensively reviewed elsewhere [21].

A key disadvantage of fusing an affinity tag on proteins of interest solely for ease of recovery are potentially undesirable *in vivo* characteristics of the tag. It is thus desirable to cleave the tag off after purification is complete. However, the use of proteases to do so is fraught with further complications including the risk of introducing another protein into the purification process and the requirement to clear both the tag and the protease before the final dosage form of the product is prepared. Proteases also bring the risk of nonspecific cleavage of the product of interest. A new technology in this area is that of using inteins as linkers between the product of interest and the affinity tag [22]. Inteins are peptides that can autocatalytically cleave themselves under appropriate conditions, for example, pH or temperature. This can provide an easy means of cleaving off the affinity tags without having to resort to the use of proteases. While this technique is still in its nascent stage, future applications to therapeutic proteins may be possible.

Affinity chromatography techniques have the huge advantage of binding specifically to the product of interest, and in many cases can enable isolation of the product of interest from fairly crude mixtures. These steps are usually employed as the capture chromatographic steps in the downstream process due to their high selectivity and ability to concentrate and separate the product quite readily. This also allows the clearance of any leached affinity ligand through subsequent polishing steps in the process. One of the main disadvantages of affinity chromatography is the high cost, due to the need for highly purified ligands. Thus, if shown to be successful, biomimetic ligands can have a significant market advantage. However, in many cases the specificity of the natural proteinaceous ligand cannot be matched by a small molecule.

Due to the high expense of these techniques, barring the routine use of Protein A chromatography for antibodies and Fc fusion proteins, affinity techniques are employed only when nonaffinity chromatographic steps do not provide the requisite purity, or do so at a significant yield trade-off. The scale of production of the final product needs to be borne in mind throughout this decision process. Affinity techniques that are not commercially established might face significant hurdles when implementation at large scale is considered.

6.4 NONAFFINITY MODES OF CHROMATOGRAPHY

These modes of chromatography are not based upon specific interactions of a domain of the target protein with a ligand; instead the interactions are to various types of amino acid residues distributed over the protein surface. Nonaffinity stationary phases are, in general, cheaper than affinity adsorbents. This, along with their applicability for all proteins has led to widespread application of these modes for preparative separations.

6.4.1 Ion Exchange Chromatography

Ion exchange (IEX) chromatography is arguably the most widely used and best understood mode of chromatographic adsorption. Ion exchangers exploit surface charges on the protein under a given pH and salt ion concentration and interact with the protein predominantly by electrostatic interactions. Ion exchangers can be positively charged (anion exchangers) or negatively charged (cation exchangers). Within these two classes, strong and weak subclasses exist, depending on the permanence of charge over a broad pH range. In general, strong ion exchangers can maintain their charge over a broader pH range than weak ion exchangers.

Often, the isoelectric point (pI) of a protein is used as a determinant of the charge on the protein surface. At pH < pI, a protein takes on a positive overall charge while at pH > pI it is predominantly negatively charged. However, it is important to recognize that proteins are ampholytes, possessing both positive and negative surface charges arising from the presence of acidic or basic amino acid residues. The pI only reflects the overall surface charge of the protein — patches of either charge still exist on the protein surface, and the protein usually interacts through these charged patches rather than through an averaged charge over its entire surface. Thus, it is still useful to screen both anion and cation exchangers for a given protein and to do so over a range of pHs. This allows the exploration of a range of conditions over which the charge on the protein surface and the charge on the surfaces of the contaminant proteins can vary, thus influencing the selectivity of the unit operation quite significantly. In fact, if the retention characteristics of the product allow it, one should screen conditions >pH 7 for cation exchange and <pH 7 for anion exchange since if the product species does bind it is likely that fewer contaminant proteins will do so under those conditions.

6.4.1.1 Modeling of Ion Exchange Chromatography

Ion exchange is based on the stoichiometric exchange of ions bound on the stationary-phase surface with the charged solute. When the solute binds to the stationary phase, it displaces a certain number of ions (equal to its characteristic charge) from the surface into solution. This exchange is given by the equation:

$$C_P + \nu Q_M \Leftrightarrow Q_P + \nu C_M \tag{6.5}$$

where ν is the characteristic charge, the subscript P denotes the protein, the subscript M denotes the mobile-phase counterion that binds to the stationary phase, Q denotes the concentration on the stationary phase, and C denotes the concentration in the mobile phase. The stoichiometric displacement model (SDM) [23] can describe linear retention of proteins on IEX by the equation:

$$\log k' = -\nu \log(C_M) + \text{const.} \tag{6.6}$$

where C_M is the concentration of the counterion in the mobile phase.

The steric mass action (SMA) model [24] extends the SDM formalism to nonlinear chromatography by accounting for the number of charged sites on the stationary-phase surface that are shielded by an adsorbed protein molecule, and hence unavailable for exchange if another protein molecule were to approach the surface. Analogous to the characteristic charge of a protein (ν), the steric factor (σ) depends not only on the protein molecule, but also on the mobile-phase conditions (i.e., pH) and stationary phase characteristics (e.g., ligand density).

Equation 6.7 provides an implicit isotherm equation for a protein molecule on IEX.

$$K = \left(\frac{Q}{C}\right)\left(\frac{C_M}{\Lambda - (\nu + \sigma)Q}\right)^{\nu} \tag{6.7}$$

where Q and C denote the concentrations of the protein on the stationary and mobile phase, respectively, and Λ is the ionic capacity of the stationary phase.

Under analytical conditions, from Equation 6.2 and Equation 6.6 one can obtain:

$$\log k' = \log(\phi K \Lambda^{\nu}) - \nu \log C_M \tag{6.8}$$

To determine SMA parameters for a given protein under a specific mobile-phase pH [25], isocratic experiments with small column loads are carried out over a range of mobile-phase salt concentrations. This enables the characteristic charge (ν) and the equilibrium constant K to be determined from Equation 6.8. Next, adsorption isotherms are measured over several different mobile-phase salt concentrations and simultaneously fit to Equation 6.7 to obtain σ. A significant advantage of the SMA formalism over the Langmuir equation is that once parameters are obtained for a given protein at a particular mobile-phase pH, Equation 6.7 can be employed to predict the isotherm at other mobile-phase salt concentrations.

The SDM/SMA formalisms provide a ready means of understanding and describing adsorption phenomena in IEX. However, they are based on the assumption that the protein interacts with point charges on the stationary-phase surface. In reality, retention in IEX is more complex and is due to the interaction of electrostatic fields of the various components of the system. The electric double layer and Donnan potential theories have been employed to develop a more fundamentally rigorous, but more complex description of IEX [26].

6.4.1.2 Resins for Ion Exchange

A large variety of preparative resins ($>50\ \mu$ particle size) are commercially available for IEX [27,28]. These differ from each other not only in terms of the strength of the functional groups (i.e., weak or strong) but also in terms of their functional group chemistry, ligand density, bead backbones, and transport properties into the beads. All of these properties contribute to quite a variation in the separation characteristics of these resins. Few ion exchangers operate solely on the basis of electrostatic interactions; a range of secondary interactions can influence selectivity quite significantly on these resins. Hence, it is recommended to screen several of these resins over a range of pH to identify resins with suitable capacity and selectivity.

Strong IEX resins retain their charge over a wide range of pH while weak resins do so over a narrower pH range. Typically strong exchangers have functional groups with very low pK (sulfopropyl in strong CEX) or very high pK (quaternary amino-type strong AEX). Table 6.1 lists some of the common IEX functional groups and their pKs.

TABLE 6.1
Functional Groups for Ion Exchange

Name	pK
Anion exchange chromatography	
Diethyl aminoethyl (DEAE)	9–9.5
Quaternary aminoethyl (Q)	>9.5
Dimethyl aminoethyl (DMAE)	9
Trimethyl aminoethyl (TMAE)	>13
Cation exchange chromatography	
Carboxymethyl (CM)	3–5
Phosphate	<2 and 6
Sulfonate (S)	2
Sulfoethyl (SE)	2
Sulfopropyl (SP)	<1

Table 6.2 lists some of the more established cation exchange (CEX) and anion exchange (AEX) resins, their manufacturers, the functional group, particle-size ranges, and information about the base matrix, were available. Cross-linked agarose based IEX resins have historically been amongst the first IEX resins to be marketed commercially by Pharmacia LKB (now part of GE Healthcare). These are available in a wide variety of functionalization chemistries. A recent addition to the family of resins based on cross-linked agarose has been the XL series of resins. In these resins, the pores of the agarose base matrix are filled with functionalized dextran chains creating a higher binding capacity. Polymeric matrices are also widely employed for IEX and include methacrylate-based chemistries, polystyrene, and other recent chemistries. For chemistries that are too hydrophobic, a hydrophilic polymer is often applied as a coating to reduce nonspecific interactions. Finally, the Hyper D series are composite stationary phases composed of a rigid macroporous support filled with a structurally weaker, but highly functionalized gel [29].

Beyond the basic bead morphologies, a variety of IEX resins have been developed with novel chemistries to either provide greater binding capacity or to improve mass transfer into the beads, allowing for higher operational flow rates. Tentacular supports (e.g., Fractogel and Fractoprep series from Merck KgGA) have polyelectrolyte chains grafted onto porous matrices. These also offer high binding capacities, since they are presumed to access more than just the immediate surface of proteins [30]. Perfusive supports have a network of larger macropores (6000 to 8000 Å) that allow convective flow (e.g., POROS resins). These connect to a network of smaller pores (500 to 1500 Å) into which the protein diffuses and binds. This has been shown to enable the use of higher flow rates leading to an increase in operational throughput [31]. The Unosphere resins from Biorad have also been shown to involve perfusive flow into the beads [32,33]. The development of agarose-based perfusive supports has also been reported [34].

6.4.1.3 Loading and Binding Capacity

Ion exchange media usually offer fairly high binding capacities for proteins (up to 100 mg/ml static binding capacities are common). Binding capacities are usually limited by the presence of other binding impurities or by the presence of an elevated salt concentration in the feed load. To maximize binding capacities, most ion exchangers are loaded at low conductivities. The concentration of the relevant counterion (positive in CEX and negative in AEX) must be controlled to obtain consistent product binding. As a result, placement of the IEX step in the process is significant. If the feed load is not low enough in conductivity, dilution with or even buffer exchange into a low salt strength buffer may be

TABLE 6.2
Some Preparative Ion Exchange Resins: Anion Exchange and Cation Exchange

Brand Name	Vendor	Bead Size (μ)	Functional Group	Base Matrix
Anion exchange chromatography				
DEAE 650 S, M and C	Tosoh	35, 65, 100	Diethylamino ethyl	Methacrylate
Fractogel DMAE	Merck KGgA	65	Dimethylamino ethyl	Methacrylate
Fractogel EMD DEAE	Merck KGgA	65	Diethylamino ethyl	Methacrylate
Fractoprep DEAE	Merck KGgA	30–150	Dimethylamino ethyl	Vinyl copolymer
ANX Sepharose FF	GE Healthcare	90	Diethylamino ethyl	Cross-linked agarose
DEAE Sepharose FF	GE Healthcare	90	Diethylamino ethyl	Cross-linked agarose
DEAE Ceramic Hyper D	Pall Corporation	50	Diethylamino ethyl	Ceramic bead filled with a hydrogel
POROS PI50	Applied Biosystems	50	Poly ethylene imine	Coated Polystyrene divinyl benzene (PSDVB) with macropores
Super Q 650 S, M, and C	Tosoh	35, 65, 100	Quaternary ammonium	Methacrylate
QAE 550C	Tosoh	100	Quaternary aminoethyl	Methacrylate
Q Sepharose FF	GE Healthcare	90	Quaternary ammonium	Cross-linked agarose
Fractogel EMD TMAE Hicap	Merck KGgA	65	Trimethyl ammonium ethyl	Methacrylate
Unosphere Q	Biorad	120	Quaternary ammonium	Polymeric
Fractoprep TMAE	Merck KGgA	30–150	Trimethyl ammonium ethyl	Vinyl copolymer
Capto Q	GE Healthcare	90	Quaternary ammonium	High flow agarose

TABLE 6.2
Continued

Brand Name	Vendor	Bead Size (μ)	Functional Group	Base Matrix
Q Sepharose XL	GE Healthcare	90	Quaternary ammonium	Dextran attached to 6% cross-linked agarose
Q Ceramic Hyper D	Pall Corporation (formerly Biosepra)	50	Quaternary ammonium	Ceramic bead filled with a hydrogel
POROS HQ50	Applied Biosystems	50	Quaternary ammonium	Coated PSDVB with macropores
Cation exchange chromatography				
CM 650 S, M, and C	Tosoh	35, 65, 100	Carboxymethyl	Methacrylate
CM Sepharose FF	GE Healthcare	90	Carboxymethyl	Cross-linked agarose
Fractogel EMD COO^-	Merck KGgA	65	Carboxymethyl	Methacrylate
CM Ceramic HyperD	Pall Corporation	50	Caboxymethyl	Ceramic bead filled with a hydrogel
SP 650 S, M, and C	Tosoh	35, 65, 100	Sulfopropyl	Methacrylate
SP 550 C	Tosoh	100	Sulfopropyl	Methacrylate
SP Sepharose FF	GE Healthcare	90	Sulfopropyl	Cross-linked agarose
Fractogel EMD SO_3	Merck KGgA	65	Sulfoisobutyl	Methacrylate
Fractogel EMD SE Hicap	Merck KGgA	65	Sulfoethyl	Methacrylate
Unosphere S	Biorad	120	Sulfo	Polymeric (proprietary vinyl copolymer)
Fractoprep SO_3	Merck KGgA	30–150	Sulfo	Vinyl copolymer
SP Sepharose XL	GE Healthcare	90	Sulfopropyl	Dextran attached to 6% cross-linked agarose

(Continued)

TABLE 6.2
Continued

Brand Name	Vendor	Bead Size (μ)	Functional Group	Base Matrix
S Ceramic HyperD	Pall Corporation	50	Sulfo	Ceramic bead filled with a hydrogel
POROS HS50	Applied Biosystems	50	Sulfo	Coated PSDVB with macropores

required. In such a case, reduced solubility of the product or impurity species in low-conductivity solutions must be considered.

While it is intuitive that higher capacities can be obtained at lower load salt strengths, this generally accepted paradigm might not always hold. Recently, it has been shown that highly basic antibodies binding to CEX actually show an optimal capacity with an initial increase in capacity with salt strength followed eventually by the expected decrease [35]. This has been explained by the strong positive charge on the monoclonal antibody that can serve to electrostatically repel other antibody molecules. As salt concentration is increased, the repulsive charges are shielded and exert a smaller influence, and the behavior eventually follows the expected pattern.

6.4.1.4 Buffers for Ion Exchange Chromatography

Common buffers used for CEX chromatography include citrate, phosphate, acetate, and MES (all of which buffer between pH 5 to 7 which is commonly used for CEX, phosphate can be used above pH 7 as well). Common buffers for anion exchange chromatography include HEPES, Tris, and borate. It is advisable to use a buffering species that does not bind to the IEX resin being employed. For this reason, the use of Tris buffers is avoided on cation exchangers even if they are being operated at pH 7 to 8 in the buffering range for Tris. Tris cations can be adsorbed by cation exchangers resulting in pH fluctuations during operation due to changes in concentration of the buffering species. Of course, sometimes stability of the product species can force the selection of a buffering species, for example, some proteins are stable only in the presence of phosphate. This can be problematic for AEX which can bind phosphate ions. In such an event, make certain a primary buffering species is present to maintain pH in addition to the one required for product stability.

Ion exchangers are typically loaded at low to moderate load conductivities (based on the product retention profile) to enable a high product binding capacity. To prevent pH fluctuations, it is important to ensure that the column is saturated with the same ionic species it will encounter during loading and beyond. For example, a cation exchanger that will be loaded, washed, and eluted with Na^+ ions should be preequilibrated with the same ions. If the column has a preponderance of H^+ ions on the surface prior to loading, displacement of these ions from the surface by Na^+ ions present in the column load material will result in a pH dip that might destabilize the product already loaded on the column. Preequilibration is typically effected by passing the same buffer used for column equilibration at a 4 to $10\times$ strength.

Another example reinforcing the importance of buffer selection for IEX chromatography is provided by Ghose et al. [36]. In this example, storage of a strong fractogel SO_3 cation exchange resin in NaOH led to the creation of weak cationic functionalities on the resin backbone, due to slow degradation of the polymethacrylate backbone forming carboxylic acid groups. This led to the creation of an ion exchanger that bound H^+ (to the COO^- functionalities) under low salt conditions and released them when a step increase in Na^+ concentration occurred, leading to a transient pH decrease during wash and elution. This was addressed by use of a higher buffer strength during wash and elution to minimize the extent of the transition.

For the most part, IEX processes are relatively robust with respect to small temperature variations. However, Tris buffers are notorious for significant changes in pH with temperature, which must be borne in mind while employing that buffer system.

6.4.1.5 Choice of Salts for Wash and Elution

While ion exchangers can be eluted by a transition in pH from binding to nonbinding conditions, this requires a change in type of buffer employed and can cause complications in the pH profile of the effluent, as mentioned above. Nevertheless, this strategy is sometimes employed, since the pH change can improve the separation over what can be obtained by a change in salt concentration. Chromatofocusing using retained pH gradients formed by simple buffering species have been shown to result in effective product elution in IEX [37]. Due to some of the robustness challenges with pH transitions, the simplest way of operating ion exchangers — with salt steps at a single operating pH — is often preferred. The use of simultaneous salt and pH transitions during IEX can often give better separations, but should be employed only if there is reasonable assurance of process reproducibility and robustness in a manufacturing scenario.

Although the type of salt ion employed for wash and elution is sometimes seen to influence the selectivity of the separation [38], this parameter is definitely less important than the type of resin selected and the pH of operation. Hence for the most part, sodium chloride is the salt of choice in ion exchange. High concentrations of sodium chloride (and other halide salts), especially at low pHs, can cause corrosion problems for stainless steel tanks used in process-scale purification suites. Accordingly, sodium chloride is often substituted with sodium sulfate or a higher concentration of the buffering species for product elution.

6.4.1.6 Impurity Clearance in Ion Exchange

Ion Exchange is useful for the removal of a wide range of impurities and is an extremely versatile chromatographic unit operation. Due to its high binding capacity, IEX is often used as the capture and concentration step in downstream processes. In such a role, precautions must be taken not to operate at the pH optimum for any proteases that can degrade the product. It is also used in polishing steps to clear trace levels of contaminants. Apart from host cell protein contaminant removal, IEX has been found to be useful for DNA, endotoxin, and viral clearance. DNA typically bears a strong negative charge and will bind strongly to AEX columns. It is expected to flow through on CEX steps but the clearance is often inferior to AEX. Complexation of DNA with the product can negatively impact DNA clearance through IEX. Endotoxin is also a strongly negatively charged moiety that can bind strongly to AEX. Techniques relying on AEX are commercially sold as flowthrough resins for endotoxin clearance (e.g., DNA Etox from Sterogene, Carlsbad, CA). Both AEX and CEX can effectively clear model viruses to typically yield between 2 and 6 logs of clearance through the chromatographic step. AEX flowthrough has yielded stellar results for monoclonal antibodies and has been validated as a generic step for viral clearance [39].

While the interaction on IEX resins is predominantly electrostatic, it must be remembered that agents that cause changes in protein conformation can also influence binding. Accordingly, chaotropes such as urea and hydrophobic competitors such as propylene and ethylene glycol have been used for washes and as load additives on IEX. Being nonionic, these agents allow binding to the IEX columns, but can remove impurities that are bound to the product or nonspecifically bound to the resin. Detergents and zwitterionic amino acids such as glycine have also been employed for modulating IEX selectivity [38].

6.4.1.7 Methodology for IEX Process Development

Typically, IEX resins are best screened for selectivity in a linear gradient format with low to moderate protein loading. Screening should be carried out over a range of pHs using relevant in-process assays to analyze both peak pools and fractions for purity. As explained in the chapter on resin screening, plots of cumulative percentage of purity vs. cumulative yield can help select the best resin and operating pH. Even though one of AEX or CEX might suggest themselves based on the protein pI, it might be prudent to screen both in case one of the modes gives a dramatically different selectivity. For some proteins, bind and elute operations can be developed on both AEX and CEX, while for others, if one of these types is operated in a bind and elute mode, the other can usually be operated only as a flowthrough operation.

Even though selectivity is usually tested in the linear gradient mode, it is often simpler to operate process-scale separations in the step gradient format. Linear gradients require more sophisticated skids for operation and the linearity of the gradient can be influenced by extra-column mixing and bed inhomogenities. Buffer consumption and operating time are usually greater in linear gradient operations. However, linear gradients do provide better separations between components with similar retention characteristics. Hence, they are often employed for variants separations. Due to the simplicity of operation and scale-up, it is worth the time to develop a step gradient operation where the separation is not negatively impacted by operating in this format.

Another important area during process development is the definition of peak pooling criteria for triggering and terminating peak collection during elution. These are typically based off the UV signal of the column effluent. The trigger is usually set at a low absorbance value while the signal for peak collection is based off a percentage of the maximum height the elution peak achieves. These criteria take on greater significance if certain impurities are present at the front- or back-ends of the elution peak. Since the protein achieves a very high concentration during elution, one must ensure that the detector is not saturated when the absorbance at peak maximum is being measured. This can be ensured by using a small path length flow cell or by moving away from 280 nm detection to another wavelength (typically 300 nm) where the extinction coefficient is lower. Paying attention to this aspect can help ensure consistent product quality through scale-up and process transfer between facilities. Since glycoforms can result in significant variations in the surface charge of proteins (both through their own charge in the form of sialilation and by shielding charged patches), different glycoforms are found at various points of the elution peak. Thus, scale-independent and well-defined peak collection criteria in IEX take on an even greater significance.

Once the key separation conditions in terms of loading, wash, and elution have been identified, standard strip, regeneration, and storage conditions can be added to complete the process. Table 6.3 lists a typical set of operating parameters for CEX and AEX as an example.

TABLE 6.3
Operating Conditions for IEX

Cation exchange chromatography
Preequilibration: 250 m*M* sodium phosphate, pH 6.0
Equilibration: 25 m*M* sodium phosphate, pH 6.0
Load: to 50 g/l
Equilibration buffer wash:
Low salt wash: 25 m*M* phosphate, 50 m*M* NaCl, pH 6.0
Elution: 25 m*M* phosphate, 200 m*M* NaCl, pH 6.0
Strip: preequilibration buffer
Regeneration: 0.5 *N* NaOH
Storage: 0.1 *N* NaOH

Anion exchange chromatography
Preequilibration: 250 m*M* Tris, pH 8.0
Equilibration: 25 m*M* Tris, pH 8.0
Load: to 50 g/l
Equilibration buffer wash:
Low salt wash: 25 m*M* Tris, 50 m*M*NaCl, pH 8.0
Elution: 25 m*M* Tris, 200 m*M* NaCl, pH 8.0
Strip: preequilibration buffer
Regeneration: 0.5 *N* NaOH
Storage: 0.1 *N* NaOH

6.4.2 Hydrophobic Interaction Chromatography

Hydrophobic interaction chromatography (HIC) is based on interactions between hydrophobic (aliphatic or aromatic) ligands on the stationary phase with hydrophobic patches on the surface of proteins. HIC is possibly the second most prevalent mode of chromatography for preparative protein separations following IEX. Since the fundamental basis for interactions between the two techniques are so different, they are often employed as orthogonal methods for protein separations. The existence of hydrophobic interactions have been noticed since 1948 in a work by Tiselius on dye retention in paper chromatography, but these interactions were only exploited for protein separations starting in 1973 with investigations by Shaltiel and Er-el [40] and by Hjerten [41]. For further reading on the fundamentals and history of HIC, the reader is referred to several excellent reviews [42–47].

Interactions of proteins on HIC are promoted by salts, especially lyotropic salts (e.g., sodium citrate, ammonium sulfate, potassium phosphate) as defined by the Hofmeister series (Figure 6.2). Chaotropic salts such as sodium thiocyanate can decrease retention on HIC. Most proteins, except for the most

Increasing lyotropic effect (salting out) →

Anions: SCN^-, I^-, ClO_4^-, NO_3^-, Br^-, Cl^-, COO^-, SO_4^{2-}, PO_4^{3-}

Cations: Ba^{2+}, Ca^{2+}, Mg^{2+}, Li^+, Cs^+, Na^+, K^+, Rb^+, NH_4^+

← Increasing chaotropic effect (salting in)

FIGURE 6.2 Hofmeister series.

hydrophobic ones, require some concentration of salt in the loading buffer to bind on HIC columns. Elution is typically effected by a decrease in the salt concentration of the mobile phase and, sometimes by employing a hydrophobic competitor such as ethylene glycol.

Since HIC is based on interactions with the protein surface, conditions that result in subtle changes in the 3D (three-dimensional) conformation of proteins can influence selectivity quite profoundly. While at first glance it may seem that mobile-phase pH is unimportant for this technique which does not involve charged interactions, in practice pH can influence retention and selectivity quite significantly by causing structural changes and by titrating charged patches that may lie in the vicinity of hydrophobic patches on the protein surface. The effect of pH on retention in HIC is quite unpredictable, hence screening experiments are usually carried out over a range of mobile-phase pHs to identify conditions with the best product retention and selectivity for impurity removal.

Binding capacity has traditionally been limited on HIC resins (<50 mg/ml resin), especially in comparison with IEX (<200 mg/ml resin). A possible explanation for this lies in the generalized interactions with the protein surface that exist on HIC, requiring a bigger footprint for successfully binding the protein. Another reason is the limitation in ligand density — very high ligand densities will render the surface too hydrophobic and can result in product denaturation or irreversible binding to the surface. Binding capacity is also related to size of the protein; very large biomolecules (such as antibodies) generally have lower binding capacities. Accordingly, HIC is often employed in a flowthrough mode in which the product of interest flows through leaving impurities bound on the column. This is especially successful if most impurities are retained higher than the product of interest and also has the added advantage of requiring exposure to a lower salt concentration.

In biopharmaceutical production, removal of aggregated forms of the product is very important to reduce any risks with eliciting an immune response. HIC occupies a unique niche in these separations since aggregates are usually higher retained on HIC. An explanation for this observation is provided by the solvophobic theory.

A characteristic feature of hydrophobic interactions is an increase in their strength with increasing temperature. In fact, HIC is one of the most temperature sensitive modes of chromatography — something that needs to be borne in mind during process transfer and scale-up. Controlling column temperature within a range is important for reproducible operation of HIC columns. At process scale, precautions should be taken to maintain buffer, load, and column temperatures within the range that is qualified during process development and characterization studies.

6.4.2.1 Physicochemical Basis for HIC

Hydrophobic interactions (i.e., fear of water) are the dominant basis behind HIC and RPC. The solvophobic theory has been one of the first theories to explain the fundamental basis of HIC [48]. From a thermodynamic perspective, the free-energy change for HIC and RPC (reversed phase chromatography) systems are dominated by the energy required to form a cavity in the mobile phase to accommodate the solute. In an effort to reduce the energy required to keep it in solution, the solute associates with the hydrophobic stationary phase. This reduces the exposed surface area for both the stationary phase and the solute, and is thus energetically favorable.

The energy required for cavity formation is proportional to the surface tension of the mobile phase and can be expressed as:

$$\Delta G^0_{\text{cav}} = -\Delta A \gamma + \text{const.} \tag{6.9}$$

where ΔA is the difference in surface area of the stationary-phase surface and the protein between the bound and the unbound states, γ is the solution surface tension.

For aqueous salt solutions, the surface tension can be estimated from the molal surface tension increment, σ and the salt concentration expressed in molal units, m.

$$\gamma = \gamma_0 + \sigma m \tag{6.10}$$

Here γ_0 is the surface tension of pure water. Kosmotropic salts from the Hofmeister series have a positive σ while chaotropic salts have a negative σ. In HIC systems which are mildly hydrophobic, a decrease in the mobile-phase salt concentration is sufficient to reduce the surface tension and cause the protein to energetically prefer the mobile phase over the adsorbed state. On the other hand, organic solvents reduce surface tension quite dramatically and can elute proteins from very hydrophobic RPC matrices. Aggregated forms of proteins being larger in size, present a larger surface area than a single protein molecule. Accordingly, they tend to require greater energy to exist in the solution phase and hence tend to be retained higher than the monomeric species on HIC.

Over a fairly broad range of salt concentrations, the relationship for linear retention in HIC is given by:

$$\log k' = \lambda m + \text{const.} \tag{6.11}$$

Thus, linear retention of proteins increases with an increase in concentration of a kosmotropic salt in HIC.

The exponentially modified Langmuir (EML) isotherm has been most widely employed for describing the adsorption capacities of proteins over a range of mobile-phase salt concentrations [49]. This is given by:

$$Q = \frac{Q_{\max} b e^{km} C}{1 + b e^{km} C} \tag{6.12}$$

where m is the mobile-phase salt molality, Q and C are the concentrations of the protein on the stationary and mobile phase, respectively, and $Q_{\max}$, b, and k are fit parameters.

Another theory to describe solute retention in HIC is the preferential interaction theory [50], which is an application of a series of papers describing protein solubility in solution [51,52] to HIC systems. This is based on the interaction of salt ions and their association with macromolecules such as proteins. Kosmotropic salts have negative preferential interaction parameters and tend to be excluded from the immediate vicinity of the protein. Thus, in their presence, proteins tend to adopt a globular conformation with minimal exposed surface area (hence the name salting-out salts) or to adsorb to a hydrophobic surface. Chaotropic salts have positive preferential interaction parameters and tend to associate with the protein surface thus increasing its solubility and decreasing retention on a HIC column. While neither the solvophobic theory nor the preferential interaction theory are used to a significant extent during process development on HIC, they provide an elegant means of understanding how HIC systems work. Models for the temperature dependence of retention in HIC systems have been described [53].

6.4.2.2 Resins for HIC

A variety of commercially available HIC resins for preparative chromatography exist (Table 6.4). Most vendor companies offer a range of resins with differing hydrophobicities based on the type of ligand employed. Ligand density on the surface also influences the overall hydrophobicity quite significantly. A case in

TABLE 6.4
Some Preparative HIC Resins

Brand Name	Vendor	Bead Size (μ)	Functional Group	Base Matrix
Ether 650 S and M	Tosoh	35, 65	Ether	Methacrylate
Butyl 650 S, M, and C	Tosoh	35, 65, 100	Butyl	Methacrylate
Hexyl 650M	Tosoh	65	Hexyl	Methacrylate
Phenyl 650 S, M, and C	Tosoh	35, 65, 100	Phenyl	Methacrylate
Phenyl Sepharose FF (low sub)	GE Healthcare	90	Phenyl	Cross-linked agarose
Phenyl Sepharose FF (hi sub)	GE Healthcare	90	Phenyl	Cross-linked agarose
Butyl Sepharose	GE Healthcare	90	Butyl	Cross-linked agarose
Octyl Sepharose	GE Healthcare	90	Octyl	Cross-linked agarose
Macroprep *t*-butyl	Biorad	50	*t*-Butyl	Methacrylate
Macroprep methyl	Biorad	50	Methyl	Methacrylate

point is the macroprep methyl resin from Biorad, which is quite hydrophobic, despite having the shortest aliphatic chain length possible. Another example is the switch in hydrophobicities between phenyl and butyl ligand containing HIC resins from GE Healthcare and Tosoh Biosciences. Phenyl sepharose 6FF is generally observed to be more hydrophobic than butyl sepharose 4FF (both from GE Healthcare) but in contrast the Tosoh butyl 650M resin is more hydrophobic than Tosoh phenyl 650M.

A general rule of thumb in HIC is to select the most hydrophobic resin that does not denature your product and still allows elution under low salt concentration conditions. Selecting a weakly hydrophobic resin requires the use of high salt concentrations for binding and thus creates the risk of precipitation of the product. Selecting too hydrophobic a resin runs the risk of denaturing the product or causing a significant yield loss from product that does not elute even under low salt conditions. In general, hydrophobic proteins are most successfully chromatographed on stationary phases with a mildly hydrophobic nature and with a relatively low salt concentration. In contrast, less hydrophobic proteins typically employ more hydrophobic stationary phases and use higher salt concentrations. Screening resins can be carried out by running linear gradients of decreasing salt concentration at a few mobile-phase pHs with analytical loads of the product species. The mass recovery in the elution peak and the peak profiles (no excessive tailing, peak splitting, or loss of product in the strip/regeneration) can be used to indicate the occurrence of denaturation [54].

Conformational changes during HIC have been studied quite extensively [42,55–57]. The determination of whether extra peaks observed during analytical linear gradient elution represent separation of an impurity/product isoform or are due to a conformational change in the product is important to distinguish. Rechromatography of the eluting peaks under the same gradient conditions can shed light on these phenomena. If the peaks chromatograph "true" (i.e., elute at the same place as in the original trace), the extra peaks are likely to represent a real separation. On the other hand, if reinjecting the main peak results in several peaks in the elution trace, the smaller peaks are likely to represent conformationally altered forms of the product species. For preparative separations, it is wise to avoid conditions (resin, salt concentration, and pH) that result in such extensive conformational changes. Conformational changes are likely to be even more extensive on RPC which is generally regarded as more hydrophobic, and hence a harsher version of HIC.

6.4.2.3 Selection of Loading Conditions

While analytical linear gradients can shed light on affinity of the product species, they do not give information about the appropriate loading salt concentration or the binding capacity. Except for very hydrophobic proteins or in case a very hydrophobic resin can be successfully employed without product denaturation, lyotropic salts will need to be present in the sample being loaded to enhance binding capacity. The first step in selecting the binding salt concentration is to generate precipitation curves for the product in solution at various mobile-phase salt concentrations and over a range of pHs. Turbidity of the solution (measured by light scattering or absorbance measurements at a high wavelength of 400 to 500 nm) can indicate the formation of precipitates. Figure 6.3 shows an example of a precipitation curve. Typically, the closer the mobile-phase pH is to the pI of the protein, the lower will be the salt concentration at which precipitation occurs. It is best to operate at a pH which is most stabilizing for the product. To maximize product loading capacity on HIC, a salt concentration just below the point of precipitation in solution is selected. Of course, care must be taken to ensure that product denaturation does not occur at this salt concentration by using the methods described in Section 6.4.2.2.

Some of the commonly employed lyotropic salts for HIC loading include ammonium sulfate, sodium sulfate, sodium citrate, and potassium phosphate. Sodium chloride is only mildly lyotropic and is employed when the product does not need much additional impetus to bind. While all of these salts are employed

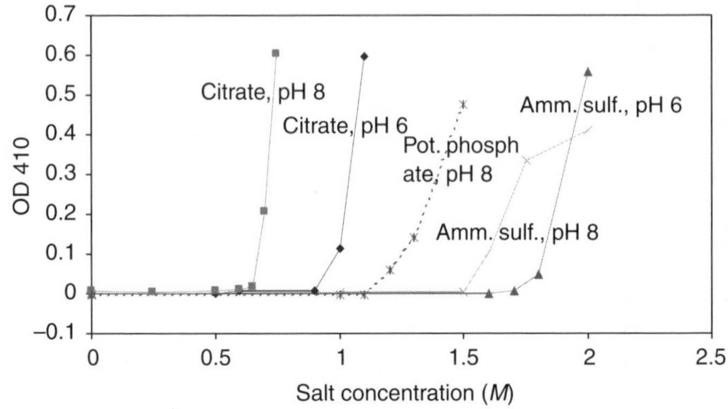

FIGURE 6.3 Precipitation curves on HIC with various salts. pI of protein is pH 8.5.

during small-scale HIC operations, disposal and corrosivity considerations become significant at process-scale when large quantities of salt containing buffers need to be stored and disposed of. Ammonium sulfate, when discharged into water bodies can release ammonia resulting in the promotion of algal growth and subsequent creation of anaerobic conditions. Thus, direct discharge of this salt is usually not permitted. Similar issues can occur for phosphate salts. High concentrations of sodium chloride can corrode stainless steel tanks despite passivation or storage at low temperatures. Accordingly, sodium citrate and sodium sulfate turn out to be the most suitable lyotropic salts for large-scale applications. The latter is limited in terms of its solubility in water ($<1.2\ M$), but has a high molal surface tension increment that can enable the use of lower solution concentrations [58].

The salt addition step to the product intermediate also requires careful development. Since a higher salt concentration solution needs to be brought in contact with the product to achieve a certain salt concentration in the column load, it must be ensured that little, if any, precipitation occurs at the high salt concentration. This is especially important at large-scale when mixing rates in tanks might lead to a longer exposure time to the high salt concentration conditions than the almost instantaneous mixing that occurs in a test-tube. It is recommended to fully characterize the effects of the high salt concentration in the dilution buffer, contact times, and hold times for the product in the load solution prior to process scale-up. When precipitation might occur at the high salt concentration necessary for obtaining a reasonable binding capacity, online mixing of the product with the dilution buffer should be considered to minimize contact time. This is easily achievable on most chromatography skids in which the load is prepared continuously just before it is actually loaded on the HIC column.

6.4.2.4 Development of Wash Conditions

Following product loading, the column is often washed to elute any weakly bound impurities. On HIC, this can be accomplished by employing the equilibration buffer (which is typically at a similar salt concentration as the load solution) or if product binding permits, by employing a lower salt concentration than the load. pH is an important variable that can be optimized for the wash buffer since protein conformation changes significantly with pH. This can enable contaminants that bound during loading to be removed by the wash. For proteins that are tightly bound, specific wash conditions can be developed using mixtures of chaotropes to greatly enhance the separation power of the HIC step. Mixtures of mobile-phase additives (urea, glycerol, and sodium thiocyanate) in a wash buffer enabled the removal of impurities that could not be removed by any of the agents when used alone even at higher concentrations [59]. Useful hydrophobic competitors include ethylene and propylene glycol and detergents. As explained by the solvophobic theory, alcohols (typically ethanol or isopropanol employed at concentrations $<10\%$ v/v) can reduce hydrophobic interactions quite significantly. However, since these agents are also denaturing in nature, they must be employed judiciously. In addition, polymeric competitors such as detergents (usually $<2\%$ Triton X100 or Tween 20/80) can bind to the column, requiring the use of organic solvents to remove them so that column lifetime does not suffer. Disposal issues for these reagents must also be considered during large-scale operation.

6.4.2.5 Selection of Elution Conditions

Elution from HIC resins is accomplished under low salt conditions. The exact concentration of lyotropic salt in the elution buffer needs to be optimized to leave more hydrophobic impurities behind on the column. Some impurities, such as product aggregates will elute at the end of the elution peak and hence, the selection of the salt concentration in the elution buffer is important to enable reliable and reproducible peak collection. Very rarely, when the association between product and resin is very strong, hydrophobic competitors or alcohols are employed for product elution. These agents can only be employed when product stability permits — otherwise selecting a less hydrophobic resin is advisable.

After a HIC step, the product usually ends up in a buffer at moderate salt concentration (only rarely can a very low salt concentration be employed for selective elution). This can create issues for placing this step in a purification process. If this step is placed prior to IEX, buffer exchange might be required to allow product binding on the IEX column. This adds an additional unit operation to the process sequence and may not be desirable for throughput and yield reasons. Accordingly, HIC steps are usually placed following IEX since the high salt concentration in the IEX eluate is compatible with the elevated salt concentrations required for HIC binding. One exception to this general paradigm is when zwitterionic lyotropes such as glycine are employed in process buffers for this step [60]. Glycine has a lower molal surface tension increment than most lyotropic salts but relatively high concentrations of this amino acid can be loaded on to IEX columns without interfering with product binding. One disadvantage of this strategy is the higher cost of glycine (or other amino acids) as a buffer component for large-scale operation.

6.4.2.6 Methodologies for Process Development

Figure 6.4 shows a flow sheet for development of a HIC step. The first step in HIC process development is to generate precipitation curves for the product with various lyotropic salts (especially sodium citrate) at several mobile-phase pHs. The precipitation curves are indicative of the salt tolerance of the product in solution, typically the product will not be able to withstand the same salt concentration in the presence of the stationary phase. Based on the salt concentration that causes precipitation, a reasonable salt concentration is selected to start the linear gradient HIC resin screening. The initial gradients typically end at a very low salt concentration in the buffer of choice, followed by a column strip with water and regeneration with 0.5 *N* sodium hydroxide. These gradient screening experiments are useful in assessing affinity of the product for HIC resins and can often be employed to eliminate resins that are completely unsuitable. Resins that exhibit very weak binding requiring the use of high salt concentrations for binding, or ones from which the product does not elute even at low salt concentrations, can be eliminated at this stage for a bind and elute process. Peak splitting or unusually broad elution peaks can also be indicative of product denaturation during these gradient experiments. If this is the case, the gradients should be repeated with a lower initial salt concentration, since most of the denaturation occurs at the high salt conditions employed for column loading.

The selectivity of HIC resins are better evaluated under preparative loadings while still using gradient elution. Fractions can be collected throughout the gradient elution and plots of cumulative yield vs. cumulative purity can be plotted as described in the chapter on resin screening. Binding capacities under dynamic or static conditions can be measured at various mobile-phase salt concentrations on the resins and pH conditions that are still under consideration. Capacity and selectivity data should help identification of a resin and pH condition to take on to the method development phase.

In the method development phase, appropriate loading, wash, and elution conditions are developed. These are combined with standard strip (water),

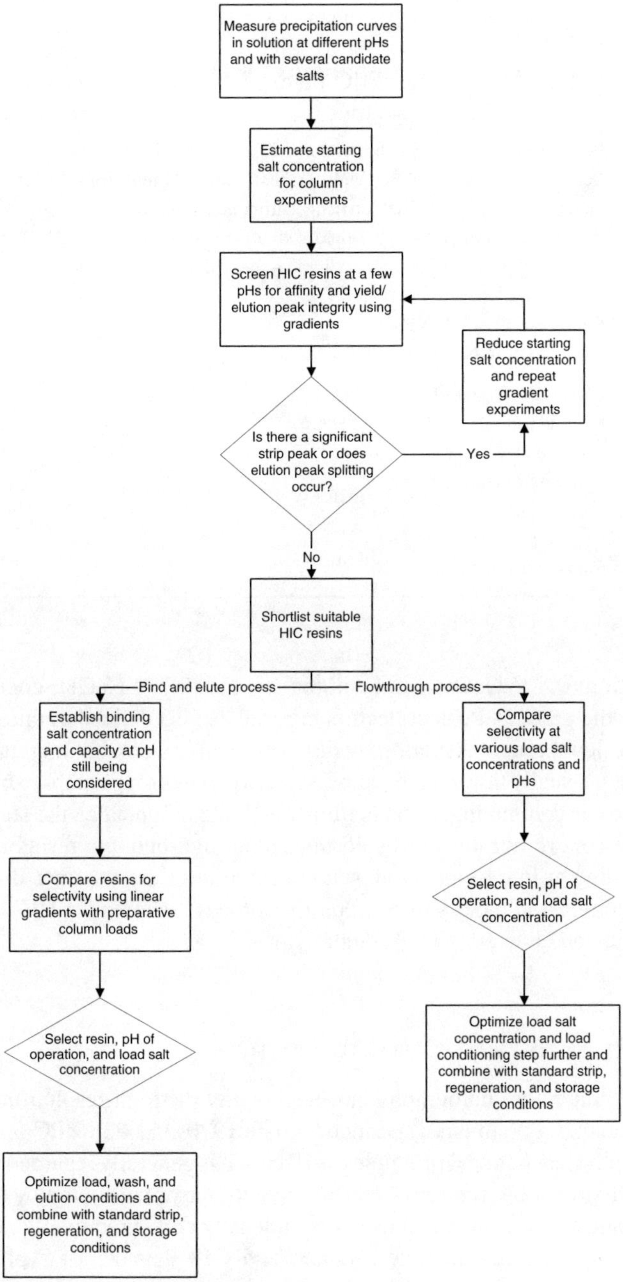

FIGURE 6.4 Process development methodology of HIC.

TABLE 6.5
Typical Operating Conditions for HIC

Bind and elute operation
Equilibration: 250 m*M* sodium citrate, 25 m*M* Tris, pH 8.0
Load: to 20 g/l after dilution with 500 m*M* citrate, 25 m*M* Tris, pH 8.0 to obtain ~250 m*M* citrate ion concentration in load
Equilibration buffer wash: 250 m*M* sodium citrate, 25 m*M* Tris, pH 8.0
Elution: 25 m*M* Tris, 50 m*M* sodium citrate, pH 6.0
Strip: water
Regeneration: 0.5 *N* NaOH
Storage: 0.1 *N* NaOH

Flowthrough operation
Equilibration: 50 m*M* citrate, pH 6.0
Load: to 50 g/l
Equilibration buffer wash:
Strip: water
Regeneration: 0.5 *N* NaOH
Storage: 0.1 *N* NaOH

regeneration (0.5 *N* NaOH), and column storage (0.1 *N* NaOH) conditions to complete the process. Peak collection criteria are also defined at this stage.

For a flowthrough operation, resins that exhibit weak binding can still be considered. Usually however, the most selective flowthrough steps exhibit weak binding to the column under the loading conditions. Typically, the least hydrophobic resins are not the most selective, more hydrophobic resins that show weak binding at low to moderate salt concentrations are the ones that are the most selective. Table 6.5 shows a possible process scheme for HIC operation in the bind and elute and flowthrough modes.

6.4.3 Reversed-Phase Chromatography

Reversed phase chromatography has been one of the high-resolution methods for peptide and protein analysis since the late 1970s [61,62]. RPC operates on the same physicochemical principles as HIC — it is generally regarded as a more strongly hydrophobic version of HIC as the stationary phases employ ligands of longer chain length and at a higher ligand density compared to HIC. Common aliphatic chain lengths include C4, C8, and C18 ligands. RPC phenomena can also be explained by the solvophobic principles used to explain HIC. For a detailed treatment of retention mechanisms in RPC, please consult Vailaya [63].

As a result of its greater hydrophobicity, RPC can be strongly denaturing to proteins. It has been shown that the RPC of papain yields two peaks, with the size of the higher retained, denatured peak being proportional to the residence time of the sample on the column [64]. This is not of consequence in analytical chromatography where the purpose is to separate analytes, but is the primary reason behind restricted usage of RPC for preparative protein purification. Another disadvantage of RPC is the high hydrophobicity of the stationary phase which requires the use of organics to elute solutes off the column. Once again, these raise disposal and handling issues at large-scale and also contribute to the denaturing influence on proteins. RPC is typically carried out in high-pressure HPLC (high performance liquid chromatography) columns — small particle-size incompressible stationary phases are employed to ensure high efficiency of the columns.

Despite these disadvantages, RPC has been employed successfully for preparative protein purification for several approved biopharmaceuticals including insulin, insulin-like growth factor I, erythropoietin, and GCSF [65]. RPC is employed during the production process for human insulin [66]. Recombinant human insulin-like growth factor (rhIGF-I) has been separated from its variants containing oxidized methionine residues by preparative purification on a C_{18} column packed with larger particle-size (15μ) beads using gradient elution with an acetonitrile–acetic acid system [67].

Several vendors supply preparative RPC stationary phases with varying ligand chain lengths and a range of particle sizes including Vydac (Grace-Vydac, Hesperia, CA), Waters (Milford, MA), and JT Baker (Mallinckrodt Baker, Phillipsburg, NJ). Most of these resins have a silica base bead because of the requirement for incompressible and mechanically stable beads under the high pressures employed for HPLC.

Most of the proteins purified by this technique are smaller in size (generally <50 kDa) since larger biomolecules tend to denature irreversibly during RPC. For these products, RPC possesses unparalleled separation capabilities to remove variants and closely related impurities. In fact, the unfolding of proteins during RPC is one of the reasons behind the high resolution this technique offers — the entire protein surface can be probed, and hence even small differences in sequence can be resolved.

At large-scale, ethanol is preferred as the solvent of choice for RPC. Ion-pairing agents are often added to lower mobile-phase pH and render the silica backbone uncharged. Ion-pairing agents also complex with charged patches on the solute and render the protein surface uniformly hydrophobic. Thus, ion-pairing agents eliminate nonspecific ionic interactions of the solute with the stationary-phase backbone and improve the separation efficiency. While analytical RPC often employs tri-fluoro acetic acid (TFA) as an ion-pairing agent, at large-scale acetic acid or triethyl ammonium acetate (TEAA) are employed.

6.4.4 Hydroxyapatite Chromatography

Hydroxyapatite (HA) is one of the few inorganic media employed for preparative protein chromatography. The chemical formula of hydroxyapatite is $Ca_{10}(PO_4)_6(OH)_2$ [68,69]. Ceramic HA (type I and type II which differ slightly in their surface distributions of Ca and phosphate groups) from Biorad is one of the few stationary phases available for preparative scale chromatography. Despite its less widespread usage, HA can provide highly selective separations, once it is adequately understood.

Two kinds of binding sites dominate the HA surface. One is predominantly positively charged and is called the C site and is comprised of one or several adjacent Ca^{2+} ions. The other is predominantly negatively charged and is called the P site since it is made of one or several adjacent phosphate ions. In addition to the cation exchange possible with the C sites, one can also have mild metal chelate interactions with the Ca^{2+} ions. Clearly, interaction mechanisms on HA are complex.

It has been shown that basic proteins interact predominantly with the P sites while acidic proteins interact with the C sites. However, the mechanism goes beyond simple ion exchange interactions since it has been shown that Cl^- ions are not very effective in eluting acidic proteins since they cannot approach C sites very closely. In contrast, phosphate ions are significantly more effective in binding to C sites and can elute acidic proteins much more effectively. In general, increasing sodium chloride concentration can elute basic proteins quite effectively since Na^+ ions can interact effectively with the P sites. To elute other proteins, phosphate buffer concentration is increased. Very few proteins remain bound to HA matrices at high phosphate buffer concentrations.

Since the matrix is inorganic and includes Ca^{2+} ions, this stationary phase is sensitive to the presence of chelators. Even low concentrations of chelators such as ethylene diamine tetraacetic acid (EDTA) or citrate ions can dissolve the matrix over time and should be avoided. Despite similar concerns, Tris buffers have been employed successfully at low concentrations to enable operation of the step above pH 7.0. In addition, the matrix is unstable at low pH conditions ($pH > 6$ is recommended) but can tolerate alkaline conditions quite well. Since phosphate buffers can reduce interactions with both P and C sites, the presence of phosphate ions should be avoided in the load material or binding capacity can suffer quite drastically.

Ceramic HA packings are quite incompressible and can be operated at elevated flow rates. However, large-scale column packing is not without its challenges. Obtaining a uniform and homogenous packed bed can be a challenge due to the high density of the packing, its tendency to settle rapidly, and the requirement for minimizing physical handling of the material to reduce risk of fracturing the particles. HA columns are often dry-packed followed by buffer flow in the upward direction to enable settling.

Metal ions present in process flow streams and buffers (usually trace levels of Fe^{3+} and Cu^{2+}) can substitute Ca^{2+} ions in the matrix at the top of the column over time. This can lead to the appearance of a gray ring in the column over time. An easy means of preventing this is to employ a smaller HA "guard column" before the process-scale column so the gray ring forms there instead. Since HA is relatively incompressible, very high flow rates can be employed on the guard column.

The first experiment on HA is to load the product on a 50 m*M* MES, pH 6.8 buffer and run a linear gradient of sodium chloride to a 1 *M* concentration, followed by a gradient of sodium phosphate to 1 *M*. This experiment helps determine if the product behaves as a basic or acidic protein in terms of behavior on HA. If the product elutes in the NaCl gradient, the product behaves as a basic protein and interacts primarily with the P sites. For such a product, it is advantageous to operate at a higher pH to maximize product affinity on the matrix and allow impurities to either flow through at lower NaCl concentrations or bind tightly and be removed in a strip containing phosphate ions. If the product does not elute in NaCl, it is likely to elute in the phosphate ion gradient that follows. This is usually the case for acidic proteins. In this case, a wash with a high concentration of NaCl can remove basic impurities. Elution can be effected by a combination of sodium chloride and sodium phosphate or by sodium phosphate alone. Table 6.6 shows a typical set of operating conditions on HA for both a basic and acidic protein. The strip buffer for HA is usually a high concentration of sodium phosphate and the storage buffer usually contains a low concentration of phosphate ions to ensure stability of the matrix.

6.4.5 Immobilized Metal Affinity Chromatography

Immobilized metal affinity chromatography (IMAC) (a.k.a metal chelate chromatography) was developed in 1975 by Porath and coworkers and since then has become quite a popular technique for lab-scale isolation of proteins as well as in several large-scale production processes [70–72]. Despite inclusion of the word affinity in its title, IMAC is not based upon biospecific affinity in the sense of interaction of a natural ligand with its binding site and is treated separately here. Nevertheless, IMAC can be quite selective for some proteins. IMAC is based upon interactions between transition metal ions immobilized on the stationary phase with amino acid residues such as His, Trp, Cys, and Lys that can chelate with metal ions. Primary interactions are through surface His residues, with multiple binding sites being involved in the interaction simultaneously [73]. In fact, a protein will show the highest affinity for arrangements

TABLE 6.6
Example of Operating Conditions on HA

Basic protein
Preequilibration: 250 m*M* sodium phosphate, pH 7.4
Equilibration: 25 m*M* sodium phosphate, 50 m*M* NaCl, pH 7.4
Load: to 50 g/l
Equilibration buffer wash:
Low salt wash: 25 m*M* phosphate, 120 m*M* NaCl, pH 7.4
Elution: 25 m*M* phosphate, 200 m*M* NaCl, pH 7.4
Strip: preequilibration buffer
Regeneration: 0.5 *N* NaOH
Storage: 0.1 *N* NaOH, 25 m*M* sodium phosphate

Acidic protein
Preequilibration: 250 m*M* sodium phosphate, pH 6.8
Equilibration: 25 m*M* MES, 50 m*M* NaCl, pH 6.8
Load: to 50 g/l
Equilibration buffer wash:
Low salt wash: 25 m*M* MES, 200 m*M* NaCl, pH 6.8
Elution: 25 m*M* MES, 100 m*M* sodium phosphate, pH 6.8
Strip: preequilibration buffer
Regeneration: 0.5 *N* NaOH
Storage: 0.1 *N* NaOH, 25 m*M* sodium phosphate

of surface sites that match its own pattern of functional groups, making this technique quite selective. However, not all proteins can be purified by this technique. Usually proteins with multiple surface His residues have the greatest affinity and thus stand the best chance of being purified by this method [74]. Alternatively, $(His)_6$ tags have been employed for laboratory scale isolation of a large number of proteins.

6.4.5.1 IMAC Resins and Metal Ions

As shown in Table 6.7, several IMAC resins are available for preparative chromatography. Most IMAC resins are sold uncharged, requiring the user to charge the column with the appropriate metal ion prior to use. Several types of ligands (Figure 6.5) can be employed to hold the metal ion on the stationary phase and their chemistry plays a major role in determining the number of spare coordination sites on the metal ions. The most common ligand is iminodiacetic acid (IDA) which is tridentate. Others include tris(carboxymethyl) ethylene diamine (TED) and nitrilo triacetic acid (NTA) which are penta- and tetra-dentate,

TABLE 6.7
Some Preparative Resins for IMAC

Resin Name	Vendor	Bead Size (μ)	Ligand	Base Matrix
Fractogel EMD Chelate	Merck KGgA	65	IDA	Methacrylate
Chelating Sepharose BB	GE Healthcare	100–300	IDA	Cross-linked agarose
Chelating Sepharose FF	GE Healthcare	100	IDA	Cross-linked agarose
Profinity IMAC	Biorad	120	IDA	Unosphere polymeric beads

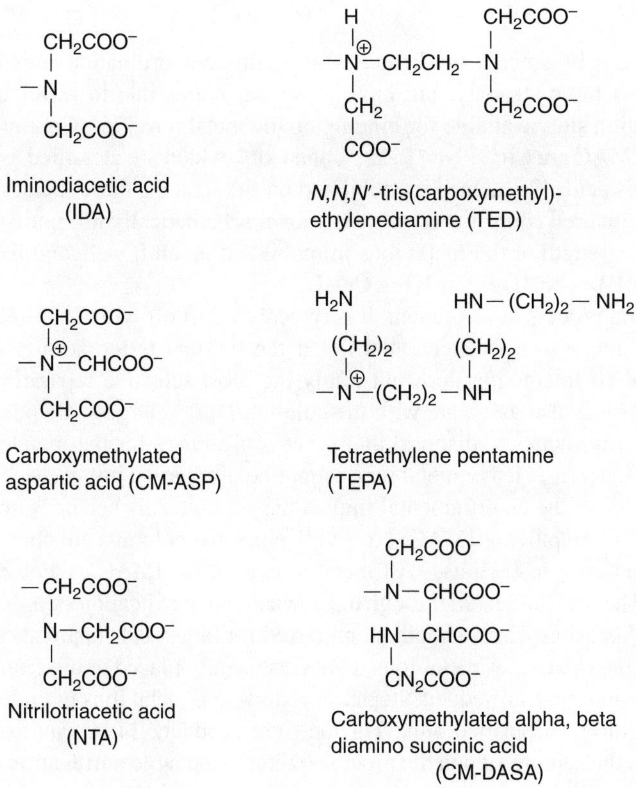

FIGURE 6.5 Chelating ligand in IMAC.

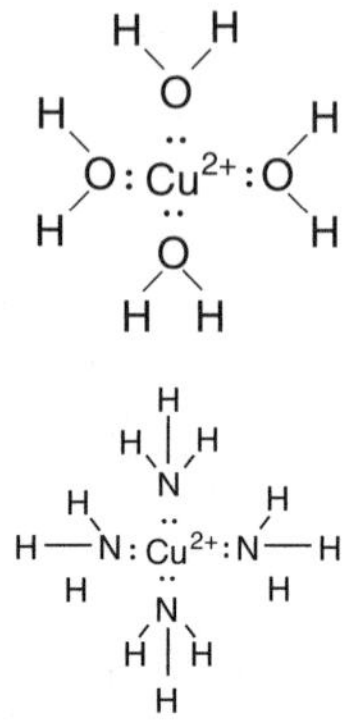

FIGURE 6.6 Metal ion coordination linkages in IMAC. (a) Metal ions are highly solvated in solution as a result of coordination with water molecules. (b) Water molecules can be replaced by a stronger base (such as His residues on the protein surface).

respectively. In general, the ligands with multiple coordination sites hold the metal ions more strongly, but have lower capacities due to fewer unpaired coordination sites available for binding on the metal ions [75]. Common metal ions for IMAC are Cu^{2+}, Ni^{2+}, Zn^{2+}, and Co^{2+} which are classified as borderline Lewis acids [76]. Once immobilized on the ligands, these metal ions still possess unpaired coordination sites as shown schematically in Figure 6.6. The retention strength of the metal ions immobilized on an IDA ligand follow the order Cu(II) > Ni(II) > Zn(II) $\sim$ Co(II).

During process development, it is typical to start off with Cu^{2+} metal ions and then move to a weaker metal ion if the binding is too strong. Usually, the strongest interaction does not imply the most selective separation, since impurities can also associate with the column. During large-scale purification of biopharmaceuticals, disposal and process clearance for the metal ions are primary concerns. Heavy metal ions cannot be discharged into water bodies or streams due to the environmental impact they cause. This becomes one of the liabilities of employing IMAC, especially since the columns are charged with metal ions prior to each use. All metal ions used for IMAC except Zn^{2+} are toxic and have to be cleared through the downstream purification steps following IMAC. If possible, Zn^{2+} should be preferred for large-scale applications. As a result of the toxicity of metal ions, IMAC is usually placed at the front-end of the process so that subsequent steps can remove the metal ions to undetectable levels in the final purified bulk. For the right product, IMAC can have a big impact as the capture step in the process owing to the large purification factor is possible on this mode of operation. However, care should be taken not to expose the IMAC column to chelating agents often present in cell culture media such as EDTA, citrate, and high concentrations of Tris ions since these can strip the metal ions off the column.

6.4.5.2 Buffers for IMAC

Immobilized metal affinity chromatography steps are typically operated at close to neutral pH to ensure that the His residues that interact with the metal ions are uncharged and can form coordination linkages. Since the ligands on IMAC resins are negatively charged (such as IDA), buffers for this technique usually include a moderate salt concentration (0.2 to 0.5 *M* NaCl or equivalent) to prevent nonspecific ionic interactions with any uncharged sites. Common buffers employed for IMAC include phosphate and acetate with low concentrations of Tris buffer also being suitable on most resins.

Elution from IMAC resins can be effected by decreasing mobile-phase pH (which causes His residues to acquire a positive charge and cease to chelate metal ions) or by employing a mobile-phase modulator such as imidazole, which can also chelate with the metal ions. Harsher methods such as using EDTA to strip the metal ions, and thus elute protein associated with them, are not used in biopharmaceutical production, since a high concentration of metal ion would end up in the elution pool. While employing low pH elution, it should be ascertained if the ligand–Me^{2+} linkage is strong enough so that metal ions do not leach out. Step or linear gradient elution with imidazole is the most commonly used elution method. However, it must be recognized that imidazole itself binds quite strongly to the metal ions and hence does not function in quite the same way as salt concentration does in IEX. There can be a significant delay in breakthrough of a step or linear gradient from the column due to retention of imidazole on the resin. Imidazole breakthrough profiles can be monitored at 230 nm in the absence of a protein load. To achieve more reproducible elution profiles, the column is often presaturated with imidazole and a low concentration (0.5 to 2 m*M*) can also be included in the column load and wash buffers, if binding capacity does not suffer substantially.

Immobilized metal affinity chromatography media can tolerate chaotropes, organic solvents, and detergents quite well. These agents can be employed as wash buffer additives to selectively remove impurities. More typically however, the wash will consist of a buffer with either lower pH than the load or with a moderate concentration of imidazole in-between that of the load and elution conditions.

Ethylene diamine tetraacetic acid is often added to the elution pool following IMAC to chelate any metal ions that may have leached off the column during purification. EDTA is included in the strip buffer to remove metal ions from the column after each use. Following the metal ion strip, the column can be sanitized and stored in sodium hydroxide.

6.4.5.3 Modeling of Interactions on IMAC

The SMA model of IEX was extended to IMAC systems [77] since this mode of chromatography is also based upon interactions of the protein with discrete binding sites on the surface and imidazole acts analogous to salt ions in IEX by binding to a single chelating site on IMAC. The key difference lies in the relatively higher affinity of imidazole for the chelation sites. Upon interaction, the protein interacts with n_P sites on the stationary phase and shields σ_P metal ion sites.

$$Q_\mathrm{m} = \frac{(\Lambda - n_\mathrm{P} Q_\mathrm{P}) K_\mathrm{m} C_\mathrm{m}}{1 + K_\mathrm{m} C_\mathrm{m}} \tag{6.13}$$

where the subscripts P and m refer to the protein and the mobile-phase modulator, respectively, n_P is the number of interaction sites of the protein with the surface, L is the bed capacity determined by imidazole binding alone, and K is the equilibrium constant.

6.4.5.4 Process Development on IMAC

The first step in process development is to screen for affinity with various metal ions using pH or imidazole gradients to elute the product. These analytical experiments are also a good time to get a preliminary evaluation of selectivity of this unit operation since this information can be valuable in deciding on a metal ion to use. Once the metal ion is selected, a comparison is made between pH and imidazole gradients to get the best separation. If imidazole is selected, a presaturation step with a high ($\sim$10 to 50 mM) concentration of imidazole is used. Following this, binding capacity for the product needs to be evaluated while including low concentrations of imidazole in the load (0 to 2 mM) to maintain the presaturation. Finally, wash and elution imidazole concentrations are selected to give the best purity and yield from the process step. If decrease in pH is chosen as the elution procedure, sequential step reductions in pH can be used to obtain the appropriate conditions for wash and elution.

Table 6.8 provides an example of operating conditions for an IMAC step. IMAC is fairly versatile in terms of its placement in a process sequence. Since, at a minimum, the load conditions include significant salt concentration to block ionic interactions both HIC and IEX eluates can be directly loaded without the need for buffer exchange (unless a citrate buffer system was used). Placement as the very first step in the process can be problematic owing to the possible presence of chelating agents in cell culture and fermentation media even though capture is where IMAC can have the greatest impact.

TABLE 6.8
Example of Operating Conditions for an IMAC Process Step

IMAC operation
Flush: 100 m*M* acetate, pH 4.0
Charging: 100 m*M* acetate, 100 m*M* zinc sulfate, pH 4.0
Flush: 100 m*M* acetate, pH 4.0
Preequilibration: 25 m*M* sodium phosphate, 200 m*M* NaCl, 50 m*M* imidazole, pH 7.0
Equilibration: 25 m*M* sodium phosphate, 200 m*M* NaCl, 2 m*M* imidazole, pH 7.0
Load: to 50 g/l
Equilibration buffer wash:
Imidazole wash: 25 m*M* sodium phosphate, 200 m*M* NaCl, 5 m*M* imidazole, pH 7.0
Elution: 25 m*M* sodium phosphate, 200 m*M* NaCl, 20 m*M* imidazole, pH 7.0
Strip: preequilibration buffer
Regeneration: 0.1 *M* EDTA
Storage: 0.1 *N* NaOH

6.4.6 Other Techniques

6.4.6.1 Thiophilic Interaction Chromatography

Thiophilic interaction chromatography (TIC) was first discovered by Porath and coworkers in the 1980s and was based on interactions between sulfur containing ligands and certain classes of proteins, especially antibodies. Further investigations showed that this technique, although promoted by kosmotropic salts was distinct from HIC. Interestingly, it has been found that the presence of sodium chloride significantly reduces the extent of thiophilic interactions [78]. Elution from TIC resins is typically achieved under low pH conditions or with hydrophobic disruptors like ethylene glycol.

The mechanism of interaction is thought to involve an electron donor–acceptor pair in close proximity to each other on the ligand and on the protein [79]. Accordingly, the ligands for TIC consist of a sulfone group and a nucleophile (typically S or N) proximal to each other. The general structure for a thiophilic ligand is: $-O-CH_2-CH_2-SO_2-CH_2-CH_2-X-R$ where X is the nucleophile and R is an alkyl or aromatic side group. A variety of heterocyclic ligands have been prepared for TIC and have been shown to bind antibodies from human serum [80].

Thiophilic interaction chromatography is largely associated with antibody purification [81] and although no investigation of the binding site on antibodies is available in the literature, it is to be presumed to interact with the Fc region of antibodies. No commercially available adsorbents based on TIC are currently available, although as explained below, HCIC (hydrophobic charge induction chromatography) resins which originated from TIC are now available.

6.4.6.2 Hydrophobic Charge Induction Chromatography

Hydrophobic charge induction chromatography was developed by Burton and Harding [82] as a means of obtaining salt independent adsorption of proteins on hydrophobic matrices. Since proteins can adsorb irreversibly on very hydrophobic resins leading to yield losses, a charge inducible ligand that can acquire a repulsive charge at low pH was employed. The heterocyclic ligand used here was very similar to those used for TIC (mercapto ethyl pyridine) except that a significantly higher ligand density was employed to impart a salt independent adsorption characteristic. Two HCIC resins are now commercially available from Ciphergen Biosystems (now part of Pall Corporation). These are MEP Hypercel (2 mercapto pyridine) and MBI Hypercel (2 mercapto 5 benzimidazole sulfonic acid) both on a cellulose base bead with a 80 to 100 μ particle size.

Proteins can bind to HCIC even under low salt conditions owing to its high hydrophobicity. Elution is effected by decreasing pH. Thus, superficially the technique bears resemblance to Protein A chromatography. Accordingly, several of the initial HCIC investigations focused on antibody purification and capture [83,84]. While good purification results were claimed for the antibodies, only SDS-PAGE analysis was employed to evaluate selectivity of the technique.

Recently, a comprehensive evaluation of the binding mechanism on HCIC was undertaken [85]. Both monoclonal antibodies and model proteins were found to bind to MEP Hypercel equally strongly and specific affinity for antibodies was ruled out. Nevertheless, the technique was found to have greater selectivity than HIC matrices with respect to host cell protein removal and may enable purification of more hydrophobic proteins than can be recovered from HIC resins. Thus, even though HCIC might not be an effective alternative to Protein A chromatography, it can have broad applications in protein purification that have only just begun to be exploited [86].

6.4.6.3 Mixed Mode Ion Exchangers and Silica

While most single mode resins inadvertently contain an element of another mode of interaction associated with them, some resins are deliberately designed to combine two or more modes of interaction together. This can help create more selective resins for certain classes of biomolecules.

Mixed cation and anion exchangers have been made on a silica backbone and are sold by JT Baker under the tradename Bakerbond ABx [87]. These resins have not become popular for preparative separations, partially due to the difficulty of column equilibration and controlling pH during operation since both anionic and cationic buffer ions can be adsorbed by the resin. Recently, a variety of multimodal ligands were prepared and screened for both AEX and CEX on an agarose backbone [88,89]. For CEX the ligands contained a hydrogen acceptor group close to a carboxylic acid while for AEX the ligands contained amine functionalities along with aliphatic or aromatic groups. At least for the model proteins studied, both these sets of ligands could be operated under typical conditions for IEX with an increase in salt concentration causing elution. None of these ligands have been introduced commercially yet.

Naked underivatized silica has also been employed for protein separations at a large scale [90]. More commonly, silica is employed as the base matrix for a wide range of preparative resins [91]. Interactions of proteins with silica have been shown to be due to a combination of electrostatic and hydrophobic interactions [92]. The authors demonstrate the capability of this stationary phase in replacing tandem IEX and HIC steps in the purification of a recombinant protein expressed in mammalian system. While being selective due to its mixed mode nature, silica suffers from the disadvantage of being unstable under alkaline conditions often employed for column regeneration and storage. Regeneration at large-scale requires simultaneous reduction of electrostatic and hydrophobic interactions through agents such as tetra-methyl ammonium chloride (TMAC) or combinations of sodium chloride and organic solvents.

6.4.6.4 Size Exclusion Chromatography

Size exclusion chromatography (SEC) separates biomolecules on the basis of their ability to penetrate a network of pores in the chromatographic stationary phase [93]. Smaller molecules that penetrate further into the network of pores elute later than larger molecules. While its unique ability to separate on the basis of molecular size and shape is useful for laboratory scale separations, SEC (a.k.a gel filtration and gel permeation chromatography) is not widely used at process-scale due to its low efficiency. Since the sample does not bind on the column, loading volumes are highly restricted (usually $<5\%$ column volumes). Due to the use of compressible media, throughput is also severely restricted. Nowadays, it is common to employ UF/DF in place of SEC for buffer exchange and to employ other techniques such as IEX and HIC for other kinds of separations. The manufacturing processes for several early biotechnology products do continue to employ SEC.

6.5 CONCLUSIONS

Preparative chromatography is the fundamental unit operation for biopharmaceutical downstream processing. The resolution from impurities that can be achieved with chromatographic techniques has so far not been duplicated in any other type of unit operation despite significant advances in membrane chromatography and liquid–liquid separations. While most of the important modes of chromatography have been in existence at least for a couple of decades, an improved understanding of the molecular level separation phenomena involved are still being obtained and in many cases resulting in a huge impact on the way process development is carried out. In addition, there are continuous improvements being made in stationary phases for preparative chromatography that are leading to a steady stream of new chromatographic products for specific classes of biomolecules. It is anticipated that both these fields of endeavor will continue into the foreseeable future lending even greater richness to the field of chromatographic separations.

REFERENCES

1. Guiochon G, Shirazi SG, and Katti A. *Fundamentals of Preparative and Nonlinear Chromatography*. New York: Academic Press, 1994.
2. Narayanan S. Preparative affinity chromatography of proteins. *Journal of Chromatography A* 1994; 658:237–258.
3. Scopes R. *Protein Purification — Principles and Practice*, 3rd ed. New York: Springer-Verlag, 1994.
4. Burnouf T. Chromatography in plasma fractionation: Benefits and future trends. *Journal of Chromatography B* 1995; 664:3–15.
5. Burnouf T and Radosevich M. Affinity chromatography in the industrial purification of plasma proteins for therapeutic use. *Journal of Biochemical Biophysics Methods* 2001; 49:575–586.
6. Curling J. Affinity chromatography — from textile dyes to synthetic ligands by design: Part I. *Biopharm* 2004; 7:34–42.
7. Curling J. Affinity chromatography — from textile dyes to synthetic ligands by design: Part II. *Biopharm* 2004; 8:60–66.
8. Gunzer G and Hennrich N. Purification of α1-proteinase inhibitor by triazine dye affinity chromatography, ion exchange chromatography and gel filtration on Fractogel TSK. *Journal of Chromatography* 1984; 296: 221–229.
9. Yarmush M, Weiss A, Antonsen K, Odde D, and Yarmush D. Immunoaffinity purification: Basic principles and operational considerations. *Biotechnology Advances* 1992; 10:413–446.
10. Burgess R and Thompson N. Advances in gentle immunoaffinity chromatography. *Current Opinion in Biotechnology* 2002; 13:304–308.
11. Lowe CR, Burton SJ, Burton N, Alderton W, Pitts J, and Thomas J. Designer dyes: Biomimetic ligands for the purification of pharmaceutical proteins. *Tibtech* 1992; 10:442–448.
12. Clonis YD, Labrou NE, Kotsira VP, Mazitsos C, Melissis S, and Gogolas G. Biomimetic dyes as affinity chromatography tools in enzyme purification. *Journal of Chromatography A* 2000; 891:33–44.

13. Lowe CR. Combinatorial approaches to affinity chromatography. *Current Opinion in Chemical Biology* 2001; 5:248–256.
14. Li R, Dowd V, Stewart D, Burton S, and Lowe CR. Design, synthesis and application of a Protein A mimetic. *Nature Biotechnology* 1998; 16:190–195.
15. Tribbick G, Triantafyllou B, Lauricella R, Rodda S, Mason T, and Geysen H. Systematic fractionation of serum antibodies using multiple antigen homologous peptides as affinity ligands. *Journal of Immunological Methods* 1991; 139:155–166.
16. Cwirla SE, Peters EA, Barrett RW, and Dower WJ. Peptides on phage: A vast library of peptides for identifying ligands. *Proceedings of the National Academy of Science USA* 1990; 87:6378–6382.
17. Maclennan J. Engineering microprotein ligands for large-scale affinity purification. *BioTechnology* 1995; 13:1180–1184.
18. Reynolds G and Millner P. Synthetic peptides as affinity ligands. In: Millner PA, Ed. *High Resolution Chromatography — A Practical Approach.* Oxford, UK: Oxford University Press, 1999, pp. 191–215.
19. Romig T, Bell C, and Drolet D. Aptamer affinity chromatography: Combinatorial chemistry applied to protein purification. *Journal of Chromatography B* 1999; 731:275–284.
20. Smith JC, Derbyshire RB, Cook E, Dunthorne L, Viney J, Brewer SJ, Sassenfeld HM, and Bell LD. Chemical synthesis and cloning of a poly(arginine) coding gene fragment designed to aid polypeptide purification. *Gene* 1984; 32:321–327.
21. Ford CF, Suominen I, and Glatz CE. Fusion tails for the recovery and purification of recombinant proteins. *Protein Expression and Purification* 1991; 2:95–107.
22. Wood D, Wu W, Belfort G, Derbyshire V, and Belfort M. A genetic system to modulate intein function for use in biotechnology. *Nature Biotechnology* 1999; 17:889–892.
23. Drager RR and Regnier FE. Application of the stoichiometric displacement model of retention to anion-exchange chromatography of nucleic acids. *Journal of Chromatography* 1986; 359:147–155.
24. Brooks CA and Cramer SM. Steric mass action ion exchange displacement profiles and induced salt gradients. *AIChE Journal* 1992; 38:1969–1978.
25. Gadam SD, Jayaraman G, and Cramer SM. Characterization of non-linear adsorption properties of dextran-based polyelectrolyte displacers in IEX systems. *Journal of Chromatography* 1993; 630:37–52.
26. Stahlberg J. Retention models for ions in chromatography. *Journal of Chromatography A* 1999; 855:3–55.
27. Arshady R. Beaded polymer supports and gels II. Physicochemical criteria and functionalization. *Journal of Chromatography* 1991; 586:199–219.
28. Boschetti E. Advanced sorbents for preparative protein separation purposes. *Journal of Chromatography A* 1994; 658:207–236.
29. Weaver LE and Carta G. Protein adsorption on cation-exchangers: Comparison of macroporous and gel-composite media. *Biotechnology Progress* 1996; 12:342–355.

30. Muller W. New ion exchangers for the chromatography of biopolymers. *Journal of Chromatography* 1990; 510:133–140.
31. Afeyan NB, Gordon NF, Mazsaroff I, Varady L, Fulton SP, Yang YB, and Regnier FE. High-performance liquid chromatographic separation of flow-through particles for the biomolecules: perfusion chromatography. *Journal of Chromatography* 1990; 519:1–29.
32. Hunter AK and Carta G. Protein adsorption on novel acrylamido based polymeric ion exchangers I: Morphology and equilibrium adsorption. *Journal of Chromatography A* 2000; 897:65–80.
33. Hunter AK and Carta G. Protein adsorption on novel acrylamido based polymeric ion exchangers II: Adsorption rates and column behavior. *Journal of Chromatography A* 2000; 897:81–97.
34. Gustavsson PE and Larsson PE. Superporous agarose, a new material for chromatography. *Journal of Chromatography A* 1996; 734:231–240.
35. van Reis R. *Charge Exclusion Phenomena in Separation Media.* American Chemical Society National Meeting, San Diego, CA, March 2005.
36. Ghose S, McNerney TM, and Hubbard B. pH transitions in ion-exchange systems: Role in the development of a cation-exchange process for a recombinant protein. *Biotechnology Progress* 2002; 18:530–537.
37. Kang X and Frey DD. Chromatofocusing of peptides and proteins using linear pH gradients formed on strong IEX sorbents. *Biotechnology and Bioengineering* 2004; 87:376–387.
38. Gagnon P, Godfrey B, and Ladd D. A method for obtaining unique selectivities in IEX by the addition of organic polymers to the mobile phase. *Journal of Chromatography* 1996; 743:51–55.
39. Curtis S, Lee K, Blank GS, Brorson K, and Xu Y. Generic/matrix evaluation of SV40 clearance by anion-exchange chromatography in the flowthrough mode. *Biotechnology and Bioengineering* 2003; 84:179–186.
40. Shaltiel S and Er-el Z. Hydrophobic chromatography: Use for purification of glycogen synthetase. *Proceedings of the National Academy of Science USA* 1973; 70:778–781.
41. Hjerten S. Some general aspects of HIC. *Journal of Chromatography* 1973; 87:325–331.
42. Shansky RE, Wu SL, Figueroa A, and Karger BL. Hydrophobic interaction chromatography of biopolymers. In: Gooding K and Regnier F, Eds. *HPLC of Biological Macromolecules*, Vol. 51. Chromatographic Science Series. New York: Marcel Dekker, 1990, pp. 95–14.
43. El Rassi Z, Lee A, and Horvath Cs. Reversed-phase and hydrophobic interaction chromatography of peptides and proteins. *Bioprocess Technology* 1990; 9:447–494.
44. Quioroz JA, Tomaz CT, and Cabral J. Hydrophobic interaction chromatography of proteins. *Journal of Biotechnology* 2001; 87:143–159.
45. Ingraham R. Hydrophobic interaction chromatography of proteins. In: Mant CT and Hodges RS, Eds. *High Performance Liquid Chromatography of Peptides and Proteins*. New York: CRC Press, 1991, pp. 425–435.

46. Wu SL and Karger B. Hydrophobic interaction chromatography of proteins. *Methods in Enzymology* 1996; 270:27–47.
47. Kato Y. High-performance HIC of proteins. In: Giddings JC and Keller RA, Eds. *Advances in Chromatography*, Vol. 1. New York: Marcel Dekker, 1987, pp. 97–115.
48. Horvath Cs, Melander W, and Molnar I. Solvophobic interaction in liquid chromatography with nonpolar stationary phases. *Journal of Chromatography* 1976; 125:129–156.
49. Antia FD and Horvath Cs. Gradient elution in non-linear preparative chromatography. *Journal of Chromatography* 1989; 484:1–27.
50. Roettger BF, Myers JA, Ladisch MR, and Regnier FE. Adsorption phenomena in HIC. *Biotechnology Progress* 1989; 5:79–88.
51. Arakawa T and Timasheff S. Preferential interactions of proteins with salts in concentrated solutions. *Biochemistry* 1982; 21:6545–6552.
52. Arakawa T and Timasheff S. Mechanism of protein salting in and salting out by divalent cation salts: Balance between hydration and salt binding. *Biochemistry* 1984; 23:5912–5923.
53. Vailaya A and Horvath Cs. Retention thermodynamics in HIC. *Industrial and Chemical Engineering Research* 1996; 35:2964–2981.
54. Gagnon P, Grund E, and Lindback T. Large scale process development for HIC Part I: Gel selection and development of binding conditions. *Biopharm* 1995; 8:21–29.
55. Wu SL, Figueroa A, and Karger B. Protein conformational effects in HIC — retention characteristics and the role of mobile phase additives and stationary phase hydrophobicity. *Journal of Chromatography* 1986; 371:3–27.
56. Mant CT and Hodges RS. Effect of HPLC solvents and hydrophobic matrices on denaturation of proteins. In: Mant CT and Hodges RS, Eds. *High Performance Liquid Chromatography of Peptides and Proteins*. New York: CRC Press, 1991, pp. 437–475.
57. Tibbs TL and Fernandez E. A lactalbumin tertiary structure changes on HIC surfaces. *Journal of Colloid and Interface Science* 2003; 259:27–35.
58. Gagnon P and Grund E. Large scale process development for HIC Part IV: Controlling selectivity. *Biopharm* 1996; 9:55–64.
59. Shukla AA, Sorge L, Peterson J, Lewis P, and Waugh S. Preparative purification of a recombinant protein by HIC: Modulation of selectivity by the use of chaotropic additives. *Biotechnology Progress* 2002; 18:556–564.
60. Gagnon P. Use of HIC with a non-salt buffer system for improving process economics in purification of monoclonal antibodies. Waterside Conference on Monoclonal and Recombinant Antibodies, Miami, FL, April 30–May 3, 2000.
61. Molnar I and Horvath Cs. Separation of amino acids and peptides on non-polar stationary phases by HPLC. *Journal of Chromatography* 1977; 142:623–640.
62. Huang J and Guiochon G. Applications of preparative HPLC to the separation and purification of peptides and proteins. *Journal of Chromatography* 1989; 492:431–469.

63. Vailaya A. Fundamentals of reversed-phase chromatography: Thermodynamic and extrathermodynamic treatment. *Journal of Liquid Chromatography and Related Technologies* 2005; 28:965–1054.
64. Benedek K, Dong S, and Karger B. Kinetics of unfolding of proteins on hydrophobic surfaces in RPC. *Journal of Chromatography* 1984; 317:227–243.
65. Lu P, Carr D, Chadwick P, Li M, and Harrison K. Process purification of polypeptides and proteins by reversed phase column chromatography: Misconceptions and reality. *Biopharm* 2001; 14:28–35.
66. Kroeff E, Owens R, Campbell E, Johnson R, and Marks H. Production scale purification of biosynthetic human insulin by reversed-phase HPLC. *Journal of Chromatography* 1989; 461:45–61.
67. Olson C, Reifsnyder D, Canova-Davis E, Ling V, and Builder S. Preparative isolation of recombinant human IGF-1 by reversed-phase HPLC. *Journal of Chromatography* 1994; 675:101–112.
68. Kawasaki T. Hydroxyapatite as a liquid chromatographic packing. *Journal of Chromatography* 1991; 544:147–184.
69. Gagnon P. Hydroxyapatite chromatography. In: Gagnon P, Ed. *Purification Tools for Monoclonal Antibodies.* Tucson: Validated Biosystems, 1996, pp. 87–102.
70. Sulkowski E. Purification of proteins by IMAC. *Trends in Biotechnology* 1985; 3:1–7.
71. Yip TT and Hutchens TW. Immobilized metal ion affinity chromatography. *Molecular Biotechnology* 1994; 1:151–164.
72. Wong JW, Albright RL, and Wang NHL. IMAC — chemistry and bioseparation applications. *Separation and Purification Methods* 1991; 20:49–106.
73. Johnson RD and Arnold FH. Multipoint binding and heterogeneity in IMAC. *Biotechnology and Bioengineering* 1995; 48:437–443.
74. Wirth HJ, Unger KK, and Hearn MTW. Influence of ligand density on the properties of metal-chelate affinity supports. *Analytical Biochemistry* 1993; 208:16–25.
75. Gaberc-Porekar V and Menart V. Perspectives of IMAC. *Journal of Biochemical Biophysics Methods* 2001; 49:335–360.
76. Ueda EKM, Gout PW, and Morganti L. Current and prospective applications of metal ion–protein binding. *Journal of Chromatography A* 2003; 988:1–23.
77. Vunnum S, Galant SR, Kim YJ, and Cramer SM. Immobilized metal affinity chromatography — modeling of nonlinear multicomponent equilibrium. *Chemical Engineering Science* 1995; 11:1785–1803.
78. Botros HG. Thiophilic interaction chromatography: Principles and applications. *IJBC* 1999; 4:209–220.
79. Porath J and Belew M. Thiophilic interaction and the selective adsorption of proteins. *Tibtech* 1987; 5:225–229.
80. Scholz G, Wippich P, Leistner S, and Huse K. Salt independent binding of antibodies from human serum to heterocyclic ligands. *Journal of Chromatography B* 1998; 709:189–196.

81. Boschetti E. The use of thiophilic chromatography for antibody purification: A review. *Journal of Biochemical Biophysical Methods* 2001; 49:361–389.
82. Burton SC and Harding D. HCIC: Salt independent protein adsorption and facile elution with aqueous buffers. *Journal of Chromatography A* 1998; 814:71–81.
83. Schwartz W, Judd D, Wysocki M, Guerrier L, Birck-Wilson E, and Boschetti E. Comparison of HCIC with Protein A affinity chromatography for harvest and purification of antibodies. *Journal of Chromatography A* 2001; 908:251–263.
84. Boschetti E. Antibody separation by HCIC. *Trends in Biotechnology* 2002; 20:333–337.
85. Ghose S, Hubbard B, and Cramer SM. Protein interactions in HCIC. *Biotechnology Progress* 2005; 21:498–508.
86. Weatherly G, Bouvier A, Lydiard D, Chapline J, Henderson I, Schrimsher J, and Shepard S. Initial purification of recombinant botulinum neurotoxin fragments for pharmaceutical production using HCIC. *Journal of Chromatography A* 2002; 952:99–110.
87. Nau DR. Optimization of mobile phase conditions for antibody purification on a mixed mode chromatographic matrix. *BioChromatography* 1989; 4:131–143.
88. Johansson BL, Belew M, Eriksson S, Glad G, Lind O, Maloisel JL, and Norrman N. Preparation and characterization of prototypes for multi-modal separation media aimed for capture of negatively charged biomolecules under high salt conditions. *Journal of Chromatography A* 2003; 1016:21–33.
89. Johansson BL, Belew M, Eriksson S, Glad G, Lind O, Maloisel JL, and Norrman N. Preparation and characterization of prototypes for multi-modal separation media aimed for capture of positively charged biomolecules under high salt conditions. *Journal of Chromatography A* 2003; 1016:35–49.
90. Reifsnyder DH, Olson CV, Etcheverry T, Prashad H, and Builder SE. Purification of IGF-1 and related proteins using underivatized silica. *Journal of Chromatography A* 1996; 753:73–80.
91. Nawrocki J. The silanol group and its role in liquid chromatography. *Journal of Chromatography A* 1997; 779:29–71.
92. Ghȯse S, McNerney T, and Hubbard B. Preparative protein purification on underivatized silica. *Biotechnology and Bioengineering* 2004; 87:413–423.
93. Hagel L. Size exclusion chromatography. In: Janson JC and Ryden L, Eds. *Protein Purification: Principles, High Resolution Methods and Applications*. New York: VCH Publishers, 1989.

7 Screening of Chromatographic Stationary Phases

Abhinav A. Shukla and Xuejun Sean Han

CONTENTS

7.1 INTRODUCTION

Chromatographic unit operations have become universal in biopharmaceutical purification processes. No other type of unit operation can compare with chromatography in terms of its ability to achieve the purities required for injectable biopharmaceuticals [1,2]. Chromatographic steps exist in a variety of modes based on the type of functional group attached to the base matrix. Further complexity is provided by the possibility of the base matrix itself interacting with solutes through secondary interactions [3], ligand density of the functional groups, spacer chemistry, pore structure, and mass transport properties of the adsorbent. Selecting the best resin not only assures a reliable separation, but can also go a long way toward assuring process robustness and creating better process economics. However, the process of resin screening can be tremendously time- and resource-intensive if one seeks to explore all possible combinations of operating parameters and chromatographic stationary phases. In addition to selectivity for a given separation, a variety of other performance attributes (e.g., pressure flow characteristics, column lifetime, cleaning, and sanitization) are considered during resin selection. Clearly, the development of effective and resource efficient strategies for resin screening is an important component of bioprocess development.

In sharp contrast to the importance of the screening stage, only a few literature references have focused on resin screening. The sequential use of binding experiments to measure capacity and linear gradient experiments to screen for selectivity has been described [4]. This experimental plan was demonstrated for anion exchange chromatographic purification of a protein derived from microbial fermentation. A variety of anion and cation exchange chromatographic media were screened for flow performance, ionic capacity, and binding capacity for a model protein [5]. Due to the large amount of experimentation involved in screening every available resin, identification of an appropriate chromatographic stationary phase for a given separation remains relatively arbitrary. Often, familiarity with an existing resin in a process for a different molecule or the identification of the first resin that works reasonably well determines which resin is selected.

This chapter focuses on heuristics for the selection of the appropriate chromatographic resin for a given separation, assuming that the appropriate mode of chromatography has already been identified. Strategies for effectively screening nonaffinity chromatographic stationary phases are provided and high throughput screening techniques are discussed. Finally, a case study is provided for the development of a cation exchange polishing step for an Fc fusion protein.

7.2 RESIN SCREENING

7.2.1 Selection of Resins

Various modes of chromatography that are based on generalized interactions with the protein surface such as ion exchange, hydrophobic interaction, hydroxyapatite, and reversed phase chromatography are commonly used in protein separations. While the specific process development activities vary depending on which mode is being employed, a key step that is common to all of them is the identification of a resin that offers the greatest selectivity and product binding capacity while maintaining a high product yield. The selection of biospecific affinity resins (such as Protein A affinity media) often follows a different set of criteria, which are described in detail elsewhere in this book.

Maximizing product binding capacity is important to enable economic processing. Dynamic capacity measurements can be employed to determine binding capacity. If multiple load conditions need to be screened, static binding measurements are often employed. These are at least indicative of trends in dynamic capacities, although it is good practice to verify dynamic binding capacities through a few independent measurements. In practice, a good rule of thumb for operational loading capacity is at least 20 g/l packed resin. The achievable capacity varies significantly with the biomolecule in question (in general larger the size, lower the capacity) and with the mode of chromatography being employed (HIC [hydrophobic interaction chromatography] resins usually display low-binding capacities that are dependent on the load salt concentration). While the dynamic binding capacity limits the achievable loading, operational load capacity may often be set even lower, depending on what impact this has on the efficiency of the separation.

One needs to begin with a clear idea of what one expects from the unit operation being considered. What impurities does one expect to remove at that stage in the downstream process? Capture-unit operations are usually focused on removal of impurities that differ substantially from the product of interest. This includes water (capture steps are often expected to result in significant concentration of the product species from a dilute feedstream) and host cell protein impurities that are present in both intra- and extra-cellularly expressed products. Further downstream, the chromatographic steps are focused on the removal of impurities that are similar to the product species and hence tougher to get rid of. This would include aggregated forms of the product, product variants (with slight differences in charge, conformation, disulfide mispairing or glycosylation) or clipped species.

The basic definition of selectivity for isocratic chromatography is provided by the classic equation:

$$\alpha = \frac{k_2}{k_1} \tag{7.1}$$

where α is defined as the separation factor and k_2 and k_1 are the dimensionless retention times for the two solutes being separated.

The dimensionless retention time for a solute under isocratic conditions is given by:

$$k = \frac{t_r - t_0}{t_0} \tag{7.2}$$

where t_r is the retention time of the solute and t_0 is the void volume of the column being used.

The resolution between two species (define what this is) also depends on the peak widths and is given by the expression [6]:

$$R_S = \frac{1}{4}(\alpha - 1)\sqrt{N}\frac{k}{1+k} \tag{7.3}$$

where N is the column efficiency.

These equations can help screen for stationary phases when a discrete binary separation is being developed. However, in the case of biomolecules this is rarely the case, since the separation is usually from a wide variety of species, many of which are not completely characterized (e.g., host cell protein impurities). In addition, even when a polishing chromatographic step is being developed to remove a specific impurity such as high molecular weight aggregate, most biomolecules (and their impurities) are so heterogenous that approximation as a single solute cannot characterize their behavior sufficiently. Hence, screening chromatographic conditions for selectivity is usually empirical and necessitates the examination of a wide range of chromatographic resins, as well as solution conditions.

Often in industrial process development, screening of resins is carried out by examining the efficacy of a separation under a starting set of operating conditions. Alternative resins are screened under the same conditions. While this ensures that the resin finally selected is compatible with process operating conditions, it does not compare the resins on an equal footing, since the operating conditions arbitrarily selected can be significantly different from the optimal conditions for a given resin. In addition, the definition of selectivity shown in Equation 7.1 is an intrinsic property of the chromatographic resin under a unique set of mobile phase conditions, and does not depend on column loading. Accordingly, selectivity comparisons need to be carried out under analytical conditions in contrast to the preparative loads that are often employed during screening. To compare resins on a more equal footing, it has been suggested to investigate separation efficacy by operating analytical linear gradient experiments over a range of mobile phase pHs and buffers [4]. Simpler gradients with only buffer strength varying with time are to be preferred. For example, salt concentration gradients on ion exchange carried out under a few different pH conditions are to be preferred to pH gradients or simultaneous pH and salt gradients during screening.

Since the affinity of the biomolecule under the test conditions is unknown, it may elute at different parts of the gradient on different resins. Accordingly, a second gradient run is sometimes conducted with gradient conditions set such that elution occurs close to the center of the gradient. Sometimes, comparing purities of the elution pools from these gradient experiments is enough to enable a choice between chromatographic resins, especially if one or two conditions provide outstanding impurity clearance. Usually however, the impurity levels in the elution pools will be too close to allow a clear choice of stationary phase on this criterion alone. To make a better selection, a retention map for the most relevant impurity must be plotted to distinguish selectivity trends between resins. For instance, while the pool purities for two resins might be similar, the impurity might be concentrated in the front part of the elution peak on one of the resins. This might allow for the development of a selective wash step on that resin, or one might initiate peak collection a little later into the elution peak to avoid collecting the impurity.

A diagnostic map for visualizing elution trends that we have often employed successfully in process development is a plot of cumulative product impurity level vs. cumulative yield, as one moves from left to right through the elution peak. Consider the elution profiles on two CEX resins in (cation exchange) Figure 7.1a and Figure 7.1b. Also plotted on the figures are the percentage of impurity for each of the elution peak fractions that were collected and analyzed by HPLC. As can be seen from the figures, in case A the impurity is skewed toward the back side of the peak while in case B it is distributed evenly through the elution peak. However, if all fractions were to be pooled, the impurity level for both the pooled peaks would be very similar. To create the diagnostic plot for each chromatogram, one moves from left to right through the fractions and calculates the percentage of product yield and the percentage of impurity for all fractions to the left of that point. This is shown schematically in Figure 7.2. These numbers can then be plotted against each other in a plot of cumulative percentage of yield vs. cumulative impurity level. Figure 7.3 shows these plots for the two resins in cases A and B mentioned earlier. From this figure, it can be clearly seen that resin A is superior to B in terms of selectivity for removing that impurity. One can obtain a significantly higher yield on resin A than on resin B for a fixed level of impurity clearance. On resin A it may be possible to terminate elution peak collection earlier to obtain a cleaner product at 85% yield.

7.2.2 High Throughput Screening Techniques

The plot of cumulative impurity level vs. cumulative product yield provides an effective means of comparing the selectivity of various chromatographic

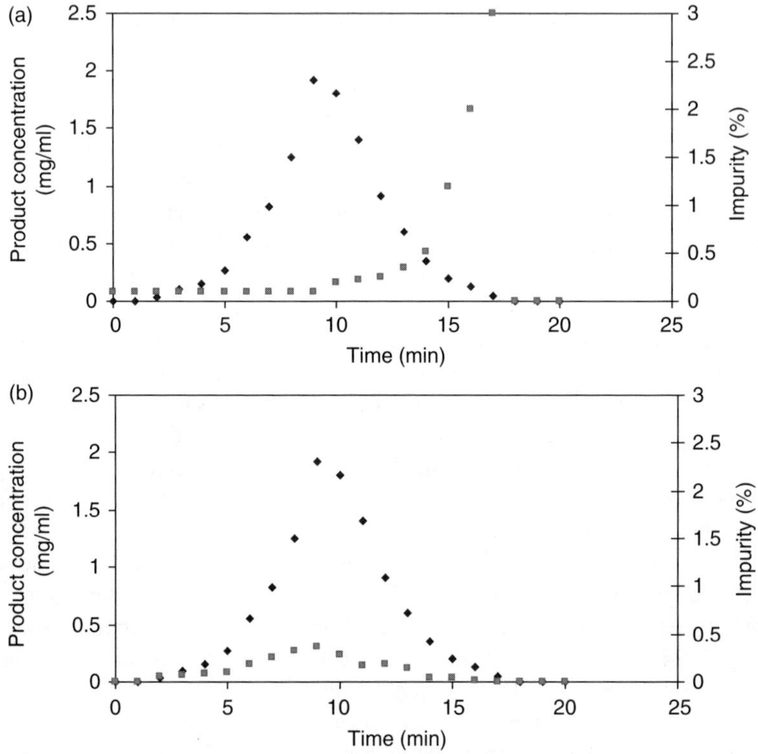

FIGURE 7.1 Chromatograms on resins A (a) and B (b) showing product elution and impurity level.

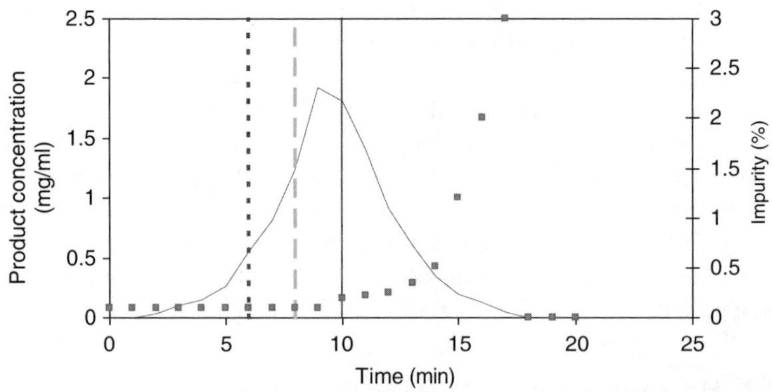

FIGURE 7.2 Calculation of cumulative percentage of impurity and cumulative product yield for a chromatogram with fraction analysis.

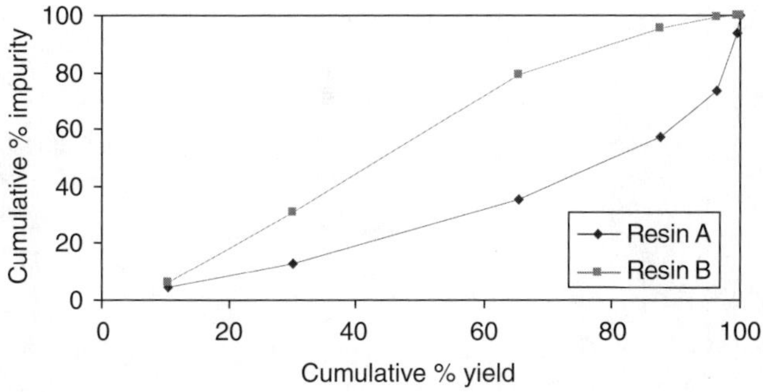

FIGURE 7.3 Plot of cumulative percentage of impurity vs. cumulative yield created from Figure 7.2.

resins under specific pH and buffer types. However, it still involves significant experimentation to screen a wide variety of resins and mobile phase conditions. In recent years, batch screening carried out in a high throughput mode has emerged as an alternative means of screening large combinations of operating conditions rapidly.

High throughput batch systems have been employed for identifying both selective and high-affinity small molecule displacers for two model proteins in a cation exchange system [7]. The percent protein displaced at low mobile-phase salt concentrations was used as the measure for comparing the efficacy of various displacer molecules, and these were then correlated to their structure using quantitative structure efficacy relationship (QSER) models. HIC conditions were optimized to remove high molecular weight aggregate from a monoclonal antibody in-process stream using high throughput screening in a 96 well plate [8]. Six different salts and eight HIC resins were rapidly screened to determine the highest salt concentration that could be used without precipitating the product. Following this screen, a subset of conditions were selected and tested for relative binding between the product and a high molecular weight aggregate species that was sought to be cleared in this process step. Eight cation exchange resins were screened at four different pHs and over a range of eluting mobile phase salt concentrations to construct pseudo chromatograms from UV measurements of the filtered supernatants [9]. This narrowed down the experimentation in the column mode, since preliminary operating conditions were available for each resin from the batch mode. Anion exchange flowthrough chromatographic operating conditions were screened for a monoclonal antibody for host cell protein and leached Protein A clearance [10]. It was determined that thc best clearance occurs under mildly retained conditions for the monoclonal antibody. In the same presentation, the authors also created contour plots of selectivity vs. sodium phosphate and sodium chloride concentration

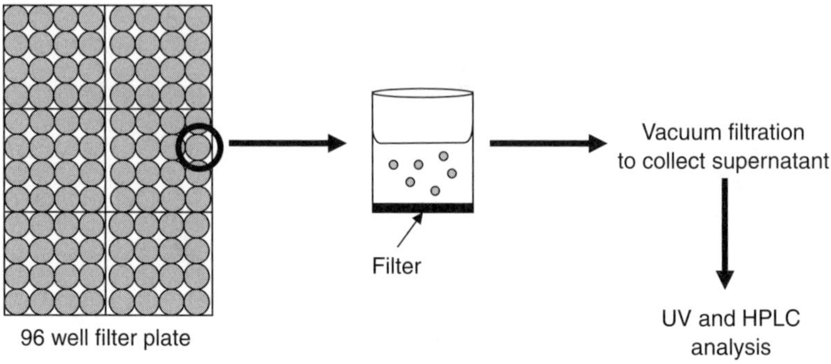

FIGURE 7.4 A high throughput screening experimental setup for chromatographic resins.

for antibody binding to a ceramic hydroxyapatite resin. Since hydroxyapatite possesses a complex chemistry with a combination of anion and cation exchange and metal chelate interactions, the two types of salts produce different effects on retention. Once again, a hydroxyapatite flowthrough step was developed under mildly retained conditions for the product. Clearly, high throughput screening using 96 well plate systems can be applied successfully for a variety of chromatographic modes and separation problems. Applying this format for screening allows one to narrow down the list of variables that need to be studied by column experiments and can be a very useful first stage in process development.

Figure 7.4 shows the schematic for a high throughput screening operation. As can be seen from the figure, a large number of conditions can be screened per plate with a significant reduction in both resin and protein usage. The rate-limiting step for the screening now becomes the analysis time for impurities and necessitates the usage of high throughput analytical chromatography and automated ELISA assays. Concentrations can be readily estimated by taking UV readings on a spectrophotometric plate reader, and this allows for the ready determination of binding properties under various mobile phase conditions.

High throughput screening results obviously do not directly translate into experimental data from column chromatographic experiments. First, batch experiments lack the resolution possible on chromatographic columns due to the presence of a large number of theoretical plates in a packed column and only a single plate in a well-mixed batch system. Second, differences in mass transport properties into beads cannot be adequately captured in the batch mode with no convective flow. Some chromatographic stationary phases have significantly better mass transport properties which translate into operational advantages such as a higher-operational flow rate. Nevertheless, the batch mode allows for the screening of situations where the stationary phase chemistry plays the dominant role in the separation.

While no commercial HTS systems are currently available for screening chromatographic systems, many users have developed a modified high Tecan system conventionally used for biomolecular screening in drug discovery. Alternatively, a system called retentate chromatography or surface enhanced laser de-ionization (SELDI) has been developed by Ciphergen Biosystems, Fremont CA (now part of Pall Corporation, Pensacola, FL). In this system, surfaces with various chromatographic functionalities are exposed to bioprocess samples under various mobile phase conditions. Proteins (either product or impurities) that bind to the surface are displaced by laser ionization and analyzed by time-of-flight mass spectrometry which provides information about the mass and percentage of dominant species. By studying the retained sample over a range of mobile phase conditions, one can rapidly gain information about a variety of chromatographic operating conditions. While the mass spectrometry analysis provides this tool with a high resolution, high throughput analytical tool that can shed light on major process impurities in addition to the product elution profile, the chip surface functionalization does not mimic that of preparative stationary phases. Since ligand density, backbone and spacer arm chemistry play an important role in determining resin selectivity, the data from retentate chromatography is expected to be indicative rather than definitive for determining process separation conditions.

7.2.3 Other Points to Consider while Selecting a Resin

Screening chromatographic stationary phases for binding capacity and selectivity are only two criteria considered while deciding which resin to use in a process application. A large and ever growing number of vendor companies sell resins for preparative biomolecule separations. If possible, it is advisable to stick with manufacturers and resin grades that one already has experience with in manufacturing processes. When dealing with a new vendor whose resins are not widely employed in bioprocess purification, auditing the vendor to examine their manufacturing and quality systems is highly recommended. Vendors are also expected to make available regulatory support files that contain manufacturing and quality control information for a given stationary phase. Lot-to-lot reproducibility is key while implementing a resin in a manufacturing process. It is important to realize that while vendor companies do have release tests for each lot of resin they produce, these criteria differ quite substantially from what one is looking for in an actual bioseparation process. For example, vendors typically test ion exchange resins for binding capacity for a model protein. However, fairly wide variations in surface ligand density might still produce identical binding capacity for a model protein, but still result in significant differences in host cell protein clearance. Very often, one does not utilize the charged functionalities themselves for impurity binding, instead these might adhere to the resin through hydrophobic association with the backbone. During process development multiple lots of stationary phase should be tested to determine if process performance changes. If a process step is found to be particularly sensitive to the incoming lots of resin and no substitution with a more robust resin is possible, one should consider instituting an in-house lot release test using an actual process feedstream. The scale of resin production at the vendor company and experience with scaling-up a new resin to larger production quantities is an important point to consider in the early stage. One obviously does not want to be in a situation in which ones biomolecule requires rapid scale-up due to dramatic clinical success, and the vendor company cannot scale-up resin production successfully. Finally, for commercial stage molecules it is wise to qualify a second source of resin as one would for other chemical raw materials. This depends on whether the production process can tolerate switching between two types of resins, and is usually more easily achievable for affinity resins where selectivity depends on the immobilized functionality and is not significantly influenced by nonspecific backbone interactions.

7.3 CASE STUDY: DEVELOPMENT OF A CATION EXCHANGE PURIFICATION STEP FOR AN Fc FUSION PROTEIN

The case study presented below demonstrates some of the concepts discussed so far in the chapter and provides a logical flow plan for experimentation in selecting a chromatographic stationary phase. The example selected here is that of an Fc fusion protein expressed in a mammalian host cell line and captured from cell supernatant by Protein A affinity chromatography. While developing polishing steps for monoclonal antibodies and Fc fusion proteins, it is common to employ a representative Protein A column eluate as the feed stock, irrespective of whether this step immediately follows capture or not. Not only does this parallel development strategy for the polishing steps save time, it is often necessary since the level of impurities after a single polishing step is often so low that meaningful screening of operating conditions for the third chromatographic step cannot be carried out. Even in such cases, employing two polishing chromatographic steps is often the practice to create some level of process redundancy and improve overall process robustness.

The CEX chromatographic step was developed primarily to reduce high molecular weight aggregate from the Protein A elution pool following low pH viral inactivation. A lesser aim was to achieve clearance of leached Protein A and host cell protein impurities.

TABLE 7.1
List of Cation Exchange Stationary Phases Screened for Purification of an Fc Fusion Protein

Resin Name	Strong/Weak Functionality	Bead Chemistry	Bead Size (μ)	Manufacturer
SP Sepharose FF	Strong	Agarose	90	GE Healthcare
CM Sepharose FF	Weak	Agarose	90	GE Healthcare
SP Sepharose XL	Strong	Agarose with dextran coupling	90	GE Healthcare
Toyopearl CM 650M	Weak	Polymethacrylate	65	Tosoh Biosep
Toyopearl SP 650M	Strong	Polymethacrylate	65	Tosoh Biosep
Fractogel SO_3 Hicap	Strong	Toyopearl 650M base bead with tentacles bearing ligands grafted	65	Merck KgGA
Fractogel SE Hicap	Strong	Same as above with sulfoethyl functionalities	65	Merck KgGA
Macroprep HS	Strong	Polymethacrylate	50	Biorad
Macroprep CM	Weak	Polymethacrylate	50	Biorad
Unosphere S	Strong	Single step polymerization	120	Biorad
Fractoprep SP	Strong	Tentacle chemistry similar to Fractogel on a proprietary base bead		Merck KgGA

7.3.1 Step 1 — Short-Listing Resins to Be Screened

Table 7.1 lists the cation exchange resins that were screened for this molecule. Selection of these resins was based on familiarity with the vendors as reliable suppliers in addition to large-scale commercial process experience with some of the resins. The resins cover both strong and a few weak cation exchangers as well as a range of bead chemistries. A common feature of all of these stationary phases is their relatively large particle size ($>50\ \mu$) that can be employed in preparative scale columns without an overly restrictive pressure drop limitation.

7.3.2 Step 2 — Batch Capacity Measurements

Static capacity measurements were carried out in the batch mode on these resins in a 25 m*M* phosphate buffer at pH 5.0 and 6.0. Figure 7.5 shows the data obtained under a subset of the conditions studied. As can be seen from the figure, most resins showed high binding capacities for the Fc fusion protein

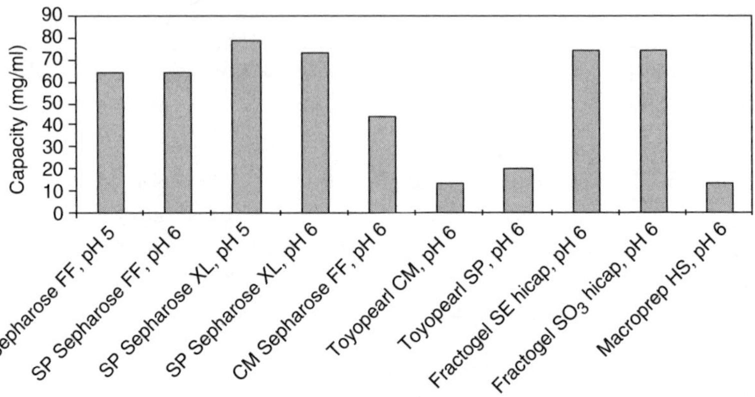

FIGURE 7.5 Batch capacity measurements on CEX.

except for Toyopearl SP650M, CM650M, and Macroprep HS. Trends in batch binding capacity matched trends from dynamic capacity measurements in the column mode (data not shown).

7.3.3 Step 3 — Binding Affinity Measurements

Linear salt gradient experiments were conducted on the cation exchangers to determine the strength of binding between the product and the resins. Salt gradients were run at pH 5.0 and 6.0 in 25 m*M* phosphate buffer with sodium chloride as the eluting salt. The salt concentration at peak maxima were read off the gradient experiments after correction for system delay time, and are plotted in Figure 7.6a and Figure 7.6b. At pH 5.0 the product did not elute even in 500 m*M* NaCl from Macroprep HS, Macroprep CM, and Unosphere S media. In general, the product showed high binding strength on all resins. Affinity on pH 5.0 was expectedly higher than at pH 6.0 due to further ionization of the protein. None of the elution conductivities were low enough to create issues in binding the Protein A column elution pool after acid inactivation and neutralization.

7.3.4 Step 4 — Peak Splitting Screening

While the Fc fusion protein had high binding capacities and strong affinity on CEX chromatography, peak splitting was observed during the linear gradient experiments described above. Some of the injected protein sample was found to be retained on the column despite operating the linear gradient to high salt concentration and instead elute during the strip with 0.5 *N* NaOH. Reinjection of

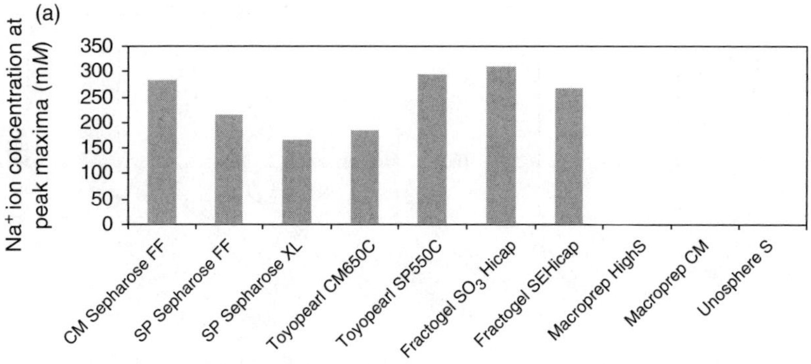

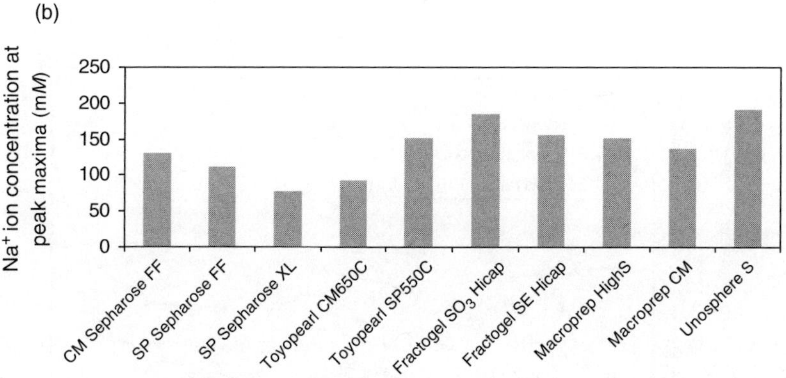

FIGURE 7.6 Binding affinity measurements on CEX: (a) affinity at pH 5.0; (b) affinity at pH 6.0.

the peak eluting during the gradient (following dilution to lower conductivity) once again exhibited this splitting phenomenon. This served to indicate that the splitting was not a separation phenomenon but instead an adverse influence of exposure of the protein to the chromatographic column. It has been shown that unfolding of proteins on binding to chromatographic surfaces can lead to denaturation and result in split peaks [11]. Since exposing the protein to a potentially denaturing condition is not advisable, the observation of peak splitting became a negative screen for further resin selection. Figure 7.7 shows the percentage of peak splitting measured at pH 5.0 and 6.0 on the CEX resins. As can be seen from the figure, significantly greater peak splitting was observed at pH 5.0 than at pH 6.0. Clearly, denaturation of the product depends on the mobile phase pH in addition to the resin chemistry. Accordingly, pH 5.0 was ruled out for process operation. At pH 6.0 low levels of peak splitting were seen on the agarose-based resins that are presumably more hydrophilic

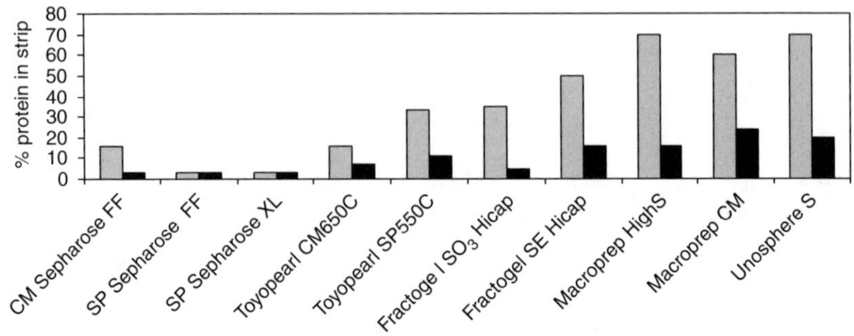

FIGURE 7.7 Peak splitting on CEX at pH 5.0 and 6.0. Black: pH 5.0, Gray: pH 6.0.

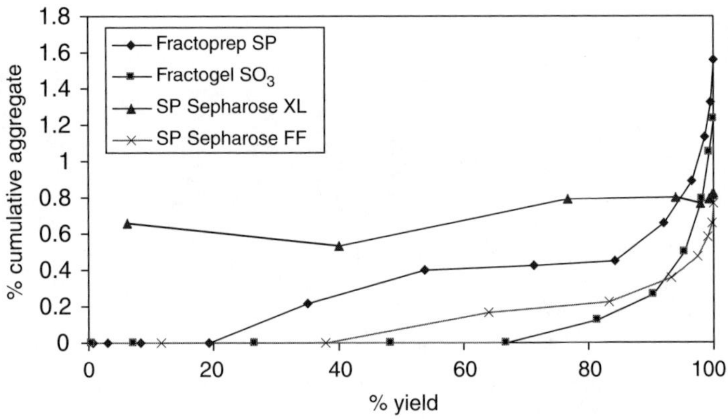

FIGURE 7.8 Plot of cumulative percentage of aggregate vs. cumulative yield on four cation exchange resins.

and on Fractogel SO_3 Hicap in which presumably interactions take place with functional groups on the tentacles rather than with the resin backbone. These four resins were selected and continued with for further screening.

7.3.5 Step 5 — Selectivity for High Molecular Weight Aggregate Clearance

Fractions were collected from analytical linear gradient experiments and analyzed by size-exclusion chromatography and quantitated by UV absorbance measurements. From this data, a plot of cumulative percentage of high molecular weight aggregate vs. cumulative yield was constructed (Figure 7.8). As can be seen from the figure, practically no aggregate clearance was obtained on SP Sepharose XL. On the other hand, both Fractogel SO_3 Hicap and SP Sepharose FF were effective in clearing aggregate. On these resins, fairly low cumulative aggregate levels (<0.5%) could be obtained at relatively high product yields (>90%). Terminating elution peak collection at 10% UV280 peak maximum was found to achieve good clearance on both resins with high product yield.

7.3.6 Step 6 — Selectivity for Leached Protein A and Host Cell Protein Removal

Using the information on elution salt strength from the analytical linear gradient experiments as a starting point, a step gradient process was set up on the two resins. The step gradient process employed the Protein A column eluate (following viral inactivation and neutralization) as its load. CHO host cell protein and leached Protein A clearance data were obtained from representative step gradient experiments loaded identically at 40 mg/ml capacity on the two resins (Figure 7.9a and Figure 7.9b). As can be seen from the figures, the Fractogel SO_3 Hicap resin was more selective than SP Sepharose FF. Based on this data as

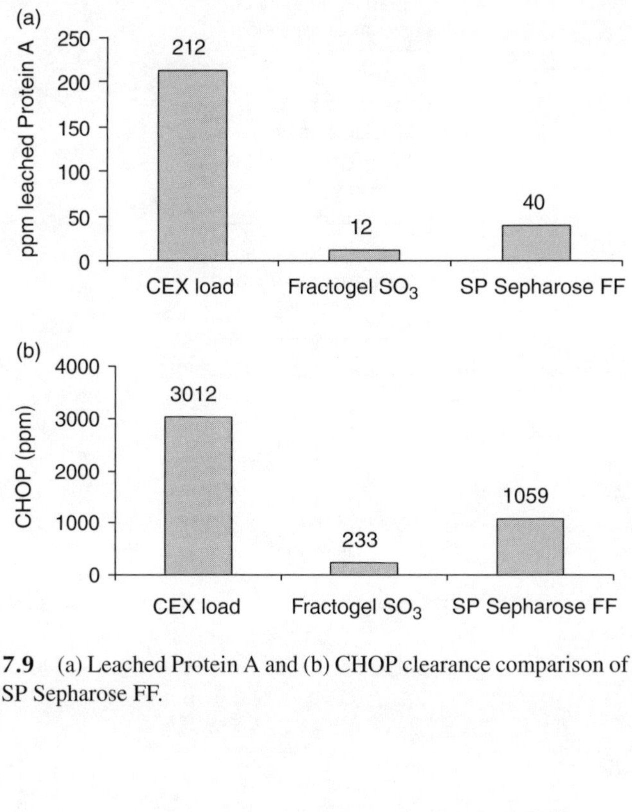

FIGURE 7.9 (a) Leached Protein A and (b) CHOP clearance comparison of Fractogel SO_3 and SP Sepharose FF.

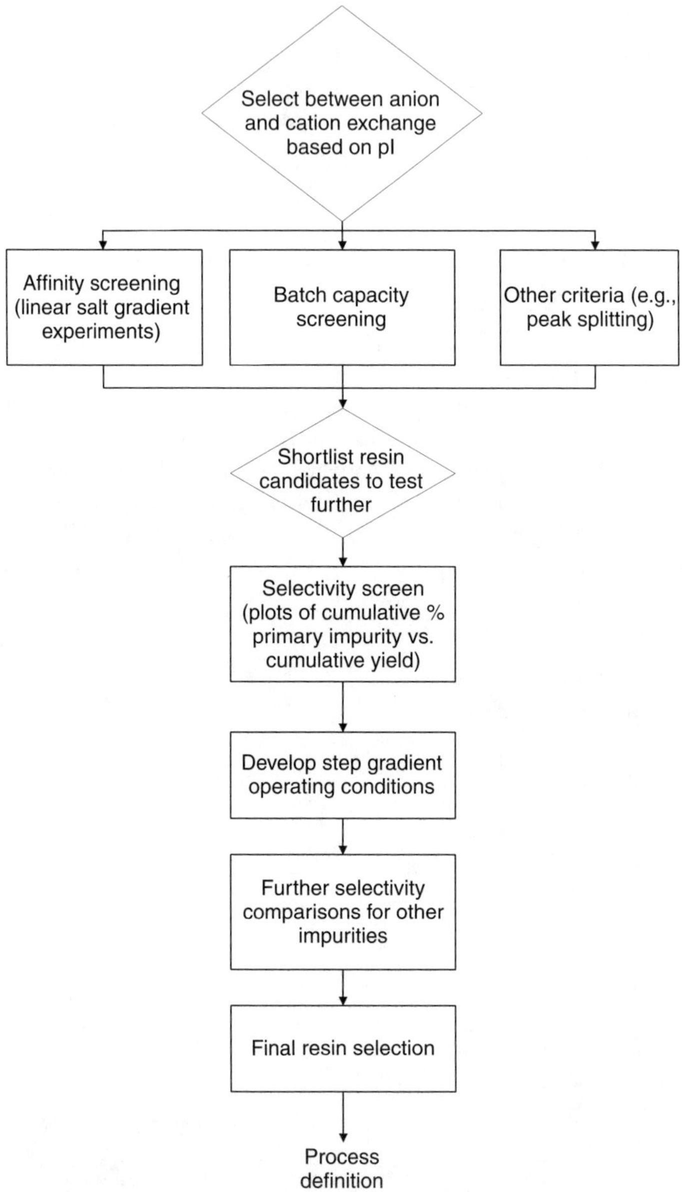

FIGURE 7.10 Flowsheet for development of a CEX chromatographic step for the Fc fusion protein.

well as higher binding capacity and greater affinity; Fractogel SO_3 was selected as the cation exchange resin for this application.

A flow sheet of experiments can be made to summarize development of the cation exchange unit operation is shown in Figure 7.10.

7.4 CONCLUSIONS

A clear, methodical plan for screening chromatographic stationary phases is highly important for successful process development in a time line-driven and resource-limited operating paradigm that dominates the biopharmaceutical industry. Tools like high throughput screening and retentate chromatography can greatly assist the rapid collection of data over a wide operating space and serve as a starting point for column experiments. It is also imperative to design and conduct column experiments in a way that allows for an equitable comparison of chromatographic stationary phases. A better understanding of selectivity and how it varies over the operating space is very useful. Also required are data interpretation tools such plots of cumulative percentage of impurity vs. cumulative product yield that can indicate trends in selectivity between resins, even when pools would not indicate significant differences.

REFERENCES

1. Jungbauer A and Boschetti E. Manufacture of recombinant proteins with safe and validated chromatographic sorbents. *Journal of Chromatography A*, 1994; 662:143–179.
2. Jungbauer A. Chromatographic media for bioseparation. *Journal of Chromatography A*, 2005; 1065:3–12.
3. Shukla AA, Bae SS, Moore JA, and Cramer SM. Structural characteristics of low-molecular-mass displacers for cation-exchange chromatography — II. Role of the stationary phase. *Journal of Chromatography A*, 1998; 827:295–310.
4. Rathore AS. Resin screening to optimize chromatographic separations. *LC GC*, 2001; 19:616–622.
5. Levison PR, Mumford C, Streater M, Brandt-Nielsen A, Pathirana ND, and Badger SE. Performance comparison of low pressure ion-exchange chromatography media for protein separation. *Journal of Chromatography A*, 1997; 760:151–158.
6. Wankat, PC. *Rate Controlled Separations*. Springer-Verlag: New York, 1994.
7. Rege K, Ladiwala A, Tugcu N, Breneman C, and Cramer SM. Parallel screening of selective and high affinity displacers for proteins in ion-exchange systems. *Journal of Chromatography A*, 2004; 1033:19–28.

8. Coffman J, Kramarczyk J, Bastek P, Molnar K, and Kelley, B. High-throughput screening of chromatographic resins and excipients to optimize selectivity. Recovery of Biological Products XI, Banff, Canada, September 2003.
9. Bastek P, Molnar K, Kelley B, and Coffman J. High-throughput screening of resins and excipients for chromatographic process development. American Chemical Society National Meeting, Anaheim, CA, March 2004.
10. Coffman J and Kelley B. High-throughput screening in downstream process development of protein therapeutics. American Chemical Society National Meeting, San Diego, CA, March 2005.
11. Wu SL, Figueroa A, and Karger BL. Protein conformational effects in HIC — retention characteristics and the role of mobile phase additives and stationary phase hydrophobicity. *Journal of Chromatography*, 1986; 371:3–27.

8 *A Priori* Prediction of Chromatographic Separations from Protein Structure Data

Asif Ladiwala, Curt M. Breneman, and Steven M. Cramer

CONTENTS

8.1 INTRODUCTION

The *a priori* prediction of chromatographic behavior directly from protein structure data has been a long-standing goal in the separations field. The availability of predictive models can decrease the uncertainty associated with most chromatographic development work, reducing the time needed to bring biological drug products to the market. Furthermore, such investigations will also enable us to gain insights into the factors influencing the affinity and selectivity of biomolecules in different chromatographic systems. This information can in turn be employed to design more efficient processes, and perhaps even enable the development of tailored chromatographic resin materials with unique selectivities for specific separation applications.

8.2 QUANTITATIVE STRUCTURE–PROPERTY RELATIONSHIPS

Researchers in drug discovery and analytical chemistry have dedicated much attention toward developing an improved understanding of the specific interactions that occur among given chemical species. Mainly, scientists recognized the need to establish relations between chemical structures of compounds and their properties. The term quantitative structure–property relationship (QSPR) is used generically to describe such correlations. The first correlation of this kind was reported in the 19th century by Brown and Fraser [1] in the area of alkaloid activity. Later, regressions were published, establishing the dependence between the structure and the equilibrium and rate constants of drugs such as antihistamines and compounds employed for anesthesia [2–4]. Hammett's correlations [3,4] are generally termed linear Gibbs free energy relationships (LFERs) because the Hammett parameter (σ) is related to thermodynamic fundamentals. LFERs have played an important role in the prediction of properties of compounds with similar structures to the molecules used to generate the correlations. Over the past few decades, researchers in drug design have actively developed and employed computational tools to accelerate the development of both new and improved therapeutics [5–9]. Regression models have been generated to relate the properties of drug compounds, such as activity, partitioning, and toxicity, to the chemical structures of the corresponding molecules [9–13]. When the activity of a drug molecule is related to its structure, the correlations are commonly referred to as quantitative structure–activity relationships (QSARs).

Structure–property modeling approaches employed in drug discovery have been extended to chromatography in an attempt to link the structure of compounds to their adsorption behavior on surfaces and resin materials. In general, when the retention of compounds in a chromatographic column is correlated with its structural properties, the models are referred to as quantitative structure–retention relationships (QSRRs) [14]. QSRRs (or more generally QSPRs) are statistically derived relationships between the chromatographic parameters determined for a representative series of analytes in given separation systems and the quantities accounting for the structural differences among analytes tested. They are essentially manifestations of LFERs and referred to as extrathermodynamic relationships (ETRs), that is, they are not necessarily a consequence of thermodynamics [15]. Extrathermodynamic approaches combine detailed models of processes with certain concepts of thermodynamics. It is well known that the thermodynamic properties of a given substance are bulk properties reflecting just the net interactive effects in that system. The magnitude of thermodynamic parameters represents the combination of individual interactions that may take place at the molecular level. Thus, classical thermodynamics fails to explain the precise molecular interactions responsible for retention and only provides an overall picture. It is therefore, inadequate for the purpose of analyzing retention in chromatographic systems. However, the development of LFERs that can predict retention, suggests the presence of a real connection between some correlated quantities, the nature of which can be subsequently identified. Some of the key goals of QSRR studies are (1) prediction of retention for a new solute; (2) identification of the most informative structural descriptors; (3) elucidation of the molecular mechanism of separation in a given chromatographic system; (4) evaluation of complex physiochemical properties of solutes, other than chromatographic for example, their hydrophobicity; and (5) estimation of relative biological activities within a set of drugs and other xenobiotics as well as the material properties of individual members of a family of chemicals [14].

Quantitative structure–retention relationship (QSRR) models have been widely reported for gas chromatography (GC) [16–18], reversed phase chromatography (RPLC) [14,19,20], and micellar electrokinetic chromatography (MEKC) [21,22] of small molecules. These reports are, however, based on the generation of models with a relatively small number of predetermined physiochemical properties of the solutes. Such an approach may not be entirely desirable, since it assumes complete knowledge of the nature of the interactions between the solutes and the stationary phase resin when selecting the molecular properties to be employed for model building. It has been shown that QSRR models can be successfully derived for small molecules in reversed phase chromatography employing a partial least squares (PLSs) modeling approach with a genetic algorithm (GA) based feature selection [23]. This approach is a departure from the traditional QSRR modeling methods because a large number of physiochemical properties were initially calculated for the molecules in the dataset and the properties that were most highly correlated with the experimental response were identified by a capacity-controlled GA feature selection routine. The selected parameters were then employed to generate the QSRR model based on PLS regression and molecules not present in the training set were employed for testing the predictive ability of the resulting model. This methodology not only results in the generation of predictive models, but also enables the investigation of the nature of the interactions between the solute and the resin in a given chromatographic system through model interpretation.

Much of the early research in this field focused on generating predictive models for the retention of small molecules under different chromatographic conditions and modes. Advances in computational chemistry and chemometrics have enabled researchers to compute physicochemical parameters of larger biological molecules such as proteins and nucleic acids for the purpose of QSPR studies. These studies can facilitate the generation of models that can predict the chromatographic behavior of biomolecules in different chromatographic modes. The case study presented in this chapter demonstrates the utility of QSPR modeling as a method development tool for chromatographic bioprocesses.

8.3 QSPR MODELING FUNDAMENTALS

In order to generate a QSPR model for a set of compounds, the physicochemical properties of molecules are numerically represented in terms of molecular descriptors or features. These molecular property descriptors become the independent variables in the model, while the chromatographic data (e.g., retention time, adsorption isotherm parameters, free energy of adsorption, etc.) constitutes the dependent variable. Regression algorithms are employed to correlate the dependent variable (i.e., response/data) with a set of relevant independent variables (i.e., descriptors) to produce the QSPR model. Details of the various aspects of QSPRs are provided in the following sections.

8.3.1 MOLECULAR DESCRIPTORS

As more researchers have become aware of the power of QSPRs, a large number of molecular property descriptors have been defined in order to accommodate particular applications. Hansch and coworkers [24] pioneered a hydrophobic term which was based on the partition coefficient of compounds between octanol and water. Although this hydrophobic term was originally employed for biomolecules, later its application became common in other areas of research [25]. Kamlet and coworkers [26–28] have developed and employed solvatochromic parameters to establish relations between solute–solvent interactions in a system in terms of linear solvation energy relationships (LSERs). However, a limitation of solvatochromic parameters and, therefore, LSERs is that they are determined experimentally, requiring much experimentation and the compounds of interest are not always available in large enough quantities. A significant breakthrough in the study of LSERs was the work published by Famini and coworkers [29–31]. They defined the theoretical linear solvation energy relationship (TLSER) parameters which enabled the elimination of the intensive experimental phase involved in the development of the LSERs. Weiner [32] developed topological indices for correlating the boiling points and surface tension of alkanes. Hall and Kier [33–35] have reported extensively on the use of connectivity indices in the field of physical chemistry as well as for drug studies. Tipker and Verloop [36] have defined parameters to describe the size and shape of functional groups. These parameters have been employed in developing pesticides as well as in drug analysis [25,37]. In addition, molecular connectivity-based two-dimensional (2D) descriptors have also been employed for both drug design and chromatography modeling [38–40].

The three-dimensional (3D) structure of a molecule determines its physical and chemical properties as well as the manner in which it can interact with other molecules and surfaces. Mechanical models have been employed to visualize molecules in 3D. These representations have been used to understand the reactivity of molecules and their biological interactions. In defining the 3D structure of a molecule, it is assumed that molecules are in their most favorable conformation when the energy state is at a minimum. In order to minimize the structure, molecular mechanics calculations are performed and the first and second derivative of the energy with respect to the distances and orientations of the atoms within the molecule are determined. These calculations require the knowledge of the potential function parameters which are generally obtained by using appropriate force fields such as MM3, AMBER, CHARMM, and MMFF94 to name a few. However, for larger molecules, these calculations can become computationally cumbersome and, therefore, empirical equations and heuristics are used to estimate the potential function parameters. Once the geometry of the molecules has been optimized 3D descriptors such as volume, radius, shape, and the van der Waals and the solvent accessible surface areas and the surface area of the molecule associated with different molecular properties can be readily computed.

8.3.2 Feature Selection

Although it might seem problematic at first, it is not unusual to begin the modeling process with more descriptors than experimental cases —

a potentially dangerous situation where over-determined models may be produced. A theoretical justification for feature selection (or descriptor removal) lies in the fact that all molecular descriptors provide information concerning some observable chemical properties, but most descriptors are not general enough or do not have truly linear relationships with observable responses to apply to all molecule behavior. Thus, by the elimination of descriptors that are not relevant to a particular property of interest, the signal-to-noise ratio is increased and superior models are produced. An important consideration when utilizing large numbers of modern molecular descriptors is finding a way to select a small number of the most important features from a set of several hundred possibilities, while safeguarding against model overtraining. Due to the difficulty involved in making the correct choices, the development of new feature selection methods is an active area of research. The literature is abounding with examples of feature selection strategies that have been successfully applied in QSPRs. These include stepwise regressions [41], forward selection [42] and backward elimination [43], simulated annealing [44], and evolutionary and genetic algorithms [45–47].

8.3.3 Modeling Techniques

To produce QSPR models, it is necessary to have good descriptors as well as robust regression/machine learning methods that can capture and exploit the chemical information encoded in the descriptors. Various methods have been utilized for building chemical property models. A few of these are listed in Table 8.1.

When small numbers of independent descriptors are used to build models, simple linear methods such as MLR may provide adequate models. In the more general case where large numbers of correlated descriptors are used, extra care must be taken to avoid producing an over-determined model that lacks predictive power. PLS [48], ANN, and SVM regression [49] work well with such large numbers of descriptors, provided appropriate feature selection procedures are used during model construction.

Early efforts toward developing predictive QSPRs for the chromatographic retention behavior of proteins in ion-exchange systems employed GA-based feature selection for variable reduction followed by PLS regression [50]. The case study presented in this chapter employs a novel, sparse SVM feature selection algorithm followed by a capacity-controlled nonlinear SVM regression algorithm [51], which was found to yield robust and generally predictive models through the use of boostrapping and bagging techniques (discussed in Section 8.5).

TABLE 8.1
Regression and Classification Techniques Commonly Employed in Structure–Property Modeling Studies

Local Learning (LL)	Similarity of molecules with parts of the training data
Multiple Linear Regression (MLR)	Multidimensional least-squares analysis
Principal Component Analysis (PCA)	Combines descriptors to find relationships that explain linear relationships in the data
Partial Least Squares (PLSs)	Like PCA, but uses the experimental data as well to develop a small set of "latent variables" that explain the data
Artificial Neural Networks (ANNs)	Nonlinear node-based learning system
Support Vector Machines (SVMs)	Linear or nonlinear classification or regression system

8.4 PROTEIN DESCRIPTOR GENERATION

Several different types of descriptors have been employed in the example presented in the following discussion. These include traditional 2D and 3D descriptors obtained using the commercially available molecular operating environment (MOE, CCG, Inc., Montreal, Canada) software package, as well as some modern electron density-derived descriptors such as those from transferable atom equivalent (TAE) calculations. Details of these descriptor types are provided in the following sections.

8.4.1 MOE Descriptors

The MOE software package provides a combination of traditional molecular property descriptor types that span several classes, including connectivity-based topological 2D and shape-dependent 3D molecular features. MOE descriptors span the following three classes of descriptors that are common to the molecular modeling community:

1. *2D.* Topological molecular descriptors are defined as numerical properties that can be calculated using a connection table representation of a molecule (e.g., elements, formal charges and bonds, but not atomic coordinates). These descriptors are computed using only atom type and connectivity information and are known to be remarkably effective in a number of applications [52]. The fact that topological descriptors carry no molecular conformation or shape

information is both a limitation and a benefit, in that their use reduces the ambiguity that can result from modeling highly flexible molecules.

2. *i3D.* Internal 3D descriptors use spatial information for each molecule, but are invariant to rotations and translations of the molecule. This class includes descriptors that incorporate quantum mechanical or empirical force field results, or rely only on the internal coordinates of each molecule. These include potential energy descriptors, surface areas, volumes, dipole moment, and bulk shape descriptors. Descriptors such as those obtained from TAE also fall into this category.
3. *x3D.* External 3D descriptors also use 3D coordinate information, but in addition, they require an absolute frame of reference (e.g., molecules docked into the same receptor). These are less commonly used in QSPR investigations, but figure prominently in CoMFA ligand/binding site investigations and other rational design techniques. x3D descriptors were not employed in the QSPR models developed in the following case study.

8.4.2 TAE/RECON Descriptors

Electron density-derived features constitute another important class of novel descriptors. A wide variety of molecular property descriptors can be derived from the electron density distributions obtained from *ab initio* calculations. However, even with the rapid advances in computer architecture and the anticipated continued growth in computational power, a direct calculation of the properties of large molecules is computationally unfeasible. Although much faster to compute, semiempirical methods are not capable of producing useful electron densities. In order to accurately obtain molecular electron density-derived descriptors with a substantial reduction in computational time, Breneman and coworkers [53] have developed the TAE method which is based on the theory of atoms in molecules (AIM) [54]. In the transferable atom equivalent/reconstruction (TAE/RECON) method, atomic contributions are used to rapidly generate whole molecule electron density-derived descriptors that have been shown to closely approximate those available through *ab initio* calculations [55]. These descriptors essentially provide information about acidity/basicity, hydrophobicity, hydrogen-bonding capacity, and polarity as well as molecular polarizability and have been successfully employed in several QSAR and QSRR studies [23,53,56]. The RECON2000 program [55] was employed to calculate TAE/RECON descriptors for the proteins in the following case study.

8.5 SVM MODELING ALGORITHM

Successful QSPR generation requires a wise choice of descriptors and robust modeling methodologies. The case study presented in this chapter uses a modeling approach which begins with a very large set of descriptive features, at which point a sparse SVM feature selection strategy is used to identify a small subset of relevant molecular property descriptors. Visualization of the resulting SVM models then allows the interpretation and further refinement of the feature subset. Finally, a nonlinear SVM regression is used to generate a predictive model. Figure 8.1 provides an overview of this process.

Feature Selection: In the present QSPR modeling strategy, a feature selection approach based on linear l_1-norm SVM regression is applied, so that a linear algorithm can be formulated for the SVM to reduce the computational cost as compared to one using a quadratic algorithm [51,57]. Within this technique, a series of linear l_1-norm SVM models are constructed for different random partitions of the training data into training and validation sets. Each different set of training proteins is called a fold and the model created using this set is used to make predictions on the validation set of proteins left out of the training set for that particular fold. This procedure is repeated x times, which results in x different training and validation subsets and the construction of x distinct, but similar models. This is termed bootstrapping in QSPR modeling literature and is known to provide better model generalization. Finally, an ensemble of these models is used so as to avoid a loss of useful information during the feature selection step. This technique is termed as bootstrap aggregation or bagging [58,59]. In each

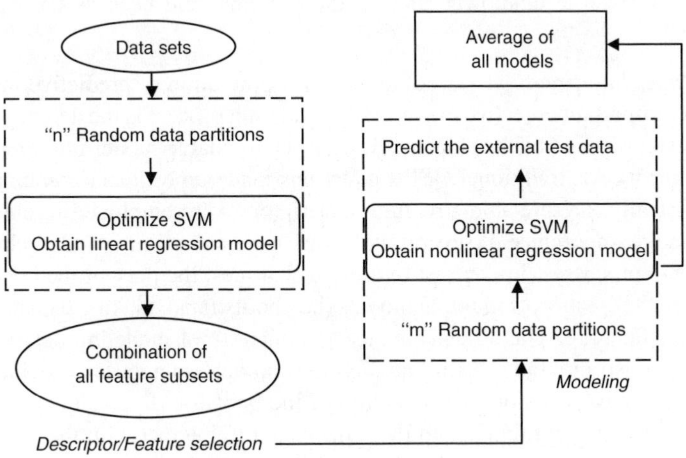

FIGURE 8.1 Flowchart of the overall SVM–QSPR modeling approach.

linear l_1-norm SVM fold, the optimal weight vector has relatively few nonzero weights with a degree of sparsity that depends on the SVM model parameters. The features with nonzero weights then become potential attributes to be used in the nonlinear SVM. The bagging approach captures important effects that might otherwise have been lost in a single model-based feature selection approach. As part of this technique, the important features for each individual linear SVM model are recorded and combined together to produce a final descriptor set that contains chemical information about the chromatographic property being modeled. Thus, the probability of inadvertently discarding useful descriptors is reduced.

Model Building: The above feature selection algorithm is incorporated into a larger scheme for property prediction. In this case, a set of training data is used to perform feature selection, after which the results are used to generate user-friendly graphics (i.e., star plots, discussed below) that can be used to evaluate and further focus the descriptor set according to chemical intuition. Finally, nonlinear SVM predictive models are constructed based on the final combined descriptor set. Comparisons between linear and nonlinear predictions show that trends are preserved, but the use of nonlinear modeling methods significantly improves the results. Again, in order to get a more robust and general predictive results, multiple QSPR models based on the same feature set are built. So instead of using a single model that is heavily and easily affected by chance correlations, the bagged average of all nonlinear model predictions is used to produce our final prediction results. The predictive quality of the models is initially determined by their performance on the validation sets, but the predictive power of the method is only revealed when predictions are made for true unknowns that is, the proteins held back as the external test set.

Model Interpretation: In addition to the generation of predictive models, another important objective of the QSPR modeling process is the determination of the significance of the selected descriptors to enable model interpretation. In earlier works, traditional QSPR equations made up of linear combinations of physically interpretable structural descriptors were employed to elucidate the relative importance of several molecular mechanisms involved in chromatographic processes. In contrast to these techniques, the present methodology relies on the creation of multiple models (i.e., bootstraps), making data analysis more challenging. The dimensionality and quantity of modeling data are too large to grasp in tabular form and simple statistical summaries provide only rudimentary information. Visualization of the bootstrap folds allows users to extract information mined from the models and to interact with the modeling process. Accordingly, a graphical visualization tool known as star plots was developed and employed to characterize the relative importance of the selected descriptors across the multiple models present in the bootstrap aggregate. In most multivariate visualization applications, star plots are generated in a multiplot format where each plot represents one variable, and each radial line represents the magnitude of a particular variable in the data matrix [60]. When the endpoints of the rays are connected together with a line, the resulting figure resembles a star.

In the star plots presented in the following example, each star corresponds to a single selected descriptor, where the radius of each spoke is the weight of that descriptor in one of the sparse SVM regression models used in the bootstrap (normalized by the magnitude of the weight of the most important descriptor in the same bootstrap fold). This technique visually represents the relative importance of the descriptors in each of the folds of the bagged model and provides a measure of the consistent importance of the descriptor over all of the bootstrap models. For each descriptor, the sum or average of all radii (or the surface area of the star) can be used to represent its overall relative importance in the ensemble model. The descriptor weights from all x bootstraps of the linear SVR models used in the bagging procedure are mapped onto the star plots in the manner shown in Figure 8.2. In the example shown in the figure, descriptor 1 is consistently important in all folds, while descriptor 2 has less uniform significance.

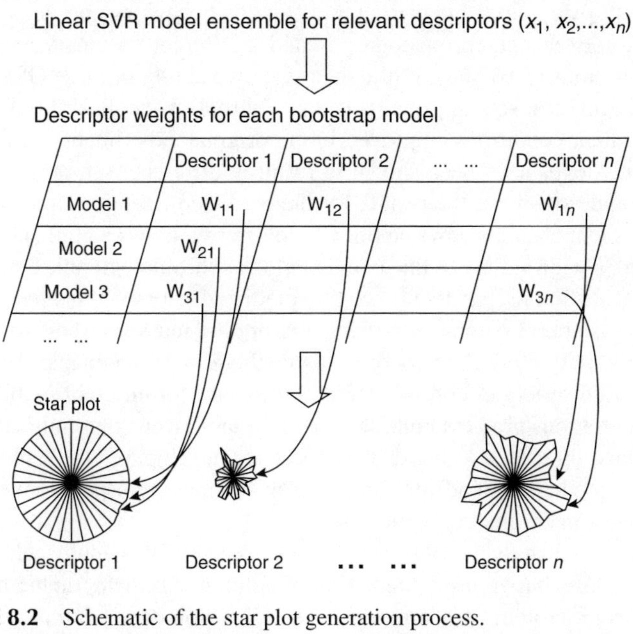

FIGURE 8.2 Schematic of the star plot generation process.

8.6 MULTISCALE MODELING FOR THE PREDICTION OF COLUMN CHROMATOGRAPHIC PERFORMANCE FROM PROTEIN STRUCTURE DATA: A CASE STUDY

Ion exchange chromatography is one of the most widely employed chromatographic steps in downstream purification of biologics. Protein retention in ion exchange chromatography is a complex function of stationary and mobile phase effects [61,62]. It has been suggested that although electrostatics is the primary mode of interaction in ion exchange systems, nonspecific interactions such as van der Waals and hydrophobic interactions can also play an important role in determining selectivity in these systems [63–66]. Despite the high level of understanding of the phenomena responsible for protein retention in ion exchange systems, the selection of appropriate chromatographic conditions for the separation of complex biological mixtures remains a challenge. Since protein selectivity can be affected by both mobile and stationary phase conditions, subtle selectivity differences can be exploited to design more efficient separations. Thus, the availability of predictive tools and models for protein binding in ion exchange systems can be very useful for accelerating process development for ion exchange chromatographic separations.

Previous work from our group has demonstrated that QSRR models can be successfully employed to predict the retention behavior of proteins in linear gradient ion exchange chromatography under different stationary and mobile phase conditions [50,67,68]. While the predictive ability of these QSRR models for external test sets of proteins is a useful tool, its applicability is limited to the gradient conditions employed in the original experiments. Differences in gradient slopes have been shown to result in differences in protein elution behavior and, therefore, the results for the predicted retention times in a particular gradient cannot always be linearly transferred to other gradients. On the other hand, a wide variety of linear and nonlinear chromatographic behavior — both lab scale and preparative — can be predicted with knowledge of protein adsorption isotherm parameters, using appropriate mass transport models for chromatography [69–72]. In theory, this methodology can enable the prediction of a wide variety of chromatographic behavior for different mobile phase salt concentrations/gradient conditions and for any given column loading condition. Thus, there is a strong driving force for developing QSPR models for the a priori prediction of adsorption isotherm and mass transport parameters of biomolecules in chromatographic systems.

In this case study we discuss the development of a multiscale modeling approach for the *a priori* prediction of column chromatographic behavior directly from protein structure data. The development of this technique accomplishes a long-standing goal in the bioseparations field to connect molecular level investigations of protein adsorption to the macroscopic/process scale. This multiscale modeling approach involves the use of protein isotherm parameters predicted from molecular scale QSPR models in concert with a macroscopic transport model for column chromatography. As proof of concept, the applicability of this approach as a method development tool is demonstrated by predicting the column separation of a mixture of test proteins directly from their crystal structure data.

8.6.1 Steric Mass Action Formalism

The steric mass action (SMA) formalism is a three-parameter model for the description of multicomponent protein–salt equilibria in ion exchange systems [73]. The multipoint binding of a protein molecule to the stationary phase is represented as a stoichiometric exchange of mobile phase protein and bound counterions as follows:

$$C_i + \upsilon_i \overline{Q_{\mathrm{s}}} \Leftrightarrow Q_i + \upsilon_i C_{\mathrm{salt}} \tag{8.1}$$

where Q and C are the solute concentrations on the stationary and mobile phases, respectively and C_{salt} is the mobile phase salt concentration. ν_i is the characteristic charge of the adsorbing solute, i and $\overline{Q_{\mathrm{s}}}$ is the concentration of sites on the stationary phase available for adsorption. The equilibrium constant for the ion exchange reaction is given by

$$K_{\mathrm{SMA}} = \left(\frac{Q_i}{C_i}\right)\left(\frac{C_{\mathrm{salt}}}{\overline{Q_{\mathrm{s}}}}\right)^{\upsilon_i} \tag{8.2}$$

The electroneutrality of the stationary phase requires

$$\Lambda = \overline{Q_{\mathrm{s}}} + \sum_{i=1}^{n} (\nu_i + \sigma_i) Q_i \tag{8.3}$$

where Λ is the total ionic capacity of the stationary phase. Thus, the SMA isotherm for a single component i is given by the implicit Equation 8.4

$$C_i = \left[\frac{Q_i}{K_{\mathrm{SMA}}}\right]\left[\frac{C_{\mathrm{salt}}}{\Lambda - (\sigma_i + \nu_i)Q_i}\right]^{\nu_i} \tag{8.4}$$

The three parameters, namely the characteristic charge (ν), the equilibrium constant (K_{SMA}), and the steric factor (σ) define the isotherm of a biomolecule in an ion exchange system. The SMA parameters, once determined, can be used to describe the adsorption of proteins at any concentration and in any salt microenvironment. This model has been shown to accurately predict ion exchange chromatographic behavior as isocratic [71], gradient [72], and displacement [69] chromatography.

8.6.2 Chromatographic Transport Models

The most complete transport model that can describe the chromatographic behavior of solutes is the general rate model. However, it is computationally expensive and therefore, employing it for optimization of preparative chromatography would be impractical. As far as possible, one would like to employ lumped rate models such as the transport- and reaction-dispersive models [74]. As outlined by Natarajan and Cramer [75], the analysis of the various dimensionless groups enables the identification of a lumped rate model appropriate for a given resin system. The overall mass balance in the column for a multicomponent system may be written as follows:

$$\frac{\partial C_i}{\partial \tau} + \beta \frac{\partial Q_i}{\partial \tau} + \frac{\partial C_i}{\partial x} = \frac{1}{\mathrm{Pe}_i} \frac{\partial^2 C_i}{\partial x^2} \tag{8.5}$$

Here, Pe_i is the Peclet number for species i and τ is dimensionless time. The lumped rate model equations, in terms of the SMA isotherm parameters, can then be written as:

1. Transport Dispersive Model:

$$\frac{\partial Q_i}{\partial \tau} = \mathrm{St}_i (Q_i^{\mathrm{equil}} - Q_i) \tag{8.6}$$

$$\mathrm{St}_i = \frac{k_{m,i} L}{u} \tag{8.7}$$

Here, C_i, Q_i, and Q_i^{equil} are the mobile-, stationary-, and equilibrium stationary-phase concentrations, respectively, and $k_{m,i}$ is a lumped mass transport coefficient for the ith component. The value of $k_{m,i}$ may be determined experimentally by an HETP analysis as outlined by Natarajan and Cramer [75]. Also, L is the column length, and u is the mobile-phase velocity. In this model, the Peclet number, Pe_i, accounts for the axial dispersion effects while the Stanton number, St_i, represents a lumped mass transport coefficient that accounts for film, pore, and surface diffusion effects (depending on the relative importance of these phenomena). This model is employed for resins wherein mass transport is the rate limiting step.

2. *Reaction Dispersive Model*:

$$\frac{\partial Q_i}{\partial \tau} = k_{\text{ads}} C_i \overline{Q}_1^{\nu_i} - k_{\text{des}} C_1^{\nu_i} Q_i \tag{8.8}$$

This lumped rate model can be employed when the kinetics of adsorption–desorption is the rate limiting resistance in a resin system.

8.6.3 Protein Dataset

Steric mass action isotherm parameters were obtained for a set of 16 proteins on SP Sepharose Fast Flow (FF) using published experimental techniques [76,77]. The crystal structures of these structurally diverse proteins were downloaded from the RSCB Protein Data Bank [78]. The PDB codes and isotherm parameter values of these proteins are presented in Table 8.2. Sybyl v6.5 (Tripos, St. Louis, MO) was used to preprocess the raw PDB files by eliminating the heteroatoms and waters of hydration present in the published protein structures.

TABLE 8.2
SMA Parameters for the Proteins Employed in the Present Case Study

#	Protein	PDB	ν	K_{SMA}	σ
1	Turkey egg lysozyme	135L	7.4	0.0329	14.8
2	Protease carlsberg	1AF4	3.0	0.0030	4.8
3	Chicken egg lysozyme	1AKI	5.6	0.0763	17.0
4	Avidin	1AVE	9.3	0.0055	46.3
5	Bovine phospholipase	1BP2	2.6	0.0824	57.1
6	α-Chymotrypsinogen A*	1CHG	3.9	0.0475	31.7
7	Protease nagarase	1CSE	2.7	0.0313	(na)
8	Horse Cytochrome C	1HRC	5.9	0.0295	15.8
9	Elastase	1LVY	4.6	0.0014	88.6
10	Pyruvate kinase	1PKN	5.9	0.0063	108.5
11	Bee phospholipase	1POC	7.3	0.0008	(na)
12	Ribonuclease A*	1RBX	5.4	0.0296	17.2
13	Bovine Cytochrome C	1RIE	5.5	0.0470	17.7
14	Trypsinogen	1TGB	4.1	0.0037	22.8
15	γ-Chymotrypsin	2GCH	5.0	0.0106	36.5
16	α-Chymotrypsin	4CHA	3.8	0.0565	21.5

Note: Proteins marked with a "*" were employed as the external test set in the QSPR models.

These structures were then employed to calculate MOE and TAE/RECON descriptors for the proteins using the appropriate software. In addition, a principal component analysis (PCA) was performed based on all calculated descriptors to identify two representative proteins (α-chymotrypsinogen A and ribonuclease A) as external test set cases to verify the predictive power of the QSPR models.

8.6.4 QSPR Model Generation

A total of 279 MOE and RECON descriptors were computed for the proteins to give a composite set of traditional 2D and 3D as well as electron density-derived TAE descriptors. The SVM modeling procedures described above were applied to this dataset and three independent QSPR models were generated for the isotherm parameters. The models showed high cross-validated R^2 values for the training data ($R^2 > 0.85$) indicating that the predicted values of the isotherm parameters were in good agreement with the experimental training data. More importantly, the QSPR models were successfully able to predict the SMA parameter values for the external test set proteins. Representative results of the QSPR model for the characteristic charge (ν) are shown in Figure 8.3 and a summary of the results for all models is presented in Table 8.3. These results clearly demonstrate the utility of the QSPR models for the a priori prediction of the SMA isotherm parameters of proteins in ion exchange systems.

8.6.5 The Multiscale Model

The multiscale modeling approach developed in the present study involves the use of protein isotherm parameters predicted from the above QSPR models in concert with a lumped rate transport dispersive model for chromatographic systems. Specifically, the SMA parameters of the two test proteins, α-chymotrypsinogen A and ribonuclease A, predicted from the QSPR models (Table 8.3) were used to simulate their chromatographic performance on an SP Sepharose FF column. The values of the lumped mass transfer coefficients for the proteins required in the transport dispersive model simulations were obtained from the literature [70]. A linear gradient elution of the binary protein mixture was carried out and the results were compared to column simulations obtained from the multiscale model. Figure 8.4 shows an overlay of the effluent protein profiles obtained in the experiment and the simulation. As seen in the figure, the simulated separation of the two proteins is in very good agreement with the experimental results. It is important to remind the reader that the isotherm parameters of the test proteins employed in this separation were not experimentally determined, but were predicted from the appropriate QSPR models. Thus, these results represent a true multiscale modeling approach,

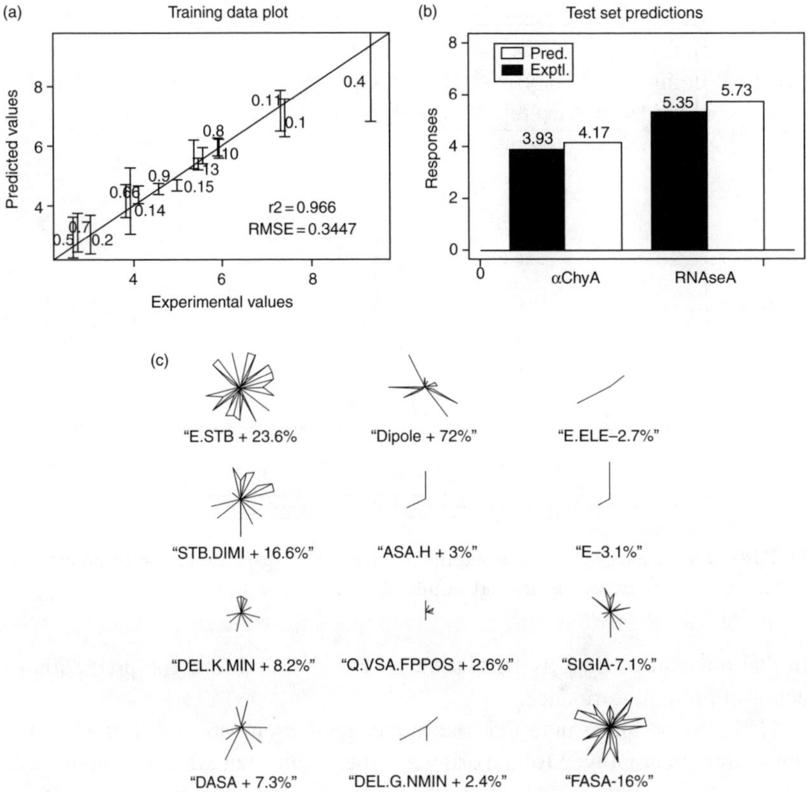

FIGURE 8.3 QSPR model for protein characteristic charge ν: (a) training data plot, cross-validated $R^2 = 0.97$, (b) external test set predictions, and (c) star plot representation of all molecular descriptors identified by the feature selection process.

TABLE 8.3
Summary of Test Set Predictions Obtained from the Individual QSPR Models

		α-Chymotrypsinogen A		Ribonuclease A	
Property	R^2_{cv}	**Exptl.**	**Pred.**	**Exptl.**	**Pred.**
Characteristic charge (ν)	0.97	3.93	4.17	5.35	5.73
Equilibrium constant (K_{SMA})	0.94	0.0475	0.0498	0.0296	0.0390
Steric factor (σ)	0.85	31.69	32.53	17.18	14.65

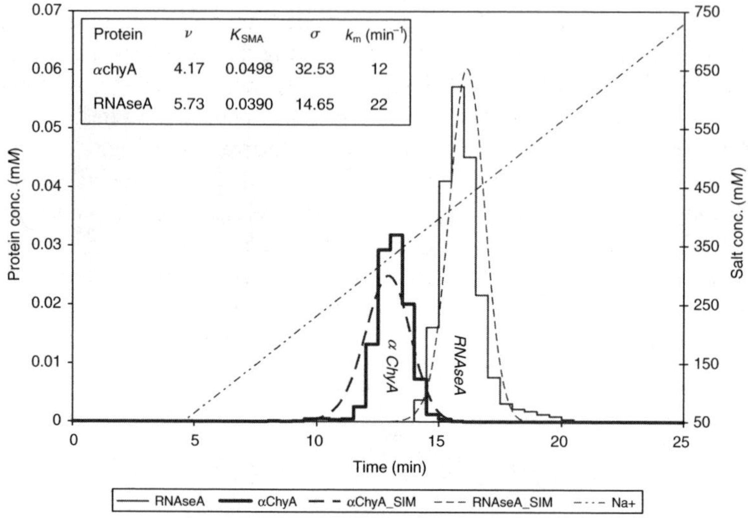

FIGURE 8.4 Linear gradient separation of α-chymotrypsinogen A and ribonuclease A: Comparison of experimental and simulated (SIM) separations.

in that one can go directly from protein crystal structure to the prediction of actual column performance.

It is important to note that the model predictions are not limited to the ionic strength employed for the original experiments, but are in fact applicable across a wide range of ionic strengths due to the explicit consideration of the salt concentration in the SMA formalism. Furthermore, since the isotherm parameters are being accurately predicted by the QSPR models, it is expected that this approach can be employed to predict column performance under a wide range of loading conditions (as shown in References 69–71). Finally, although a specific experimental system was employed to demonstrate the approach, the multiscale protocol may be readily applied to other chromatographic systems and modes of operation.

8.6.6 Summary of Case Study

The results presented in this case study indicate that it is indeed possible to generate predictive QSPR models of protein SMA parameters in ion exchange systems using a SVM regression technique. The primary focus of this case study was on the predictive ability of QSPRs and, therefore, details of the model interpretation are not provided in this section. The reader is referred to the literature for this information [76].

The ability to predict protein isotherm parameters can have direct implications for various ion exchange processes. As proof of concept, a multiscale modeling approach was developed and employed for predicting the chromatographic separation of a test set of proteins using their isotherm parameters predicted by the QSPR models. The simulated separation showed good agreement with the experimental data. While the results presented in this example have focused on a single gradient separation for the purpose of demonstrating the approach, it is important to note that once the SMA parameters are predicted by the QSPR models, they can in theory be employed to predict any preparative chromatographic separation (e.g., linear gradient, step gradient, displacement chromatography). This ability to predict chromatographic behavior of proteins directly from their crystal structures (or eventually from protein primary sequence information) can have significant implications for a range of biotechnology processes.

8.7 QSPR AS A BIOPROCESS DEVELOPMENT TOOL

The above example demonstrates the utility of the QSPR modeling approach for the *a priori* prediction of chromatographic separations of proteins. The practical application of this technique in a typical downstream bioprocessing setup would involve the generation of models to predict the chromatographic behavior of the product of interest and the key impurities in a given biological mixture. Using these models, computational experiments may then be carried out by varying different operational parameters and using the appropriate QSPR models to predict the resolution of the resultant separation. Thus, this strategy can enable the *in silico* design and optimization of chromatographic separations of bioprocess mixtures. The following discussion outlines some of the key steps that would be involved in the use of QSPRs as a design tool for downstream purification processes. It also describes some of the potential hurdles that may be encountered during this process and presents some generic solutions to these issues.

1. The first step in the QSPR modeling process is the development of predictive models using a representative training set of molecules. Once developed, these models may be utilized for different product campaigns without the need for additional experimentation and model development for every new bioproduct. Accordingly, the goal is to build models that are capable of predicting the chromatographic behavior of a diverse set of test molecules that may be encountered in typical bioprocesses. At the same time, it is desirable to have independent QSPR models that focus on the different classes of molecules (e.g., small proteins, antibodies, nucleic acids, etc.) for better predictive ability. Tools such as PCA and clustering [79,80] and other similarity metrics [81] based on calculated molecular property descriptors may be employed for the identification of representative

training data sets, so that the resultant models can yield reliable predictions over a broad chemical space.

2. Once a training set of molecules has been identified, crystal structures may be determined in-house or obtained from an online database such as the RCSB Protein Databank (www.pdb.org) or the ExPASy proteomics server (www.expasy.org). Structures obtained from databanks must be checked for their completeness and accuracy and often require preprocessing to remove heteroatoms (e.g., water molecules, ligands, etc.) and redundant chains. Commercially available molecular modeling/visualization software packages such as MOE (CCG Inc., Montreal, Canada), Sybyl (Tripos Inc., St. Louis, MO), and Insight II (Accelrys, San Diego, CA) can be employed for this purpose. Where crystal structure data is not available, primary sequence information may be input into the molecular modeling programs to generate a peptide chain for the protein, which may be subsequently used to calculate shape-independent 2D descriptors. If 3D structure is important to model the binding, homology models (described below) may be employed to predict the 3D structure from the primary sequence. Molecular modeling packages such as MOE:Homology and Modeler (Accelrys) have homology modeling and structure optimization tools built-in that can enable the identification of suitable template molecules and the subsequent generation of 3D structures from sequence data.
3. With molecular structures in hand, a wide variety of physicochemical parameters of the molecules that is, descriptors can be computed using one or more different software packages such as MOE, SYBYL, and RECON2000 [55] to name a few.
4. Experimental data (retention data, isotherm parameters, etc.) is then obtained for the training set molecules using standard experimental techniques. In-house historical data or published data obtained under similar experimental conditions may also be employed for this purpose. Data must be obtained for all training set molecules for each experimental system (i.e., chromatographic mode, resin, mobile phase condition) of interest for the generation of independent QSPR models for these different systems.
5. Using the experimental data in concert with the calculated descriptors, QSPR models can be generated using one of many available feature selection and model building techniques. AI Trilogy™ (Ward Systems Group, Frederick, MD) (GA/ANN), SIMCA-P (Umetrics, Umeå, Sweden) (PLS), Analyze/StripMiner™ (ANN) (Prof. Mark Embrechts, RPI, Troy, NY), and MOE:QuaSAR (PLS) are some examples of software packages that may be employed for model

building. Typically 80 to 90% of the data is used for training the QSPR models, while the remaining molecules are employed as an external test set to verify their predictive ability. The quality of the trained model is usually estimated as the coefficient of determination (or goodness of fit, R^2), while the predictive power of the model is measured by the goodness of prediction parameter (Q^2). Table 8.4 outlines some of the hurdles that may be encountered at the model building stage.

In addition to predictive validation, other statistical analyses (e.g., cross-validation, response permutation testing, etc.) are often carried out to ensure the reliability of the resultant models. The reader is referred to a recent review by Eriksson et al. [82] for a summary of the techniques employed for rigorous model validation.

6. Once robust QSPR models are available, they can be used to predict chromatographic separations for new biologics from bioprocess mixtures. The models may be employed either directly to predict gradient separations under specific conditions, or incorporated into a more powerful multiscale modeling platform (as described in the case study), which can enable the design and optimization of chromatographic processes over a wider range of operating conditions.
7. Finally, at the end of the development cycle, experimental data obtained for the new molecules may be added to the training data of the QSPR models to enhance their predictive ability for future applications.

8.8 ADVANCES IN QSPR MODELING TECHNIQUES AND FUTURE DIRECTIONS

8.8.1 PHYSICALLY INTERPRETABLE DESCRIPTORS

The MOE and RECON descriptors employed in the example discussed above are generic in nature and represent common physicochemical properties of molecules. Clearly, the generality of these descriptors poses some challenges during the model interpretation process. Thus, there is a strong motivation for the development of physically interpretable descriptors for use in QSPRs involving biological molecules.

Most natural proteins are derived by the combination of twenty commonly occurring amino acids. This relatively small set of building blocks provides a unique opportunity to develop residue-based protein-specific descriptors for use in QSPR modeling. The use of protein-specific descriptors can result in the generation of better predictive models with fewer selected descriptors, which in

TABLE 8.4
Common Problems Encountered at the Model Building Stage: Causes and Solutions

Problem	Possible Causes	Solutions
Low R^2	Insufficient training data	• Add more training molecules • Focus the model on a smaller region of chemical space, that is, a subset of chemometrically similar molecules
	Irrelevant descriptors (i.e., low signal-to-noise ratio)	• Include relevant descriptors on the basis of knowledge of a particular chromatographic system
	Poor model building algorithm (e.g., linear model built for inherently nonlinear data)	• Examine other feature selection and regression techniques (e.g., linear vs. nonlinear regression)
Low Q^2 (but high R^2)	Nonrepresentative training data	• Use a model that focuses on the respective subclass of biomolecules • Add more training molecules
	Over-determined model, which may be due to	
	1. Over-fitting during training	• Perform cross-validation during training • Use algorithms that better control the training process (e.g., back propagation algorithm for ANNs)
	2. Many selected descriptors	• More rigorous feature selection algorithms or other intelligent feature selection techniques driven by knowledge of the separation process
	3. Highly nonlinear modeling algorithm	• Use a modeling algorithm that minimizes over-fitting by controlling the degree of model nonlinearity (e.g., capacity controlled SVM)

turn minimizes the possibility of model overdetermination. Furthermore, these descriptors offer the advantage of improved model interpretability since they can be easily linked to protein structural features. The use of residue-based descriptors also opens up the possibility of utilizing state-of-the-art graphical tools to visualize the distribution of residues associated with selected descriptors directly on the protein molecules. This can maximize the amount of physicochemical information regarding the nature and relative importance of various interactions responsible for biomolecule adsorption in a given chromatographic system, which can be extracted via model interpretation.

8.8.2 QSPR Models from Primary Sequence Information

The results presented in the above case study successfully demonstrate the ability to predict chromatographic column separations directly from protein crystal structure data. The application of this approach for chromatographic process design and optimization relies on the availability of crystal structure data of the biomolecule of interest as well as the key impurities in a given feed mixture. However, crystal structure information is sometimes not available for molecules encountered in industrial bioprocesses. Thus, there is a strong driving force for refining the present multiscale modeling strategy so as to ensure its success as a method development tool for the biotechnology industry.

One possible solution to this problem is the generation of predictive QSPR models using topological 2D descriptors which can be computed from the primary sequence of the molecule, without the need for 3D structure information. As described above, the MOE package computes a large number of 2D descriptors based on the connection table representation of a molecule (e.g., elements, formal charges and bonds, but not atomic coordinates). These include physical properties of the molecule (such as molecular weight, log P, molar refractivity, partial charge), subdivided van der Waals surface area of atoms associated with specific bin ranges of these physical properties, various atom and bond counts, and some pharmacophore feature descriptors. While this approach may be very useful for modeling some chromatographic systems, it could result in significant model degradation in systems where molecular size and shape factors are important.

Recent advances in the molecular modeling field have resulted in the development and refinement of homology modeling [83,84] and threading techniques [85,86] that can be employed to estimate the 3D structure of a protein from its primary sequence information. These techniques offer an excellent opportunity to overcome the drawbacks of using 2D descriptors alone in QSPR model generation. Homology modeling relies on the identification of a structurally conserved region (SCR) for a family of homologous molecules. Once an SCR is identified, appropriate loops based on the unaccounted gaps in the primary sequence of the target molecule are identified from available databases and added onto the SCR. Finally, the side chains of all amino acid residues are incorporated into the structure followed by an energy minimization procedure to yield the final predicted structure of the protein. On the other hand, threading algorithms are based on the premise that there are a limited number of unique folds found in proteins. It involves determination of the appropriate fold for a given amino acid sequence by comparing the query sequence against a database of folds. The degree of similarity is given by the Z-score calculated for each sequence/profile pair and the structure–sequence match is validated by energy calculations. Homology modeling and threading methods are often used together and may be combined with other protein folding algorithms that have been extensively researched by several groups [87–90].

The above techniques can somewhat mitigate the overdependence on crystal structure data for generating predictive QSPR models for proteins. Clearly, the development of efficient strategies for building QSPR models based on protein primary sequence information can greatly enhance the general applicability of the multiscale modeling protocol to industrial bioprocesses.

8.9 CONCLUSIONS

Protein affinity in chromatographic systems is a result of the complex interplay of several physicochemical effects arising from the multicomponent adsorption of different species under various operating modes and conditions. The example presented in this chapter and other prior publications from our group demonstrate the utility of structure–property modeling techniques to predict and understand the factors responsible for the binding affinity of biomolecules in different chromatographic modes. The ability to predict column chromatographic separations of biomolecules using the multiscale modeling approach can have significant implications for bioprocess design and optimization. Furthermore, the availability of strategies for rigorous model validation and physically interpretable descriptors for proteins can significantly increase the robustness, reliability, and interpretability of QSPRs and the resultant multiscale models. These developments represent the state-of-the-art in structure–property modeling as applied to chromatography and can have a significant impact on the way in which the biotechnology industry carries out downstream process development.

ACKNOWLEDGMENTS

This work was supported by NSF Grant BES-0418413, NSF Grant BES-0214183, and GE Healthcare (Uppsala, Sweden). The Support Vector Machine (SVM) Regression program was developed independently in Professor

Kristin Bennett's group in the Department of Mathematics at Rensselaer Polytechnic Institute. Special thanks to Qiong Luo, Nagamani Sukumar, Dechuan Zhuang, Kaushal Rege, and Matthew Kuure-Kinsey for their valuable discussions and suggestions.

REFERENCES

1. Brown, A.C. and Fraser, T.R. On the connection between chemical constitution and physiological action. part I. On the physiological action of the salts of the ammonium bases, derived from Strychnia, Brucia, Thebaia, Codeia, Morphia, and Nicotia. *Transactions of Royal Society of Edinburgh* 1869; 25:151–203.
2. Burkhardt, G.N. Influence of subsituents on organic reactions: a quantitative relationship. *Nature* (*London*) 1935; 136:684.
3. Hammett, L.P. The effect of structure upon the reactions of organic compounds: Benzene derivatives. *Journal of American Chemical Society* 1937; 59:96–103.
4. Hammett, L.P. Reaction rates and indicator acidities. *Chemical Reviews* 1935; 17:67–79.
5. Xu, J. and Stevenson, J. Drug-like index: a new approach to measure drug-like compounds and their diversity. *Journal of Chemical Information and Computer Sciences* 2000; 40:1177–1187.
6. Desjarlais, R.L., Sheridan, R.P., Seibel, G.L., Dixon, J.S., Kuntz, I.D., and Venkataraghavan, R. Using shape complementarity as an initial screen in designing ligands for a receptor-binding site of known 3-dimensional structure. *Journal of Medicinal Chemistry* 1988; 31:722–729.
7. Bravi, G., Green, D.V.S., Hann, M.M., and Leach, A.R. PLUMS: a program for the rapid optimization of focused libraries. *Journal of Chemical Information and Computer Sciences* 2000; 40:1441–1448.
8. Meng, E.C., Shoichet, B.K., and Kuntz, I.D. Automated docking with grid-based energy evaluation. *Journal of Computational Chemistry* 1992; 13:505–524.
9. Hansch, C. The QSAR paradigm in the design of less toxic molecules. *Drug Metabolism Reviews* 1985; 15:1279–1294.
10. Wessel, M.D., Jurs, P.C., Tolan, J.W., and Muskal, S.M. Prediction of human intestinal absorption of drug compounds from molecular structure. *Journal of Chemical Information and Computer Sciences* 1998; 38:726–735.
11. Arulmozhiraja, S. and Morita, M. Structure–activity relationships for the toxicity of polychlorinated dibenzofurans: approach through density functional theory-based descriptors. *Chemical Research in Toxicology* 2004; 17:348–356.
12. Norinder, U. Support vector machine models in drug design: applications to drug transport processes and QSAR using simplex optimisations and variable selection. *Neurocomputing* 2003; 55:337–346.
13. Subramanian, G. and Kitchen, D.B. Computational models to predict blood-brain barrier permeation and CNS activity. *Journal of Computer-Aided Molecular Design* 2003; 17:643–664.

14. Kaliszan, R. Quantitative structure-retention relationships applied to reversed-phase high-performance liquid-chromatography. *Journal of Chromatography A* 1993; 656:417–435.
15. Sjostrom, M. and Wold, S. Linear free-energy relationships — local empirical rules — or fundamental laws of chemistry. *Acta Chemica Scandinavica Series B — Organic Chemistry and Biochemistry* 1981; 35:537–554.
16. Mnuk, P., Feltl, L., and Schurig, V. Gas chromatographic study of the inclusion properties of calixarenes. 2. Selective properties of cyclic tetra- to octamers derived from phenol, and some problems associated with the use of calixarenes in capillary gas chromatography. *Journal of Chromatography A* 1996; 732:63–74.
17. Li, S.Y., Sun, C., Wang, Y., Xu, S.F., Yao, S.C., and Wang, L.S. Quantitative structure retention relationship studies for predicting relative retention times of chlorinated phenols on gas chromatography. *Journal of Environmental Sciences (China)* 2002; 14:418–422.
18. Gao, Y.H., Wang, Y.W., Yao, X.J., Zhang, X.Y., Liu, M.C., Hu, Z.D., and Fan, B.T. The prediction for gas chromatographic retention index of disulfides on stationary phases of different polarity. *Talanta* 2003; 59:229–237.
19. Kiridena, W. and Poole, C.F. Structure-driven retention model for solvent selection and optimization in reversed-phase thin-layer chromatography. *Journal of Chromatography A* 1998; 802:335–347.
20. Tan, L.C. and Carr, P.W. Study of retention in reversed-phase liquid chromatography using linear solvation energy relationships — II. The mobile phase. *Journal of Chromatography A* 1998; 799:1–19.
21. Yang, S.Y., Bumgarner, J.G., Kruk, L.F.R., and Khaledi, M.G. Quantitative structure-activity relationships studies with micellar electrokinetic chromatography — Influence of surfactant type and mixed micelles on estimation of hydrophobicity and bioavailability. *Journal of Chromatography A* 1996; 721:323–335.
22. Salo, M., Siren, H., Volin, P., Wiedmer, S., and Vuorela, H. Structure-retention relationships of steroid hormones in reversed-phase liquid chromatography and micellar electrokinetic capillary chromatography. *Journal of Chromatography A* 1996; 728:83–88.
23. Breneman, C.M. and Rhem, M. QSPR analysis of HPLC column capacity factors for a set of high-energy materials using electronic van der Waals surface property descriptors computed by transferable atom equivalent method. *Journal of Computational Chemistry* 1997; 18:182–197.
24. Hansch, C., Maloney, P.P., and Fujita, T. Correlation of biological activity of phenoxyacetic acids with Hammett substituent constants and partition coefficients. *Nature* 1962; 194:178.
25. Patrick, G.L. *An Introduction to Medicinal Chemistry*. 2nd ed. Oxford University Press, Oxford, UK 1997.
26. Kamlet, M.J. and Taft, R.W. Solvatochromic comparison method. 1. Beta-scale of solvent hydrogen-bond acceptor (Hba) basicities. *Journal of the American Chemical Society* 1976; 98:377–383.

27. Taft, R.W. and Kamlet, M.J. Solvatochromic comparison method. 2. Alpha-scale of solvent hydrogen-bond donor (Hbd) acidities. *Journal of the American Chemical Society* 1976; 98:2886–2894.
28. Kamlet, M.J., Abboud, J.L., and Taft, R.W. Solvatochromic comparison method. 6. Pi-Star scale of solvent polarities. *Journal of the American Chemical Society* 1977; 99:6027–6038.
29. Lowrey, A.H., Cramer, C.J., Urban, J.J., and Famini, G.R. Quantum-chemical descriptors for linear solvation energy relationships. *Computers and Chemistry* 1995; 19:209–215.
30. Lowrey, A.H. and Famini, G.R. Using theoretical descriptors in quantitative structure–activity-relationships — HPLC capacity factors for energetic materials. *Structural Chemistry* 1995; 6:357–365.
31. Wilson, L.Y. and Famini, G.R. Using theoretical descriptors in quantitative structure–activity-relationships — some toxicological indexes. *Journal of Medicinal Chemistry* 1991; 34:1668–1674.
32. Wiener, H. Structural determination of paraffin boiling points. *Journal of the American Chemical Society* 1947; 69:17–20.
33. Kier, L.B. and Hall, L.H. Molecular connectivity in structure-activity analysis. Research Studies Press Ltd., Hertfordshire, England and John Wiley & Sons, New York, 1986.
34. Kier, L.B. and Hall, L.H. Molecular connectivity analyses of structure influencing chromatographic retention indexes. *Journal of Pharmaceutical Sciences* 1979; 68:120–122.
35. Hall, L.H. and Kier, L.B. The nature of structure–activity relationships and their relation to molecular connectivity. *European Journal of Medical Chemistry* 1977; 12:307–312.
36. Tipker, J. and Verloop, A. Use of Sterimol, Mtd, and Mtd-Star steric parameters in quantitative structure-activity-relationships. *ACS Symposium Series* 1984; 255:279–296.
37. Gaudio, A.C., Korolkovas, A., and Takahata, Y. Quantitative structure–activity-relationships for 1,4-dihydropyridine calcium-channel antagonists (Nifedipine Analogs) — a quantum chemical/classical approach. *Journal of Pharmaceutical Sciences* 1994; 83:1110–1115.
38. Topliss, J.G.e. *Quantitative Structure–Activity Relationships of Drugs.* Academic Press, New York, 1983.
39. Toropov, A.A. and Roy, K. QSPR modeling of lipid-water partition coefficient by optimization of correlation weights of local graph invariants. *Journal of Chemical Information and Computer Sciences* 2004; 44:179–186.
40. Moon, T., Chi, M.W., Park, S.J., and Yoon, C.N. Prediction of HPLC retention time using multiple linear regression: using one and two descriptors. *Journal of Liquid Chromatography and Related Technologies* 2003; 26:2987–3002.
41. Cronin, M.T.D., Aptula, A.O., Duffy, J.C., Netzeva, T.I., Rowe, P.H., Valkova, I.V., and Schultz, T.W. Comparative assessment of methods to develop QSARs for the prediction of the toxicity of phenols to *Tetrahymena pyriformis. Chemosphere* 2002; 49:1201–1221.

42. Maran, U. and Slid, S. QSAR modeling of genotoxicity on non-congeneric sets of organic compounds. *Artificial Intelligence Reviews* 2003; 20:13–38.
43. Votano, J.R., Parham, M., Hall, L.H., Kier, L.B., and Hall, L.M. Prediction of aqueous solubility based on large datasets using several QSPR models utilizing topological structure representation. *Chemistry and Biodiversity* 2004; 1:1829–1841.
44. McClelland, H.E. and Jurs, P.C. Quantitative structure-property relationships for the prediction of vapor pressures of organic compounds from molecular structures. *Journal of Chemical Information and Computer Sciences* 2000; 40:967–975.
45. Kubinyi, H. Variable selection in QSAR studies. 1. An evolutionary algorithm. *Quantitative Structure–Activity Relationships* 1994; 13:285–294.
46. Leardi, R., Boggia, R., and Terrile, M. Genetic algorithms as a strategy for feature-selection. *Journal of Chemometrics* 1992; 6:267–281.
47. Xu, L. and Zhang, W.J. Comparison of different methods for variable selection. *Analytica Chimica Acta* 2001; 446:477–483.
48. Wold, H. Partial least squares, in *Encyclopedia of Statistical Sciences*. John Wiley & Sons, New York, pp. 581–591, 1985.
49. Vapnik, V.N. *The Nature of Statistical Learning Theory*. Springer-Verlag, New York, 1995.
50. Mazza, C.B., Sukumar, N., Breneman, C.M., and Cramer, S.M. Prediction of protein retention in ion-exchange systems using molecular descriptors obtained from crystal structure. *Analytical Chemistry* 2001; 73:5457–5461.
51. Bi, J.B., Bennett, K.P., Embrechts, M., Breneman, C., and Song, M.H. Dimensionality reduction via sparse support vector machines. *Journal of Machine Learning Research* 2003; 3:1229–1243.
52. Hall, L.H. and Kier, L.B. The E-state as the basis for the molecular structure space definition and structure similarity. *Journal of Chemical Information and Computer Sciences* 2000; 40:784–791.
53. Breneman, C.M., Thompson, T.R., Rhem, M., and Dung, M. Electron-density modeling of large systems using the transferable atom equivalent method. *Computers and Chemistry* 1995; 19:161.
54. Bader, R.F.W., Carroll, M.T., Cheeseman, J.R., and Chang, C. Properties of atoms in molecules — atomic volumes. *Journal of the American Chemical Society* 1987; 109:7968–7979.
55. Breneman, C.M. and Sukumar, N., *RECON 2000*, in *RECON 2000 Software, RPI, Troy, NY*, Troy, NY, 2000.
56. Lavine, B.K., Davidson, C.E., Breneman, C., and Katt, W. Electronic van der Waals surface property descriptors and genetic algorithms for developing structure–activity correlations in olfactory databases. *Journal of Chemical Information and Computer Sciences* 2003; 43:1890–1905.
57. Song, M.H., Breneman, C.M., Bi, J.B., Sukumar, N., Bennett, K.P., Cramer, S., and Tugcu, N. Prediction of protein retention times in anion-exchange chromatography systems using support vector regression. *Journal of Chemical Information and Computer Sciences* 2002; 42:1347–1357.

58. Breiman, L. Using iterated bagging to debias regressions. *Machine Learning* 2001; 45:261–277.
59. Breiman, L. Bagging predictors. *Machine Learning* 1996; 24:123–140.
60. Chambers, J., Cleveland, W., Kleiner, B., and Tukey, P.A., *Graphical Methods for Data Analysis*, ed., Wadsworth International Group, W.I.l. Duxbury Press, Belmont, CA, 1983.
61. Kopaciewicz, W. and Regnier, F.E. Mobile phase selection for the high-performance ion-exchange chromatography of proteins. *Analytical Biochemistry* 1983; 133:251–259.
62. DePhillips, P. and Lenhoff, A.M. Determinants of protein retention characteristics on cation-exchange adsorbents. *Journal of Chromatography A* 2001; 933:57–72.
63. Shukla, A.A., Bae, S.S., Moore, J.A., and Cramer, S.M. Structural characteristics of low-molecular-mass displacers for cation-exchange chromatography — II. Role of the stationary phase. *Journal of Chromatography A* 1998; 827:295–310.
64. Roth, C.M. and Lenhoff, A.M. Electrostatic and van der Waal's contributions to protein adsorption — computation of equilibrium-constants. *Langmuir* 1993; 9:962–972.
65. Sheng, Y.J., Tsao, H.K., Zhou, J., and Jiang, S.Y. Orientation of a Y-shaped biomolecule adsorbed on a charged surface. *Physical Review E* 2002; 66:011911:011911–011915.
66. Bowen, W.R., Pan, L.C., and Sharif, A.O. Predicting equilibrium constants for ion exchange of proteins — a colloid science approach. *Colloids and Surfaces A* 1998; 143:117–131.
67. Ladiwala, A., Rege, K., Breneman, C.M., and Cramer, S.M. Investigation of mobile phase salt type effects on protein retention and selectivity in cation-exchange systems using quantitative structure retention relationship models. *Langmuir* 2003; 19:8443–8454.
68. Tugcu, N., Song, M.H., Breneman, C.M., Sukumar, N., Bennett, K.P., and Cramer, S.M. Prediction of the effect of mobile-phase salt type on protein retention and selectivity in anion exchange systems. *Analytical Chemistry* 2003; 75:3563–3572.
69. Gadam, S.D., Gallant, S.R., and Cramer, S.M. Transient profiles in ion-exchange displacement chromatography. *AIChE Journal* 1995; 41:1676–1686.
70. Natarajan, V., Ghose, S., and Cramer, S.M. Comparison of linear gradient and displacement separations in ion-exchange systems. *Biotechnology and Bioengineering* 2002; 78:365–375.
71. Gallant, S.R., Kundu, A., and Cramer, S.M. Modeling nonlinear elution of proteins in ion-exchange chromatography. *Journal of Chromatography A* 1995; 702:125–142.
72. Gallant, S.R., Kundu, A., and Cramer, S.M. Optimization of step gradient separations — consideration of nonlinear adsorption. *Biotechnology and Bioengineering* 1995; 47:355–372.

73. Brooks, C.A. and Cramer, S.M. Steric mass-action ion-exchange — displacement profiles and induced salt gradients. *AIChE Journal* 1992; 38:1969–1978.
74. Guiochon, G., Golshan Shirazi, S., and Katti, A.M. *Fundamentals of Preparative and Non-linear Chromatography*. Academic Press, New York, 1994.
75. Natarajan, V. and Cramer, S. A methodology for the characterization of ion-exchange resins. *Separation Science and Technology* 2000; 35:1719–1742.
76. Ladiwala, A., Rege, K., Breneman, C., and Cramer, S.M. A priori prediction of adsorption isotherm parameters and chromatographic behavior in ion-exchange systems. *Proceedings of the National Academy of Sciences of the United States of America* 2005; 102:11710–11715.
77. Shukla, A.A., Bae, S.S., Moore, J.A., Barnthouse, K.A., and Cramer, S.M. Synthesis and characterization of high-affinity, low molecular weight displacers for cation-exchange chromatography. *Industrial and Engineering Chemistry Research* 1998; 37:4090–4098.
78. Berman, H.M., Westbrook, J., Feng, Z., Gilliland, G., Bhat, T.N., Weissig, H., Shindyalov, I.N., and Bourne, P.E. The Protein Data Bank. *Nucleic Acids Research* 2000; 28:235–242.
79. Potter, T. and Matter, H. Random or rational design? Evaluation of diverse compound subsets from chemical structure databases. *Journal of Medicinal Chemistry* 1998; 41:478–488.
80. Taylor, R. Simulation analysis of experimental-design strategies for screening random compounds as potential new drugs and agrochemicals. *Journal of Chemical Information and Computer Sciences* 1995; 35:59–67.
81. Sheridan, R.P., Feuston, B.P., Maiorov, V.N., and Kearsley, S.K. Similarity to molecules in the training set is a good discriminator for prediction accuracy in QSAR. *Journal of Chemical Information and Computer Sciences* 2004; 44:1912–1928.
82. Eriksson, L., Jaworska, J., Worth, A.P., Cronin, M.T.D., McDowell, R.M., and Gramatica, P. Methods for reliability and uncertainty assessment and for applicability evaluations of classification- and regression-based QSARs. *Environmental Health Perspectives* 2003; 111:1361–1375.
83. Goldsmith-Fischman, S. and Honig, B. Structural genomics: Computational methods for structure analysis. *Protein Science* 2003; 12:1813–1821.
84. Marti-Renom, M.A., Stuart, A.C., Fiser, A., Sanchez, R., Melo, F., and Sali, A. Comparative protein structure modeling of genes and genomes. *Annual Review of Biophysics and Biomolecular Structure* 2000; 29:291–325.
85. Madej, T., Gibrat, J.F., and Bryant, S.H. Threading a database of protein cores. *Proteins–Structure Function and Genetics* 1995; 23:356–369.
86. Panchenko, A., Marchler-Bauer, A., and Bryant, S.H. Threading with explicit models for evolutionary conservation of structure and sequence. *Proteins-Structure Function and Genetics* 1999:133–140.
87. Sun, S.J., Thomas, P.D., and Dill, K.A. A simple protein-folding algorithm using a binary code and secondary structure constraints. *Protein Engineering* 1995; 8:769–778.

88. Yuan, X., Shao, Y., and Bystroff, C. Ab initio protein structure prediction using pathway models. *Comparative and Functional Genomics* 2003; 4:397–401.
89. Znamenskiy, D., Chomilier, J., Le Tuan, K., and Mornon, J.P. A new protein folding algorithm based on hydrophobic compactness: rigid unconnected secondary structure iterative assembly (RUSSIA). I: Methodology. *Protein Engineering* 2003; 16:925–935.
90. Znamenskiy, D., Le Tuan, K., Mornon, J.P., and Chomilier, J. A new protein folding algorithm based on hydrophobic compactness: rigid unconnected secondary structure iterative assembly (RUSSIA). II: Applications. *Protein Engineering* 2003; 16:937–948.

9 Membrane Chromatography: Analysis of Breakthrough Curves and Viral Clearance

Mark R. Etzel and William T. Riordan

CONTENTS

9.1 INTRODUCTION

Most chromatographic separations utilize columns packed with beads. The bead diameter is an important factor: small beads result in fast diffusion times and large numbers of plates, but also high pressure drops. Large beads are used in process-scale separations to allow for increased flow rates without incurring high pressure drops and the resulting bed compression and eventual plugging. However, large beads have long diffusion times, low plate numbers, and low dynamic capacities. In 1988, membrane chromatography was first introduced as a means to overcome the limitations of column chromatography [1]. Microporous membranes containing immobilized ligands were used as the chromatographic media. Because the membranes were thin ($\sim$0.1 mm), pressure drop limitations were not significant. Diffusional limitations were eliminated because solute was transported through the pores of the membrane by convection, not diffusion. The first devices were hollow fiber membranes where the surface was activated for affinity ligand attachment. Membrane chromatography has evolved since 1988. Several reviews of membrane chromatography spell out the evolution of the technology over the years [2–7]. Single-layer and hollow-fiber devices were abandoned because of poor performance. Affinity chromatography gave way to ion exchange chromatography as the primary ligand type. Vendor promotion turned away from protein purification to purification of large biomolecules such as plasmid DNA, viruses, and very large proteins ($>$250 kDa), where chromatography beads have low capacity. Applications such as viral clearance and purification of gene therapy vectors are examples. Three primary vendors have emerged for membrane chromatography products: Millipore Corporation (Bedford, MA, USA, Intercept™), Pall Biopharmaceuticals (East Hills, NY, USA, Mustang™), and Sartorius AG (Goettingen, Germany, Sartobind™). In this chapter, the principles and experimental methods applicable to membrane chromatography will be presented, and two applications will be offered as examples.

9.2 PRINCIPLES OF MEMBRANE CHROMATOGRAPHY

Two key advantages of membrane chromatography over columns packed with beads are (1) mass transfer limitations are reduced or eliminated leading to fast binding of the solute to the ligand sites on the membrane surface and (2) low transmembrane pressure drop. For the target solute to be captured by the binding sites on the membrane surface, the solute must flow into the pore structure, diffuse to the wall of the pore, and bind to the ligand. The result of this process is that the solution passing out of the membrane (effluent) is less concentrated in the solute than is the feed solution. The breakthrough curve (BTC) is a plot of the effluent concentration vs. time or effluent volume. Ideally the BTC is sharp, meaning no solute comes out in the effluent solution until the membrane reaches saturation, at which point the effluent solution is the same concentration as the feed solution. The extent to which this is not the case is a measure of the impact of slow adsorption kinetics, slow mass transfer, and mixing in the flow system. The faster the flow rate, the more likely the BTC will be broad. The following sections will present the principles of mass transfer, adsorption kinetics, and mixing in the flow system in the context of describing the sharpness of the BTC.

9.2.1 Adsorption Kinetics

A simple algebraic model of the BTC can be derived for the case of irreversible adsorption in the absence of axial dispersion in the membrane, mass transfer limitations, and mixing in the flow system [8]. This model was derived from the continuity equation using Langmuir adsorption kinetics as the constitutive relation:

$$C = \frac{1}{1 + (1 - e^{-n})e^{n(1-T)}} \tag{9.1}$$

where $C = c/c_0$, c is the effluent concentration, c_0 is the feed solution concentration, n is the dimensionless number of transfer units, and T is the dimensionless throughput. Axial dispersion in the membrane is typically negligible, and irreversible adsorption is often a good approximation for process-scale protein purification, because the equilibrium dissociation constant (K_d) is small for tight binding, and c_0 is large. Therefore the ratio c_0/K_d approaches infinity, and adsorption is essentially irreversible. Mass transfer limitations and mixing in the flow system are discussed in subsequent sections.

The parameter T for irreversible adsorption ($c_0/K_d \gg 1$) is given by:

$$T = \frac{\varepsilon c_0}{(1 - \varepsilon)c_l}(\tau - 1) \tag{9.2}$$

where ε is the void fraction of the membrane, and c_l is the total ligand capacity of the membrane based on the solid volume of the membrane. The throughput parameter is a measure of the loading of the membrane. It is the ratio between the amount of solute loaded into the membrane via the feed solution and the maximum amount of solute that can bind to the membrane. The dimensionless time $\tau = vt/L$, where v is the interstitial liquid velocity, L is the membrane thickness, and t is time.

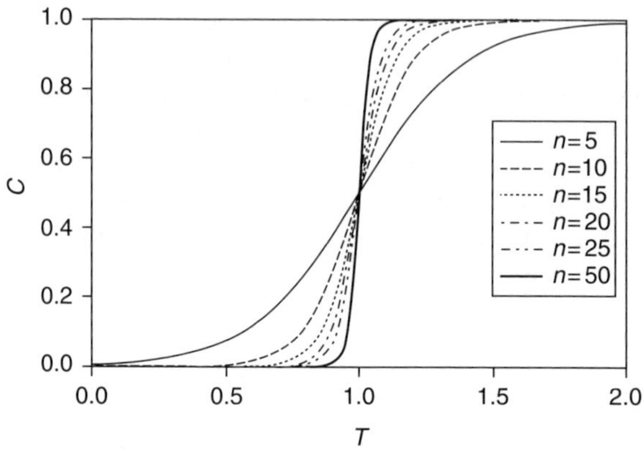

FIGURE 9.1 Breakthrough curve predicted using Equation 9.1 for different values of the number of transfer units *n*.

The parameter *n* is given by:

$$n = \frac{(1-\varepsilon)k_a c_l L}{\varepsilon v} \tag{9.3}$$

where k_a is the association rate constant of the solute with the ligand. The parameter L/n is the height of a transfer unit, comparable to the height equivalent to a theoretical plate (HETP) commonly found in the chromatography literature. When *n* is large, or HETP is small, breakthrough curves and elution peaks are sharp.

Equation 9.1 is plotted for various values of *n* in Figure 9.1. The BTC is reasonably sharp when $n = 20$–25. Not much is gained by going to $n = 50$ and beyond. Increasing *n* requires a high capacity (c_l), a fast association rate constant (k_a), and a long residence time in the membrane (L/v). If a high flow rate is desired, as is usually the case, then one or more of the other parameter values must have a large value. Thus, most chromatographic membranes use ion exchange binding (high k_a), a high ligand density (high c_l), and several layers (high *L*) to achieve sharp BTCs at high flow rates (high *v*).

The assumption of irreversible adsorption made in the derivation of Equation 9.1 is valid for values of c_0/K_d approaching infinity, as mentioned above. The practical cut-off for when c_0/K_d is large enough was determined to be $c_0/K_d > 60$, set by the criteria that Equation 9.1 fall within 95% of the exact solution at $C = 0.1$ for finite values of c_0/K_d. In other words, the exact solution for $C = 0.1$ was used to find *T*, and then the value of *C* from Equation 9.1 at that *T* had to be within 95% of the exact solution.

9.2.2 Mass Transfer

To eliminate mass transfer effects, the residence time in the membrane (L/v) must be much greater than the time scale for diffusion from the center of the membrane pore to the wall:

$$L/v \gg d_p^2/4D \tag{9.4}$$

where d_p is the diameter of the pore and D is the diffusion coefficient of the solute. This situation is frequently not the case when the membrane is thin (small L), the pores are large (large d_p), and operation is at high flow rate (large v). Most membrane chromatography systems are operated at residence times of 1 to 10 sec. Membrane pore sizes of less than 1 μm eliminate mass transfer limitations for large proteins when residence times are about 1 sec or longer. However, some membranes have a pore size of about 5 μm, and then residence times of about 100 sec or longer are required to obtain sharp BTCs for large proteins. For very large biomolecules such as plasmid DNA and viruses, even longer residence times are needed because D is smaller. As a rule of thumb, D is approximately proportional to the inverse of the molecular mass raised to the one-third power. Therefore, systems separating small proteins such as alpha-lactalbumin (14.4 kDa, $D = 1.1 \times 10^{-6}$ cm^2/sec) can be operated at higher flow rates than systems separating large proteins such as thyroglobulin (660 kDa, $D = 2.5 \times 10^{-7}$ cm^2/sec).

A few examples will illustrate the use of Equation 9.4. BTCs were sharp when α-lactalbumin and thyrogobulin were captured onto a chromatographic membrane having a pore size $d_p = 0.65$ μm, a stack thickness $L = 0.098$ cm, and operated at velocity $v = 4.9 \times 10^{-3}$ cm/sec [9]. In this case, the time scales for diffusion (4 msec for thyroglobulin and 1 msec for α-lactalbumin) were much smaller than the residence time in the membrane ($L/v = 20$ sec). On the other hand, BTCs were broad when thyroglobulin was captured onto a chromatographic membrane having a pore size of 5 μm, a stack thickness of 0.06 cm, and operated at a velocity of 4.2×10^{-2} cm/sec. In this case, the time scale for diffusion (0.25 sec) was too close to the residence time in the membrane ($L/v = 1.4$ sec). Even at a residence time of 14 sec the BTC was not sharp for this system, which indicates that the residence time in the membrane needs to be much greater than the time scale for diffusion to obtain a sharp BTC.

9.2.3 Mixing in the Flow System

Broad BTCs can result solely from liquid mixing in the pump, tubing, fittings, membrane holder, stack of membranes, and detector system. For example,

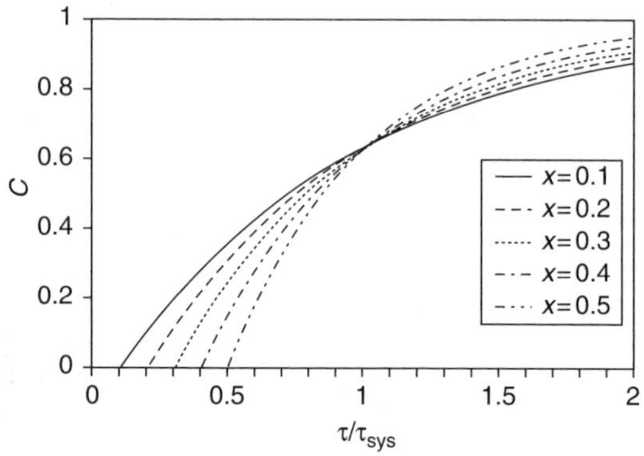

FIGURE 9.2 Breakthrough curve for a nonbinding tracer predicted using Equation 9.5 for different values of x = fraction unmixed volume, where τ_{sys} is the system mean residence time.

if the liquid flowing through the membranes has different residence times, for example, shorter times through the center and longer times through the edges, then it will broaden the BTC. The simplest model found to describe mixing in the flow system in membrane chromatography is the serial combination of a continuously stirred tank reactor (CSTR) and an ideal plug flow reactor (PFR) [10]:

$$C = 1 - \exp\left(\frac{x - (\tau/\tau_{sys})}{1 - x}\right) \tag{9.5}$$

where τ_{sys} is the dimensionless mean residence time in the system, and x is the fraction PFR volume ($x = \tau_{PFR}/\tau_{sys}$). After the delay time $= x\tau_{sys}$ from the dead volume, Equation 9.5 can be used to predict the BTC for a nonbinding tracer. Prior to that time ($\tau \leq x\tau_{sys}$) $C = 0$ (Figure 9.2). Typically, mixing in the flow system is not a significant factor in determining the shape of the BTC, because $x\tau_{sys}$ is small compared to the values of τ at the point of breakthrough, defined as when $C = 0.1$.

9.3 EXPERIMENTAL DESIGN AND DATA ANALYSIS

The following example will be used to illustrate how to conduct an experiment and analyze the results. Data were taken from the literature for capture of a small protein (α-lactalbumin) by an anion exchange membrane [9].

9.3.1 Experimental Procedure

Flat-sheet polyvinylidene difluoride membranes (acylimidazole activated Durapore membranes, Millipore, Bedford, MA) were reacted with 2-aminoethyltrimethylammonium chloride to make the anion exchange membranes. These membranes were 140 μm thick and had a pore size of 0.65 μm, an internal surface area of 155 cm^2/cm^2 of frontal area, and a void fraction of $\varepsilon = 0.7$. A 7-layer stack of these 25 mm diameter membranes sandwiched between 2 blank membranes upstream and downstream (11 membrane discs total) was placed into a membrane holder. The blank membranes aided in flow distribution. Protein solution (0.05 g/l α-lactalbumin in 50 m*M* Tris, pH 8.3) was loaded into the membrane stack at a flow rate of 1 ml/min, and the absorbance at 280 nm of the effluent solution measured vs. time. Mixing in the flow system was measured by loading a nonbinding tracer (0.05 g/l α-lactalbumin in 50 m*M* Tris, 2 *M* NaCl, pH 8.3).

9.3.2 Mixing in the Flow System

The response to loading a nonbinding tracer was fit using Equation 9.5 resulting in a fraction PFR volume of $x = 0.67$ and a dimensionless residence time for the system of $\tau_{sys} = 9.4$ (Figure 9.3). To generate this plot from the raw data, the voltage signal from the detector was determined for the baseline (V_{BL}) using only buffer without protein, and the feed solution (V_{FS}) while bypassing the membrane holder. Then the voltage signal from the BTC was converted to C using the equation $C = (V - V_0)/(V_{FS} - V_0)$. This conversion assumes that absorbance is linearly related to protein concentration, which is a good assumption for dilute protein solutions ($c < 2$ g/l), as was the case in this

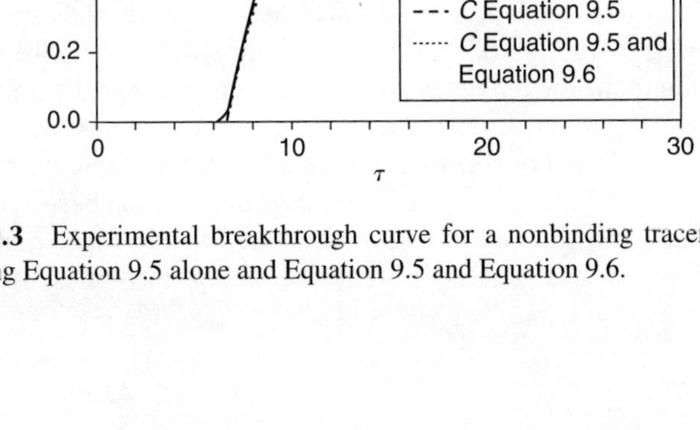

FIGURE 9.3 Experimental breakthrough curve for a nonbinding tracer, and fitted curves using Equation 9.5 alone and Equation 9.5 and Equation 9.6.

experiment ($c_0 = 0.05$ g/l). The x-axis was obtained by converting time to dimensionless time τ (= vt/L) using the values of $v = 4.85 \times 10^{-3}$ cm/sec ($v = Q/\varepsilon A$ where $Q = 1$ ml/min, $\varepsilon = 0.7$, and $A = 4.91$ cm^2) and $L = 0.098$ cm (=7 × 140 μm).

The values of x and τ_{sys} mentioned above were obtained using the SOLVER function in Excel to minimize the sum of the square of the difference between the model and the data (least squares method). Another perhaps more accurate method is to obtain τ_{sys} from the first moment of the data using the equation:

$$\tau_{sys} = \int_0^{\infty} (1 - C)\mathrm{d}\tau = \int_0^1 \tau \mathrm{d}C \tag{9.6}$$

Then this calculated value of τ_{sys} is used along with Equation 9.5 to fit the data by using x as the only fitted parameter value in Excel. Using this method, $\tau_{sys} = 10.3$ and $x = 0.638$. This result is also plotted in Figure 9.3, and is nearly identical to the first method.

Frequently, rather than reporting liquid volumes directly, the volumes are normalized by dividing by the membrane volume. This makes the results dimensionless and independent of scale. The volumes are then referred to in terms of membrane volumes. For example, to normalize the effluent liquid volume and express it in terms of membrane volumes, divide it by the membrane volume: (effluent volume)÷(membrane volume) = $\varepsilon\tau$. When the system volume is normalized and expressed in terms of membrane volumes, it is equal to: $\varepsilon\tau_{sys} = 7.2$ membrane volumes for these data [9]. Of this, $x\varepsilon\tau_{sys} = 4.6$ membrane volumes is the PFR portion, which includes 1 membrane volume for the stack of 7 membranes, and $(1 - x)\varepsilon\tau_{sys} = 2.6$ membrane volumes is the CSTR portion. One membrane volume equals 0.481 ml in this experiment. In conclusion, if the value of τ at the point of breakthrough ($C = 0.1$) is much greater than $x\tau_{sys} = 6.3$ to 6.6, then mixing in the flow system can be neglected. This criterion will be checked in a subsequent section.

9.3.3 Mass Transfer

To ignore mass transfer effects, Equation 9.4 must be satisfied. For the experimental system described, $L/v = 20$ sec, and the RHS of Equation 9.4 is 1 msec [$=(0.65 \times 10^{-4}\ \text{cm})^2/4(1.1 \times 10^{-6}\ \text{cm}^2)$]. Therefore, the time scale for convection in the membrane is 20,000 times greater than the time scale for boundary layer mass transfer to the wall of the pores, and mass transfer can be safely neglected. Based on this calculation, a greater flow rate than 1 ml/min, perhaps even 200 ml/min, could have been used and still not have a mass transfer limitation. Thus, although the flow rate used was 125 membrane volumes per hour, it might have been possible to use 25,000 membrane volumes per hour without encountering a mass transfer limitation. Column chromatography using beds of packed beads typically operates at flow rates of 30 column volumes per hour, much lower than the flow rates possible using membrane chromatography.

9.3.4 Adsorption Kinetics and the Breakthrough Curve

The experimental BTC for α-lactalbumin is shown in Figure 9.3. The point of breakthrough ($C = 0.1$) occurred at $\tau = 93.6$. This value is 14 to 15 times greater than $x\tau_{sys}$, which means that mixing in the flow system can be neglected as a factor in determining the shape of the BTC. The point of breakthrough occurred at 66 membrane volumes ($=\varepsilon\tau$). The dynamic binding capacity of the membrane is then $\varepsilon\tau c_0$ or 3.3 mg/ml expressed as mg bound per ml of membrane.

To fit Equation 9.1 to the BTC, values of the two unknowns (k_a and c_l) were assumed temporarily, allowing calculation of T using Equation 9.2 and n using Equation 9.3. The other parameter values (ε, c_0, v, L, and τ) are already known. Using the temporary values of T and N, Equation 9.1 was used to calculate C. Then SOLVER in Excel was used to minimize the square of the differences between the calculated and observed values of C using k_a and c_l as fitted parameters. The solution found was $k_a = 1900/M$sec and $c_l = 0.00085\,M$. The value of n was 14. Example calculations are shown in Table 9.1.

The fitted value for $c_l = 0.00085\,M$ is expressed as moles of α-lactalbumin bound per L of membrane solid volume. The solid volume of the membrane divided by the total volume of the membrane equals $(1 - \varepsilon)$. Therefore, the fitted value of the membrane capacity is 3.7 mg/ml when expressed on a mass and total-membrane-volume basis ($=(1 - \varepsilon)c_l$). This value corresponds closely to the value of 3.3 mg/ml determined from the point of breakthrough as mentioned above. In conclusion, the fitted and observed binding capacities match, which provides validation of the model and the fitted parameter values.

The BTC was not symmetric. Instead, the BTC first rose sharply toward $C = 0.6$ to 0.8, and then rose slowly, but never reached $C = 1.0$ (Figure 9.4). Even after loading 500 membrane volumes ($=\varepsilon\tau$) of feed solution the BTC rose to only $C = 0.986$, whereas it rose to $C = 0.5$ in only 72 membrane volumes. In contrast, the washing and elution curves rapidly approached baseline [9]. The washing curve fell to $C = 0.1$ at $\tau = 78$. The elution curve was sharp and symmetric. It emerged at $\tau = 6.2$, about the dead time of the flow system, peaked at $\tau \approx 8$, and reached 95% of the total amount eventually eluted at $\tau = 27$.

One explanation for the observed behavior is that during elution protein binding is quickly and completely disrupted in the elution buffer, and mass

TABLE 9.1
Example Calculations for Fitting Equation 9.1 to the Breakthrough Curve Data

τ	C_{exptl}	T	C_{model}	$(C_{exptl} - C_{model})^2$	
0.0	0.013	0.00	0.000	1.78E-04	
36.6	0.012	0.34	0.000	1.39E-04	
73.7	0.017	0.70	0.015	6.10E-06	$c_l = 0.00085\ M$
86.1	0.016	0.82	0.073	3.24E-03	$k_a = 1900/M\text{sec}$
92.3	0.063	0.88	0.152	7.93E-03	$n = 14$
98.5	0.280	0.94	0.292	1.39E-04	
104.7	0.562	1.0	0.486	5.67E-03	
110.9	0.746	1.1	0.685	3.67E-03	
117.1	0.800	1.1	0.833	1.11E-03	
123.2	0.843	1.2	0.920	5.93E-03	
135.6	0.875	1.3	0.984	1.19E-02	
148.0	0.888	1.4	0.997	1.19E-02	
185.1	0.919	1.8	1.000	6.51E-03	
222.2	0.947	2.1	1.000	2.83E-03	
259.4	0.951	2.5	1.000	2.38E-03	
296.5	0.956	2.8	1.000	1.90E-03	
333.6	0.962	3.2	1.000	1.48E-03	
370.7	0.966	3.6	1.000	1.15E-03	
407.8	0.970	3.9	1.000	9.15E-04	

The fitted parameter values c_l and k_a were determined by least squares regression analysis.

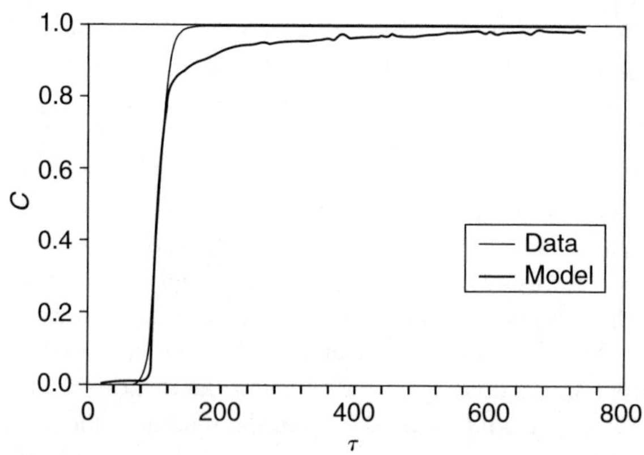

FIGURE 9.4 Experimental breakthrough curve for α-lactalbumin and fitted curve using Equation 9.1.

transfer is not rate limiting as shown before. Mixing in the flow system is the primary cause of broadening of the elution curve. For the BTC on the other hand, slow adsorption kinetics could be the rate-limiting factor. The asymmetry in the BTC may result from surface crowding effects wherein the protein adsorption rate slows as the membrane approaches saturation. The net effect of crowding is that the adsorption rate constant k_a is not constant, but falls as the surface fills up. This effect causes the shape of the BTC to transition from initially sharp, when adsorption is fast and surface crowding is not a factor, to a slowly inclining shape as the adsorption rate slows and saturation is approached. This effect is sometimes referred to as the car parking problem, because random car parking can leave many spaces that are each too small to park a new car even though the aggregate area of all these spaces is adequate.

The shape of the washing curve is determined by the removal of unbound protein from the void volume of the system, plus dissociation of any weakly bound protein. In contrast to the elution buffer, the washing buffer does not disrupt the binding of the protein. Therefore, weakly bound protein will dissociate from the membrane surface during washing, but not during elution. If only unbound protein was removed from the membranes, then the washing curve would have dropped to baseline at about the mean residence time of the flow system ($\tau = 9.4$ to 10.3). Instead, it took about 8 times longer than that to wash to $C = 0.1$. Thus, some dissociation of weakly bound protein probably occurred during washing.

9.3.5 Scale-Down and Scale-Up

Successful scale-down and scale-up of membrane chromatography systems require an accurate, scientifically based model. Equation 9.1 to Equation 9.6 can be used for this purpose. To obtain equal BTC performance (C vs. time is the same), the values of n and T must match at each time point for the small- and large-scale, and mixing in the flow system (x and τ_{sys}) must be either the same or small enough to be negligible. When the same membrane material and feed stream are used at large and small scale, parameters such as c_0, ε, c_l, k_a, d_p, and D will most likely be constant. However, v, L, x, and τ_{sys} may not be constant, because the flow rate, number of layers in the membrane sack, and extent of mixing in the flow system may increase with increasing scale. However, if L/v is kept constant, and mixing in the flow system is verified to be negligible, then equal performance at different scales should be expected. The impact of potential deviations in operating parameters (c_0 and v), and membrane chromatography device parameters (ε, c_l, k_a, d_p, and L) can then be evaluated using the model, and be used to steer clear of regions where performance is too sensitive to normal variation.

9.4 VIRAL CLEARANCE USING MEMBRANE CHROMATOGRAPHY

The potential for contamination of therapeutic proteins produced in cell culture by viruses is a regulatory concern. Steps are included in downstream processing specifically to meet regulatory requirements; redundant and complementary unit operations are included that clear any potential viral contaminant from the protein product. For viral clearance applications, performance is measured by the log reduction value (LRV), which is simply LRV $= -\mathrm{Log}_{10}(C)$. Typical LRV values for anion exchange column chromatography are LRV $= 4$ to 6 [11].

9.4.1 Adaptation of the Model to Viral Clearance Applications

The assumption of irreversible adsorption made in the derivation of Equation 9.1 is valid for values of c_0/K_d approaching infinity, as mentioned above. This is a good assumption for the BTC in process-scale protein separations, where the feed solution is concentrated. In viral clearance operations, the feed solution can be very dilute (pM to nM). Therefore, depending on the value of K_d, we may have two limiting cases (1) $c_0/K_\mathrm{d} \gg 1$ and irreversible adsorption, and (2) $c_0/K_\mathrm{d} \ll 1$ and linear adsorption. These two cases will be considered in the following sections.

9.4.1.1 Irreversible Adsorption Case

For irreversible adsorption, where c_0/K_d approaches infinity, the practical cutoff when c_0/K_d is large enough was found to be $c_0/K_\mathrm{d} > 30$, determined by setting the criteria that Equation 9.1 fall within 95% of the exact solution at LRV $= 4$. The mathematical relationship between LRV, T, and n for irreversible adsorption can be derived from Equation 9.1:

$$\mathrm{LRV} \approx \frac{n(1-T)}{\ln(10)} \tag{9.7}$$

Using Equation 9.7, we find that there is a linear decline in LRV with increasing T. The slope of this plot is approximately $-n/\ln(10)$, and the y-intercept is approximately $n/\ln(10)$.

For irreversible adsorption, Equation 9.2 can be rearranged to find the number of membrane volumes processed ($\varepsilon\tau$) at any value of the parameter T when

$\tau \gg 1$:

$$\varepsilon\tau\bigg|_{\text{irreversible}} \approx \frac{T(1-\varepsilon)c_1}{c_0} \tag{9.8}$$

The parameter T in Equation 9.8 is a dimensionless measure of the relative amount of material loaded into the membrane. $T = 0.0$ corresponds to the point where the feed solution has just started to emerge at the exit of the membrane. $T = 1.0$ corresponds to the point where the total mass loaded into the membrane equals the total membrane capacity. For an infinitely sharp BTC ($n \to \infty$), $T = 1.0$ also corresponds to 100% saturation of the membrane. However, this is impractical. A practical target for operation can be found by examination of Equation 9.7. We seek to obtain a LRV = 4 while also attaining a large loading capacity. For example, we can attain LRV = 4 at $T = 0.08$ and $n = 10$, or at $T = 0.90$ and $n = 90$. Therefore, it is desirable to have a large value of n because we can achieve a much larger throughput (greater T) while still attaining LRV = 4.

If we set a target to attain 90% of the saturation capacity ($T = 0.9$) at LRV = 4.0, then we find from Equation 9.7 that this target corresponds to attaining a value of $n = 92$.

From Equation 9.3, we see that attaining $n = 92$ requires a high capacity (c_1), thick membrane stack (L), low flow rate (v), and fast adsorption rate constant (k_a). For example, for the membrane system analyzed in Section 9.3, the invariant membrane parameters are $k_a = 1900$/Msec, $c_1 = 0.00085\ M$, and $\varepsilon = 0.7$. Therefore, to attain the above target (LRV = 4 at $T = 0.9$) requires $L/v = 133$ sec. This residence time is much longer than the time used in the experiment ($L/v = 20$ sec). This example illustrates a general rule of thumb: it is easier to obtain a sharp BTC for protein purification than it is to achieve a target LRV for viral clearance.

9.4.1.2 Linear Adsorption Case

For the linear adsorption case where $c_0/K_d \ll 1$, Equation 9.1 is not valid. In this case, the BTC is given by:

$$C = 1 - \exp(-nT)\int_0^n \exp(-\eta)I_0(2\sqrt{\eta nT})\,\mathrm{d}\eta \tag{9.9}$$

where I_0 is the modified Bessel function of the zero order. Values of n and T that result in LRV = 4 were calculated from Equation 9.9 and are listed in Table 9.2. In general, when LRV = 4 for any given value of n, the corresponding value of T is smaller in the linear adsorption case than the irreversible adsorption case.

TABLE 9.2
Values of *n* and *T*

n	*T*
10	0.008
30	0.26
50	0.39
70	0.46
90	0.52

LRV = 4 from Equation 9.9 for the case of linear adsorption.

In other words, as in the irreversible adsorption case of Equation 9.7, LRV for the linear adsorption case is a function of only n and T, but the values of LRV for the linear adsorption case are generally smaller at a given value of n and T. Only when $T = 0$ is the LRV for the linear adsorption case equal to the LRV for the irreversible adsorption case. This is because when $T = 0$, Equation 9.9 reduces to $C = \exp(-n)$, because $I_0(0) = 1$, and $\mathrm{LRV} = n/\ln(10)$, which is the same result as Equation 9.7 when $T = 0$.

The definition of T is different for the linear adsorption case:

$$T = \frac{\varepsilon K_\mathrm{d}}{(1-\varepsilon)c_1}(\tau - 1) \tag{9.10}$$

where K_d is the dissociation equilibrium constant. Equation 9.10 can be rearranged to calculate the membrane volumes of feed solution processed at any value of T when $\tau \gg 1$:

$$\varepsilon\tau\Big|_{\text{linear}} \approx \frac{T(1-\varepsilon)c_1}{K_\mathrm{d}} \tag{9.11}$$

We can see from Equation 9.11 that the volume of feed solution processed at a given value of T is not at all related to the feed solution concentration for the linear case, whereas for the case of irreversible adsorption, it was inversely related to the feed solution concentration as in Equation 9.8. Also, because $K_\mathrm{d} \gg c_0$ for the linear adsorption case, throughput expressed as $\varepsilon\tau$ or T is going to be lower than for the irreversible adsorption case.

9.4.2 Use of the Model for Design

From a regulatory perspective, if a membrane chromatography product was shown to attain LRV = 4 for a particular feed solution at a fixed concentration (c_0), loading volume ($\varepsilon\tau$), and residence time (L/v), then the LRV should exceed 4 for a smaller loading volume, longer residence time, or more dilute feed solution. Validation of a membrane chromatography system for viral clearance should utilize measuring the LRV of effluent fractions over time rather than the entire effluent pool, and the trend of LRV vs. T can be determined to aid in setting allowable operating limits.

From a membrane design point of view, we have set the above target (LRV = 4.0 and $T = 0.9$), but need to set some additional constraints to fully define the problem. For example, what flow rate and volumetric throughput will be attractive compared to competing technologies? One approach to answering this question is to take values for the flow rate and volumetric throughput from the commercially successful viral filtration systems. It should be noted that viral filtration removes viruses by a sieving mechanism, which is totally different than the adsorption mechanism used in membrane chromatography. Nevertheless, we can use the performance capabilities of viral filtration membranes as a target for membrane chromatography systems too. A commercially successful viral filtration system is the Viresolve™ filters from Millipore (Bedford, MA), which can achieve LRV = 4.0 for the bacteriophage ϕX174 when operated at a flow rate of 150 l/m^2-h, a throughput of 300 l/m^2, and a pressure drop of 2.0 bar [12]. This flow rate and throughput target corresponds to $\varepsilon v_{\text{min}} = 4.2 \times 10^{-3}$ cm/sec and $\varepsilon\tau_{\text{min}}L_{\text{min}} = 30$ cm, respectively, for a membrane chromatography system. One advantage of membrane chromatography is a lower pressure drop. At 2.0 bar, the membrane system analyzed in Section 9.3 would attain the target flow rate when $L = 6.2$ cm based on the reported hydraulic permeability [13]. Therefore, pressure drop is not a limitation.

We can use Equation 9.7 and Equation 9.8 for the irreversible adsorption case, and Equation 9.3 to calculate the minimum L under conditions constrained by meeting the targets for flow rate ($\varepsilon v_{\text{min}} = 4.2 \times 10^{-3}$ cm/sec), throughput ($\varepsilon\tau_{\text{min}}L_{\text{min}} = 30$ cm), and viral clearance (LRV = 4.0) as set above. We use data from Section 9.3 to illustrate these calculations. The value of L_{min} needed to meet the flow rate and viral clearance targets is found from substitution of LRV = 4.0 and $T = 0.9$ into Equation 9.7 to obtain $n = 92$, which is then substituted into Equation 9.3 along with $\varepsilon v_{\text{min}} = 4.2 \times 10^{-3}$ cm/sec to solve for L_{min}. The value of L_{min} needed to meet the throughput target is found from substitution of $\varepsilon\tau_{\text{min}}L_{\text{min}} = 30$ cm into the LHS of Equation 9.8. An example calculation is shown in Table 9.3 using values of c_1, c_0, and k_a taken from Section 9.3. To meet the throughput requirement we require $L_{\text{min}} = 0.46$ cm. However, this value is too thin to meet the viral clearance target of LRV = 4.0.

TABLE 9.3
Example Calculation for Use of the Model to Design a Membrane Chromatography System for Viral Clearance

Known	Calculation of L_{min} Using Equation 9.7, Equation 9.3[a], and Equation 9.8[b]
$\varepsilon = 0.7$ $\varepsilon v_{min} = 4.2 \times 10^{-3}$ cm/sec $\varepsilon \tau_{min} L_{min} = 30$ cm $k_a = 1900/M$sec $c_l = 0.00085\ M$ $c_0 = 3.5 \times 10^{-6}\ M$	$L_{min} = \frac{92(\varepsilon v_{min})}{(1-\varepsilon)k_a c_l} = \frac{(92)(4.2 \times 10^{-3}\ \text{cm/sec})}{(1-0.7)(1900/M\text{sec})(8.5 \times 10^{-4}\ M)}$ $= 0.8\ \text{cm}^{a}$ $L_{min} = \frac{30\ \text{cm}}{\varepsilon \tau_{min}} = \frac{30\ \text{cm}}{(T(1-\varepsilon)c_l)/c_0}$ $= \frac{30\ \text{cm}}{(0.9(1-0.7)8.5 \times 10^{-4}\ M)/3.5 \times 10^{-6}\ M} = 0.46\ \text{cm}^{b}$

LRV = 4 at a flow rate of 150 l/m^2-h and a throughput of 300 l/m^2.

For that, we require $L_{min} = 0.8$ cm. Thus, a membrane stack thicker than $L = 0.8$ cm would exceed the targets set above. In conclusion, the principles outlined above can be used to design membrane chromatography systems for viral clearance. Desirable system parameters include (1) high membrane capacity c_l, (2) thick membrane stack L, (3) dilute feed solution c_0, and (4) fast association rate constant k_a. This is in the case of irreversible adsorption.

The solution is slightly different for the case of linear adsorption. In that case, we need to know K_d to use Equation 9.11, and the feed solution concentration does not affect performance. In addition, we cannot realistically attain the above target (LRV = 4 at $T = 0.9$). From Table 9.2, we see that $T < 0.9$ when LRV = 4 for all reasonable values of n. Thus, the throughput T is less at a given value of n and LRV, and the LRV is less at a given value of n and T for the linear adsorption case, compared to the irreversible adsorption case. In the linear adsorption case, we can choose a value of n and determine the value of T when LRV = 4 from Table 9.2. The value of L_{min} to meet the viral clearance target is calculated from Equation 9.3, and the value of L_{min} to meet the flow rate target is calculated from Equation 9.11.

9.4.3 Comparison to the Literature

The model can be used to analyze data taken from the literature [13], where the LRV was measured for a membrane chromatography system similar to the one in Section 9.3. The effect of throughput ($=\varepsilon\tau$) on LRV for ϕX174 is shown in

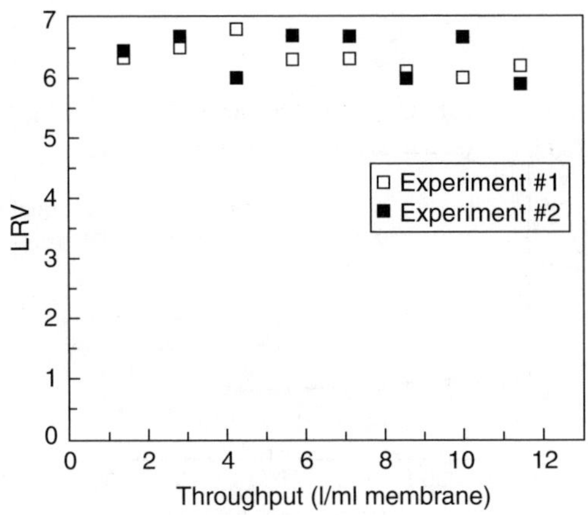

FIGURE 9.5 Effect of throughput on the LRV for ϕX174 in 25 m*M* Tris, pH 8.1 using a flow rate of 3400 membrane volumes per hour. (From Phillips M, Cormier J, Ferrence J, Dowd C, Kiss R, Lutz H, and Carter J. *J Chromatogr A* 2005; 1078:74–82. With permission.)

Figure 9.5. The feed solution in this experiment was very dilute: 1.5×10^7 pfu/ml ($c_0 \approx 1 \times 10^{-13} M$). The membrane capacity was reportedly $c_1 = 0.0058\ M$, measured using tosyl glutamic acid, and $L = 0.1$ cm, and $\varepsilon = 0.7$. From these values, the parameter T in Equation 9.2 can be calculated: $T \approx 4 \times 10^{-11}$. In essence, $T \approx 0$, and LRV $= n/\ln(10)$ from Equation 9.7. Therefore, the LRV is not a function of throughput T, and this may be why no dependency on T is observed in Figure 9.5.

The effect of linear velocity (εv) on the LRV for endotoxin, herring sperm DNA, host cell protein (HCP), and the bacteriophage ϕX174 is shown in Figure 9.6. Because the feed solutions in these experiments were also very dilute: 2000 endotoxin units/ml, 1 μg/ml DNA, 700 to 1000 ng/ml HCP, and 1.5×10^7 pfu/ml ϕX174, a value of $T \approx 0$ was assumed in Equation 9.7. The LRV should have decreased with increasing linear velocity because n is inversely proportional to εv in Equation 9.3, and LRV is proportional to n in Equation 9.7. This did not occur. Possible explanations for this discrepancy are (1) heterogeneity in the feed solution caused some nonbinding HCP, endotoxin, or ϕX174 to pass through the membrane while all other material bound, (2) a miniscule amount of channeling or bypassing of membrane, (3) k_a in Equation 9.3 is proportional to flow rate because the system is boundary layer mass transfer limited not kinetically limited, and (4) mixing in the flow system dominates performance and not kinetics or mass transfer. For example, regarding point (1), the virus particles may not all have a negative net charge at pH 8.1, which might render those particles unable to bind to the positively

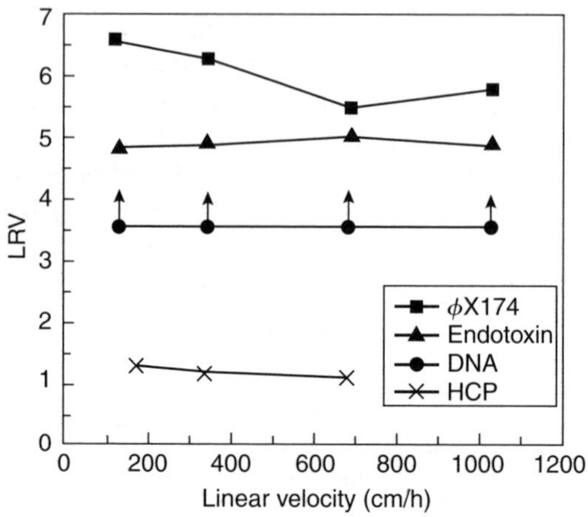

FIGURE 9.6 LRV for ϕX174, DNA, endotoxin, and HCP vs. linear velocity. Upward arrows indicate that DNA was not detected in the effluent. (From Phillips M, Cormier J, Ferrence J, Dowd C, Kiss R, Lutz H, and Carter J. *J Chromatogr A* 2005; 1078:74–82. With permission.)

charged ion exchange membrane. The Henderson-Hasselbalch equation for acid base equilibrium can be used to illustrate this point. If the value of pH–pK_a for an acid is 4.0, then 99.99% of the acid exists as a negative ion, and 0.01% remains uncharged. If only 0.01% of the virus was not charged negative at pH 8.1, and passed through the membrane without binding, then we would find that LRV = 4, even though the LRV for the virus particles that were capable of binding would be much higher than 4. For comparison, the virus ϕX174 has a pI = 6.6, making pH–pI only 1.5 for the experiment of Figure 9.6.

These discrepancies between the observations and the model point out other possible topics for exploration, and the utility of combining mathematical models of the basic principles with experimental observations to discover what we know and do not know about the mechanism of action of a unit operation. Furthermore, the model can be used to identify the critical and noncritical operating parameters that determine system performance. Future work should examine the reason for the lack of decline in LRV with increasing flow rate and should verify the expected decline in LRV with increasing throughput.

9.5 CONCLUSIONS

Membrane chromatography is a new technology designed to overcome the flow rate and diffusion limitations of traditional columns packed with beads. Advances in the theoretical and experimental understanding of the performance of membrane chromatography has led to new and improved designs, and refocused the target applications toward new areas including viral clearance, gene therapy, and very large proteins, whereas in the past the target has been purification of small proteins. This chapter presents the principles of membrane chromatography and some simple mathematical models of performance, and explains step-by-step how to use these models to analyze laboratory and pilot plant data. The use of these models for scale-up and scale-down of membrane chromatography, is also presented. The models are adapted to the application of membrane chromatography to viral clearance. Critical operating parameters are identified using the models, and the design of membrane chromatography systems specifically for viral clearance applications is explained using the models. Viral clearance data from the literature is used to illustrate the application of the models to data analysis. Some points of agreement and some discrepancies were found when comparing the predictions from the model to the data from the literature. It is clear that we do not yet understand everything about the science of membrane chromatography, especially when applied to viral clearance operations. Some suggestions were made for future research. It is hoped that this chapter will lead to developing a more solid, scientific understanding of the mechanism of action of membrane chromatography and its increased adoption by the biopharmaceutical industry.

REFERENCES

1. Brandt S, Goffe RA, Kessler SB, and O'Connor JL. Membrane-based affinity technology for commercial scale separations. *Bio/Technology* 1988; 6:779–782.
2. Etzel MR. Layered stacks. In: Svec F, Tennikova TB, and Deyl Z, Eds. *Monolithic Materials: Preparation, Properties and Applications*. Amsterdam: Elsevier, 2003, pp. 213–234.
3. Ghosh R. Protein separation using membrane chromatography: opportunities and challenges. *J. Chromatogr. A* 2002; 952:13–27.
4. Zeng X and Ruckenstein E. Membrane chromatography: Preparation and applications to protein separation. *Biotechnol. Prog.* 1999; 15:1003–1019.
5. Charcosset C. Purification of proteins by membrane chromatography. *J. Chem. Technol. Biotechnol.* 1998; 71:95–110.
6. Roper DK and Lightfoot EN. Separation of biomolecules using adsorptive membranes. *J. Chromatogr. A* 1995; 702:3–26.
7. Thömmes J and Kula MR. Membrane chromatography — an integrative concept in the downstream processing of proteins. *Biotechnol. Prog.* 1995; 11:357–367.
8. Heister NK and Vermeulen T. Saturation performance of ion-exchange and adsorption columns. *Chem. Eng. Prog.* 1952; 48:505–516.
9. Yang HW, Viera C, Fischer J, and Etzel MR. Purification of a large protein using ion-exchange membranes. *Ind. Eng. Chem. Res.* 2002; 41:1597–1602.

10. Fogler HS. *Elements of Chemical Reaction Engineering*. 2nd ed. Englewood Cliffs, NJ: Prentice-Hall, 1992.
11. Curtis S, Lee K, Blank GS, Brorson K, and Xu Y. Generic/matrix evaluation of SV40 clearance by anion exchange chromatography in flow-through mode. *Biotechnol. Bioeng.* 2003; 84:179–186.
12. Millipore Technical Brief, Viresolve® NFP Filters Predictably Scale, TB1020EN00, Bedford, MA, 2002.
13. Phillips M, Cormier J, Ferrence J, Dowd C, Kiss R, Lutz H, and Carter J. Performance of a membrane adsorber for trace impurity removal in biotechnology manufacturing. *J. Chromatogr. A* 2005; 1078:74–82.

10 Ultrafiltration Process Design and Implementation

Herb Lutz and Bala Raghunath

CONTENTS

Ultrafiltration is a membrane-based separation technology widely employed in biopharmaceutical manufacturing to concentrate and separate biological molecules such as proteins. Figure 10.1 shows several ultrafiltration steps used in the downstream purification of a monoclonal antibody.

This chapter is organized into the following sections: process requirements, technology fundamentals, commercial products, development of a phase 1 process, scale-up, equipment selection, process validation, troubleshooting, advanced topics, and a guide to the literature.

10.1 ULTRAFILTRATION PROCESS REQUIREMENTS

Ultrafiltration (also commonly called UF/DF for ultrafiltration/diafiltration) uses polymeric membranes to retain a biologic product while allowing low molecular weight solutes and water to pass through the membrane. Ultrafiltration is widely used to (1) concentrate (or dewater) the product and (2) remove low molecular weight impurities or buffer components while replacing them with a fresh buffer. Size exclusion chromatography (SEC), was used in the past for buffer exchange but has since been superseded due to its high cost and

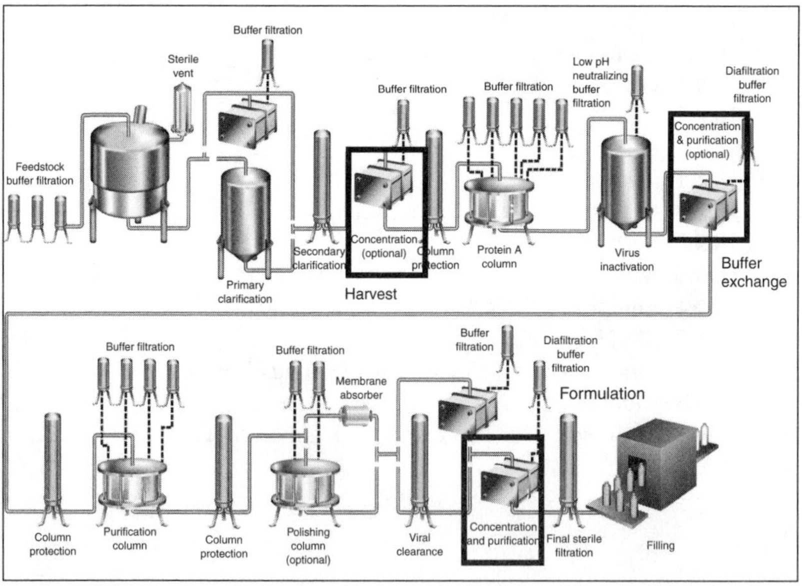

FIGURE 10.1 Location of ultrafiltration steps in MAb downstream processing.

TABLE 10.1
Ultrafiltration Step Process Objectives and Considerations

Step	Harvest	Buffer Exchange	Formulation
Process objectives	Reduce capture column size by reducing the batch volume and/or desalting	Put the product into a loading buffer optimized for high selectivity	Concentrate the product and put it in formulation buffer for fill and finish
Feed	Cell and colloid free (0.2 μm filtered) with lipids, 2–20 Kl batch size, 0.1–1 g/l protein	Column eluate or neutralized inactivation step typically at high salt	Column eluate or dilute virus filter permeate, 0.1–4 g/l protein, 0.2–6 Kl batch size
Product	1–5 g/l product ready to load on a capture column	>95% exchange into new buffer	Vial ready at: 5–20 g/l protein, >99% formulation buffer, with low extractables, endotoxin, and multimers
Key considerations	Fouling, large batch volume	Precipitation	Product quality

protein concentration limitations.[1] Table 10.1 summarizes these ultrafiltration applications shown in Figure 10.1.

Each application needs to be designed to meet its process requirements in a robust manner while meeting process constraints and having its performance optimized. General process constraints include considerations of product stability, possible limitations on cleaning agents, and process duration to balance the production line. Hardware and software constraints involve operator safety and integration with the rest of the plant. Optimized performance considerations include consistency, ease-of-use, economics, and product yield.

10.2 ULTRAFILTRATION TECHNOLOGY FUNDAMENTALS

Ultrafiltration is typically operated in tangential flow filtration (TFF) mode as shown in Figure 10.2. TFF involves passing a permeate fluid through the membrane (with a velocity component perpendicular, or normal to the membrane) and passing fluid across the membrane surface (with a velocity component tangent to the membrane).

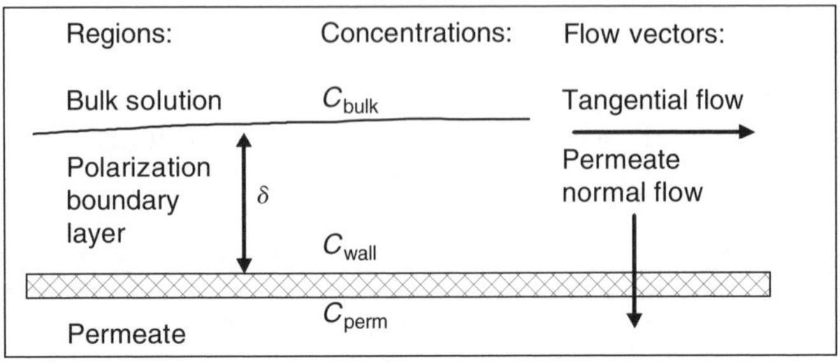

FIGURE 10.2 Tangential flow filtration.

10.2.1 Surface Polarization

Polarization is a fundamental phenomenon occurring in ultrafiltration wherein retained solutes concentrate at the membrane surface. Understanding polarization and controlling its effects are essential to implementing a good process. Solutes entrained by the permeate flow are retained by the membrane. They accumulate on the membrane surface and form a region of high concentration called the polarization boundary layer. A steady state is reached where Brownian diffusion helps the retained solute migrate away from the membrane surface while tangential convective flow carries antibody along the membrane surface, and normal convective flow carries it toward the membrane. The back transport leading to steady state operation gives TFF a high capacity. The elevated membrane surface concentration is called C_{wall} (as visualized by Vilker et al.[2] and McDonogh et al.[3]). Neglecting tangential convection allows a one-dimensional (1D) mass balance derivation for the single solute, polarization equation[4] where k is defined as the mass transfer coefficient (ratio of Brownian antibody diffusivity D to the boundary layer thickness δ):

$$\text{Polarization} \quad C_{\text{wall}} - C_{\text{perm}} = (C_{\text{bulk}} - C_{\text{perm}}) \cdot \exp(J/k) \qquad (10.1)$$

The mass transfer coefficient k and the boundary layer thickness are dependent on the cross flow. At high flux rates the wall concentration can significantly exceed the bulk concentration with potential impacts on protein aggregation and membrane fouling (Figure 10.3). The protein mass held up in the polarization layer can be on the order of 1.5 g/m^2 depending on concentrations, mass transfer, and flux. Analysis of multicomponent systems is more complex and must include van der Waals interactions between polarized species.[5,6]

An empirical gel model is obtained from Equation 10.1 by taking $C_{\text{perm}} = 0$ and setting C_g (or gel concentration) $= C_{\text{wall}}$ to get Equation 10.2. Figure 10.4

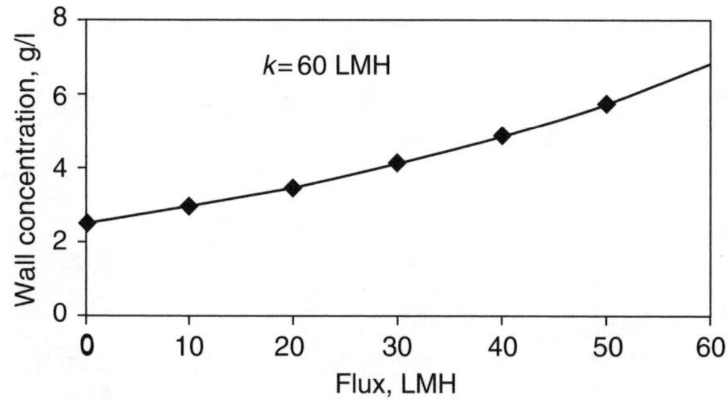

FIGURE 10.3 Wall concentration.

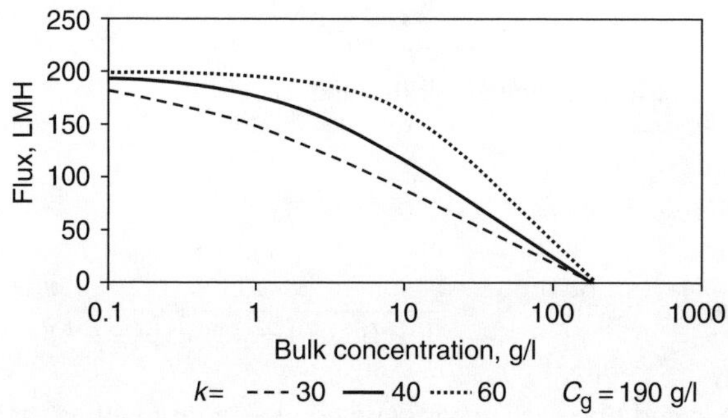

FIGURE 10.4 Gel model.

shows that Equation 10.2 can provide a good fit to data over a range of concentrations but breaks down at lower concentrations where the flux is determined by membrane permeability. It should be noted that C_g values obtained from such plots do not correspond to a separate gel phase.

$$\text{Gel model} \quad J = k\ln(C_g/C_{\text{bulk}}) \tag{10.2}$$

10.2.2 Sieving and Retention

The intrinsic membrane sieving, also called passage or transmission, is defined as $S_i(= C_{\text{perm}}/C_{\text{wall}})$ while intrinsic membrane retention or rejection is defined as $R_i(= 1 - S_i)$. The intrinsic sieving is inherent to the membrane and solute, while an observed sieving as $S_o(= C_{\text{perm}}/C_{\text{bulk}})$ varies with polarization. Equation 10.1 is rearranged to show that observed sieving depends on intrinsic sieving

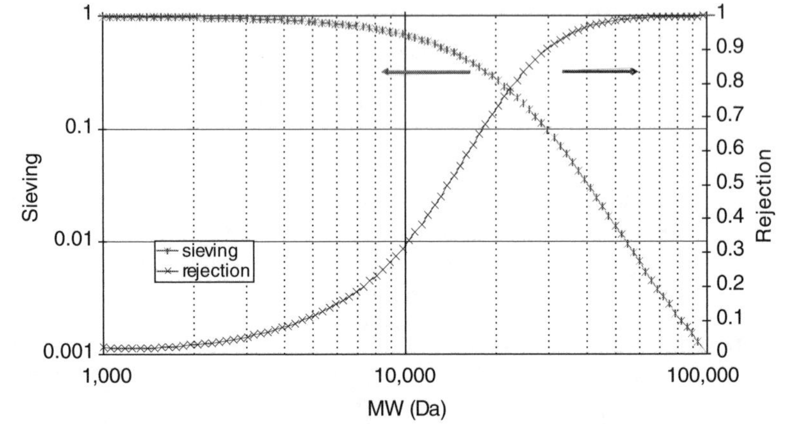

FIGURE 10.5 Intrinsic membrane sieving and retention.

and polarization:

$$\text{Observed sieving} \quad S_o = \frac{1}{1 + ((1/S_i) - 1) \cdot \exp(-J/k)} \tag{10.3}$$

While intrinsic and observed sieving are equivalent at low flux, at high fluxes the wall concentration can increase significantly as a result of polarization. This causes the observed sieving to increase and approach 100%, regardless of the intrinsic sieving.

The intrinsic sieving characteristics of a UF membrane can be characterized by using a polydisperse nonadsorbing solute such as dextran[7] as shown in Figure 10.5. Retention is based on hydrodynamic size, not molecular weight, so linear chain dextrans show a higher sieving than globular proteins of the same molecular weight. The designation of ultrafiltration membranes are considered to fall within the 1 k to 1,000 kDa range with tighter membranes considered to be reverse osmosis and more open membranes microfiltration.

10.2.3 Flux

Figure 10.6 shows that flux flattens out at high pressures. It has been shown that retained antibodies at a wall concentration of 191 g/l have an osmotic pressure Π of 30 psi.[8] That is, an elevated pressure of 30 psig must be applied to the protein-rich retentate side of a water permeable membrane containing 191 g/l of antibody in order to prevent water back-flow from the permeate side of the membrane containing water at 0 psig. This diminishes the driving force for flow

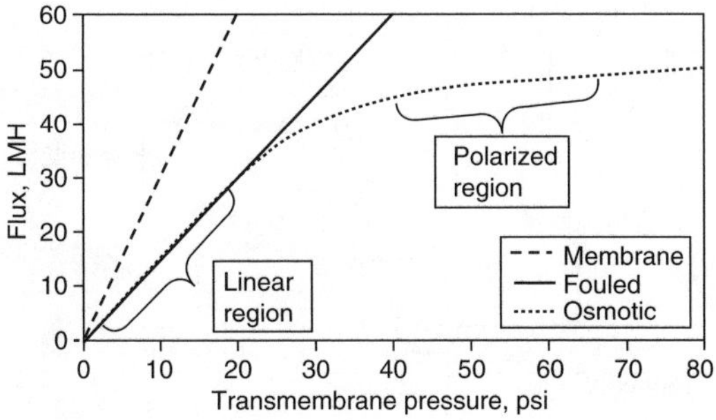

FIGURE 10.6 Ultrafiltration flux behavior.

and leads to the mechanistic-based osmotic flux model:[9]

$$\text{Polarization flux model} \quad J = \frac{\text{TMP} - R \cdot \Delta\Pi(C_{\text{wall}})}{\mu \cdot (R_{\text{membrane}} + R_{\text{fouling}})} \tag{10.4}$$

where TMP (transmembrane pressure) is the pressure difference across the membrane, R is the intrinsic membrane retention, delta Π is the osmotic pressure at the wall concentration, μ is the permeate viscosity, and R_{membrane} and R_{fouling} are the hydraulic resistances of the membrane and fouling layer respectively. An empirically based flux model can also be defined by omitting the osmotic term and adding a compressible polarization resistance term in the denominator.[4]

Equation 10.4 requires a mass transfer coefficient k to calculate C_{wall} and a relation between protein concentration and osmotic pressure. Pure water flux obtained from a plot of flux vs. pressure is used to calculate membrane resistance (typically small). The LMH/psi slope is referred to as the NWP (normal water permeability). The membrane plus fouling resistances are determined after removing the reversible polarization layer through a buffer flush. Note that in a device where the feed pressure varies along the feed channel, the TMP is calculated as an average: $\text{TMP} = (P_{\text{feed}} + P_{\text{retentate}})/2 - P_{\text{permeate}}$. To illustrate the components of the osmotic flux model, Figure 10.6 shows flux vs. TMP curves corresponding to just the membrane in buffer ($R_{\text{fouling}} = 0$, $C_{\text{wall}} = 0$), fouled membrane in buffer ($C_{\text{wall}} = 0$), and fouled membrane with osmotic pressure.

The region at low flux/low TMP is called the linear region and is dependent on TMP but independent of cross flow and bulk concentration. The region at high flux/high TMP is called the polarized region and is independent of TMP but dependent on cross flow and bulk concentration. This extremely counterintuitive result is the consequence of polarization. In between these two regions lies what

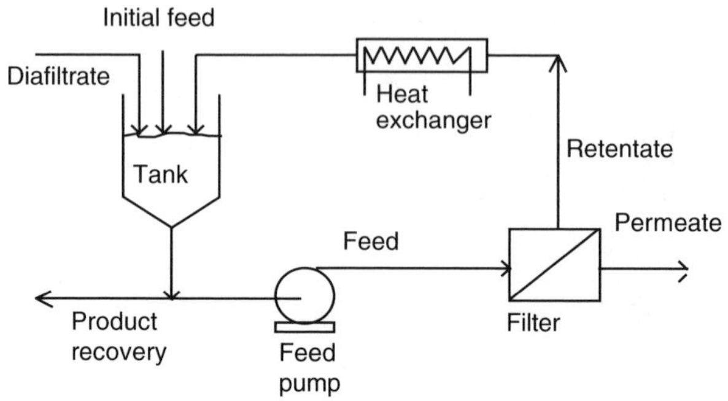

FIGURE 10.7 Ultrafiltration process schematic.

is termed the knee of the flux curve. Increasing pressure beyond the knee gives diminishing return in improving flux.

10.2.4 Processing

Figure 10.7 shows the components in a batch ultrafiltration process. Processing involves charging the feed tank with the protein product solution, recirculating this feed using the feed pump, and withdrawing product-free permeate waste. As permeate is withdrawn, the tank volume drops and the concentration of retained product in the tank increases. This concentration step proceeds until the product concentration meets the target formulation concentration. The buffer formulation is then modified during a diafiltration step where buffer, or diafiltrate is added at the same rate as permeate is withdrawn. The tank volume and retained protein concentration remains constant during this step. After sufficient buffer exchange has taken place, the retained product may be recovered from the retentate. This sequence of concentration and diafiltration is referred to as the diafiltration strategy (see Section 10.4.2).

Figure 10.8 and Equation 10.5 show the relationship between retained product concentration and the volume reduction factor X = (initial volume)/(final volume) for different membrane retention characteristics and a starting concentration of 0.1 g/l.[4] For a fully retained product ($R = 1$), a tenfold volume reduction ($X = 10$) produces a tenfold more concentrated product at 1 g/l. However, if the product is only partially retained, the volume reduction does not proportionately increase the final concentration due to losses through the membrane. Depending on the feed concentration and the product sieving, a 5- to 50-fold increase in product concentration may be required

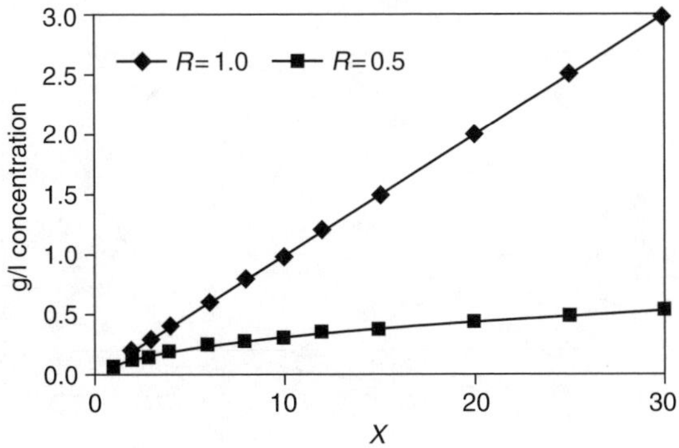

FIGURE 10.8 Concentration mode operation.

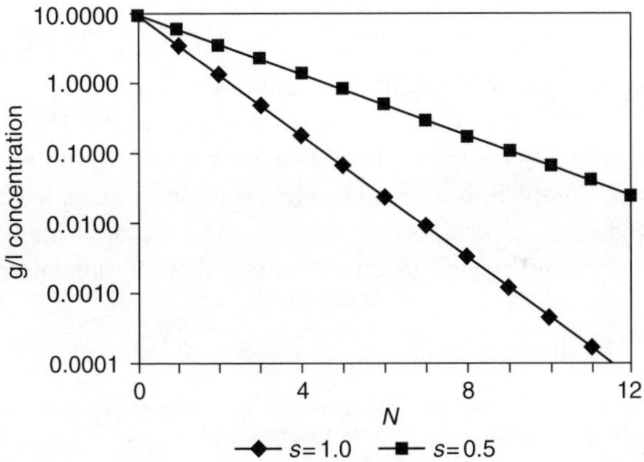

FIGURE 10.9 Diafiltration mode operation.

corresponding to a 5- to 50-fold volume reduction

$$\text{Concentration mode operation} \quad C = C_0^* X^{\mathrm{R}} \tag{10.5}$$

Figure 10.9 and Equation 10.6 show the relationship between retentate concentration and diavolumes N = (buffer volumes added)/(fixed retentate volume) for different membrane passage characteristics and a starting solute concentration of 10 g/l.[4] For a fully passing solute ($S = 1$) such as a buffer, the retentate concentration decays 10-fold with each 2.2 diavolumes. Partially retained solutes do not decay as quickly and require more diavolumes to reach a final concentration target. A fully retained solute ($S = 0$) maintains

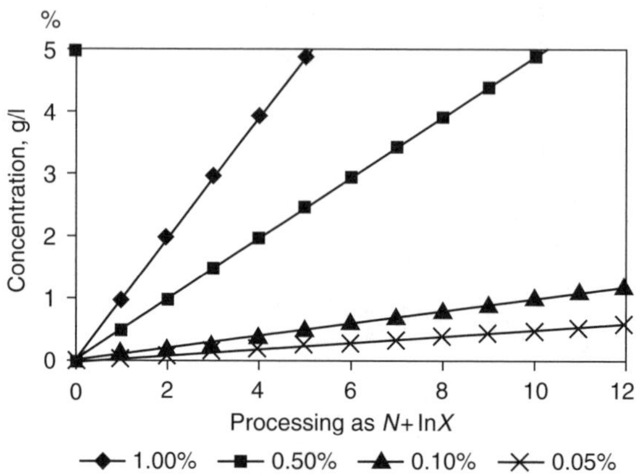

FIGURE 10.10 Permeate losses.

its retentate concentration constant at the initial value. For fully sieving solutes, >4.5 diavolumes are needed to achieve a specification of <1% of the original buffer components. It is common to add an extra 1 to 2 diavolumes as a safety factor to ensure complete buffer exchange. Note that incomplete mixing (due to dead legs and liquid droplets on tank walls) becomes significant at high diavolumes (<14) and causes the curves in Figure 10.9 to flatten out.

$$\text{Diafiltration mode operation} \quad C = C_o^* \exp(-S^* N) \qquad (10.6)$$

The formulas for concentration and diafiltration can be combined for the entire process to derive an expression for the loss of product in the permeate. This loss is shown in Figure 10.10 and Equation 10.7 for different levels of processing and sieving characteristics. Note that a membrane with 1% sieving (99% retention), can have a process yield losses much higher than 1% because the protein is repeatedly cycled past the membrane during the entire process with losses at every pass. A high yielding process (<1% product loss) requires sieving of <0.1% (retention of >99.9%).

$$\text{Permeate losses} \quad L = 1 - \exp[-S^*(N + \ln X)] \qquad (10.7)$$

The purification of a product p from an impurity i by an ultrafiltration process is shown in Figure 10.11 and Equation 10.8 where C_{i0} and C_{p0} are the initial g/l concentrations of the two components, Y_p is the yield of product in the retentate, and ψ is the selectivity($= S_i/S_p$), the ratio of sieving. High yields are obtained in purifying out small solutes (high selectivity) but are compromised

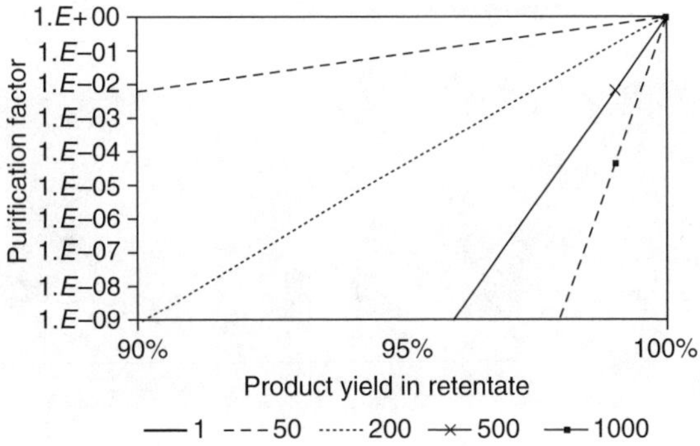

FIGURE 10.11 Purification vs. yield.

in removing larger impurities with similar sieving to the product

$$\text{Purification factor} \quad \text{PF} = \frac{(C_\text{i}/C_\text{i0})}{(C_\text{p}/C_\text{p0})} = \left(\frac{\text{ppm\_in\_product}}{\text{ppm\_in\_feed}}\right) \cdot Y_\text{p}^{\psi-1} \quad (10.8)$$

10.3 COMMERCIAL PRODUCTS

10.3.1 ULTRAFILTRATION MEMBRANES

Early ultrafiltration membranes had thin surface retentive layers with an open structure underneath as shown in Figure 10.12. These membranes were prone to defects and showed poor retention and consistency. Composite membranes have a thin retentive layer cast on top of a microfiltration membrane. These composites demonstrate consistently high retention. Membranes are also surface modified to make them lower binding and fouling resistant. This means consistently higher fluxes and less product losses through adsorption to the membrane.

Table 10.2 compares properties of different commercially available UF membranes. Membrane selection is based on experience with vendors, molecular weight rating for high yields, chemical and mechanical robustness during product processing and clean-in-place (Extractables, Adsorption, Swelling, Shedding, Class VI), flux (LMH) for sizing and costing, and the Quality/Consistency (ISO, cGMP) of the vendor and the membrane. Regenerated cellulose is often selected due to its low fouling property that improves consistency over the process, increases fluxes, makes cleaning easier, and improves yield.

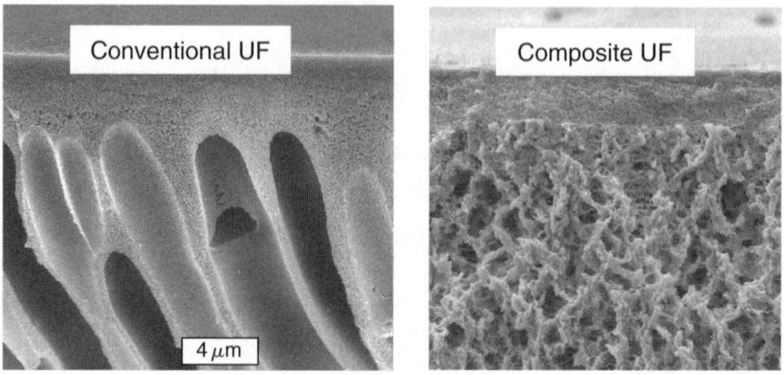

FIGURE 10.12 Ultrafiltration membranes.

TABLE 10.2
Ultrafiltration Membrane Properties

	Modified Polyether Sulfone	Regenerated Cellulose
Composite form	Yes	Yes
Available MW ratings	5–100 kDa	1–1000 kDa
Oxidants compatibility	OK	Low chloride tolerence
pH compatibility	1–4	2–12
Extractibles	Moderate	Moderate
g/M2 protein binding	0.2–0.5	0.1
Fouling	Moderate	Very low
Class VI	Pass	Pass
Strength	High	Moderate
Temp. range	4–50°C	4–50°C
Flux	Moderate	High

Membranes with low molecular weight cut-off ratings provide higher retention but have corresponding low flow rates, requiring more membrane area to achieve the separation. This leads to larger pumps with large holdup volumes and potential negative impact on product quality. A rule of thumb for selecting membrane NMWL (nominal molecular weight limit) is to take 0.2 to 0.3 of the product MW or 30 to 50 kDa membranes for high antibody retention at reasonable flux rates. Note that this rating is a nominal value and the specific retention properties vary among membranes and vendors depending on the marker solute selected (protein, dextran in a particular buffer), and the level of retention selected for the marker solute.

10.3.2 Ultrafiltration Modules

Table 10.3 compares properties of commercially available ultrafiltration modules shown in Figure 10.13 and Figure 10.14. Module selection is made on the basis of experience with the vendor, chemical and mechanical robustness in the process fluid and CIP (extractables, adsorption, swelling, shedding, class VI), vendor support, feed channel plugging, holdup/working volume, mass transfer efficiency (affecting pump and area sizing), packing density (area/volume), scalability, ability to integrity test, and ease-of-use. Flat sheet modules (i.e., cassettes) have become the dominant module format in the biopharmaceutical industry due to their high mass transfer efficiency (high fluxes at low tangential flows) and linear scale-up for reliability and speed of implementation. High mass transfer is achieved using fine screens in the feed channel that act as a turbulence promoter, and reduce protein polarization at the membrane surface. High mass transfer efficiency translates into less membrane area and smaller

TABLE 10.3
Commercial Ultrafiltration Module Performance

	Spiral	Fiber	Cassette
Screens/spacers	Yes	No	Yes/No
Typical # in series	1–2	1–2	1–2
Packing density M2/M3	800	1,000–6,000	500
Feed flow LMH	700–5,000	500–18,000	400
Feed pressure drop psi/module	5–15	1–5	10–50
Channel height mm	0.3–1	0.2–3	0.3–1
Plugging sensitivity	High	Moderate	High
Working volume L/M2	1	0.5	0.4
Holdup volume L/M2	0.03	0.03	0.02
Module cost $/M2	40–200	200–900	500–1,000
Ruggedness	Moderate	Low-moderate	High
Module areas M2	0.1–35	0.001–5	0.05–2.5
Membrane types	RO–UF	RO–UF–MF	UF–MF
Relative mass transfer efficiency[a]	6	4	10
Ease of use	Moderate	High	Moderate
Scalability[b]	Fair	Moderate	Good

[a] Qualitative, based on relative fluxes.

[b] Cassettes keep retentate path length constant and require lower feed flow rates.

FIGURE 10.13 Commercial ultrafiltration modules.

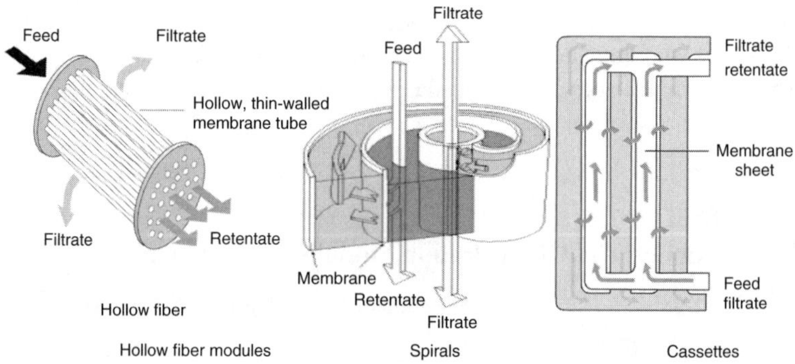

FIGURE 10.14 Module flow paths.

pump size. This means small systems with high recovery of the valuable protein product and less pump passes, minimizing the risk of damage to the protein product. For relatively dilute proteins, a fine-feed channel turbulence promoter or screen is recommended to give high efficiency without causing excessive feed channel pressure drops. For higher concentration protein solutions with significant viscosity, a coarse screen is recommended. Note that the presence of turbulence promoters modifies the simple picture shown in Figure 10.2 but a mass transfer coefficient can still be defined.[10]

10.4 DEVELOPMENT OF AN EARLY CLINICAL PHASE PROCESS

10.4.1 OBJECTIVES AND METHODS

Development of a robust, consistent and optimized TFF step may be achieved by following a systematic process development methodology. The rigor and

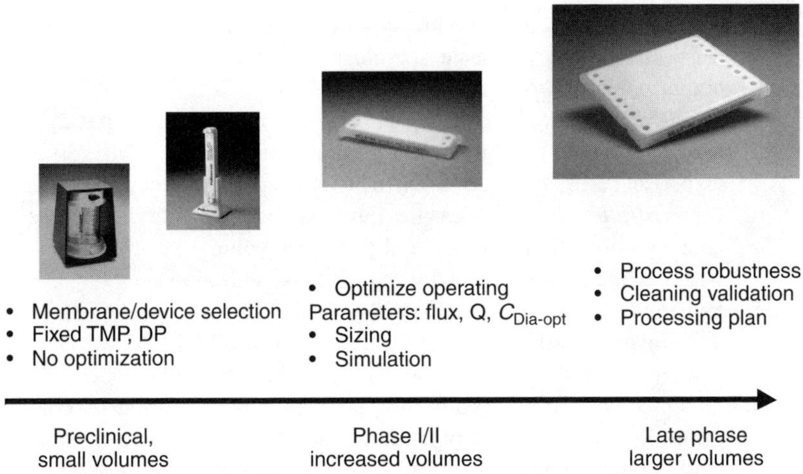

FIGURE 10.15 Process development timeline.

detail required in the method is influenced by (1) phase of drug development, (2) volume of product available, (3) relative novelty of the product, (4) development timelines. Figure 10.15 depicts the focus of process development activity as a function of a product development timeline.

The objectives of early phase process development are to determine if it is feasible for the UF membrane/device to meet the target concentration and buffer exchange specifications while producing an acceptable quality product. As the drug progresses through development, process development focuses on generating additional data to optimize, confirm, and validate performance (product yield, aggregation, system sizing).

Candidate membranes can be screened using a low volume, stirred cell device to check yields, fluxes, and fouling characteristics even though the flux performance in these devices does not scale. If larger volumes (>50 ml) of feedstock are available, a linearly scalable cassette is preferred. Other candidate modules with different membrane chemistries, MW ratings, and feed channel screens can also be tested if feed volume and time permits.

A typical method for UF process development and optimization is as follows:

1. *Module verification*: initial flush to wet the membrane; measure normalized water permeability (NWP) as LMH/psi corrected to 25°C; measure air diffusion through wetted membrane as cc/min-m^2 and compare to specification to verify membrane integrity. Flush with buffer to precondition the membrane.

2. *Initial dynamics*: Add product solution and measure flux and passage vs. time in permeate recycle mode to allow for membrane conditioning in product solution.
3. *Sensitivity to initial operating conditions*: measure flux, passage, feed to retentate pressure drop (ΔP), and turbidity vs. TMP (30 to 50 psi) and cross flow (ΔP of 10 to 30 psi) in permeate recycle mode.
4. *Concentration mode*: Measure flux, passage, turbidity, retentate concentration, temperature, and permeate volume vs. time while withdrawing permeate to achieve target concentration.
5. *Sensitivity to final operating conditions*: measure flux, passage, feed to retentate pressure drop (ΔP), and turbidity vs. TMP (30 to 50 psi) and cross flow (ΔP of 10 to 30 psi) in permeate recycle mode.
6. *Diafiltration mode*: Measure flux, passage, turbidity, impurity concentration in retentate, temperature, and permeate volume vs. time while withdrawing permeate and adding diafiltrate to achieve target impurity concentration.
7. Recover retentate product by depolarizing and using a plug-flow flush (see data analysis). Measure yield.
8. *Clean membrane under vendor recommended conditions*: Flush with buffer, then WFI, add 0.1 *N* NaOH solution, recirculate for 45 min at 20°C in permeate recycle mode, and measure NWP.

Additional testing can include repeatability evaluation, cleaning optimization, and further exploration of particular conditions. For high final concentrations and a final formulation buffer leading to significantly different osmotic pressures, it is worth repeating both the operating conditions step #3 and the concentration mode step #4 in the new buffer after step #6.

10.4.2 Data Analysis

The module used should be integral and show NWP within the normal device range. The initial dynamics should show asymptotic approach to steady-state flux and retention. Continued decline in performance indicates a membrane compatibility issue.

Operating condition data should mirror the trends shown in Figure 10.16. Behavior contrary to these trends may indicate faulty experimental procedures or an unusual new effect, that should be confirmed and understood.

The behavior of the TMP data is largely explained by the phenomenon of concentration polarization. As described in the discussion on mechanism, the bend in the flux curve is due to osmotic pressure. The drop in observed retention (increase in observed passage) is due to a constant intrinsic passage with higher wall concentrations. Additionally, high wall concentrations can lead

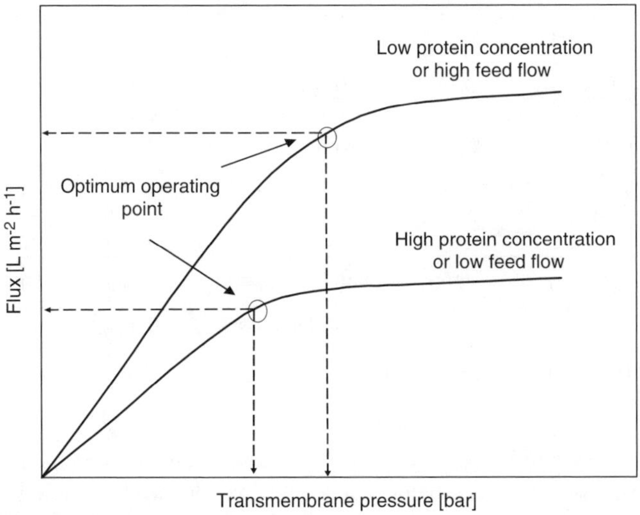

FIGURE 10.16 Flux trends.

to protein aggregation and a rise in the turbidity of the solution. The optimum TMP is located at the knee of the flux curve to obtain reasonable flux and avoid formation of aggregates. If there is a difference in the flux curves between the starting solution and the diafiltered solution, a conservative approach for polarization would be to select the lower of the two TMP values.

Higher cross flow reduces polarization by increasing the mass transfer coefficient. As a result, higher cross flows decrease C_{wall}, thereby decreasing the osmotic pressure and increasing the flux. Lower C_{wall} also increases retention and lowers aggregation. If polarization is a significant source of aggregate formation and yield loss, operation of the UF step using the method of C_{wall} control should be explored.[11]

Pumping can damage protein and pump shear is often cited as a source of degradation.[12] It has been claimed that variations in fluid velocity (shear rate) subject proteins to forces that overcome the 1 to 2 kcal of energy holding them in their normal three-dimensional (3D) conformation. Careful studies in tubes have shown that high shear rates (or velocity gradient at the wall) of 10^6 sec^{-1} degrade protein only in the presence of air interfaces.[13,14] Protein degradation through pumps and valves primarily occurs through a mild cavitation effect where gas microbubbles effervesce in low-pressure regions of fluid flow. This effect is not severe enough to cause noise or pitting in the pump like conventional cavitation but the large surface area of many microbubbles can provide a surface for denaturation. Figure 10.17 and Equation 10.9 show pumping-induced degradation characterized by the number of pump passes used,[15] where CR

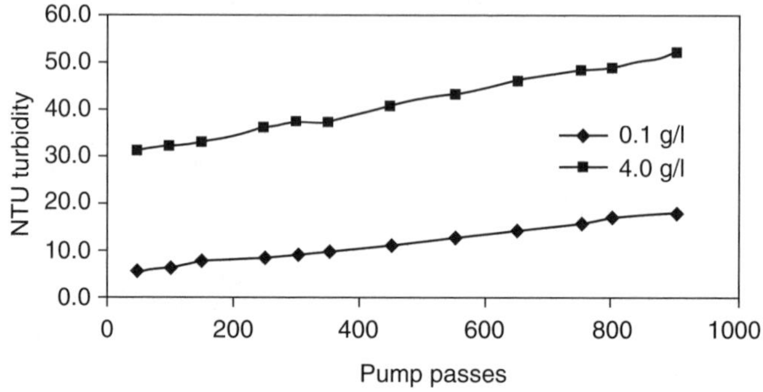

FIGURE 10.17 Pump degradation.

is the conversion ratio (permeate flow/feed flow). Vendor recommendations of optimum cross flow rates provide reasonable flux rates while avoiding increases in turbidity (see equipment selection). Final conditions may be determined on a case-by-case basis

$$\text{Pump passes} \quad \text{PP} = [N + \ln X]/\text{CR} \tag{10.9}$$

Protein quality can be measured by turbidity, SEC, or by the tendency of aggregates to plug a downstream sterile grade filter. Filter plugging is the most sensitive assessment of aggregate levels and is characterized by V_{max}.[16]

Product concentrations should be plotted vs. volume reduction factor to show agreement with expectations during the concentration step (as in Figure 10.7). Solute concentrations should be plotted vs. diavolumes to show agreement with expectations during the diafiltration step (as in Figure 10.8). Deviations in performance from expected behavior indicate problems in experimental procedure (e.g., poor tank mixing) or unusual product properties requiring further study (e.g., product precipitation at high concentrations).

A general processing sequence or diafiltration strategy could include an initial concentration step, followed by a diafiltration step with a final concentration step. One chooses the product concentration at which diafiltration is performed by the degree of concentration in the first step. One analysis[17] using the gel flux model (Equation 10.2) showed that a minimum system area and processing time is obtained by diafiltering at a protein concentration of c_g/e. For human plasma derived IgG, c_g values of 191.4 g/l have been reported, yielding an optimum diafiltration concentration of 68 g/l.[8] While these high values are found in plasma IgG processing, final recombinant antibody concentrations have generally been much smaller, in the 5 to 20 g/l range, where it is convenient to diafilter at the final formulation concentration.

TABLE 10.4
Sources of Product Loss

Source	Permeate	Adsorption	Holdup	Inactivation
Magnitude (%)	0–1	0–5	1–10	0–20
Causes	Membrane leakage, operation	Membrane adsorption	Poor system design, poor recovery method	Gas interfaces, high temperatures, polarization, pumping
Correction	Lower MW membrane, sealing issue	Different membrane material	Eliminate dead legs, improve recovery method	Check tank foaming and pump, C_{wall} control

Overall product recovery in the final retentate should be >95% and yields of >99% are common. High losses of product require further investigation and corrective action in the procedures. Sources of yield losses in the UF step are shown in Table 10.4 along with corrective actions. The largest sources of yield losses, hold-up and inactivation, are affected by system design (in Section 10.5). Inactivation can also occur through extensive process times or hold steps, high temperature spots (nonisolated pump from motor), unflushed cleaning fluid remaining in a dead leg, and contact of the protein solution with buffer solutions that promote instability (i.e., diafiltering through an isoelectric point).

Processing times, permeate fluxes, and permeate volumes are recorded during the scale-down experiments to enable scale-up system sizing (see below). Fluxes scale directly and volumes scale proportionately with the feed volume. A safety factor is generally built into the scale-up design so that process times upon scale-up will be reasonable from a manufacturing standpoint.

Feasibility requires demonstration of repeatable performance. The adequacy of the cleaning step is determined by the recovery of at least 80% of the initial normalized water flux[18] (Figure 10.18). While some variability in water flux is typical, any consistent decline reflects an inadequate cleaning procedure. Additional verification of consistency can involve measuring batch-to-batch yields (e.g., data in Table 10.5), buffer passage, process flux (Figure 10.18), and air integrity. The typical module change-out frequency is 1 year, or 50 runs, or when performance (retention, flux, integrity) drops below preset specifications.

The data should establish the feasibility of the formulation step in meeting antibody quality and concentration, buffer composition, robustness (consistency, worst case feed), antibody recovery, and economical operation

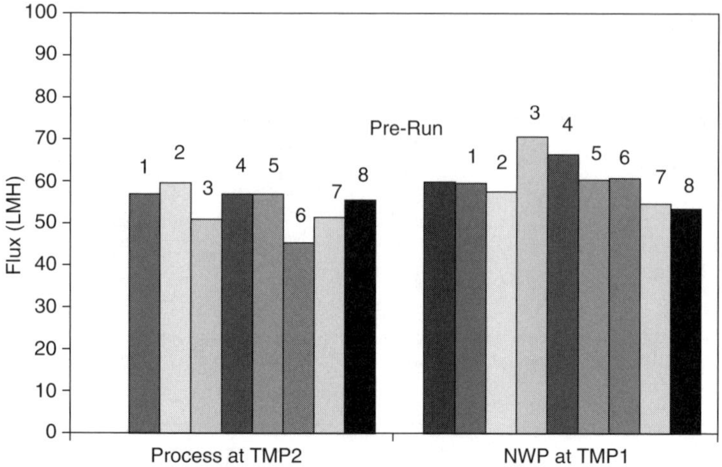

FIGURE 10.18 Membrane cleaning performance.

TABLE 10.5
Product Loss in Composite Permeate for Each Run (%w/v)

Run no.	Ultracell PLCGC	Biomax 10
1	0.00	0.01
2	0.00	0.00
3	0.00	0.00
4	0.00	0.00
5	0.00	0.00
6	0.00	0.00
7	—	0.00
8	—	0.00
9	—	0.00

goals. Data consistent with the trends shown here should establish confidence in the results. Further optimization may be appropriate for large-scale operation.

10.5 SCALE-UP

10.5.1 Sizing

The simplest scaling strategy is linear scale-up.[19] This involves system scaling based on consistent capacity (l/m^2) from small scale to large scale. Permeate, buffer, and retentate volumes also scale linearly with feed volume. TFF system performance can be sensitive to hardware design and layout. It is usually not possible to find scale-down hardware that precisely mimics large-scale versions. However, hardware functionality needs to be preserved at both scales. One should also identify potential changes in the feed material in moving from one scale to the next and conduct scale-down testing with worst case feed.

Figure 10.19 shows the concentration and diafiltration steps for an example of a 600 L feed undergoing a 21× volume reduction and 7× diafiltration. While this volume is on the high side for a final formulation step, it may be the result of pooling several fermentation batches. The input buffers and permeate waste volumes for this case are calculated from mass balances.

The system area is calculated from the average flux (Equation 10.10) over each step, permeate volumes, and the target processing time (Table 10.6).

The average flux over each step is calculated by integrating experimental flux data.

$$\text{Average flux} = \frac{\text{Liters of experimental total volume permeated}}{\text{m}^2 \text{ device area} \times \text{hours total processing time}} \tag{10.10}$$

An area-time term can be calculated for each step as the ratio of the manufacturing scale permeate volume over the average experimental flux. For the concentration step shown in the table this is (571 L)/(40 LMH) = 14 m^2-h. The area-time for each step is then added to give the total area-time for the process,

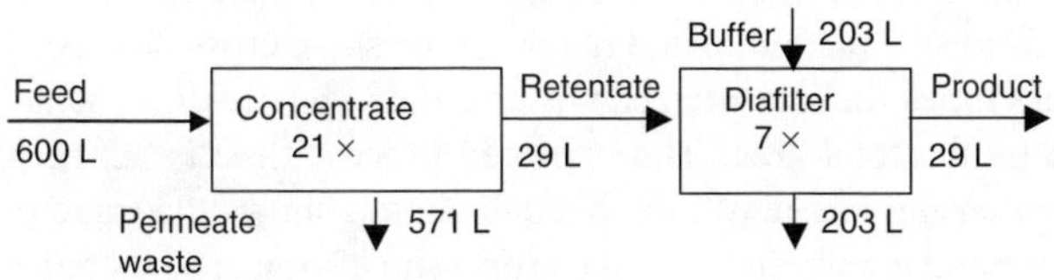

FIGURE 10.19 Ultrafiltration process sequence.

TABLE 10.6
Ultrafiltration System Sizing

Step	Concentrate	Diafilter	Total
Average flux	40 LMH	30 LMH	
Permeate volume (l)	571	203	774
Area-time (m^2/h)	14	7	21
Time (h)	2.3	1.2	3.5

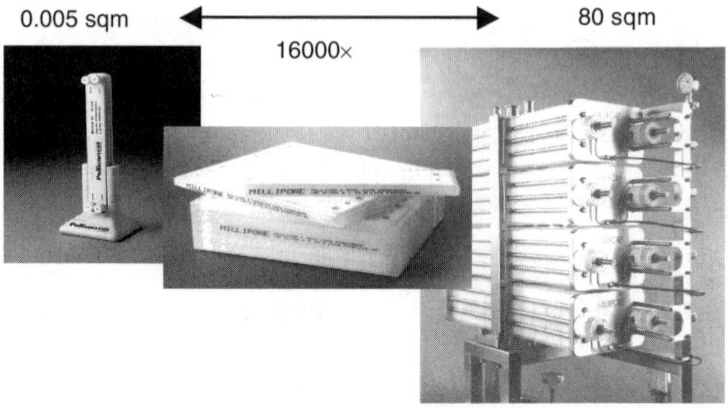

FIGURE 10.20 Scale-up.

here 21 m^2-h. When the total area-time is divided by the total processing-time, this gives a total area of $(21\ m^2\text{-h})/(4h) = 5.3\ m^2$. This total area corresponds to roughly 2 cassettes of 2.5 m^2 which can be arrayed on a single-level process holder. A 4-level holder containing 32 cassette modules is shown in Figure 10.20. It is recommended that an area safety factor be included to ensure timely process scale operation with some allowance for process variations (feed volumes, process delays, etc.). In this case, a 40% area safety factor boosts the area to 6.0 m^2, or 12 cassettes of 0.5 m^2 each. These can be arrayed as 6 cassettes on each side of a fully loaded single level process holder. As a general rule of thumb, at least a 20% safety factor is recommended. The system area can be used to size the pump and maintain the same cross flow per unit area.

Process times for each step can be calculated by dividing the area-time for each step by the total area. The expected process time is 3.5 h, well within the 4 h processing target. While this calculation shows the trade-off between process time and membrane area, the processing time can be a small component of the time it takes to turn around the UF operation (including setup, recovery, and cleaning).

Although system sizing assumes equivalent behavior between manufacturing scale and scale-down testing, increasing areas show larger hydraulic variations within the system.[19] Beyond sizing of 20 m^2 per level and 4 levels high, one should consult with the vendor.

The final stage of TFF processing entails product recovery (Table 10.7). Product recovery is the process of removing the product from the TFF system into a vessel appropriate for storage or further downstream processing. It is critical to devise an efficient recovery step in order to maximize product yield. The bulk of the product, which is typically in the recycle tank, is pumped out using the feed pump. However, some product may get held up in the piping

TABLE 10.7
Product Recovery Methods

Recovery Method	Gravity Drain	Blowdown	Plug Flow Flush	Recirculation Flush
Holdup Loss (%)	5–20	2–10	0.5–2	0.1–0.5
Notes	Design sensitive	Design sensitive		Dilutes product

and the modules. A simple gravity drain of the system through some low point near the tank outlet or feed pump facilitates additional product recovery. Other techniques may be utilized to augment product recovery. Blowdown uses compressed air introduced at a high point of the system in the retentate line with the product collection at the lowest point. Care must be taken to gently introduce the air in order to avoid foaming and product denaturing.

Additional recovery is obtained by slowly introducing buffer at a high point in the system. This will progress through the wetted components and displace held up product. The buffer will eventually exit the low point drain where a UV monitor can readily detect the transition from product to buffer.[20] The transition is generally compact and this procedure is termed a plug-flow flush in accordance with the standard residence time distribution model in chemical engineering.[21] One can also use the more extensive recirculation flush where a minimum tank volume of buffer is introduced and recirculated to allow product diffusion out of any dead legs and nooks and crannies in the wetted flow path. The additional recovery using this procedure may not warrant the product dilution effect of using this approach.

The onset of tank foaming will impose a maximum volume reduction limit, typically 40×. Any further reduction in tank volume can cause the recirculating retentate to entrain air and denature the product. The flow sheet modifications shown in Figure 10.21 can be used to extend the range of concentration to 100× and to provide some flexibility in processing a variety of batch volumes in a single skid. For a fed-batch operation, the retentate is returned to a smaller tank, not the large feed tank. Feed is added to the small retentate tank as permeate is withdrawn. The smaller retentate tank can allow a smaller working volume without foaming. A bypass line can also be used to return the retentate directly into the pump feed. Fluid from the feed tank is added slowly into this recirculation loop. This allows a holdup volume consisting of just the recirculation loop. This configuration has also been referred to in the literature as a feed-and-bleed configuration.[22]

Retentate concentration over the course of the process is shown in Figure 10.22 where the tank ratio $= V_o/$retentate tank volume, C_o is the feed

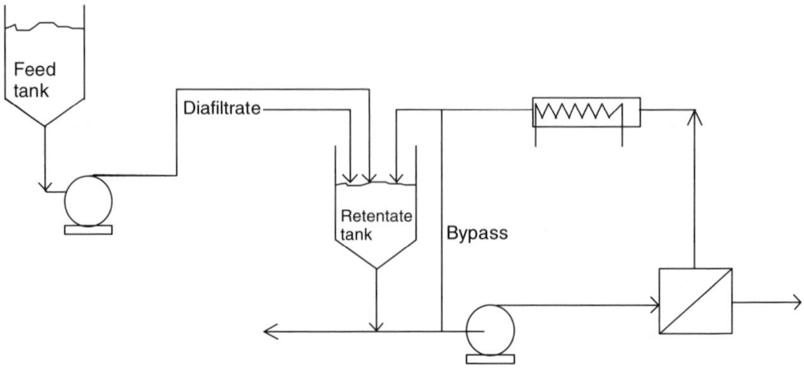

FIGURE 10.21 Alternative ultrafiltration process configurations.

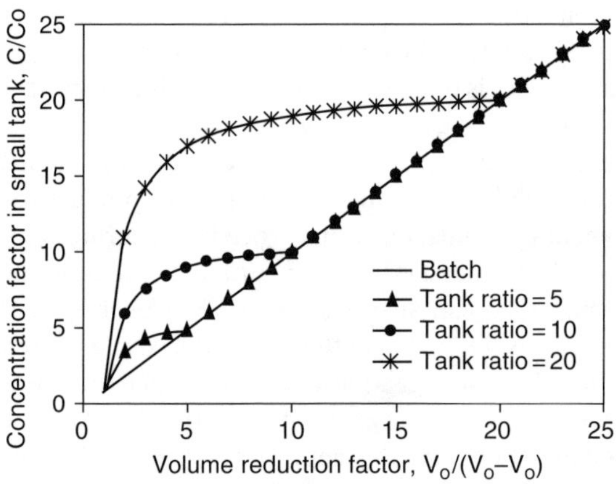

FIGURE 10.22 Fed-batch process concentrations.

concentration, and V_0 is the feed volume. The benefits sought from higher concentrations can however lead to other problems such as reduced fluxes, larger area and pumps, possible denaturation, and extra lines that may have issues with cleaning and product recovery. This has caused not only significant commissioning and validation delays but has also led to the scrapping of a process skid as unworkable. The number of pump passes will also be higher, leading to more potential protein degradation. Fed batch and bypass should be used only when necessary.

Special considerations for scale-up may also be required when processing protein solutions to high final concentrations ~150 to 200 g/l. This is particularly relevant to antibodies where higher concentration formulations facilitate subcutaneous drug administration. The effect of TMP and cross flow on flux should be measured at both feed and final retentate concentrations. It is important to generate the flux vs. concentration data in both in the initial and final (diafiltration) buffer to determine the optimum concentration for diafiltration.[23] Particular attention also must be paid to the pressure drop across the membrane module towards the end of the concentration step as it may rise due to high viscosities. The process may require operation at a lower pressure drop (or cross flow) at the expense of flux to ensure stable operation conditions.

A well-designed, robust ultrafiltration process is characterized by the following performance parameters:

Yield (Overall)	$\geq$95 to 98%
Process flux consistency	$\sim \pm$10% run-to-run
Product retention (membrane)	$\geq$99.9%
NWP recovery (run-to-run)	$\sim \pm$20% cellulose membranes $\sim \pm$20 to 35% for PES membranes
Typical flux	$\sim$Product and process specific; for example, 30 to 120 LMH for 30 kDa membrane with MAb
Typical sizing	$\sim$Product and process specific; for example, 5 to 10 m^2/kl for MAb

10.5.2 Operating Procedure

The following procedure is representative of a typical operation. As much as possible, plant SOPs (standard operating procedures) should be written to allow flexibility in operating conditions while ensuring consistent performance in meeting product specifications.

1. *WFI flush*: Installation of cassettes with proper attention to holder compression 15 min at process flows to permeate assay spec. (20 to 25 l/m^2), NWP measurement, drain.
2. *Integrity test*: Pressurize to 30 psi with sterile air, measure air flow and compare with vendor specification.
3. Buffer rinse preconditioning.
4. *Process fluid*: Fill tank, concentrate to volumes or retentate concentration, and diafilter to buffer or permeate volume.
5. *Recover product*: Drain, blow down, plug-flow flush.
6. *Cleaning*: WFI Flush, 0.1 *N* NaOH at 20°C for 45 min, WFI Flush, NWP End Pt. Spec., LAL test, Drain.
7. *Storage*: 0.05 *N* NaOH in holder up to 12 months.

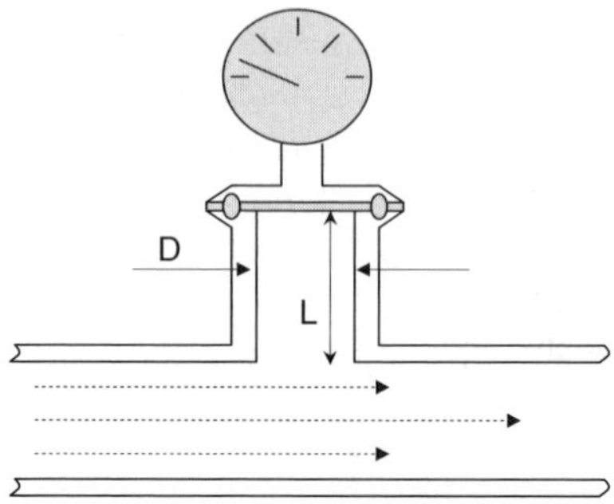

FIGURE 10.23 Dead legs.

8. *Change out*: Based on failing specs on NWP, DP, Integrity, Yield, # Cycles.
9. Allow for reprocessing (if desired).

10.5.3 System Considerations

A TFF system must be able to implement all the steps in the SOP, deliver the desired product, and meet any process constraints. At minimum, a system contains the membrane modules and holder, tank, feed pump, retentate valve, and pressure sensors for the feed and retentate lines. Heat exchanger is sized to maintain uniform temperature of the process to counteract heat added by the pump and ambient heating. Open systems may be acceptable for early clinical phase production but closed systems with hard plumbed lines are needed for validated consistency, short cycle times, and bioburden control manufacturing. Each step of the process is controlled by switching valves to introduce/discontinue the appropriate solutions to the system (air, buffer, diafiltrate, cleaning fluid, etc.).

Multiproduct facilities process a wide range of batch volumes from clinical trials to marketed products. While fed batch configurations allow flexibility to handle these wide volume ranges, one can encounter issues with cleaning and product degradation. Separate pilot systems are recommended for small batches to avoid these validation issues and speed time to market.

The monitoring and control strategy use sensors to measure permeate flow (water permeability measurement, processing, cleaning), retentate flow, air flow during integrity testing, pressures (feed, retentate, permeate), temperature, tank level or weight, permeate composition (UV absorbance as an indicator of protein content, pH, conductivity as desired), process temperature, and coolant temperature. A good design practice is to minimize the number of sensors and identify their optimal placement. These sensor readings may be displayed, logged, used to trigger alarms, and used to trigger a subsequent step in a process. Alarms may be triggered by abnormal pressure excursions (plugged feed channel, valve froze shut), low tank volumes, high permeate UV absorbance, and high temperatures. The duration of each step is determined by time, permeate volume, or fluid height in the feed tank. Process data typically is logged for GMP, trending, and diagnosis of any unusual excursions.

The flow paths that are active during each step of the SOP should be traced out on the P&ID to show valve configurations along with the relevant sensors used for monitoring and alarms. This includes the ability to vent, drain, SIP, flush the system, and respond to system upsets. It is particularly important to ensure that all flow paths are active during the cleaning step, including sample valves. In some cases, the modules may be removed from the system during cleaning. Care should also be taken to make sure that product is not inadvertently flushed to waste through automation or the use of transfer panels.

10.6 HARDWARE

10.6.1 EQUIPMENT SELECTION

Table 10.8 and Table 10.9 show equipment requirements and specifications. Regulatory requirements ensure that processing objectives are consistently met. Economical requirements include lifecycle costs such as capital, operation, validation, maintenance, cost of replacement (and revalidation if needed), staffing required to support ongoing operation (e.g., programming, calibration), and the scope of the supporting documentation and service from the vendor. Standard, easy-to-use, designs and components are less prone to failure with lower labor costs and maintenance. Components must conform to the standards of the country and operating facility (e.g., metric, voltage) to ensure compatibility and reduce the inventory of spare parts. Additional selection criteria include experience with components and vendors, and scalability. Long lead components such as specially designed pumps, sensors, and tanks must be designed and ordered early.

Operator safety involves chemical hazards (explosive solvents, biohazards, and toxic or corrosive chemicals), physical hazards (high pressures, moving parts, temperature extremes, use of steam, and obstacles to operation and maintenance), and electrical hazards (high voltages and currents, and inadequate grounding). Safeguards and alarms should be tested to make sure they work in an emergency. Safety reviews of designs are recommended along with formal hazard and operability analyses (HAZOPs) and control hazard and operability analyses (CHAZOPs).

Integration of skids with each other and the plant may impose additional constraints. Skids require that the hardware matches. Retrofit into an existing

TABLE 10.8
Equipment Requirements

Component	Regulatory	Economic	Safety
Wetted surface materials	Non-toxic, cleanable/ sanitary, consistent, chemically compatible with all process fluids (no swelling or reacting), non-shedding, non-leaching, non-adsorbing, closed system	Availability, can fabricate	
Wetted volume	Low holdup, drainable, ventable, minimize dead legs, cleanable, flushable		Pressure and temperature rating, sealing
Piping, valves, heat exchanger		Compact, can fabricate, drainable, available in a variety of formats	Operator protection from moving parts
Vessels	handles volume range, mixing, avoid foaming	Availability	
Pumps, filter holders	low protein degradation, consistent, thermal isolation		Operator protection from moving parts
Sensors/sampling	Reliable, accurate, insensitive to environmental effects (temperature, pressure changes), calibrate in closed system	Design in optimum number — not to excess	Electrical shock, closed system
Skid frame	compatible with cleaners and sanitizers	Compact	Supports load, ease-of-operation
Display	Capture and store data		Legible
All components	Documentation, easy-to-validate, quality certification	Low cost, reliable, maintainable, conform to plant standards, proven designs, spare parts	Conform to country standards

TABLE 10.9
Equipment Specifications

Hardware Component	Specifications
Filter module	Cassette, 30 kDa, Ultracel, 2 × 2.5 m^2
Holder	316L ss, Ra 0.5 μm, 1 high
Closure hydraulics	308 ss
Piping	316L ss, Ra 0.5 μm, Connections, labeled with flow direction, sized for 3–10 ft/sec process and cleaning, weld documentation
Feed Pump	Rotary lobe/Progressing cavity, Seal, 316L ss, Ra 0.5 μm, Flow, Pressures
Piping connections	Tri-Clamp® (Tri-Clover Corporation)
Frame	304 ss, Finish, Casters, Dimensions
Valves	Weir diaphragm, EPDM, PTFE, Silicone or Viton
Heat exchanger	Shell-and-tube, 316L ss, Ra 0.5 μm, Rating
Tank	Bottleneck design, Size, 316L ss, Ra 0.5 μm, well mixed/agitator, operate over volume range, avoid foaming, cleanable/spray balls, ASME stamp, no dead legs, sight glass
Air filter	Sterilizing grade, I-line, 316L ss, Ra 0.5 μm
Sensors	Type, Range, Output, reliability, thermocouples in wells, load cell
Sampling	Radial diaphragm valve
Electrical-motor	Enclosure, Power
Control box	History-Vendor & Mfr; Availability, Cost; Reliability, Accuracy; ISO/cGMP Certification, Explosion Proof

facility may impose limitations on skid dimensions for the production floor and for access to the facility.

Dead legs shown in Figure 10.23 are spaces or pockets in contact with the product that are difficult to vent, flush, and drain.[24] They arise from connecting components to the piping system (e.g., sensors, sampling ports, rupture disks), within wetted components (e.g., pumps, housings, valves, heat exchangers, tanks) or as surface roughness. Flow visualization,[25] simulations, and testing indicates that the efficiency of cleaning a dead leg is affected by the ratio L/D, the average fluid velocity in the pipe, and presence of air pockets in the dead leg.[26] While elimination of dead legs is desirable, current ASME BPE Guidelines currently recommend $L/D < 2$ (based on the internal dimensions of the dead leg).

The presence of air in piping systems can prevent fluids from wetting the internal surfaces. Cleaning effectiveness can be compromised by inadequate contact with cleaning and sanitizing fluids. Fluid velocities of 5 ft/sec are required to displace air from dead legs or to complete flood a vertical pipe.[26] Avoid vertical pipe bends that create sections where air or solids can accumulate. Liquid retention in undrainable sections represent product loss, growth areas for bioburden, or batch carryover. Venting and draining is aided by a pipe slope of at least 1/16 in. drop per foot of pipe length. Clean steam lines are self-sanitizing and may be plumbed without a vertical slope. A low point drain is required.

The high CIP flows of 5 ft/sec required for wetting internal surfaces are typically more demanding than process flows and form the basis for pump sizing. An economic analysis shows that there is a tradeoff between capital and operating costs for piping systems with optimum velocities in the range of 3 to 10 ft/sec.[27] Velocities >3 ft/sec are also recommended for cleaning and flushing as part of the 3A standards. These velocities help ensure that air bubbles in dead legs are flushed out of the piping system so that all the internal surfaces are accessible for cleaning.[26]

Although pumps do not scale consistently, they are selected to meet the flow, pressure, and pulsation requirements of the scale of operation and avoid damaging protein. Rotary lobe designs recommended for feed pumps should be run below 500 rpm and be mounted in a vertical position to allow product and cleaning solutions to drain out easily. Adding more lobes will reduce pulsation effects but increase protein degradation. Sanitary centrifugal or peristaltic pumps can be used for buffer or CIP solution transfer and for WFI loop recirculation.

10.6.2 Skid Layout

Skid layout criteria include minimizing holdup volume and dead legs, as well as allowing for flushing, cleaning, venting, draining, mixing, sanitizing/steaming, operating and servicing ergonomics, safety, etc. Holdup volume and the presence of dead legs in the wetted fluid path impact product recovery, separation efficiency (e.g., poor buffer exchange at high diavolumes), fluid volumes required (cleaning, flushing, processing), system cost and required floor space, and the ease of cleaning and sanitizing. Minimization involves reducing line lengths, employing the full 3D space using CADCAM to explore design alternatives, and using compact components (e.g., valve assemblies).

Systems must be designed to minimize product degradation by controlling physical and chemical stresses (e.g., hot spots, excessive shear, air interfaces, cavitation, local concentrations, etc.). Mixers are used to eliminate concentration gradients in tanks. Inadequate mixing caused by the addition of diafiltrate in a Tee between the tank and feed pump will reduce performance. In-line mixing

FIGURE 10.24 Ultrafiltration process skid.

(two fluids are pumped into a Tee connection and blended while flowing in a pipe) for buffer preparation from concentrates reduces waiting time, tankage, and floor space.[28]

The 3D layout of process and remote monitoring and control systems should allow operators to setup, operate, and turn around each process without undue strain. In addition, maintenance and service personnel should be able to conduct routine operations (e.g., calibration, gasket replacement) without undue strain. This requires proper orientation of displays and enough space around a processing skid to gain access. Computer 3D models, constructed using CADCAM systems, should be examined for ergonomics during design reviews (Figure 10.24).

The design, construction, commissioning, and validation of a process skid often involves a team of experts from the biopharmaceutical manufacturer, the skid supplier, and an A&E (architect and engineering) firm. This is facilitated through the use of a process with clear UF step performance requirements, roles of team members, and milestones/reviews.[29]

10.7 ULTRAFILTRATION PROCESS VALIDATION AND COMMISSIONING

Initial process validation follows Section 10.4 using scale-down devices:

- Define process objective(s) or claims (e.g., concentration, buffer exchange, yield, product quality, process time).

- Identify primary operating parameters affecting these process objectives (e.g., cross flow, TMP, temperature, concentration factor X, diavolumes N, feed volume/membrane area, pump passes, integrity test) and establish initial ranges for each. Note that the number of parameters and the width of their operating ranges selected should be sufficient to ensure process consistency, yet loose enough to allow flexibility in plant operation without over specifying requirements.
- Qualify the process by running repeated process simulations with different feed batches and membrane modules within the initial parameter ranges to track performance parameters (e.g., flux, bioburden, endotoxin, recovery, cleaning effectiveness, and NWP).
- Use filter validation guides supplied by vendors to evaluate compatibility, flushing requirements, leachables, and bioburden that may impact product quality.

Process validation for process scale[30,31] follows Section 10.5 and Section 10.6:

- Design qualification (DQ) verifies that the design meets clearly defined process objectives and constraints.
- Installation qualification (IQ) verifies that the equipment components meet the design requirements in terms of compatibility, ranges, sensitivity, finish, etc. Certificates of quality are supplied with each filter to ensure consistency.
- Operational qualification (OQ) verifies that the process skid can deliver the sequence of processing steps at the specified operating ranges, and has the appropriate alarms and monitors. This step is often part of the factory acceptance test (FAT) performed at the vendor site.
- Commissioning involves the installation and start-up of the system at the manufacturing plant. Cassettes are installed and cleaned to remove preservatives and measure extractables and bioburden. Volumes and flow rates used for flushing and cleaning need modification from the scale-down system to account for the holdup in the process scale skid.
- Performance qualification (PQ) verifies that the ultrafiltration process can repeatedly process the biologic feedstock and deliver the claimed performance. Attention should be paid to the preparation of buffers and cleaning agents used in the process. Product processing involves measuring key performance parameters (e.g., flux, yield, purity, NWP, integrity test airflow, bioburden, and endotoxin). The FDA requires three consistency batches (or conformance lots) while

the EMEA requires five. A blank run (running pure buffer after running a batch) also can be run to evaluate the amount of carryover from batch-to-batch. Module lifetimes and storage procedures need to be established in consultation with the vendor.

- Process monitoring involves tracking key performance parameters from batch-to-batch to demonstrate the process remains consistent and within control. This can impact module lifetime specifications. Documented plans for preventative maintenance and servicing are a cGMP requirement.
- Revalidation may be needed to accommodate process changes arising from vendor changes or second source qualification, manufacturing facility changes that can alter the feedstock, or processing skid modifications. This is often handled by the use of comparability protocols.[32]

10.8 TROUBLESHOOTING

TABLE 10.10
Troubleshooting Guide

Symptom	Root Cause(s)	Recommended Action(s)
Low flux	Fouling, improper cleaning, low cross flow, hardware shedding	Modify cleaning procedure, replace modules, check TMP, replace hardware components
Low yields	Poor recovery, tank foaming, leaky membrane, poor assay or sampling	Modify recovery, integrity test, check tank and retentate flow
Failed integrity	Leaky module, improper installation	Reinstall and replace modules, tighten hydraulic spec
Failed bioburden	Contaminated feed, buffers, or equipment	Sanitization, check buffers and upstream steps, swab equipment
Inadequate buffer exchange/solute removal	Insufficient diavolumes, poor mixing, poor passage	Increase diavolumes, improve mixing, check for membrane fouling or solute binding to retained solutes
External leaks	Seals	Integrity test, replace seals

10.9 ADVANCED TOPICS

New developments in ultrafiltration are found in journals and trade magazines (e.g., *Journal of Membrane Sciences*, *BioPharm International*), vendor communications (e.g., websites), patent filings, and conference presentations (e.g., annual ACS or NAMS meetings, IBC conferences). Areas of active research include new membrane polymers and surface modification with accompanying diagnostic methods (to reduce fouling, increase flux and retention, improve consistency), new module designs (to improve flux, cleanability, ease-of-use, scalability, reliability), new processing skids (better components, recovery, less holdup, better mixing, disposability, software for automated processing and archiving), new processing methods (diafiltration strategies, turbulence enhancements), and new applications (e.g., protein–protein separations, plasmids). Another chapter describes the use of charged membranes.

REFERENCES

Useful overall references include Cheryan,[4] Ho,[33] Eykamp,[34] Millipore,[23,35] and Zeman.[36] A review of membrane retention mechanisms is found in Deen.[37]

1. Kurnik RTA, Yu W, Blank GS, Burton AR, Smith D, Athalye AM, and van Reis R. Buffer exchange using size exclusion chromatography, countercurrent dialysis, and tangential flow filtration: models development and industrial application. *Biotechnol. Bioeng.*, 1995; 45:149–157.
2. Vilker VL, Colton CK, and Smith KA. Concentration polarization in protein ultrafiltration. Part I: an optical shadowgraph technique for measuring concentration profiles near a membrane-solution interface. *AIChE J.*, 1984; 27:632–637.
3. McDonogh RM, Bauser H, Stroh N, and Grauscoph U. Experimental in situ measurement of concentration polarization during ultra- and micro-filtration of bovine serum albumin and dextran blue solutions. *J. Membr. Sci.*, 1995; 104:51–63.
4. Cheryan M. *Ultrafiltration Handbook*. Pennsylvania: Technomic Publishing Company, Inc., 1986.
5. Zydney AL. Concentration effects on membrane sieving: development of a stagnant film model incorporating the effect of solute–solute interactions. *J. Membr. Sci.*, 1992; 68:183.
6. Saksena S. Protein Transport in Selective Membrane Filtration. Ph.D. Dissertation, U. Delaware, Fall 1995.
7. Tkacik G and Michaels S. A rejection profile test for UF membranes and devices. *Biotechnology*, 1991; 9:941–946.
8. Mitra G and Lundblad JL. Ultrafiltration of immune serum globulin and human serum albumin: regression analysis studies. *Sep. Sci. Technol.*, 1978; 13:89–94.

9. Vilker VLC, Colton CK, Smith KA, and Green DL. The osmotic pressure of concentrated protein and lipoprotein solutions and its significance to ultrafiltration. *J. Membr. Sci.*, 1984; 20:63–77.
10. Schwinge J, Wiley DE, and Fletcher DF. Simulation of the flow around spacer filaments between channel walls. 2. Mass-transfer enhancement. *Ind. Eng. Chem. Res.*, 2002; 41:4879–4888.
11. van Reis R, Goodrich EM, Yson CL, Frautschy LN, Whiteley R, and Zydney AL. Constant Cwall ultrafiltration process control. *J. Membr. Sci.*, 1997; 130: 123–140.
12. Charm SE and Wong BL. Shear effects on enzymes. *Enzyme Microb. Technol.*, 1981; 3:111–118.
13. Thomas CR and Dunnill P. Action of shear on enzymes: studies with catalase and urease. *Biotechnol. Bioeng.*, 1979; 21:2279–2302.
14. Maa Y-F and Hsu CC. Effect of high shear on proteins. *Biotechnol. Bioeng.*, 1996; 51:458–465.
15. Virkar PD, Narendranathan TJ, Hoare M, and Dunnill P. Studies of the effects of shear on globular proteins: Extension to high shear fields and pumps. *Biotechnol. Bioeng.*, 1981; 23:425–429.
16. Millipore Corporation. Filter Sizing Methods. Lit No AN1512EN00, 2000.
17. Ng P, Lundblad L, and Mitra G. Optimization of solute separation by diafiltration. *Sep. Sci.*, 1976; 2:499–502.
18. Millipore Corporation. Techniques for Demonstrating Cleaning Effectiveness of Ultrafiltration Membranes. Lit No TB1502EN00, 2000.
19. van Reis R, Goodrich EM, Yson CL, Frautschy LN, Dzengeleski S, and Lutz H. Linear scale ultrafiltration. *Biotechnol. Bioeng.*, 1997; 55:737–746.
20. Frenz, J. Genentech. Personal communication. 1998.
21. Levenspiel O. *Chemical Reaction Engineering*. 2nd ed. New York: John Wiley & Sons, 1972, p. 97.
22. Mir L, Michaels SL, Goel V, and Kaiser R. Crossflow microfiltration: Applications, design, and cost. In: Ho WSW and Sirkar KK, Eds. *Membrane Handbook*. New York: Van Nostrand Reinhold, 1992.
23. Millipore Corporation. Protein Concentration and Diafiltration by Tangential Flow Filtration. Lit No TB032 Rev. B. 1999.
24. American Society of Mechanical Engineers. Bioprocess Equipment Guidelines. 1997.
25. Van Dyke M. *An Album of Fluid Motion*. Stanford, CA: Parabolic Press, 1982.
26. DeLucia D. Fundamentals of CIP Design. ASME Bioprocessing Seminars. 1997.
27. Peters M and Timmerhaus KD. *Plant Design and Economics for Chemical Engineers*. 2nd ed. New York: McGraw-Hill, 1968.
28. Ogez, J. Increasing Plant Capacity Using Buffer Concentrates & Linked Unit Operations. IBC 2nd Annual Recovery & Purification of Biopharmaceuticals, San Diego, CA, November 15, 2001.
29. ISPE. *Baseline Guide Vol 5, Commissioning and Qualification*. Tampa, FL: International Society for Pharmaceutical Engineering, 2001.

30. PDA. Industrial Perspective on Validation of Tangential Flow Filtration in Biopharmaceutical Applications. Technical report No. 15, Bethesda, MD: Parenteral Drug Association, 1992.
31. Petrone J, Erdenberger T, and Esenther C. Process Validation for Monitoring the Performance of Reusable Tangential Flow Filtration Membrane Devices. ACS Annual Meeting New Orleans, LA, March 2003.
32. US FDA. *Guidance for Industry: Comparability Protocols-Chemistry, Manufacturing, and Controls Information, Draft Guidance*. Rockville, MD, Feb. 2003, www.fda.gov/cder/guidance/index.htm.
33. Ho WSW and Sirkar KK. *Membrane Handbook*. New York: Van Nostrand Reinhold, 1992.
34. Eykamp W. Section 22, Membrane separation processes. In: Perry RH and Green DW, Eds. *Perry's Chemical Engineers' Handbook*, 7th ed. New York: McGraw-Hill, 1997.
35. Millipore Corporation. Maintenance Procedures for Prostak™ Modules. Lit No P17513. 1990.
36. Zeman LJ and Zydney AL. *Microfiltration and Ultrafiltration: Principles and Applications*. New York: Marcel Dekker, 1996.
37. Deen WM. Hindered transport of large molecules in liquid-filled pores. *AIChE J.*, 1987; 33:1409–1425.

11 Virus Filtration Process Design and Implementation

Michael W. Phillips, Glen Bolton, Mani Krishnan, John J. Lewnard, and Bala Raghunath

CONTENTS

11.1 BACKGROUND

Virus filtration is one component in the overall strategy to minimize transmission of infectious pathogens for biological products. Recombinant products such as monoclonal antibodies are expressed by mammalian, bacteria or yeast cells in fermentors, and within ascites fluid, or fluids from transgenic mammals. Mammalian cell lines may contain endogenous viruses that are generated in the bioreactor. Endogenous retroviruses are expressed because the retroviral genome is integrated into the cell line and cannot be screened-out during the creation of the Master Cell Bank. This causes retrovirus like particles, or RVLP's, to be produced within the bioreactor. Products can also become contaminated by adventitious viruses which enter the process streams.

Although regulatory agencies do not mandate the inclusion of specific viral clearance technologies, they do mandate the safety standard for final doses, and require full validation of viral clearance steps by the manufacturer. Regulations and good manufacturing practice (GMP) require two orthogonal and robust steps for endogenous viral clearance [1]. For adventitious viruses, these requirements typically result in the use of multiple viral clearance steps [2]. Orthogonal steps rely upon different mechanisms to achieve virus clearance. For example, heat and low pH are two independent mechanisms for inactivating viruses.

The effectiveness of a process step for viral clearance is quantified in terms of its log removal value (LRV), defined as:

$$\text{LRV} = \log\left(\frac{C_{\text{feed}}}{C_{\text{perm}}}\right) \tag{11.1}$$

where C_{feed} and C_{perm} are the outlet and inlet virus concentrations, respectively. Several commercial virus clearance technologies and typical virus LRVs are summarized in Table 11.1. Filtration removes viruses based on size exclusion, since the pores of the filter are smaller than the virus. Filtration is considered a robust operation because the removal efficiency is insensitive to normal variations in process conditions. Consequently, most well-designed downstream processes include a virus filtration step.

There are many commercially available virus filters, each with their own strengths and weaknesses. Filter selection should be based upon the nature of the application (product and process) and the performance of the filter as demonstrated in both qualification and validation studies. A systematic methodology

TABLE 11.1
Commercially Available Virus Clearance Technologies

Technology	Primary Mechanism	Capabilities	Comments
Filtration	Size exclusion	3–4 LRV parvo 6+ LRV MuLV	Robust; Easily plugged
Low pH incubation	Inactivation	6+ LRV MuLV	Robust; Denatures some proteins
Heat	Inactivation		Robust; Mandated for Albumin (Human) and Plasma Protein Fraction (Human)
Chromatography	Adsorption or exclusion	0–6 LRV	Nonrobust
Solvent/detergent	Inactivation by lipid dissolution	6+ LRV MuLV	Only enveloped viruses
Membrane adsorber/membrane chromatography	Adsorption/size exclusion	0–6 LRV	Buffer dependence; Issues with integrity testing
Ultraviolet inactivation	Inactivation	>2–6 LRV	Masking due to protein; highly dose-dependent; virus-dependent
Gamma irradiation	Inactivation	>3–6 LRV	Batch process; effect on protein needs to be determined

for developing a robust virus removal step includes:

- Virus filter selection
- Process design and optimization
- Process sizing and simulation
- Virus validation studies
- Manufacturing implementation

This chapter discusses each of these steps, providing guidance for the general practitioner.

11.2 VIRUS FILTER SELECTION

Although at first glance the selection of a virus filter can seem daunting, selecting a virus filter often involves similar considerations used for selecting any other filter media — product performance and process compatibility/system integration. Filter manufacturers classify virus clearance filters into two broad categories based on the removal needs of the biotech industry — filters that are capable of removing viruses 50 nm or larger (retroviruses) and filters that can remove both small (~20 nm Parvoviruses) and large viruses.

Performance related criteria for selecting a virus filter include virus retention capabilities, protein product transmission/product recovery, product throughput requirements, and overall process economics. These criteria tend to be protein-product-specific, and among other things, are dependent upon virus size/LRV needs, protein concentration, protein history (freeze/thaw, prefiltration, location in downstream process train), processing time, flow rates and pressure differential, and general solution conditions (pH, ionic strength, etc.). To effectively evaluate the impact of these variables, in-house testing is often required. The impact of many of these variables on virus filter performance is discussed in Section 11.3.

Less obvious considerations for selecting a virus filter revolve around process compatibility and system integration issues. All materials of construction (including filter matrix, chemical modifications and additives used to enhance filter wettability, preservatives, filter housing, *o*-rings, support material, etc.) should be chemically compatible both with the protein product as well as all relevant processing conditions (solvents, buffer constituents, etc.). Additionally, thermal and hydraulic stress resistances, extractables, and cleaning/sterilization/sanitization attributes should be evaluated to determine if they are consistent with the proposed implementation scheme. These topics will not be further discussed in this chapter. However, additional information on these topics can be found in the PDA Technical Report No. 41 on Virus filtration [2] or from the various vendors listed in Table 11.2.

11.3 PROCESS DESIGN AND OPTIMIZATION

To properly optimize a virus filtration process and establish process robustness, it is important to consider all processing variables that impact virus retention (LRV), product recovery, and product throughput. Additionally, from an economic point of view, process optimization is extremely important. For large volume processes, such as monoclonal antibodies, virus filtration can be one of the most expensive unit operations. Virus filtration is more expensive than sterile filtration due to both higher filter costs and lower product throughputs. Table 11.3 provides a typical range for cost and performance parameters for virus and sterile filters.

Virus filtration can either be run in a normal flow filtration (NFF) mode or a tangential flow filtration (TFF) mode [3]. Historically, TFF systems were

TABLE 11.2
Vendors with Commercially Available Virus Filtration Products

Company Type	Product Name	Virus	Sizes
Millipore NFF	NFP NFR	>4 log ΦX-174 bacteriophage >6 log retrovirus	Scale-down 3.5 cm^2; process modules 0.08–1.5 m^2
Millipore TFF	Viresolve® 70	>4 log polio; >7 log retrovirus	Scale-down 150 and 1000 cm^2; process modules 0.75–1.4 m^2
	Viresolve 180	>3 log polio; >6 log retrovirus	
Sartorius NFF	Virosart CPV	>4 log PP7 bacteriophage; >6 log retrovirus	Scale-down module 5 and 20 cm^2; process modules 0.7–2.1 m^2
Pall NFF	DV20	>3 log PP7 bacteriophage; >6 log PR772 bacteriophage	Scale-down 14 and 140 cm^2; process modules 0.07–6 m^2
	DV50	>6 log PR772 bacteriophage	
Asahi TFF/NFF	Planova® 15N	>6.2 log parvovirus; >6.7 log poliovirus	Scale-down modules 10 and 100 cm^2; process modules 0.12–4.0 m^2
	Planova 20N	>4.3 log parvovirus; >5.4 log Encephalomyocarditis	
	Planova 35N	>5.9 log Bovine viral diarrhea virus; >7.3 HIV	

more common, but recent improvements in NFF filters have lead to their predominance. Because the process optimization strategy, test design, and scale-up considerations differ for the two operating modes, they are discussed separately in the following sections.

11.3.1 Normal Flow Operation

In NFF, also referred to as a dead-end filtration, fluid flows perpendicular to the filter membrane surface. Figure 11.1 shows a schematic representation of the NFF process. The fluid upstream of the filter is the feed; the downstream product is the filtrate or permeate. NFF processes can be run either under constant flow operation or constant pressure operation.

During operation, protein products or other components can accumulate at the top of the membrane or adsorb to internal surfaces. These two fouling mechanisms will reduce the hydraulic permeability of the membrane and may

TABLE 11.3
Comparison of Virus Filter Costs, Process Fluxes, and Capacities Compared to Sterilizing Grade Filters

	Filter Type			
		Virus		
Parameter	**Sterile**	**NFF 20 nm Virus Filter**	**NFF 50 nm Virus Filter**	**TFF Virus Filter (>20 nm)**
Unit filter cost, $\$/m^2$	200–300	>2000–4000	>1000	>2000–4000
Typical process flux, $l/m^2/h/psi$	200–500	0.5–6	20–30	5–15
Typical design capacity, l/m^2	>2000–4000	60–500	800–1500	250–500

NFF — normal flow filtration, TFF — tangential flow filtration.

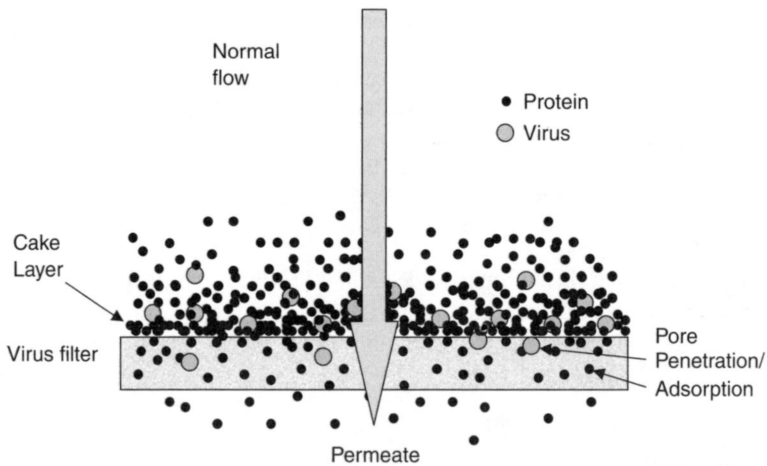

FIGURE 11.1 Normal flow filtration mode of operation.

impact virus retention. Fouling results in a decrease in filtrate flow rate with time for constant pressure operations or an increase in upstream pressure with time for constant flow operations.

The primary advantage of NFF is its ease of use. In contrast to TFF, there is no recirculation of the feed on the upstream side of the filter. As a result, NFF avoids product degradation due to shear. However, NFF is more susceptible to fouling than TFF, and thus requires cleaner process fluids.

11.3.1.1 NFF Virus Filters

Normal flow filtration virus filters are available in a variety of formats and filter areas. Small area filters are generally available for process development and optimization studies as well as virus validation studies. They typically range in size from 3.5 to 140 cm^2, and are available either as cut disks tested in stainless steel holders or as fully encapsulated plastic devices. Stainless steel holders used for testing cut disks should be cleaned, sanitized, or sterilized between uses while most self-encapsulated devices are pre-sterilized and fully disposable.

For pilot and production scale operations, commercial vendors offer individual virus filters ranging in size from 0.07 to 6 m^2. Additionally, multiple filter cartridges or module assemblies can be configured to achieve even larger filtration areas. For these large scale normal flow virus filtration operations, typical formats include pleated cartridges and hollow fiber modules that are designed to be fully disposable.

It is important to note that most virus filters cannot be sterilized by steaming in place (SIP), though there are some exceptions. However, most NFF filter capsules are autoclavable and are usually available pre-sterilized. Commercial filters that are currently available for virus removal applications in the NFF mode are summarized in Table 11.2.

11.3.1.2 NFF Test Equipment and Protocol

A schematic representation of the experimental setup to conduct NFF experiments at constant pressure is shown in Figure 11.2. The apparatus consists of a feed container to hold the protein solution for processing, along with necessary valves, the scale-down filter device, a balance, and a collection vessel for the filtrate. The scale-down test may be carried out in either a constant pressure or a constant flow mode. The constant pressure setup is often simpler and the testing is easier to execute, in that the setup does not require a pump to drive the filtration process. However, scale-down testing may also be carried out in a constant flow mode if the projected large scale operation mode is expected to be at constant flow. Both modes represent valid scale-down methodologies and the choice is often determined by large-scale process needs, operation philosophy, or individual preference.

For the constant pressure test mode, the feed container in Figure 11.2 is pressurized using air pressure. Alternately, a small positive displacement pump may be substituted for constant flow experimentation. Multiple setups in parallel

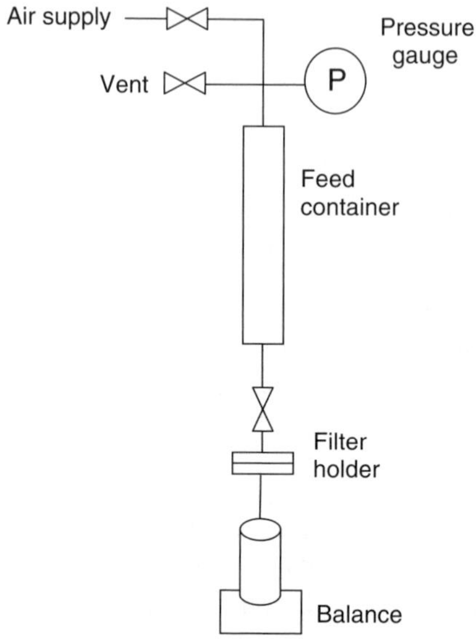

FIGURE 11.2 Experimental set-up for NFF mode of operation.

with input to a data acquisition program and a computer may be employed to evaluate performance over a range of process operation parameters.

The typical steps employed in a scale-down NFF process evaluation are described in the following general test protocol:

1. *System set-up*: Set-up equipment as shown in Figure 11.2. Record membrane lot, serial number for the test, and any other pertinent information.
2. *Water flush:* Fill the feed container with water-for-injection (WFI) quality water and flush the filter following a standard flush procedure provided by the vendor. The flush step serves to both wet the filter as well as reduce extractable levels to a predetermined low level. At the end of the flush step, measure the normalized water permeability (NWP)

$$\mathrm{NWP} = \frac{Q}{A_{\mathrm{filter}}\Delta P} \tag{11.2}$$

where Q is the measured flow rate corrected to a standard temperature to account for the temperature dependency of the water viscosity [4]. A_{filter} is the available filtration area, and ΔP is the differential pressure across the filter. Typical units for NWP are $\mathrm{l/m^2/h}$.

3. *Installation check:* Conduct a pressure-hold test to check installation integrity of the devices. Test pressure and guidelines will be specific to each filter and may be obtained from the filter vendor. (This step is optional for process development experiments, but is typically used in virus validation studies.)
4. *Buffer conditioning:* Condition the filter by flowing an appropriate buffer through the filter at a prescribed pressure to a certain loading (l/m^2). The pressure is usually identical to that used in the product filtration step.
5. *Product filtration:* Prefilter the feed material through a 0.2 μm sterilizing grade filter. Fill the feed container with the prefiltered feed solution and carry out the filtration process under a prescribed set of operating conditions (pressure and flow rate, concentration, pH, etc.). During filtration, filtrate volume (V) collected is measured and recorded at various filtration times (t). Filtration time may vary between 45 and 120 min. Assay filtrate sample for product concentration.
6. *Recovery:* When the filtration is complete, carry out a short buffer recovery step to maximize recovery of the protein from within the holdup volume using similar processing conditions employed during the product filtration step. Assay buffer flush sample for product concentration and calculate product recovery.
7. *Installation check:* Flush the filter with water and carry out a pressure-hold test to confirm installation integrity of the devices. (This step is optional for process development and is typically carried out in virus validation studies.)
8. *Calculate filter capacity and initial flux:* The filter capacity and the initial flux are generally obtained by fitting the experimental data (V vs. t) to the gradual pore plugging model [5–8] as follows:

$$\frac{t}{V} = \frac{t}{V_{\max}} + \frac{1}{Q_{\mathrm{i}}} \tag{11.3}$$

   A plot of t/V vs. t should yield a straight line with slope of $1/V_{\max}$ and a y-intercept of $1/Q_{\mathrm{i}}$, where $V_{\max}$ is the filter capacity and Q_{i} is the initial flux. Typical units for filter capacity and initial flux are l/m^2 and $l/m^2/h$, respectively. Although the gradual pore plugging model is the most widely used model, other models [9,10] are available that may provide alternate and more rigorous analyses of the filtration data.
9. *Calculate minimum required filter area*: Once the filter capacity and initial flux values are calculated, the minimum required filter area

can be calculated as:

$$A_{\min} = \frac{V_{\rm B}}{V_{\max}} + \frac{V_{\rm B}}{Q_i\, t_{\rm b}} \tag{11.4}$$

where $V_{\rm B}$ and $t_{\rm b}$ are the proposed batch volume and desired batch processing time, respectively. This value should be used only as a comparative tool during process optimization studies. The final design filter area will be determined during process simulation and virus validation studies.

11.3.1.3 NFF Process Optimization

Optimization of a virus filtration process involves evaluating the effect of a variety of process parameters to arrive at optimum conditions that would ensure robust, consistent, and scalable operation. Some of the key process development parameters that impact process performance are described in more detail below. A generic approach to optimization is schematically represented in Figure 11.3.

11.3.1.3.1 Impact of Location in Downstream Process Train

There are typically several choices of where to implement a normal flow virus filtration step within a given downstream process. As shown in Figure 11.4 for a typical monoclonal antibody process, the NFF step could conceivably be implemented at three distinct locations within the downstream process train: following the low pH inactivation step, following the intermediate chromatographic operation, or after the final chromatography step. Since protein concentration, impurity concentration, and process volumes vary dramatically throughout the downstream process train, it should come as no surprise that the actual filtration requirements can be highly dependent upon where in the process the virus filtration step is located. As seen in Figure 11.5, the required filter area can be a strong function of location placement within the downstream process train. It should be noted, however, that the results depicted in Figure 11.5 are for a particular feedstream. The interplay of specific feed and virus filter properties can lead to location dependencies either higher or lower than those depicted in Figure 11.5.

11.3.1.3.2 Impact of Feed Concentration

Feed solution concentration may impact the virus filtration process by reducing product throughput (as measured by capacity and flow). The significance of the impact may depend on the structure and morphology of the virus filter. Figure 11.6 shows the potential impact of protein concentration on the performance of a small virus filter. In general, higher protein concentrations reduce the

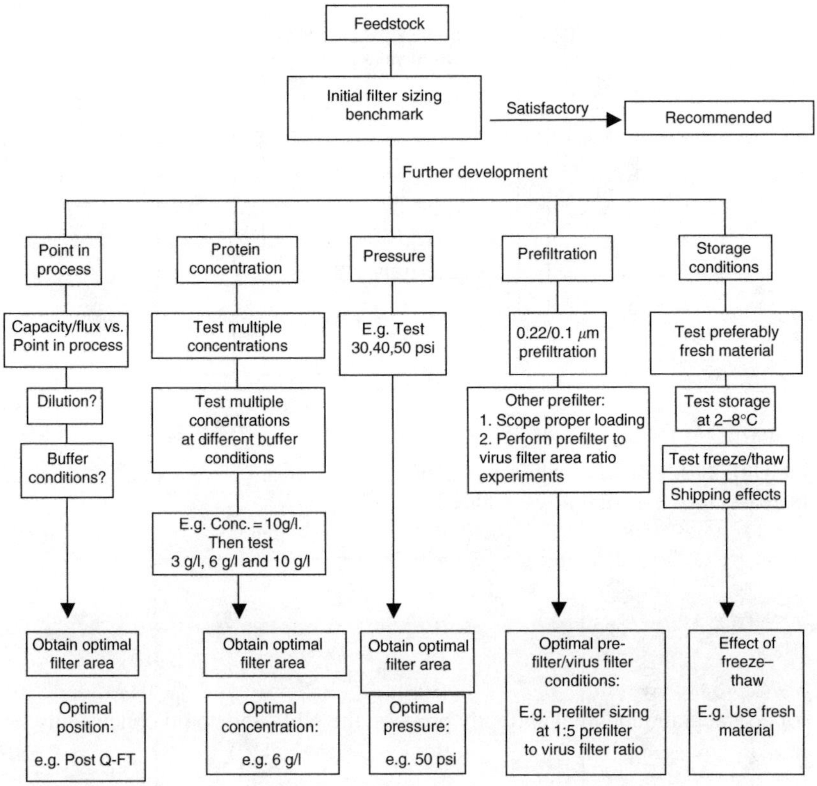

FIGURE 11.3 Example of a decision tree for optimizing NFF performance.

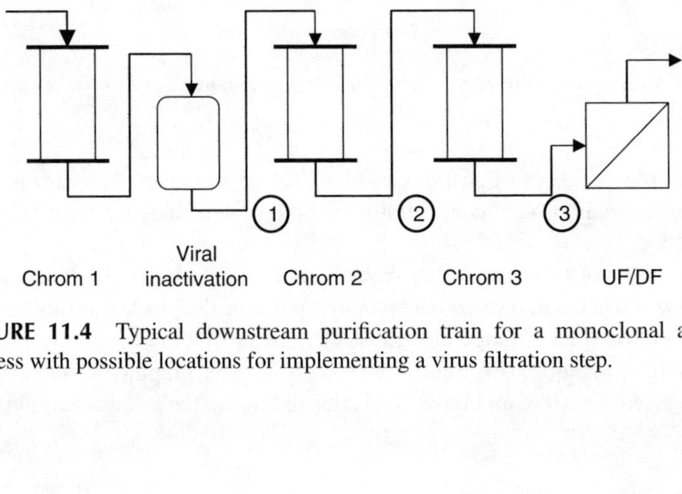

FIGURE 11.4 Typical downstream purification train for a monoclonal antibody process with possible locations for implementing a virus filtration step.

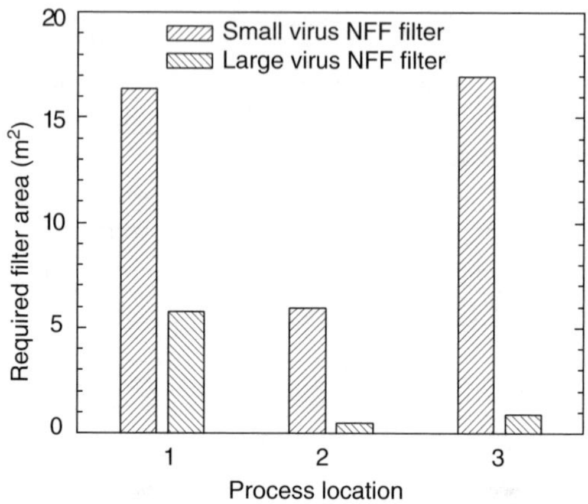

FIGURE 11.5 Typical impact of location in downstream process train on required filter area for a normal flow virus filter.

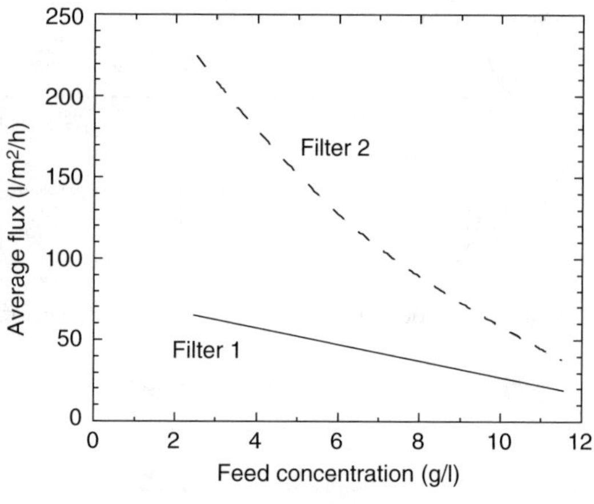

FIGURE 11.6 Typical impact of feed concentration on average flux for normal flow virus filters.

average process flux through the virus filter. The dependence of average process flux to changes in protein concentration is both protein specific and virus-filter specific.

The effect of increasing filter capacity and flow at lower product concentrations is offset by an increase in process volume as the product is diluted. The interplay of these two competing effects can often result in a feed concentration that minimizes required filtration area [11]. This is graphically depicted in Figure 11.7. As seen in Figure 11.7, an optimum feed concentration may

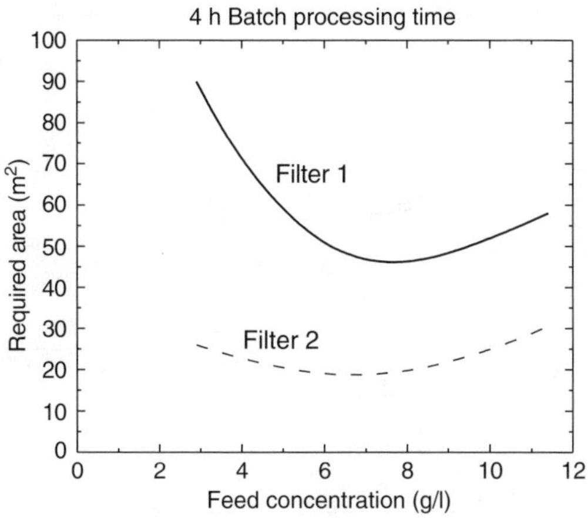

FIGURE 11.7 Typical impact of feed concentration on required filter area for normal flow virus filters.

exist that maximizes filtration performance (minimizes filtration area [m^2] and maximizes productivity [$g/m^2/h$]). For high concentrations (>10 to 15 g/l), it may be advantageous to dilute the product to improve filterability. The need to dilute the feed material prior to virus filtration may suggest that the preferred placement of the virus filtration step is immediately upstream of a UF–DF step — either before an intermediate UF–DF step or prior to the final formulation.

11.3.1.3.3 Impact of Operating Pressure

The effect of filtration pressure is often best determined by conducting an excursion study to evaluate filter capacity and flow as a function of pressure. It is customary to evaluate pressure effects in the 10 to 50 psi range. Keep in mind, however, that the maximum pressure evaluated must be within the manufacturer's pressure limit specifications. The effect of pressure may be significant, based on the feed nature and filter morphology. Figure 11.8 shows the typical impact of operating pressure on the minimum required filter area. In general, higher operating pressures increase the average process flux and decrease the required filter area [11]. The magnitude of this impact is dependent upon several factors, including feed product, feed concentration, impurity profile, and virus filter.

11.3.1.3.4 Impact of Prefiltration

Prefiltration of the feed solution can have a dramatic impact on filter performance. Prefiltration is targeted to remove various impurities or contaminants such as protein aggregates, DNA and other trace materials. While larger-size impurities can be removed by prefiltering with a 0.2 or 0.1 μm microfilter, smaller

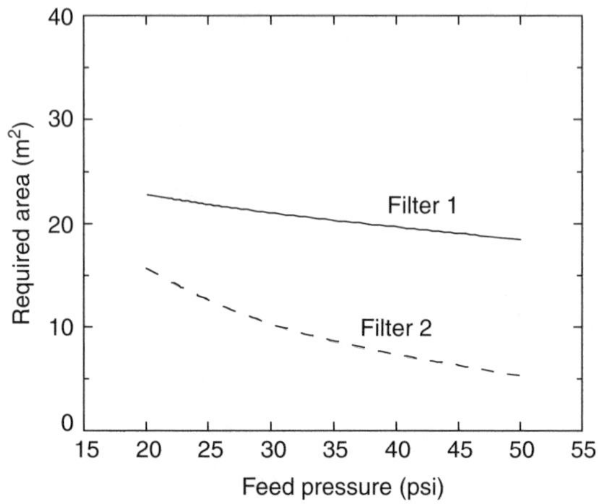

FIGURE 11.8 Typical impact of feed pressure on required filter area for normal flow virus filters.

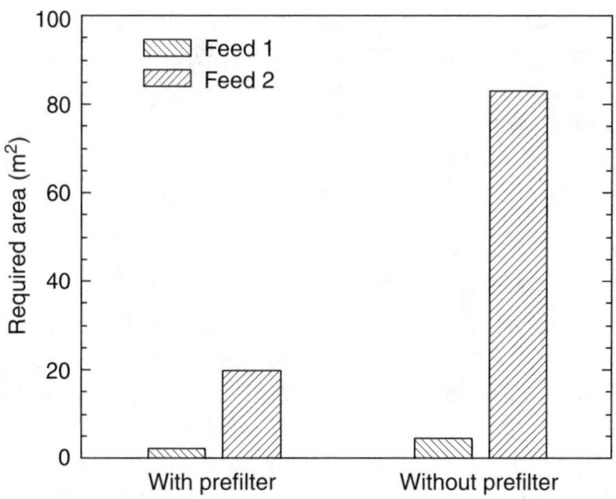

FIGURE 11.9 Typical impact of virus filter prefiltration on required filter area for normal flow virus filtration.

impurities such as protein aggregates that may only be marginally larger in size compared to the protein product, are not easily amenable to size-based removal methods. Prefiltration through adsorptive depth filtration has been observed to provide significant protection for certain virus removal filters [12]. As seen in Figure 11.9, the impact of prefiltration can be quite dramatic; with up to ten-fold reductions in required filter area sometimes achievable. As these filters work by nonspecific multimode adsorption, product recovery should be confirmed to ensure good yield.

11.3.1.3.5 Impact of Freeze/Thaw

For some protein solutions, the freeze–thaw of a material can have a significant impact on filtration performance. In fact, in some instances, it has been observed that the required filter capacity is five- to six-fold higher when measured using material that has been previous frozen compared to fresh feed [11]. While the actual purification process may not have a freeze–thaw step, feed samples required for virus validation testing are often conveniently submitted in a frozen form due to material stability/availability considerations. In such situations, if freeze–thaw is observed to produce an adverse impact of filtration performance, a prefilter is often used to restore performance similar to the unfrozen material.

11.3.2 Tangential Flow Operation

Tangential flow filtration is similar to NFF, but includes a sweep flow across the feed side of the membrane. The sweep flow tangential to the surface removes foulants, allowing both higher and more stable flow rates through the membrane. As with NFF, the portion of the feed that flows through the membrane is called the filtrate or permeate. The feed that passes tangentially across the filter surface is the retentate, and is recycled to the feed tank. As with NFF, the operation can be run either at constant transmembrane pressure (TMP) or constant permeate flux.

The TFF system has more components than the NFF system, including the retentate line with a control valve, a feed pump, a heat exchanger for removing feed pump heat, and associated monitoring and control components. As a result, it is more capital intensive and complicated than NFF. The advantage of TFF is higher flow rates and throughputs since the tangential flow reduces polarization and cake formation [4]. Figure 11.10 shows a schematic representation of a TFF process.

In TFF operation, the bulk of the feed is recirculated as retentate. With time, the volume of the recirculating retentate decreases, so that contaminants become concentrated in the retentate. To enhance product recovery, the TFF system operation often involves a diafiltration step whereby product is flushed through to the permeate via continuous buffer addition on the retentate side.

11.3.2.1 TFF Virus Filters

The filter membranes used in TFF filters can be asymmetric or symmetric ultraporous or microporous membranes. TFF filters are generally available either as flat sheet cassettes or as hollow fiber modules. TFF virus filter modules/cassettes

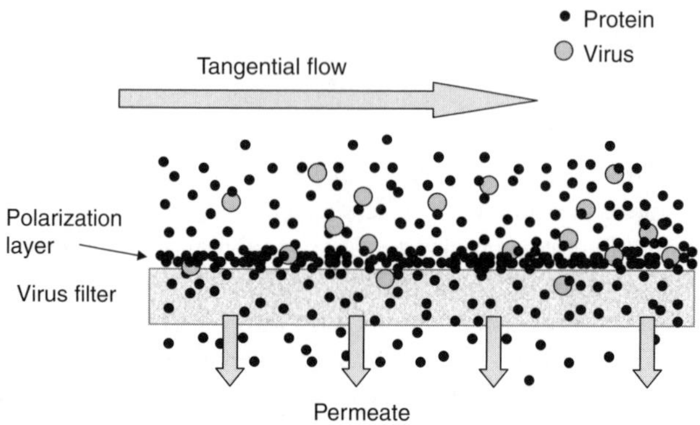

FIGURE 11.10 Tangential flow filtration mode of operation.

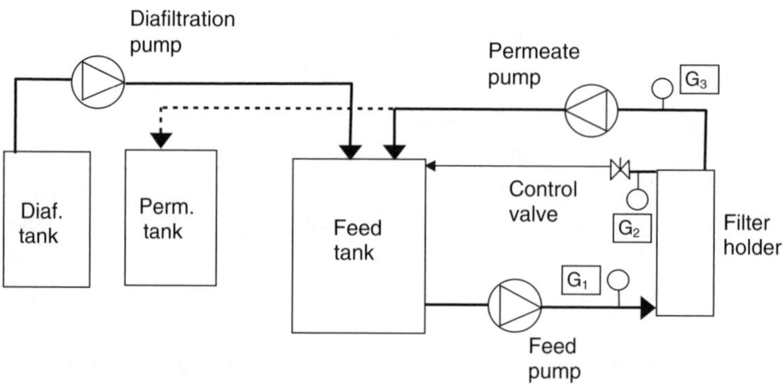

FIGURE 11.11 Experimental setup for TFF mode of operation.

are available in areas ranging from small-scale laboratory devices to high area modules for large volume applications. Cassettes or modules are usually installed in reusable housings, and can be stacked or manifolded together to provide the filtration area required for large-scale applications. Commercial filters are listed in Table 11.2.

11.3.2.2 TFF Test Equipment and Protocol

A typical experimental set-up that may be employed to carry out tangential flow virus filtration evaluation is shown in Figure 11.11. The equipment set-up generally consists of

- Feed container to hold the protein solution
- Feed, permeate and diafiltration pumps

- Virus filter holder
- Inlet (feed), outlet (retentate) and permeate pressure gauges
- Feed/permeate flow meters
- Control valves

A systematic method to evaluate the various parameters that affect virus filter performance is necessary for successful process development and scale-up. For TFF, these parameters are typically: feed concentration, feed cross-flow rate, and permeate flux or (transmembrane pressure). The following test protocol describes the typical steps in a scale-down TFF process development:

1. *Filter Installation:* Load the virus filtration membrane modules in the filter holder following the standard installation procedure provided by the filter vendor. Record the filter lot and serial number and any other pertinent information.
2. *Water Flush:* Fill the feed container with WFI quality water and flush the filter following a standard flush procedure provided by the vendor. Measure NWP toward the end of the flush step.

$$\text{NWP} = \frac{Q}{A_{\text{filter}}\text{TMP}} \tag{11.5}$$

   This equation is similar to Equation 11.2 with the exception that ΔP is now replaced with the TMP. As before, the water flow rate Q must be properly normalized for viscosity variations due to temperature.
3. *Installation Check:* Conduct a pressure hold test to check installation integrity of the devices. Test pressure and guidelines will be specific to each filter and may be obtained from the filter vendor. (This step is optional during process development, but is typically used during virus validation studies and actual production runs.)
4. *Buffer Conditioning:* Condition the filter by circulating an appropriate buffer through the filter under total recycle mode (retentate and permeate lines returned to feed reservoir) and standard conditions for permeate and cross-flow rates recommended by the vendor.
5. *Product Filtration*
   5.1 Fill the feed container with the feed solution after prefiltering it through a 0.2 μm sterilizing grade filter.
   5.2 Set-up the system in a total recycle mode.

5.3 Begin recirculation of the feed solution to a desired cross-flow rate using the feed pump.
5.4 Set and control the retentate pressure (e.g. 5 to 6 psi) using a control valve.
5.5 Slowly ramp up the permeate flow to a desired value using the permeate pump.
5.6 After equilibration, record transmembrane pressure. Collect feed and permeate samples to measure protein concentrations and protein passage.
5.7 Repeat steps 5.3 to 5.6 for various permeate and cross-flow settings.

Note: A typical operating TMP range is 0 to 10 psig with permeate fluxes ranging between 35 and 50 $l/m^2/h$ for product concentrations in the range of 2 to 10 g/l.

6. *Data Analysis:* The above data may be analyzed to select operating conditions (i.e., permeate flux and cross-flow rate) that optimize the productivity or mass flux of the recovered product, measured in grams of product recovered in the permeate per unit membrane area per unit time ($g/m^2/h$).
6.1 The mass flux may be mathematically expressed as:

$$M = J\,C_p = J\,C_f\sigma \tag{11.6}$$

where M is the mass flux in $g/m^2/h$ (gmh), J is the volumetric flux in $l/m^2/h$ (lmh), C_f and C_p are the feed and permeate concentrations of the product and σ is the product sieving coefficient or passage of product through the membrane.
6.2 During process simulation, diafiltration buffer is added at a desired flow rate through the diafiltration pump. An optimal diafiltration strategy that is commonly employed in tangential flow virus filtration is called differential diafiltration [13], where the ratio of diafiltration to permeate flow is maintained equal to the value $1 - \sigma$, where σ is the passage of product.

11.3.2.3 TFF Process Optimization

A systematic method to evaluate the various parameters that affect mass flux is necessary for successful process development and scale up. These parameters are: feed concentration, feed cross-flow rate, permeate flux or (transmembrane pressure).

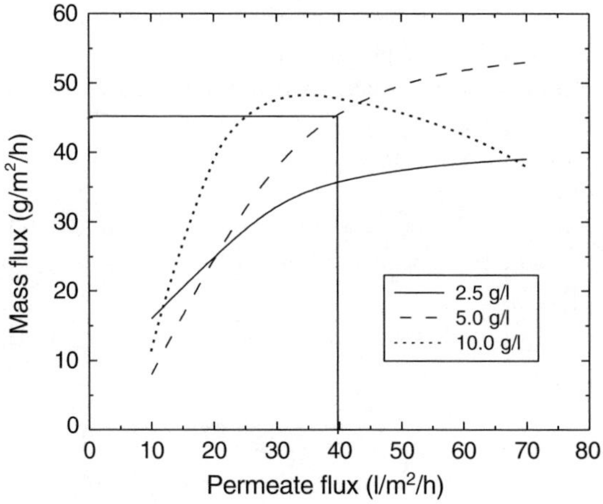

FIGURE 11.12 Typical impact of permeate flux and feed concentration on measured mass flux for TFF virus filtration.

11.3.2.3.1 Impact of Feed Concentration

The equation for mass flux suggests that operation at higher feed concentration tends to increase the mass flux. Many times, situations arise where the feed concentration of the fluid is too high for practical operation of the tangential flow process. In such cases, the feed needs to be diluted to a more appropriate concentration. As it is difficult to *a priori* know this concentration, the feed is diluted to various (lower) concentration levels, and the mass flux profile is evaluated at each concentration. The feed concentration that gives the highest stable mass flux is selected. As a rule of thumb, the permeate flux at approximately 80% of the maximum mass flux is selected for stable process operation. For the example illustrated in Figure 11.12, a feed concentration of 5 g/l at a mass flux of 45 g/m^2/h would represent a stable operating condition.

11.3.2.3.2 Impact of Feed Cross-Flow Rate and Permeate Flux

The feed cross-flow rate and permeate flux control the wall concentration, C_{wall} of the retained and partially retained solutes. This wall concentration within the concentration polarization layer ultimately impacts process performance by influencing product passage and mass flux. In order to optimize the selection of these parameters, the mass flux is evaluated at a number of different cross-flow rates and permeate fluxes. The cross-flow rate beyond which there is no significant improvement in mass flux is then selected. For the example shown in Figure 11.13, cross-flow rate 2 would typically be selected as further increases in the cross-flow rate appear to have a negligible impact on improving mass flux.

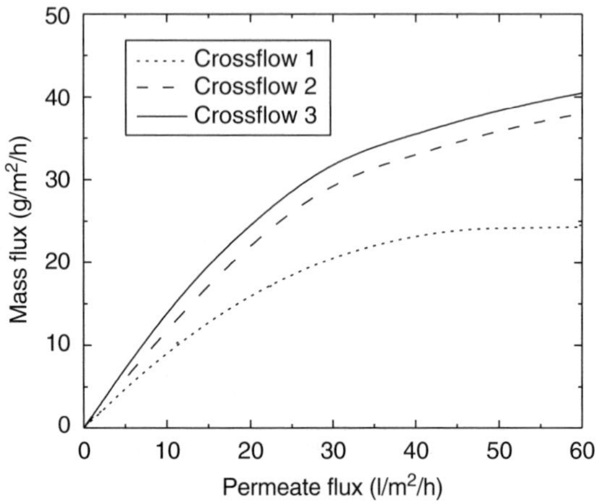

FIGURE 11.13 Typical impact of permeate flux and cross-flow rate on measured mass flux for TFF virus filtration.

11.4 PROCESS SIMULATION AND SCALE-UP

11.4.1 NORMAL FLOW FILTRATION

As mentioned previously, a gradual pore-plugging model as described by Equation 11.3 and Equation 11.4 (V_{max} model) is often used to project filtration requirements. Equation 11.4 indicates that filter sizing is impacted by the filter capacity, V_{max}, the initial flow rate, Q_i, and the batch time, t_b. Typical predictions from the V_{max} model are shown in Figure 11.14. As these examples illustrate, for typical processing times <4 h, a higher flux membrane with a corresponding lower capacity often results in lower filtration area compared to a high capacity/low flux filter. For processing times >18 h, the high capacity/low flux filter would result in a process with lower filter area. It should be noted, however, that shorter processing times have the added benefit of allowing for the possibility of in-line processing with other purification steps as well as mitigating potential product stability issues.

Once the optimum filtration conditions have been determined, it is recommended that a simulation study be performed. This would initially be performed at the small-scale (3.5 to 14 cm^2), then repeated at a larger scale as the process is scaled-up. This would involve running the filtration to the desired endpoint, which may be a specific filtration time, volume/area ratio, or percent flux decline.

One of the outcomes of successful process development is a process that is robust and easy to implement when scaled-up to manufacturing. In order to demonstrate scalability of the process, it is recommended that pilot scales

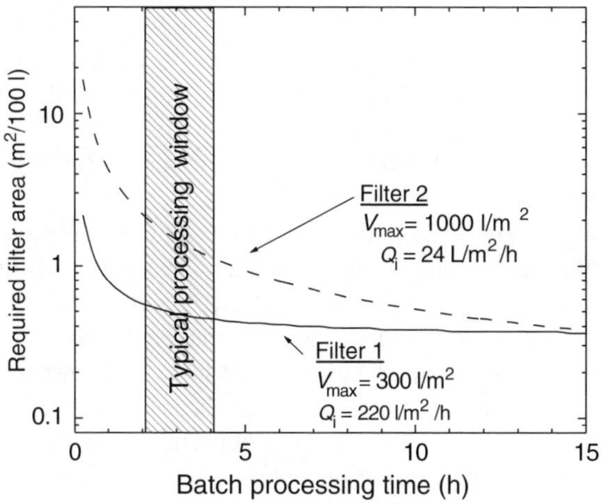

FIGURE 11.14 Typical impact of batch processing time on required filter area for NFF virus filtration for two hypothetical filters.

studies be conducted using devices containing 100 to 1000 cm^2 of filter area. This scale represents a 10- to 300-fold scale-up from the initial simulation studies.

The objectives of the pilot-scale studies are two-fold. A first objective is to obtain confirmation that the process parameters (process loading, time, flux or pressure, and yield) are with predicted ranges and estimated bounds. Second, the pilot-scale studies are used to obtain information on the entire operation (installation, flushing, sterilization/sanitization, integrity testing, process, product recovery, etc.) so as to enable drafting of SOPs and batch records for cGMP manufacturing. Information obtained during the pilot scale studies can also be used to establish appropriate performance limits for water permeability (NWP), integrity testing, and other secondary operations related to the virus filtration.

11.4.2 Tangential Flow Filtration

For TFF, process simulation refers to a scale-down simulation of the large-scale process at the proposed operating conditions of cross flow rate, loading, and permeate flux. The results of the mass flux excursion tests are used to select the optimum operating conditions. Once the optimum operating conditions are selected, the process operation, using a differential diafiltration technique is typically conducted. The objective of the differential diafiltration step is to maintain a stable, high mass flux operation throughout the entire process. Samples of the retentate and permeate are collected at several points during the diafiltration process and the protein concentration (typically using absorbance at 280 nm) is measured. At the end of the process, a composite permeate and retentate sample is collected for measurement to ensure confirmation of mass balance. After completion of the simulation run, the product is drained and the system is flushed with WFI/buffer. This step may then be followed up by a cleaning procedure before discarding the filters.

During process simulation, it is often discovered that the protein passage is not constant during the run even though the concentration may be maintained constant. This may be due to the fact that there are other components in the solution which may be rejected by the membrane and hence, polarize on the membrane surface. Changes to the polarization layer are known to affect protein passage in a complex manner. Lutz [13] describes an optimization strategy for diafiltration that considers the dependence of protein passage on concentration factor during the process. In addition, during process simulation, confirmation of product loading with respect to membrane throughput (capacity) is also established.

11.5 VIRUS VALIDATION STUDIES

The purposes of the virus validation studies are to confirm the LRV claims for the filtration step and to verify the filter sizing established in the scale-up phase. These tests are run at a small scale, maintaining critical parameters such as pressure, flux, and loading capacity at their commercial operation values while mimicking other operating procedures such as preprocessing WFI flush-outs and buffer equilibration. The tests are typically run concurrent with the manufacturing-scale consistency/validation batches. Due to the handling and assay requirements for virus studies, the tests are typically conducted at specialized labs.

The filter is challenged with representative feedstock containing a virus spike. The concentration of the virus in the feed and the pooled permeate is measured to calculate the LRV. Parallel control assays are run to correct for virus losses due to artifacts such as dilution, concentration, filtration, and storage of samples before titration. Based on the measured LRV, the filtration operation will be classified as effective (LRV > 4), moderately effective ($1 <$ LRV < 4) or ineffective (LRV < 1). A moderately effective step is recognized as contributing to the overall process LRV. No LRV credit is given for an ineffective step. Typically, manufacturers will place a lower limit of 3 on the reduction factors that will be combined to yield the overall reduction factor for the manufacturing process.

To accurately represent manufacturing settings, the test feedstock must be identical to the commercial-scale feedstock. Shipping or storage constraints may require freezing feedstock, which can result in protein aggregates. Aggregates can cause premature filter plugging that may alter scaling parameters such as loading capacity. The problem can be obviated by either removing aggregates with microfiltration or generating fresh feedstock at the site of the virus spiking study.

The viruses used in the spiking studies depend on the specifics of the process and virus contaminants. The regulations recognize relevant model viruses that represent endogenous viruses, and nonspecific model viruses to validate general viral clearance for adventitious contamination. Murine Leukemia Virus (MuLV) is the generally accepted RVLP model for endogenous virus tests. If use of a relevant virus is not possible, the manufacturer chooses the best specific model virus to serve as a model for the relevant virus. To satisfy the general viral clearance objective, the study sponsor will generally evaluate two or three additional viruses. The nonenveloped parvoviruses (~20 nm) are often accepted as a worst case for filtration. The test thus comprises a four- or five-virus panel that represents viruses of different genomes (DNA and RNA), sizes and surface properties (enveloped and nonenveloped).

Higher virus titers allow larger LRV claims to be demonstrated while providing a more rigorous challenge to the filtration operation. For this reason, regulatory guidance states that "the amount of virus added to the starting material for the production step which is to be studied should be as high as possible" [14]. However, so as not to unacceptably alter the product composition, the volume of spike should be kept below 10%, and typically below 5%. The guidance also voices concerns over virus aggregation that could be induced by deliberately concentrating the virus. The use of aggregated virus could lead to underestimation of inactivation effectiveness and overestimation of size-exclusion effectiveness [15]. In this vein, there is no quantitative regulatory guidance for the acceptable level of virus aggregation or specific means to evaluate virus aggregation. The common practice is to use size-based prefiltration to remove virus aggregates from a spiked feed stream prior to performing the clearance study.

Impurities contained within the virus spike may also foul the membrane, preventing the tests from reaching important scaling parameters such as loading capacity. Methods of generating highly pure virus preparations are increasingly being used to prevent fouling due to spike impurities [16,17]. Another alternative procedure is to determine the relationship between amount of spike and loading capacity. A series of tests with progressively larger amounts of virus spike is run to determine the impact on capacity. The size of the virus spike is then adjusted to be consistent with the target process loading capacity.

Virus retention has been observed to decline with fouling for a variety of filters [11,16,18–20]. If the virus spike required to achieve the target LRV causes excessive fouling, alternative validation methods may be used to determine LRV at higher throughput values [16]. These alternative methods may be used for validation after consulting with the appropriate regulatory agencies.

11.6 PROCESS IMPLEMENTATION

Once the virus clearance step has been optimized and virus validation studies completed, an implementation strategy is required for robust process operation. After determining the filter capacity (l/m^2) required for a process during process simulation/scale-up and virus validation studies, the filter area required for processing a given batch volume can be calculated. Various filter configurations are made available by manufacturers to facilitate large scale implementation. When multiple filter modules are required to process a given batch volume, the modules may be installed in parallel within a multiround housing or multiple filter modules can be installed separately in parallel.

11.6.1 Hardware Considerations

Normal flow virus filters are operated either in constant pressure mode or in constant flow mode. Tangential flow filters are typically operated at a constant feed/retentate flux or at a constant permeate flux. Typical factors to be considered during large-scale virus filtration system design include:

- Minimum and maximum batch volume
- Minimum and maximum flow rate; it is important to consider flow rates during pre- and post-use water flush and for post-use system cleaning/sanitization
- Maximum operating pressure and differential pressures across the prefilter and the virus filter
- If in-line dilution is needed to maintain constant feed concentration, appropriate dilution and mixing hardware
- Minimum and maximum concentration and appropriate instrumentation to span the range
- Appropriate hardware and connections to enable the filters and the system to be steamed, autoclaved, or chemically sanitized
- Filter housing configuration — individual filters in parallel or a multifilter housing
- System holdup volume versus validated post-process buffer rinse
- Pre- and post-use integrity tests

In the case of normal flow virus clearance filters, pressure vessels are typically employed for constant pressure operation. However, when very large process volumes are involved, it may be easier to use a pump with a pressure feedback loop to carry out constant pressure filtration.

11.6.2 Operating Sequence

A typical sequence of operations in a virus filtration process includes the following steps, each of which will be discussed in further detail:

- Filter installation
- Filter flushing
- Measurement of normalized water permeability
- Sterilization/sanitization
- Pre-use integrity testing
- Buffer preconditioning
- Processing and product recovery
- Postproduction integrity testing

11.6.2.1 Filter Installation

Filter modules should be installed into the housings as per manufacturer's instructions. In the case of normal flow filtration cartridges where feed-to-filtrate barrier is achieved with *o*-rings, it is important to use caution to ensure that the *o*-rings are not damaged during installation. Wetting the *o*-rings with WFI can greatly reduce friction between the *o*-rings of a cartridge filter and the steel housing base.

11.6.2.2 Filter Flushing

Most filter manufacturers recommend preuse flushing in order to flush out residual extractables from the filters. Flush volume and flushing conditions (differential pressure or flux) that are recommended by the filter manufacturers should be employed. If a prefilter is used, it may be necessary to flush the prefilter independently prior to flushing the virus clearance filter since some prefilters may have higher level of extractables when compared to virus filters.

11.6.2.3 Measurement of Normalized Water Permeability

A pre-use water permeability measurement is recommended to ensure that the filter is fully wetted and the entire installed filter area is used during filtration. Some filter users monitor the extent of flux decline during the filtration process in order to ensure that the flux decline during manufacturing does not exceed the flux decline that was observed during virus clearance validation studies.

In such cases, water or buffer flow rate through a clean filter is sometimes used as the initial flow rate, or Q_i, to calculate the extent of flux decline.

In order to measure water permeability, measure and record water flow rate, inlet/outlet pressures and water temperature. NWP can then be calculated using either Equation 11.2 for NFF operation or Equation 11.5 for TFF operation. If the NWP value for the filter assembly is outside the recommended range, ensure that the filters are fully vented, check the water temperature, pressure, and ensure that there are no flow restrictions on the downstream side of the filter assembly (narrow bore tubing/piping, sticky valves, etc.) and repeat the NWP measurement. It is important to note that the actual NWP value will depend on many factors such as housing configuration, parasitic pressure drops in the pipes, etc.

In the case of some virus clearance filters, a post-use integrity test is carried out using liquid–liquid porosimetry methods. Some liquid–liquid intrusion tests require that the membrane be thoroughly cleaned post-use prior to testing. In such cases, the preuse NWP can serve as a benchmark to determine the extent of cleanliness post-use.

11.6.2.4 Sanitization/Sterilization

In a typical downstream purification process, virus clearance filters are employed downstream of a chromatography column and upstream of an ultrafiltration/diafiltration step. Neither of these steps is considered an aseptic unit operation. However, there appears to be an industry trend to sterilize the virus filter to reduce the bioburden, if not to render it aseptic. Some end users have adopted chemical sanitization as a means of reducing bioburden. Some of the virus filters are available presterilized which eliminates the need for a sterilization/sanitization step for the filter. However, one would still need to sanitize the equipment/system.

11.6.2.4.1 Steam-In-Place

Most virus filters are ultrafiltration membranes that have air–water bubble points well in excess of 100 psi [21]. Due to the very high bubble points of these filters, (steam-in-place) SIPing a wet virus filter will usually result in very high pressure differentials across a hot filter. If the pressure is not carefully controlled during the steaming and the cool down phase, the filter can be severely damaged and the filter integrity compromised. Points to consider prior to SIPing a virus filter include:

- Manufacturer's recommended procedure for sterilization.
- Maximum temperature the filter is rated for and exposure time at this temperature.

- Maximum forward and reverse pressure differential that the filter can be subjected to at the maximum temperature.
- Is the SIPing process designed to demonstrate sterility or bioburden clearance?
- Filter cool-down procedure prior to processing.
- Since filters are relatively less resistant to reverse pressure, is there a chance of reverse pressurizing the filter during the cool-down phase?
- What data is available from the filter manufacturer to demonstrate virus retention capability post-SIPing?
- Is a post-SIP integrity test possible prior to processing?
- Is a post-SIP water or buffer flush possible prior to processing to flush out extractables resulting from thermal cycling?
- Is a post-SIP flush possible to remove any air trapped between the layers of a multilayer filter?

11.6.2.4.2 Autoclaving

Autoclaving can also be used to reduce bioburden from a filter assembly. Some virus filters may need to be autoclaved wet, and care must be taken to ensure that the autoclaving process does not dewet the membrane pores. In such cases, the autoclaving procedure must be carefully designed, particularly during the post-autoclave cool down phase when vacuum pulsing is employed. Some manufacturers recommend using a liquid cycle in order to prevent water from flashing off a hot membrane during the cooling phase.

Since autoclaving is not easy to carry out on very large filter installations, one novel way to achieve bioburden reduction is by steaming and cooling a wet filter from both the feed and the filtrate ports. This technique, termed as Autoclave-in-Place, ensures that the filter pores remain wet during the autoclaving process and that very low-pressure differentials are maintained during steaming and cool down phase.

11.6.2.4.3 Hot Water Sanitization

Some filter users have considered hot water for sanitization of filters. Typically, hot WFI at 80 to 90°C for about 30 to 60 min is employed for this purpose. As with steaming, it is important to consider the effects of high-pressure differential during hot water sanitization as well. Filters that are typically rated for 80 psi operation at 25°C have much lower ΔP ratings at 80°C. Check manufacturer's recommendations for sanitization conditions using hot WFI. For any sterilization or sanitization step, time and temperature will need to be validated for a given filter configuration.

TABLE 11.4
Alternative Chemical Sanitizing Agents

Sanitizing agent	Typical concentration
Chlorine dioxide gas	50–500 ppm
Peracetic acid	100–300 ppm
Sodium hydroxide	>0.1 N, pH 12–13
Sodium hypochlorite	20–50 ppm, pH between 6 and 8

11.6.2.4.4 Chemical Sanitization

When chemical sanitization is employed to achieve bioburden reduction, it is important to first ensure that the filters are chemically compatible with the solutions and conditions used during the sanitization step. Follow manufacturer's recommendations for chemical concentration, temperature, and maximum exposure time. The filtration system should be carefully designed to ensure that the chemicals are completely flushed from the system. Typical sanitizing agents and concentrations are shown in Table 11.4.

11.6.2.4.5 Post-Sterilization/Sanitization Flush

Some filter manufacturers may recommend a flush after sterilization/sanitization in order to wash out any extractables from the filter materials. It is important to follow manufacturer's recommendations.

11.6.2.5 Pre- and Post-Use Integrity Testing

To ensure that virus clearance is consistent with manufacturer's claims and results obtained during virus validation studies, filter integrity should be checked both pre- and post-use. To facilitate this, filter manufacturers have developed a variety of destructive and nondestructive physical integrity tests that are related to virus retention. Ultimately, the objectives of properly designed physical integrity testing are three-fold:

- Confirmation that the virus removal filter is properly installed
- Confirmation that the filter is free from gross defects and damage
- Confirmation that the filter removes viruses consistent with both manufacturers' specifications and end-user virus validation studies

To properly satisfy these requirements, a series of tests may need to be performed to ultimately confirm filter integrity. Fortunately, various tests run by filter manufacturers either as QC release tests or 100% integrity tests can often minimize the number of integrity tests that the end-user must perform. Periodic auditing of manufacturers to verify that these tests ensure filter performance is recommended. Filter manufacturers should be able to provide evidence that integrity test methods and acceptance criteria correlate to retention of viruses in the targeted size range under standard conditions.

The complexity of integrity testing virus filters should not be overlooked when selecting the virus removal filter for manufacturing. Key integrity test considerations include performance (integrity test sensitivity, robustness, and performance), safety (explosion hazards, product contamination), logistics (ease-of-use, pre- and post-use, cycle time, nondestructive), validation (test robustness and repeatability, raw material requirements, equipment calibration, proper positive controls) and regulatory (filter manufacturer support data, regulatory filings) [2].

11.6.2.5.1 Classes of Virus Filter Integrity Tests

Various destructive and nondestructive methods are available to integrity test virus filters. Currently available integrity tests for virus removal filters can generally be classified into the following categories [3,21]:

- Particle challenge tests (dextran retention, gold particle retention)
- Gas–liquid porosimetry tests (bubble point, leak, forward/diffusive flow, pressure hold/decay)
- Liquid–liquid porosimetry

Some of these tests are better suited for confirming proper filter installation, whereas others may be better suited for detecting gross defects or subtle changes in filter-pore-size distribution. A more detailed summary of the various tests along with troubleshooting techniques can be found in the PDA TR41 [2].

11.6.2.5.2 Pre and Post-Use Integrity Testing

While only nondestructive tests can be used preuse, either type of test can be used for post-use testing. Nondestructive tests are either gas–liquid or liquid–liquid porosimetry tests. In general, a gas–liquid porosimetry test such as diffusion test or pressure hold test is recommended to complement liquid–liquid porosimetry test or particle challenge test to check for gross defects in the system.

Pre-use integrity testing can be performed either before or after sterilization/sanitization. Post-sterilization integrity tests are particularly useful since they ensure that the filters are not damaged during the sterilization process. However, in an aseptic process, one must maintain system sterility during filter wetting and integrity testing steps. An automated integrity tester may be useful for post-sterilization integrity testing.

11.6.2.5.3 Automated Integrity Testers

Automated integrity testers generally offer many advantages over manual testing, including:

- Less operator variability
- Better reproducibility of results
- Ability to carry out tests post-sterilization while maintaining the system aseptic
- Ability to interface the test results with plant-wide data acquisition/control system

It is important to qualify the automated integrity testers for testing virus clearance filters. Since many virus clearance filters are composed of multiple filter layers, it may be necessary to include a longer pretest stabilization step to demonstrate comparability of manual and automated test results.

11.6.2.5.4 Integrity Testing Multifilter Assemblies

Liquid–liquid porosimetry tests and particle challenge tests are generally carried out on individual filter elements. However, it is possible to perform gas–liquid diffusion tests on multiple filter elements installed in parallel. Several different approaches, summarized in Table 11.5, can be employed for integrity testing multifilter assemblies.

When multiple filter elements are installed within a larger housing, a statistical diffusion flow rate limit for the multifilter installation can be calculated. This minimizes the risk of passing the filter housing integrity test with one or more out-of-specification filter element. The statistical diffusion limit F_S for a multi-filter assembly can be calculated as follows:

$$F_S = F + \sqrt{n(f_S - f)^2} \tag{11.7}$$

where F_S is the statistical limit for a multifilter installation, $F = nf, n$ is the total number of filters in the assembly, f is the mean of the population for the filter elements, and f_S is the diffusion specification for a single filter. It is important to note that the multifilter limit, F_S, is only a recommendation and not a specification. Most filter vendors only provide integrity test specifications for individual filter units. If the multifilter diffusion flow rate exceeds F_S, each cartridge should be tested independently and the diffusion flow rate compared

TABLE 11.5
Available Options for Integrity Testing Multifilter Assemblies

Option	Advantages	Disadvantages
Test each filter individually and then install in the housing	Will confirm integrity of each device	• Will not confirm integrity of the filter assembly • Labor intensive • Needs to be carried out pre-sterilization/sanitization
Use engineered housing that can test filters individually or in small banks Install filters in individual housings/capsules and test each filter	• Will confirm integrity of each device • Can be carried out post-sterilization or sanitization	• Engineered housing can be expensive • Time consuming • Can potentially add holdup volume to the system • Too many connections — cause integrity failures
Multiply diffusion specification per filter by the number of filters and use as gas flow rate limit for housing	• Easy to implement • Can be done with automated testers without additional documentation • No chance of false failures due to filters (may still have false failures due to system leaks)	• It is possible to have an out-of-specification filter and still pass the multifilter housing limit • Will necessitate testing individual filters against specs if housing exceeds limit
Multifilter diffusion limit is statistically derived from filter specification and mean diffusion flow rate for a population of devices (see Equation [11.7])	• Significantly reduces the possibility of having an out-of-specification filter and still pass the multifilter housing limit	• Increased risk of false failures • Will necessitate testing individual filters against specs if housing exceeds limit

against manufacturer's specifications in order to determine if there is a true failure.

11.6.2.6 Protein Processing and Product Recovery

Prior to protein processing, a buffer flush is generally recommended in order to displace WFI with the appropriate buffer. The buffer flush can be carried out using the conditions that are employed during protein filtration (same ΔP, TMP or filtrate flux). About 10 l of buffer per m^2 of filter area is a reasonable volume of buffer.

After the buffer flush, the system is ready for protein processing. The protein product should be processed using the process conditions and operating window established during the virus validation studies. For NFF filters, important operating parameters include:

- ΔP (feed-permeate pressure) for constant pressure operation
- Feed flux for constant flow operation
- Maximum allowable flux decline
- Feed concentration, pH, ionic strength, and temperature
- Prefilter and virus filter loading (l/m^2 or g/m^2)
- Volume of buffer flush for protein recovery

Likewise, for TFF filters, the important operating parameters include:

- Cross-flow rate or ΔP (feed-retentate pressure)
- Minimum and maximum permeate flux or transmembrane pressure
- Feed concentration, pH, ionic strength and temperature
- Differential diafiltration concentration
- Filter loading (l/m^2 or g/m^2)
- Final retentate volume or volumetric concentration factor
- Number of diafiltration volumes

In the case of normal flow filters, protein recovery may be enhanced with a buffer rinse. The buffer rinse can be carried out using the conditions that are employed during protein filtration (same ΔP or TMP or filtrate flux). Flush volume depends on the upstream volume of the system and desired protein yield. About 10 l/m^2 is a reasonable flush volume for a well-engineered system.

REFERENCES

1. FDA. Draft points to consider (PTC) in the manufacture and testing of monoclonal antibody products for human use. CBER 1994.
2. PDA. PDA Technical Report No. 41 Virus Filtration 2005.
3. Levy RV, Phillips MW, and Lutz H. Filtration and the removal of viruses from biopharmaceuticals. In Meltzer TH and Jornitz MW, Eds. *Filtration in the Biopharmaceutical Industry.* New York: Marcel Dekker, 1998, pp. 619–646.
4. Perry RH and Green DW. *Perry's Chemical Engineers' Handbook*, 6th ed. New York: McGraw-Hill, 1984.
5. Grace HP. Structure and performance of filter media. *AIChE J.* 1956. 2:307.
6. Hermia J. Constant pressure blocking filtration laws-application to power-law non-newtonian fluids. *Trans. IchemE* 1982; 60:183–187.
7. Hermans PH and Bredee HL. Principles of the mathematical treatment of constant-pressure filtration. *J. Soc. Chem. Ind.* 1936; 55T:1–4.

8. Badmington F, Wilkins R, Payne M, and Honig ES. Vmax testing for practical microfiltration train scale-up in biopharmaceutical processing. *Biopharm* 1995; 8(7):46–52.
9. Bolton G, LaCasse D, and Kuriyel R. Combined models of membrane fouling: development and application to microfiltration and ultrafiltration of biological fluids. *J. Memb. Sci.* in press.
10. Ho C and Zydney AL. A combined pore blockage and cake filtration model for protein fouling during microfiltration. *J. Coll. Inter. Sci.* 2000; 232:389–399.
11. Ireland T, Lutz H, Siwak M, and Bolton G. Viral filtration of plasma-derived human IgG. *Biopharm. Int.* 2004; 17:38–44.
12. Siwak M. Process for prefiltration of a protein solution. U.S. Patent US20030201229 A1, 2003.
13. Lutz H. Membrane filtration with optimized addition of second liquid to maximize flux. U.S. Patent 5,597,486. 1997.
14. CPMP EMEA. Note for guidance on virus validation studies: the design, contribution and interpretation of studies validating the inactivation and removal of viruses 1996. CPMP/BWP/268/95.
15. Brorson K, Krejci S, Lee K, Hamilton E, Stein K, and Xu Y. Bracketed generic inactivation of rodent retroviruses by low pH treatment for monoclonal antibodies and recombinant proteins. *Biotechnol. Bioeng.* 2003; 82:321–329.
16. Bolton G, Cabatingan M, Rubino M, Lute S, Brorson K, and Bailey M. Normal flow virus filtration: detection and assessment of the endpoint in bioprocessing. *Biotechnol. Appl. Biochem.* 2005; 42:133–142.
17. Previsani N, Fontana S, Hirt B, and Beard P. Growth of the parvovirus minute virus of mice MVMp3 in EL4 lymphocytes is restricted after cell entry and before viral DNA amplification: cell-specific differences in virus uncoating *in vitro*. *J. Virol.* 1997; 71:7769–7780.
18. Hirasaki T, Noda T, Nakano H, Ishizaki T, Manabe S, and Yamamoto N. Mechanisms of removing Japanese Encephalitis Virus (JEV) and gold particles using cuprammonium regenerated cellulose hollow fiber (i-BMM or BMM) from aqueous solution containing protein. *Polym. J.* 1994; 26:1244–1256.
19. Omar A and Kempf C. Removal of neutralized model parvoviruses and enteroviruses in human IgG solutions by nanofiltration. *Transfusion* 2002; 42:1005–1010.
20. Carter J and Lutz H. An overview of viral filtration in biopharmaceutical manufacturing. *Eur. J. Parenter. Sci.* 2002; 7:72–78.
21. Phillips MW. Integrity testing virus-retentive membranes. *Proceedings from the PDA Fourth International Congress*. Vienna, Austria, 1996.

12 Product Recovery from Transgenic Sources

Chenming (Mike) Zhang and
Kevin E. Van Cott

CONTENTS

12.1 INTRODUCTION

Transgenic sources (plants and animals) for biopharmaceutical production offer numerous advantages over bioreactor-based production, and the most important ones are the ease and the associated low cost for large-scale production. It has been estimated that the cost of producing a recombinant drug from transgenic plants is only 10 to 20% of the cost of using fermentation [1]. For example, depending on the scale, the total production cost of monoclonal antibodies (MAbs) via mammalian cell culture ranges from \$100/g to more than \$300/g [2], among which 40% are incurred for production of the protein and the rest for its recovery and purification. In contrast, the estimated cost is for producing a drug in a transgenic crop 12 to \$15/g [1]. Transgenic plants offer additional advantages, such as plants do not carry human pathogens; this is an advantage not only from the process economy point of view but also from a regulatory and safety perspective. Processes for protein purification from expression systems such as cell culture and transgenic animal product usually include extra steps for clearance of pathogens, such as chromatographic or membrane separation methods [3–5], which add to the total process cost. Many proteins have been targeted for production in transgenic sources; these proteins range from monoclonal antibodies (MAbs), enzymes, blood proteins, to various subunit vaccines.

The use of plants for recombinant protein production has been ongoing for almost two decades [6, 7]. The plant species used include tobacco, carrot, tomato, maize, potato, alfalfa, soybean, rice, canola, and spinach, and the list continues to grow. Each of these plants has specific advantages. Some commonly used plants have been reviewed in great detail elsewhere [8, 9]. However, the appearance of the unapproved Starlink *Bacillus thuringiensis* (Bt) toxin from recombinant corn in the food supply has caused tremendous public outcry. Hence, it is increasingly obvious that, for producing pure recombinant proteins (edible vaccines are excluded), among all the plants, tobacco as a nonfood and nonfeed crop will probably face the least regulatory resistance and public scrutiny; and thus it may be the most promising plant candidate in future plant-based biopharmaceutical production.

The list of recombinant proteins expressed in transgenic plants is growing day by day. However, strictly speaking, only two plant-made recombinant proteins, β-glucuronidase [10] and avidin [11], are currently on the market, and both are research proteins produced by ProdiGene (College Station, TX) and are marketed by Sigma Chemical Company (St. Louis, MO). Significant efforts are being devoted to developing plants as viable systems for biopharmaceutical production, although most of the efforts are still concentrated in academic environments or closely associated start-up companies. As summarized in Table 12.1, these efforts can be testified to by the numerous biopharmaceutical candidates that are in various stages of development and clinical trials.

Despite the promises and the expanded efforts, there are still no plant-made-biopharmaceuticals (PMBs) on the consumer market. There are two primary reasons for this lack of success. One is at the molecular level, that is, how to construct the recombinant gene cassette, including a promoter and a targeting sequence, to obtain the highest recombinant protein accumulation. In addition, the difference in the glycosylation patterns between plants and mammals may be responsible for the lack of success, particularly for glycoproteins (proteins that have attached sugar chains). Different strategies have been studied to humanize

TABLE 12.1
Selected Plant Biotechnology Companies and Their Plant-Made-Biopharmaceutical (PMB) Candidates Currently in Development

Major PMB Candidates	Primary Plant Host	Stage of Development	Responsible Company	Company Info
1. Alpha-galactosidase A 2. Lysosomal Acid Lipase	Tobacco	1. Preclinical studies completed 2. Preclinical studies completed	Large-Scale Biology Corporation	Vacaville, CA www.lsbc.com
Human monoclonal antibodies (mAbs)	Corn	Unknown	Epicyte Pharmaceutical Inc.[a]	San Diego, CA www.epicyte.com
Unknown	Moss	Unknown	Greenovation Biotechnology GmbH	Freiburg, Germany www.greenovation.com
None	Various	Contract manufacturer	Dow Chemical	San Diego, CA www.dowplantpharma.com
None	Tobacco	Contract manufacturer	Phytomedics Inc.	Dayton, NJ www.phytomedics.com
1. Various vaccines 2. Aprotinin 3. Trypsin	Corn	1. Various animal trials 2. Marketed for research 3. Marketed for research	ProdiGene	College Station, TX www.prodigene.com
1. Lactoferrin 2. Lysozyme	Rice, barley	Unknown	Ventria Bioscience	Sacramento, CA www.ventria.com

(Continued)

TABLE 12.1
Continued

Major PMB Candidates	Primary Plant Host	Stage of Development	Responsible Company	Company Info
1. Insulin 2. Apolipoprotein	Seeds from undisclosed plants	Both are in research phase	SemBioSys	Calgary, AB, Canada www.sembiosys.ca
None	Potato	Contract manufacturer	Axara Consulting	Frechen, Germany www.axara-consulting.com
Alpha interferon	Lemna (Duckweed)	Ongoing clinical trial	Biolex	Pittsboro, NC www.biolex.xom
1. Human serum albumin 2. Interferon	Tobacco, chloroplast	Research phase	Chlorogen	St. Louis, MO www.chlorogen.com
1. Gastric lipase 2. Human lactoferrin	Corn, tobacco	1. Phase I and IIa have been completed 2. Phase I completed	Meristem Therapeutics	Clermont-Ferrand, France www.meristem-therapeutics.com
1. Antibody (CaroRx) 2. Antibody (RhinoRx)	Tobacco	1. Phase II clinical trial 2. Phase I clinical trial	Planet Biotechnology, Inc.	Hayward, CA www.planetbiotechnology.com
Monoclonal antibodies	Alfalfa	Research phase	Medicago, Inc.	Sainte-Foy, Quebec, Canada www.medicago.com

[a]The company was acquired by Biolex in May 2004.

plant glycans, and a recent paper by Gomord and Faye [12] provides a good overview of those strategies. The other challenge is the engineering challenge, that is, how to economically recover and purify the expressed protein, which will be the focus here.

Currently, the benchmark techniques used for protein purification from transgenic plants include various chromatographic methods [13–17]. However, to recover and purify a large quantity of protein from plant material, directly utilizing chromatographic methods would be economically prohibitive. For example, to extract 1 kg of recombinant protein from a batch of transgenic plant material (for earlier stages of clinical trials), which contains 3%wt total soluble protein (TSP) and the protein accumulation level is 1% of the TSP, if 10:1 extraction buffer to biomass ratio is used and the extracting efficiency is 90%, 3,700 kg of plant material and subsequently 37,000 kg (or liter) of plant extract will need to be processed. Since the plant extract, particularly green tissue extract, is extremely heterogeneous containing fine particulates, directly using membrane-based techniques could be inefficient, as would be using chromatographic columns. Expanded bed chromatography has been shown to be effective for protein recovery from plant sources such as canola [18, 19], but its for product recovery directly from green tissue homogenate will likely be difficult. Thus, developing nonchromatographic techniques, at least for protein recovery and the early stages of protein separation and concentration, will probably be the key to answering the engineering challenge for molecular farming of recombinant proteins from transgenic plants. Therefore, we will focus on discussing how to apply two commonly used nonchromatographic techniques, aqueous two-phase extraction (ATPE) and polyelectrolyte precipitation, in protein recovery and purification from transgenic plants, particularly leafy crops like tobacco. We anticipate that these techniques can be readily applied to other plant systems for protein separation.

Production of recombinant proteins in transgenic animal milk has also been ongoing for about two decades [20,21]. Despite this long research and development time, no transgenic animal-derived proteins have been approved for sale in the United States. Companies that are currently developing transgenic animal-derived proteins include GTC Biotherapeutics (Framingham, MA), Pharming Group N.V. (Netherlands), and Progenetics LLC (Blacksburg, VA). GTC Biotherapeutics recently announced that their recombinant anti-thrombin III (ATryn®) produced in transgenic goat milk is undergoing Market Authorization Application (MAA) review with the European Medicines Evaluation Agency (EMEA), and that the United States FDA has agreed to initiation of clinical trials for patients with hereditary anti-thrombin III deficiency. Other proteins in development by GTC Biotherapeutics include human alpha-1-antitrypsin, human albumin, a malaria vaccine protein, and monoclonal antibodies. Pharming Group N.V. is developing C1 Inhibitor protein in transgenic rabbit milk, and recombinant human fibrinogen and lactoferrin in transgenic cattle. Progenetics LLC is developing recombinant human Factor IX and Factor VIII in transgenic pigs.

Several technical challenges remain to be solved before transgenic animals are accepted as routine bioreactors for production of therapeutic proteins. Possibly one of the more significant technical challenges is the evaluation of how different transgenic animal species posttranslationally modify complex proteins [22]. Compared to immortalized animal cells grown in stainless steel bioreactors (e.g., CHO, HK293, etc.), there has not been as great an effort to perform this characterization on the mammary epithelial cells of different livestock species. Another aspect important in the development of transgenic animals is the upfront cost and time involved in generating founder animals and evaluating their offspring for suitability as members of a production herd. For goats and pigs, the generation times are approximately 1 year, but for cattle it is 2 years. Thus there is a degree of risk involved in beginning a protein production effort [23]. However, once a production herd is established, capital equipment and operating costs are expected to be lower than costs for mammalian cell culture bioreactors. As we will review below, the technical and financial aspects of purification are not significantly different from other biotechnology feedstocks. At the time of publication, GTC Biotherapeutics is the only company to submit a transgenic animal-derived protein for regulatory approval. On February 22, 2006 the Committee for Medicinal Products for Human Use (CHMP) of the European Medicines Agency (EMEA) issued a negative opinion for ATryn. Major concerns included (1) not enough patients enlisted in the clinical trial, (2) insufficient long-term data on patients' immune response to ATryn or potential contaminating goat milk proteins, and (3) changing of the purification process during the clinical trial – a nanofiltration viral reduction step was added to the process after material was made for the clinical trial. GTC Biotherapeutics has stated that they will appeal the decision and has provided initial responses to these concerns in corporate conference calls after the EMEA's decision. We expect that the lessons learned during this process may enable a faster development and approval time-line for the other proteins being commercially developed.

12.2 INITIAL RECOVERY AND SEPARATION OF RECOMBINANT PROTEIN FROM TRANSGENIC LEAFY CROP

The economic pressure on the production of biopharmaceuticals is ever increasing, and this pressure lies in the demand for increasing product yield, reducing the process time, and cutting down of running costs and capital expenditure during downstream processing [24]. The overall economy of a protein purification process is largely dependent on the cost of the protein recovery from biomass and the initial purification steps. Considering the characteristics of protein extract from transgenic leafy crops, such as tobacco, in order to lower the overall process cost of a protein purification process, it is important to develop techniques that not only can separate and concentrate a target protein from the native plant components but also are capable of performing protein separation with the presence of fine solid particulates.

Aqueous two-phase extraction has been shown to be such a technique. ATPE has a low intrinsically associated running cost, and scale-up processing is readily obtainable from lab-scale experiments. More importantly, the protein concentrate from this method has been shown to be compatible with subsequent chromatographic techniques, such as ion exchange chromatography [25], immobilized metal affinity chromatography [26], and size exclusion chromatography [27]. In addition, APTE offers the potential to stabilize proteins that are vulnerable to protease degradations [28]. On the other hand, polyelectrolyte precipitation is a technique straightforward for scale-up and has a low associated cost. This technique may need to work with clarified protein extract, but compared to chromatographic methods, it still represents a significant cost reduction.

12.2.1 ATPE in Protein Separation from Transgenic Tobacco

12.2.1.1 Background and Practical Considerations

Aqueous two-phase extraction has been widely used for protein recovery and purification [29,30]. ATPE has the potential to produce a concentrated and purified product in one step, replacing a number of steps in conventional downstream protein processing such as product recovery, solid clarification by filtration or centrifugation, and initial purification [31]. Thus, it can potentially significantly reduce the cost of a protein purification process. This technique has been successfully used in both lab- and pilot-scale protein separation [31,32].

In ATPE, two immiscible phases are formed when a polymer such as poly(ethylene glycol) (PEG) is mixed with another polymer such as dextran, or salts such as potassium sulfate in particular concentrations. PEG–salt–water two-phase systems have certain advantages over polymer–polymer–water systems, such as low viscosity and lower cost [33,34], and thus are most promising in gaining widespread industrial applications in protein separations. In ATPE, a protein's distribution between the two phases depends on its surface properties, such as charge and hydrophobicity and the physicochemical properties of the two phases [35]. Although a preexisting-phase diagram is not absolutely necessary for developing a proper ATPE system for protein separation, it can provide valuable information for process development and experimental design. If a phase diagram is not available but is critical in process development, the

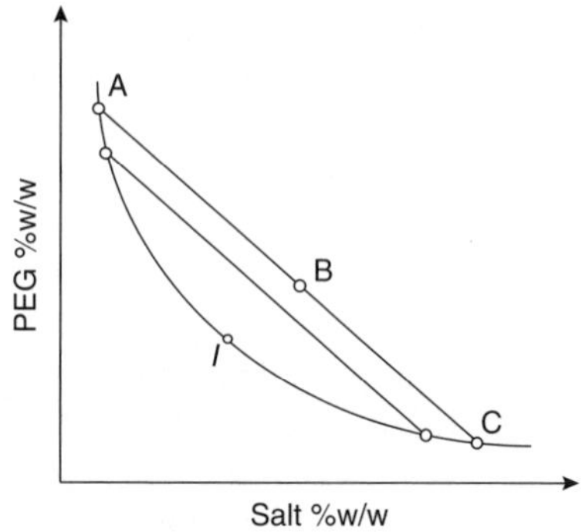

FIGURE 12.1 An aqueous two-phase system phase diagram. The binodial curve separates the homogenous region from the regions where two phases are immiscible. Point B represents the total composition of the mixture, which separates into two phases. The compositions of the two phases are represented by point A, the top polymer-rich phase, and point C, the bottom salt-rich phase (for a polymer-salt water two-phase system). Points A and C, are called nodes are located on the binodial curve. Line AC is called a tie line. When the tie line decreases in length, the two nodes will converge at point D, which is called the critical point.

commonly used and quick method to generate the phase diagram is the "cloud point" method [36]. Many phase diagrams have been reported in the literatures [37]; a simple phase diagram and its important features are illustrated in Figure 12.1. The partition coefficient of a protein, K_P, in an ATPE system is defined as

$$K_P = \frac{C_{top}}{C_{bottom}} \tag{12.1}$$

where C_{top} is the protein concentration in the top phase at equilibrium, and C_{bottom} is the corresponding protein concentration in the bottom phase. In most cases, the protein partition coefficient depends on the overall phase composition. However, if the phase composition falls anywhere on a particular tie-line (line AC in Figure 12.1), the protein partition coefficient will be the same. Nevertheless, changing the overall phase composition on a particular tie-line is important in practical process design for protein recovery and purification.

When the overall phase composition (point B in Figure 12.1) varies on a tie-line, the amount (volume) of the top- or bottom-phase varies accordingly. The volume of the individual phase is directly proportional to the length of the segment on the tie-line: top phase to BC and bottom phase to AB, respectively. The ratio of the length of BC to AB defines the phase ratio, ϕ,

$$\phi = \frac{\text{Volume of top phase}}{\text{Volume of bottom phase}} = \frac{\text{BC}}{\text{AC}} \tag{12.2}$$

When point B moves towards point C on line AC, the top-phase volume will decrease. If the protein is targeted to be partitioned in the top phase, the protein recovery or yield will also decrease, because

$$\text{Recovery} = C_{\text{top}} \cdot (\text{Volume of top phase})/M_0 \tag{12.3}$$

where M_0 is the mass of the target protein in the initial sample. However, the loss of protein recovery is not necessarily an engineering failure during protein separation. What matters to the process development engineers/scientists is the combination of protein recovery and the purification factor. As defined by Balasubramaniam et al. [31], in parallel to protein separation in chromatography, the selectivity of a target protein, α, over the contaminant proteins is defined by

$$\alpha = K_P/K_C \tag{12.4}$$

where K_C is the lumped partition coefficient of all proteins excluding the target protein. Since the protein partition coefficients are the same on a tie-line, the selectivity is also a constant on the tie-line. This offers opportunities to maximize the protein enrichment factor with limited sacrifice protein recovery [31]. The relationship between the recovery and the enrichment factor for a target protein is illustrated in Figure 12.2.

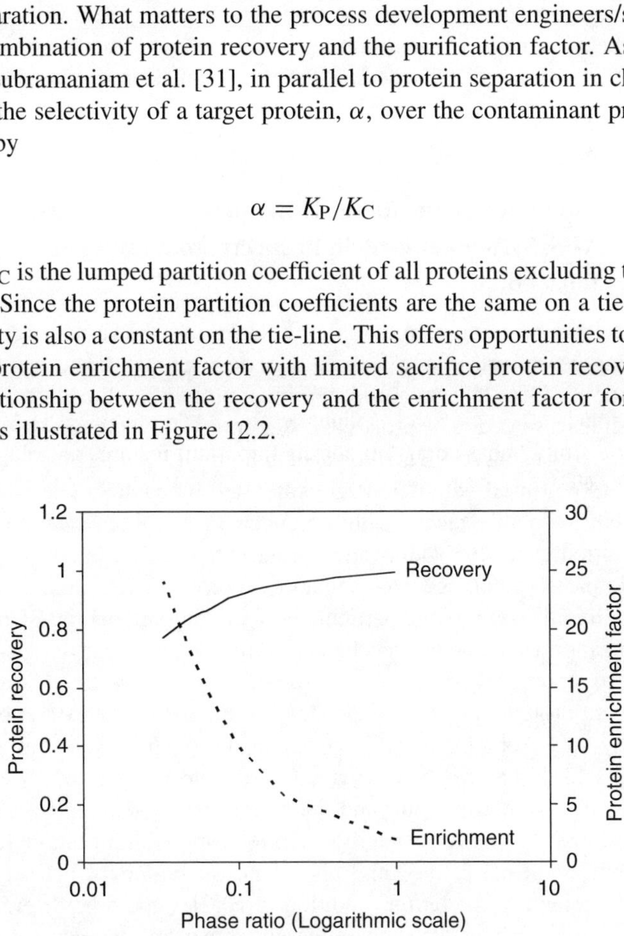

FIGURE 12.2 The relationship between protein recovery and the enrichment factor. A selectivity of 57 is used for the separation of a model protein, lysozyme, from tobacco extract in particular ATPE systems. Sacrificing tolerable protein recovery could result in a significant increase in the protein enrichment factor.

Furthermore, the overall performance of an ATPE system can be optimized by adjusting the system factors such as the type of polymer, the type of phase forming salt, ionic strength, and pH for the separation of a target protein. The fact that many factors can be adjusted to optimize the recovery of a protein presents opportunities and challenges at the same time. Since the protein partitioning mechanism in ATPE is still not well understood, process development is still largely trial and error, and there is hardly any chance to extrapolate one method from a particular system to another. To obtain the optimized conditions for a target protein separation from a particular system is thus difficult and can be extremely time-consuming. However, with the assistance of the Design of Experiment (DOE) method, the amount of effort may be significantly reduced.

12.2.1.2 Experiment Protocol for Developing an Optimized ATPE System for Protein Recovery from Transgenic Tobacco

Although it has been reported that alfalfa leaves may be stored twelve weeks after drying before processing without experiencing significant target protein degradation [38], for large-scale recombinant protein production from a leafy crop, fresh leaves need to be processed. A common flow of unit operations for protein recovery using ATPE is illustrated in Figure 12.3. The freshly cut leaves first need to be rinsed with deionized water and blown dry. The dried leaves are then cut to smaller pieces with a blender (or a leaf shredder for a large-scale operation), and after adding appropriate extraction buffer, the large pieces of leaf biomass are homogenized to facilitate protein extraction, producing a mixture of liquid and fine solid particulates, including cell debris. Blending and homogenization exert violent mechanical forces to the plant tissue, but there has been no report of detrimental effect on protein. It is worth mentioning that the extracted protein content is dependent on the pH of the extraction buffer, and it may vary from 1.0 to 1.6% (w/w) of the fresh biomass, as shown in Figure 12.4 [31]. The pH of the extraction should be selected based on the properties of the target protein, and the extract buffers should be made with the highest possible buffer capacity, such as using sodium citrate and citric acid for buffers at pH 3 to 5, and phosphate for buffers at pH 6 to 8 [31]. The ionic strength of the buffer (addition of NaCl) should be determined for individual cases. Our results showed that the effect of NaCl is pH dependent, and, at pH 7, the addition of NaCl does not significantly improve the extraction of native tobacco proteins (unpublished data). Moreover, phenolic compounds are abundant in plants, particularly in tobacco, and they have been known to interact with proteins to form less soluble and even insoluble complexes.

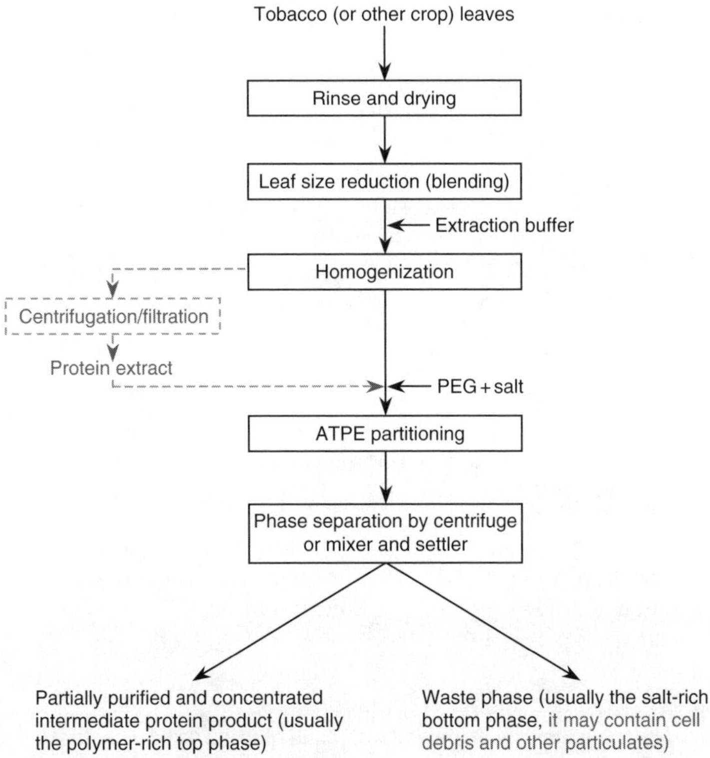

FIGURE 12.3 Unit operations (in boxes) and material flow for protein recovery by ATPE from leaf biomass. Centrifugation/filtration after homogenization and the intermediate protein extract may be omitted (in gray). However, if phase-forming agents are directly added to the homogenate, separation of the phases after ATPE may be more complicated.

In order to prevent the influence of phenolic compounds, additives such as soluble or insoluble polyvinylpyrrolidone (PVP) can be added to the extraction buffer [39].

After separating the solid residue from the supernatant, the phase forming agents including PEG and a salt such as potassium phosphate, ammonium sulfate, or sodium sulfate, may be added either as solid or stock solutions to the desired overall composition. The solution is then thoroughly mixed, and the phase separation can be expedited by centrifugation. A mixer-settler type of equipment, which relies on gravity for phase separation, may be used here for any scale of operation, particularly for large-scale operations [40]. However, phase separation by gravity may take 30 to 90 min and even longer to ensure complete phase separation [41]. Then, protein concentrations may be analyzed

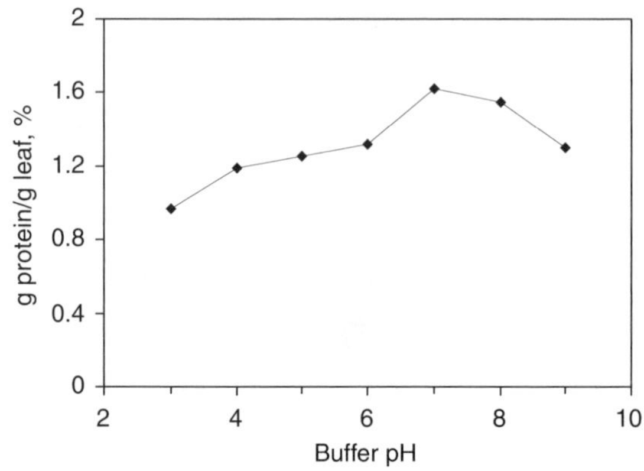

FIGURE 12.4 Percentage of g protein extracted per g tobacco leaf (flue-cured) vs. buffer pH. Buffers used: pH 3 to 5, 50 m*M* sodium citrate-citric acid; pH 6 to 8, 50 m*M* sodium phosphate; pH 9, 50 m*M* Tris base.

to determine the partition coefficients. The following sections describe in detail the separation of a model protein, egg white lysozyme, from tobacco extract by ATPE in lab-scale.

12.2.1.2.1 Tobacco Extract Preparation

Tobacco leaves cleaned by deionized water and blotting dried were first cut using a Warring blender, and the required amount of blended leaves was weighed into a 50-ml conical tube. Buffer with appropriate pH in the ratio of 10 ml for every gram of leaf was added to the conical tube (buffer to biomass ratio $= 10 : 1$). The leaves were then homogenized using a Power Gen 700 (Fisher Scientific). The homogenized mixture was allowed to stand for 20 min at room temperature ($\sim$20°C) and then centrifuged at 4°C, 12857 $\times$ *g* for 15 min. The extract was recovered by decanting the supernatant into a new conical tube. The volume of the extract recovered was recorded. The extract was then filtered using a syringe filter (45 μm) before ATPE studies. The protein concentration of the cleared extract can be determined by Bicinchoninic acid assay (BCA) [42] or Bradford assay [43] using bovine serum albumin as the standard. Lysozyme concentration in the extract can be determined by lysozyme activity assay [31].

12.2.1.2.2 ATPE Experiments and Determination of the Optimal Conditions for Lysozyme Recovery by DOE Methods

PEG-sodium sulfate systems were investigated for lysozyme recovery and separation from tobacco extract [31]. The stock solutions used were PEG

TABLE 12.2
Factors and Levels for Factorial Study of Lysozyme and Tobacco Protein Partitioning in ATPE

Factor	Low Level	High Level
PEG molecular mass	3400	8000
PEG concentration (% w/w)	10	15
Salt concentration (% w/w)	13	18
NaCl concentration (*M*)	0.1	1.2
pH	6	8

50% (w/w) in water and 30% (w/w) sodium sulfate in a buffered solution at appropriate pH. Systems of 5 g total mass containing the required amounts of PEG, sodium sulfate, sodium chloride, and tobacco extract were prepared from appropriate stock solutions. For factorial study, all the systems contained the same amount of tobacco extract, and the total mass was balanced by addition of varying amount of water. The 5 g systems were thoroughly mixed, then centrifuged at 1157 $\times$ g at room temperature for 10 min to expedite the phase separation. The bottom phase was carefully removed by aspiration and weighed. The mass of the top phase was calculated by subtracting the bottom phase from the total (5 g). In order to calculate the phase volumes, the density of each phase was estimated by measuring the mass of 100 μg of each phase in a preweighed microcentrifuge tube. The total protein concentration in each phase was determined by BCA and the lysozyme by activity assay.

To screen for the most important factors for lysozyme separation from tobacco extract, factorial study was carried out. Since interactions among the different factors are insignificant [31], a half factorial study was carried out, including the factors and correspondent levels as shown in Table 12.2. Sixteen experiments (2^4) were conducted, and the analysis response is the lysozyme selectivity. Alternatively, it is equally effective to carry out experiments evaluating the partitioning coefficients of tobacco native protein and lysozyme (target protein) separately and then combine the results [31]. From the statistical analysis (MINITAb, version 13), sodium sulfate and sodium chloride concentrations were determined to be the most important factors for further optimization.

In the response surface study, three of the five factors studied above were held constant, and their specific values were determined from the main effect plots obtained in the factorial study and by considering the limits associated

TABLE 12.3
Central Composite Design for Response Surface Study of the Effect of Sodium Sulfate Concentration and Sodium Chloride Concentration on the Selectivity of Lysozyme from Tobacco Extract by ATPE

	Coded Levels		Real Values	
Run Order	Na_2SO_4 Concentration	NaCl Concentration	Na_2SO_4 Concentration (%w/w)	NaCl Concentration (*M*)
1	−1	−1	9	0.4
2	1	−1	15	0.4
3	−1	1	9	1.4
4	1	1	15	1.4
5	−1.414	0	7.8	0.9
6	1.414	0	16.2	0.9
7	0	−1.414	12	0.2
8	0	1.414	12	1.6
9	0	0	12	0.9
10	0	0	12	0.9
11	0	0	12	0.9
12	0	0	12	0.9
13	0	0	12	0.9

PEG molecular mass: 3400
PEG Concentration: 10 %w/w
pH: 7

with the phase diagram, such as the concentrations of PEG. The PEG molecular mass was set as 3400, its concentration at 10%, and system pH at 7.0. Thirteen experiments according to central composite design were conducted, and they correspond to four cube, four axial, and five center points, as shown in Table 12.3. Sodium sulfate concentration levels were chosen as far apart as possible based on the phase diagram, and the sodium chloride levels were altered (compared with that in Table 12.2) to see if a higher concentration would increase the selectivity. The response surface is shown in Figure 12.5. The conditions at which lysozyme selectivity over the native tobacco protein was highest were determined using the response optimizer provided by MINITAB (Version 13) software.

As shown in Figure 12.5, the global solution of the response surface study predicted a lysozyme selectivity value of 57 when the sodium sulfate concentration was maintained at 16.2% w/w and the sodium chloride concentration

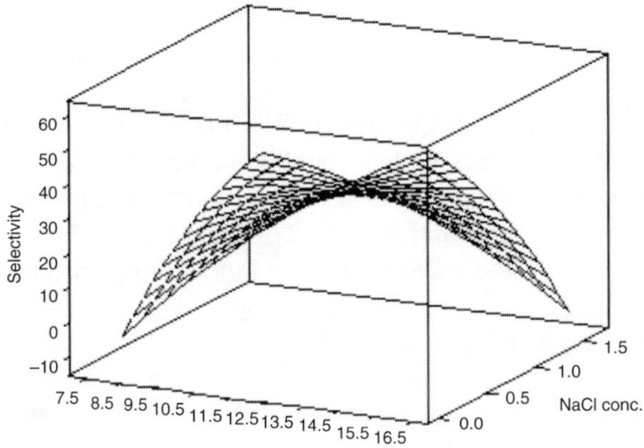

FIGURE 12.5 Response surface study of egg white lysozyme separation from tobacco extract by ATPE. Lysozyme selectivity was used as the response estimate. Other conditions used in the study: PEG 3400 at 10%w/w and pH 7.

at 0.19 *M* with PEG 3400 at 10%w/w at pH 7. Two experiments at the globally selected conditions were carried out and yielded an average selectivity of 47. The difference between the tested and predicted selectivity values probably is caused by the inaccuracy of the lysozyme activity assay for the bottom phase. Since almost all lysozyme is partitioned into the top phase, the activity assay is not sensitive enough to accurately determine the amount of lysozyme remaining in the bottom phase. Besides the lysozyme activity assay, another factor that could contribute to the difference between the theoretically predicted and the experimental selectivity values is the amount of extract added to the experimental systems.

12.2.2 Protein Purification from Transgenic Tobacco by Polyelectrolyte Precipitation

12.2.2.1 Background and Practical Considerations

Protein recovery by precipitation is often used in the early stages of protein purification processes because it is a simple unit operation, relatively inexpensive and straightforward for scaling-up. Although chromatography has been the workhorse for protein purification, precipitation can be used to fractionate the protein sample and improve the efficiency of chromatography [44]. In fact, as much as 80% of published protein purification protocols have been reported to include at least one precipitation step [45]. Among the various methods of precipitation such as salting out, addition of organic solvents or nonionic polymers such as PEG, affinity precipitation, isoelectric precipitation, and polyelectrolyte precipitation, protein precipitation by polyeletrolyte may be the most attractive one because it is more economical than other specific precipitation methods meanwhile more selective than the nonspecific methods. Polyelectrolyte precipitation is based on the electrostatic interaction between a charged polymer and the oppositely charged proteins, and, more importantly, the precipitated protein maintains its bioactivity and can be readily resuspended in aqueous solutions at higher concentrations [46]. For a more general discussion about protein precipitation, the readers can refer to a recent publication by Kumar et al. [45].

Even though it seems quite straightforward, developing a successful method for a target protein precipitation by polyelectrolyte from transgenic plant sources is still challenging. Once decided what type of polyelectrolyte is to be used for a target protein (selecting the polymer with opposite charge), a significant amount of work needs to be done to screen for the particular polymer and to determine the chain length and the dosage of the polymer in the operation [47–49]. The most commonly used polyanionic agents to precipitate positively charged protein (at certain pH) are polyacrylic acid (PAA), Glass H (polyphosphate), and carboxymethylcellulose (CMC); the most widely used polycationic polymer is polyethyleneimine (PEI). All polymers are commercially available. Moreover, the complex nature of the protein extract from transgenic sources may complicate the development of protein precipitation, as Zaman et al. [47] have reported that lysozyme precipitation by PAA from canola extract is much more inefficient than from egg white extract [50].

Protein precipitation from green leaf extract is further complicated because of the presence of polyphenolic compounds. Zhang et al. [51] showed that none of the three polymers used (PAA, CMC, Glass H) were effective to precipitate egg white lysozyme at pH 7, presumably because of the interaction between polyphenolic compounds and lysozyme, which could induce the formation of more hydrophobic and relatively more acidic complexes (lower pI) [52]. However, as shown in Figure 12.6, the effectiveness of PAA–lysozyme precipitation can be improved tremendously if the extract was obtained at a lower pH at 5, at which the amount of polyphenolic compounds extracted is thought to be decreased. Even with improved efficiency, however, the recovery of lysozyme from tobacco was low at only about 53%. The addition of polyvinylpyrrolidone (PVP) in the extraction buffer may further improve the efficiency by removing some phenolics, but it is yet to be seen whether or not the recovery can be significantly improved to the range of 80 to 90% to make this technique a feasible one in practical applications.

While, intuitively, it is a common practice to directly engage the target protein in the protein–polymer interaction during precipitation process development, but negative precipitation, that is, keeping the target protein

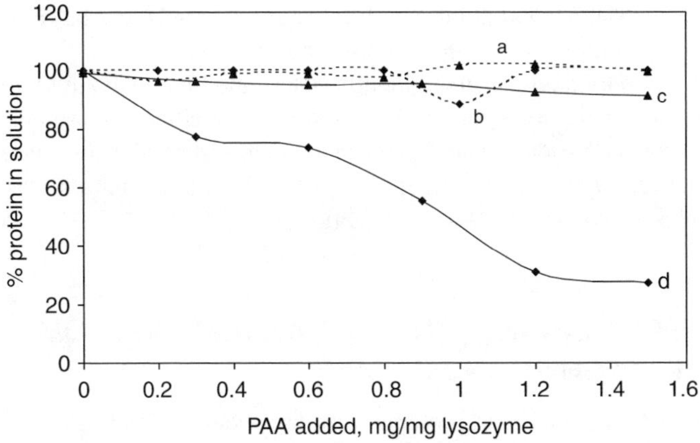

FIGURE 12.6 Lysozyme precipitation by PAA from tobacco extract. The *y*-axis indicates the weight percentage of protein remaining in the solution after precipitation. Line a, total protein at pH 7; b, lysozyme at pH 7; c, total protein at pH 5; d, lysozyme at pH 5. The lysozyme data was normalized to 100% when no PAA was added.

soluble throughout the precipitation process and to precipitate the impurities, should not be overlooked. At times, negative precipitation may well be the method of choice. For example, Jerala et al. [53] used PEI to clear up nucleic acids and of other 90% contaminating proteins during the first step isolation of a basic protein, cysteine proteinase inhibitor stefin B. One advantage of using the negative precipitation method is that one takes no risk of losing some of the target protein in the precipitate during resuspension, which could potentially be problematic. Finally, compared to ATPE, the development of a precipitation protocol is relatively straightforward since there are fewer variables involved.

12.2.2.2 Experimental Protocol for Lysozyme Precipitation by PAA from Tobacco Extract

Tobacco extract at appropriate pH can be obtained by methods outlined in Section 12.2.1.2.1. After determining the protein concentration of the extract, PAA can be added as a stock solution. In order not to dilute the protein solution extensively, the stock solution should be made as concentrate as possible. If the amount of PAA to be added can be weighed accurately for large-scale operation, directly adding solid will be a good alternative. The amount of PAA added can be calculated using the estimated lysozyme in the extract as a base. Figure 12.6 indicates the range of the polymer/lysozyme ratio in the paper by

Zhang et al. [51]. After addition of polyelectrolyte, samples can be vortexed and allowed to precipitate for 15 min or longer and then centrifuged. The amount of protein precipitated can be indirectly determined by analyzing the protein concentration in the supernatant. Because most of the tobacco proteins stay in the solution, high protein enrichment factors can usually be obtained. Zhang et al. [51] reported an enrichment factor of 8 when precipitation was carried out at pH 5 for lysozyme-PAA precipitation.

12.3 RECOMBINANT PROTEIN PURIFICATION FROM TRANSGENIC ANIMAL MILK

Purification of recombinant proteins from transgenic animal milk is in many respects similar to purification of proteins from complex multiphase feedstocks such as blood plasma and disrupted cells. Milk is a multiphase mixture that must first be clarified before it can be passed through a conventional chromatography column. The aqueous phase of milk contains soluble proteins, salts, and low molecular weight carbohydrates. The lipid phase accounts for approximately 3 to 4% of the total milk volume for bovine and caprine milk, and approximately 6 to 8% of total volume for porcine milk. The solid phase consists of casein micelles, somatic cells, and cellular debris. The casein micelles are aggregates of caseins, the most abundant milk protein, and calcium phosphate and other salts. Excellent reviews of milk protein composition from the different animal species used in the production of recombinant proteins can be found in several sources [54–56]. Most recombinant proteins produced in milk to-date partition into the aqueous phase, and thus downstream processing usually begins with steps to separate the solids and lipids.

One of the most important steps in developing an efficient purification process is the design of initial recovery and capture steps — where the feedstock is clarified and concentrated to achieve the first target isolation step. The value of designing a capture and recovery process that is scalable and achieves a significant degree of volume reduction and purification cannot be overemphasized. Most journal articles on transgenic animal-derived proteins are focused on initial purification and characterization of the protein, where bench-scale recovery and capture steps are typically used. These methods can include precipitation of caseins by lowering the pH to about 4.5 and centrifugation. In other cases, the casein micelles were solubilized by the addition of EDTA, which chelates the calcium of the micelles and breaks up the micelles [57]. Centrifugation can then be used to remove the lipids. An example of a complex recovery and capture process was reported by Drohan et al. [58] for purification of recombinant Protein C from transgenic pig milk. After using centrifugation to skim the milk, a series of polyethylene glycol precipitation steps were used for volume-reduction and purification before chromatographic processing. The overall yield of the polyethylene glycol precipitation steps was approximately 50%. A process such as this, that includes multiple precipitation/centrifugation/solubilization steps would be time-consuming, capital equipment intensive, and labor intensive if it were to be scaled-up to the hundreds of grams to kilogram production scale. However, for producing gram quantities at the bench-scale, this process was effective for purifying and characterizing recombinant Protein C in the early phases of protein characterization.

New recovery and capture processes have been developed in the past decade that represent dramatic improvements in scalability. Morcol et al. [59] have published a method based on the initial disruption of the casein micelles by adding EDTA to liberate the protein product that is partitioning within the micelles. This procedure is followed by the reprecipitation of caseins by addition of colloidal calcium phosphate particles, which are reported to selectively precipitate the caseins and not the protein product. This methodology was demonstrated with four model proteins (human albumin, bovine albumin, human alpha-1-antitrypsin, and human insulin), and $\geq$90% yields were obtained in the clarified supernatant for each protein. The particles were removed by centrifugation in the published work, and it is conceivable that a filtration step or expanded bed chromatography (see below) could also be used to make the process more amenable to scale-up.

Aqueous two-phase extraction has also been reported as a potential capture step for proteins in transgenic milk. Cole et al. [60] were able to obtain an enriched fraction of recombinant Protein C from transgenic pig milk by adjusting pH, polyethylene glycol molecular weight, and ammonium sulfate concentration. The caseins precipitated at the interface of the two phases. Even though the recombinant Protein C existed as a collection of subpopulations that varied in the nature and extent of glycosylation and γ-carboxylation, it partitioned into one phase. Capezio et al. [61] used ATPE with a model mixture of individual whey proteins (no caseins), and found that polyethylene glycol molecular weight and pH were significant factors in the enrichment of spiked alpha-1-antitrypsin.

The use of expanded bed chromatography in recovery and capture offers the potential for solids separation combined with a significant degree of purification. GE Healthcare (formerly Amersham Biosciences) offers expanded bed matrices with a variety of ion exchange and affinity ligands. Published reports that developed expanded bed chromatographic methods for processing transgenic milk have focused primarily on using the process to remove precipitated solids from the feed stream. Degener et al. [62] used the addition of zinc to precipitate caseins, and then selectively adsorb functionally active recombinant Protein C subpopulations onto the sorbent; inactive Protein C subpopulations were co-precipitated with the caseins. Ozyurt et al. [63] also used zinc precipitation combined with expanded bed chromatography to capture antithrombin III from goat milk. This process is somewhat analogous to the calcium phosphate-induced reaggregation of caseins presented by Morcol et al. [59] and appears to be generally applicable to the capture and recovery of proteins from different species. As demonstrated by Degener et al. [62], careful optimization of precipitation conditions can also be used to significantly enhance the functional purity of the expanded bed product stream. In the preceding cited reports, the milk was clarified by addition of EDTA and skimmed by centrifugation prior to expanded bed adsorption. The combination of expanded bed adsorption and delipidation of milk was reported by Gardner [64]. Removal of milk lipids was performed by first incubating lipophilic hydrogel chromatographic matrices (Sephadex LH-20 or hydroxyalkoxypropyl dextran) with whole milk, and then loading the milk/sorbent mixture onto the expanded bed column. The product protein adsorbed to the expanded bed matrix, and the smaller lipophilic sorbent particles with adsorbed lipids flowed out the top of the column during loading.

While the above recovery and capture methods have their place in process development, possibly the most promising technology based on ease of scale-up and technical transfer to cGMP manufacturing processes is membrane filtration. Several entities have developed membrane materials and the processing configurations and conditions to remove milk lipids and casein micelles while concentrating and purifying the product protein. Recently, Parker et al. [65] reported a dual tangential flow filtration process for recovery of recombinant human α-fetoprotein from transgenic goat milk. Whole milk was first passed through a 0.2 μm ceramic microfilter that removed casein and lipid particles, and then the product protein was concentrated with a 30 kDa nominal molecular weight cutoff polymeric membrane. Georges Belfort's group at Rensselaer Polytechnic Institute has published several articles modeling membrane processing of milk [66–68]. They confirmed their models experimentally by processing transgenic goat milk containing monoclonal antibodies and obtained yields up to 95%, and they also found that using a helical hollow fiber module resulted in a threefold improvement in performance over linear hollow fiber modules [69]. On the commercial side, NCSRT Inc. (Apex, NC) has developed cross-flow and tangential-flow membrane filtration systems specifically for large-scale processing of transgenic milk.

Once the product protein is recovered from the transgenic milk, further downstream processing is generally no different from recombinant proteins derived from mammalian cell culture. Pathogen reduction steps are incorporated to reduce the risk of viral transmission, just as they are for plasma-derived proteins or recombinant proteins from mammalian cell culture [70,71]. Pathogen reduction processes include solvent/detergent treatment [72], gamma-irradiation [73], and nanofiltration [74]. Proactive and preventative methods to reduce the risk of pathogen transmission are recommended in the *Points to Consider in the Manufacture and Testing of Therapeutic Products for Human Use Derived from Transgenic Animals* document from the FDA. These include maintaining specific pathogen-free (SPF) production herds at more than one geographical location, and maintaining a high level of biosecurity each production herd. Biosecurity measures include managing people flow through the facility, managing animal flow, incorporation of sentinel animals and routine testing for known pathogens, managing potential insect/rodent infestations, and carefully controlling animal feed sources. These methods are analogous to the characterization and biosecurity measures taken for maintenance of Master and Working Cell Banks in the mammalian cell culture and fermentation industries. The combination of these measures results in multiple barriers to potential pathogen transmission by the purified protein product.

The downstream purification and polishing will be highly specific to the individual protein. If the majority of the caseins are removed during the capture and recovery steps, this will greatly simplify the purification process, as caseins are otherwise present at a high concentration driving force and will tend to nonspecifically adsorb to chromatographic matrices [75]. The chromatography modes used to purify proteins from transgenic milk are as varied as the number of product proteins. Some recent examples of chromatographic processes used include cation exchange, anion exchange, hydrophobic interaction, and heparin affinity [65, 75–77]. The use of monoclonal antibodies in purification has fallen out of favor because they introduce another animal-derived biologic into the process and the final product. Recently, Pedersen et al. [78] have published a work on the chromatographic behavior of milk proteins. Through use of these types of studies combined with knowledge of the chromatographic behavior of the product protein reference standard, one can greatly simplify process development and develop a library of general strategies for removal of contaminant proteins.

Two particular complications may exist in the purification of proteins expressed at high levels in milk. First, for proteins that have multiple post-translational modifications, such as glycosylation, proteolytic processing, or γ-carboxylation, multiple subpopulations that differ in biological activity and clinical properties may exist [75, 79]. Also, as recently shown the glycosylation of recombinant C1 Inhibitor produced in transgenic rabbit milk and recombinant antithrombin II in goats, glycosylation can vary from animal to animal and change as the lactation progresses [80, 81]. Thus a purification process may need to be designed that can take a heterogeneous mix of proteins and select for the subpopulations having the best clinical properties. Second, it may be a challenge to remove the animal's endogenous version of the product protein: purification of recombinant human lactoferrin from endogenous bovine lactoferrin [77] and recombinant human serum albumin from bovine serum albumin (a joint venture between Genzyme Transgenics and Fresenius) are two examples. Recombinant human lactoferrin was purified from bovine lactoferrin by cation exchange chromatography. Purification of bovine serum albumin from recombinant human serum albumin was accomplished by affinity chromatography.

12.4 CONCLUSIONS

Production of recombinant therapeutic proteins from transgenic sources, both plants and animals, is gaining momentum. Many therapeutic candidates, particularly from transgenic plants, are at various stages of development, and several are deep into clinical trials. However, several challenges await the establishment of transgenic plants and animals as legitimate alternatives to current widely used microbial fermentation and mammalian cell culture for recombinant protein production in the biopharmaceutical industry. For transgenic plants, it is an engineering challenge to develop economical processes for protein recovery and purification from a large quantity of biomass, which could account for up to 90% of the total production cost. ATPE and polyelectrolyte precipitation are two techniques that could play important roles at early stages of protein recovery and purification from transgenic plants, especially green leaves. Both methods are straightforward for scale-up and with low associated costs. ATPE can be more versatile because of the number of adjustable variables, but for that very reason, identifying the optimal conditions for protein separation could be the bottleneck of a process development effort. Design of Experiment methods could be extremely helpful. However, to get ATPE widely used in industrial processing, the linchpin lies at understanding the mechanism of protein partitioning. Polyelectrolyte precipitation, on the other hand, can be more easily developed. How to improve the recovery of the target protein during precipitation by minimizing the interference of the polyphenolic compounds may determine if this technique can be successfully applied in protein separation from transgenic leafy plants.

Recombinant protein production from transgenic animals presents different challenges. The protein expression level may not be the most pressing issue, but the heterogeneous nature of the product, particularly glycoproteins from milk, may complicate the downstream processing. From the engineering point of view, the processes for protein purification from milk are similar to those from microbial or mammalian cell culture systems, including the reduction of viruses and other pathogens.

REFERENCES

1. Crosby L. Commercial production of transgenic crops genetically engineered to produce pharmaceuticals. *BioPharm Int* 2003; 15:60–67.
2. Hood E, Woodard SL and Horn ME. Monoclonal antibody manufacturing in transgenic plants — myths and realities. *Curr Opin Biotechnol* 2002; 13:630–635.
3. Graf EG, Jander E, West A, Pora H, and Aranha-Creado H. Virus removal by filtration. In: Brown F, and Cartwright T, Horaud F, Spieser JM. Eds. *Animal Sera, Animal Sera Derivatives and Substitutes used in the Manufacture of Pharmaceuticals: Viral Safety and Regulatory Aspects*, Vol. 99. Dev Biol Stand. Basel: Karger, 1999; pp. 89–94.
4. Adcock WL, MacGregor A, Davies JR, Hattarki M, Anderson DA, and Goss NH. Chromatographic removal and heat inactivation of hepatitis A virus during manufacture of human albumin. *Biotechnol Appl Biochem* 1998; 28:85–94.
5. Grun JB, White EM, and Sito AF. Viral removal/inactivation by purification of biopharmaceuticals. *BioPharm* 1992; 5:22–30.
6. Fraley RT, Rogers SG, Hrosch RB, Sanders PR, Flick JS, Adams SP, Bittner ML, Brand LA, Fink CL, Fry JS, Galluppi GR, Goldberg SB, Hoffmann NL, and Woo SC. Expression of bacterial genes in plant cells. *Proc Natl Acad Sci USA* 1983; 80:4803–4807.
7. Willmitzer GC, Depicker M, Dhaese P, De Greve H, Hernalsteens JP, Holsters M, Leemans J, Otten L, Schroder J, Schroder G, Zambryski P, van Montagu M, and Schell J. The use of Ti-plasmids as plant-directed gene vectors. *Folia Biol* 1983; 29:106–114.
8. Menkhaus TJ, Bai Y, Zhang C, Nikolov ZL, and Glatz CE. Considerations for the recovery of recombinant proteins from plants. *Biotechnol Prog* 2004; 20:1001-1014.
9. Fischer R, Stoger E, Schillberg S, Christou P, and Twyman RM. Plant-based production of biopharmaceuticals. *Curr Opin Plant Biol* 2004; 7:152–158.
10. Witcher DR, Hood EE, Peterson D, Bailey M, Bond D, Kusnadi A, Evangelista R, and Nikolov Z. Commercial production of β-glucuronidase (GUS): A model system for the production of proteins in plants. *Mol Breed* 1998; 4:301–312.
11. Hood EE, Witcher DR, Maddock S, Meyer T, Baszezynski C, Bailey M, Flynn P, Register J, Marshall L, Bond D, Kulisek E, Evangelista R, Nikolov Z, Wooge C, Mehigh RJ, Hernan R, Kappel WK, Ritland D, Li CP, and Howard JA. Commercial production of avidin from transgenic maize: Characterization of transformant, production, processing, extraction, and purification. *Mol Breed* 1997; 3:291–306.
12. Gomord V and Faye L. Posttranslational modification of therapeutic proteins in plants. *Curr Opin Plant Biol* 2004; 7:171–181.
13. Desai UA, Sur G, Daunert S, Babbitt R, and Li Q. Expression and affinity purification of recombinant proteins from plants. *Protein Expr Purif* 2002; 25:195–202.

14. Vaquero C, Sack M, Chandler J, Drossard J, Schuster F, Monecke M, Schillberg S, and Fischer R. Transient expression of a tumor-specific single-chain fragment and a chimeric antibody in tobacco leaves. *Proc Natl Acad Sci USA* 1999; 96:11128–11133.
15. Salmon V, Legrand D, Slomianny MC, Yazidi IE, Spik G, Gruber V, Bournat P, Olagnier B, Mison D, Theisen M, and Mérot B. Production of human lactoferrin in transgenic tobacco plants. *Protein Expr Purif* 1998; 13:127–135.
16. Krishnan R, McDonald KA, Dandekar AM, Jackman AP, and Falk B. Expression of recombinant trichosanthin, a ribosome-inactivating protein, in transgenic tobacco. *J Biotechnol* 2002; 97:69–88.
17. Gazaryan IG and Lagrimini LM. Purification and unusual kinetic properties of a tobacco anionic peroxidase. *Phytochemistry* 1996; 41:1029–1034.
18. Bai Y and Glatz CE. Bioprocess considerations for expanded-bed chromatography of crude canola extract: Sample preparation and adsorbent reuse. *Biotechnol Bioeng* 2003; 81:775–782.
19. Bai Y and Glatz CE. Capture of a recombinant protein from unclarified canola extract using streamline expanded bed anion exchange. *Biotechnol Bioeng* 2003; 81:855–864.
20. Hennighausen L. The prospects for domesticating milk protein genes. *J Cell Biochem* 1992; 49:325–332.
21. Houdebine LM. Use of transgenic animals to improve human health and animal production. *Reprod Domest Anim* 2005; 40:269–281.
22. Van Cott KE and Velander WH. Transgenic animals as drug factories: A new source of recombinant protein therapeutics. *Expert Opin Invest Drugs* 1998; 7:1683–1690.
23. Dyck MK, Lacroix D, Pothier F, and Sirard MA. Making recombinant proteins in animals — different systems, different applications. *Trends Biotechnol* 2003; 21:394–399.
24. Schügerl K and Hubbuch J. Integrated bioprocesses. *Curr Opin Microbiol* 2005; 8:1–7.
25. Persson J, Andersen DC, and Lester PM. Evaluation of different primary recovery methods for *E. coli*-derived recombinant human growth hormone and compatibility with further down-stream purification. *Biotechnol Bioeng* 2005; 90:442–451.
26. Li Y and Beitle RR. Protein purification via aqueous two-phase extraction (ATPE) and immobilized metal affinity chromatography. Effectiveness of salt addition to enhance selectivity and yield of GFPuv. *Biotechnol Prog* 2002; 18:1054–1059.
27. Srinivas ND, Rashmi KR, and Raghavarao KSMS. Extraction and purification of a plant peroxidase by aqueous two-phase extraction coupled with gel filtration. *Process Biochem* 1999; 35:43–48.
28. Zhang C, Medina-Bolivar F, Buswell S, and Cramer CL. Purification and stabilization of ricin B from tobacco hairy root culture by aqueous two-phase extraction. *J Biotechnol* 2005; 117:39–48.

29. Albertsson PA. *Partition of Cell Particles and Macromolecules*, 3rd ed. New York: Wiley, 1986.
30. Kaul RH. Eds. *Aqueous Two-Phase Systems, Methods and Protocols*. Clifton, NJ: Humana Press, 2000.
31. Balasubramaniam D, Wilkinson C, Van Cott K, and Zhang C. Tobacco protein separation by aqueous two-phase extraction. *J Chromatogr A* 2003; 989:119–129.
32. Boland MJ, Hesselink PGM, Papamicheal N, and Hustedt H. Extractive purification of enzymes form animal tissue using aqueous two phase systems: Pilot scale studies. *J Biotechnol* 1991; 19:19–34.
33. Salabat A. The influence of salts on the phase composition in aqueous two-phase systems: Experiments and predictions. *Fluid Phase Equilib* 2001; 187/188:489–498.
34. Diamond AD and Hsu JT. Aqueous two-phase systems for biomolecule separation. *Adv Biochem Eng Biotechnol* 1992; 47:89–135.
35. Brooks DE, Sharp, KA and Fisher D. Theoretical aspects of partitioning. In: Walter H, Brooks DE, and Fisher D. Eds. *Partitioning in Aqueous Two-Phase Systems: Theory, Methods, Uses, and Application to Biotechnology*. Orlando, FL: Academic Press, 1985, pp. 11–84.
36. Kaul A. The phase diagram. In: Hatti-Kaul R. Ed. *Aqueous Two-Phase Systems: Methods and Protocols*. Totowa, NJ: Humanna Press, 2000, pp. 11–21.
37. Zaslavsky BY. *Aqueous Two-Phase Partitioning*. New York: Marcel Dekker, 1995.
38. Khoudi H, Laberge S, Ferullo JM, Bazin R, Darveau A, Castonguay Y, Allard G, Lemieux R, and Vezina LP. Production of a diagnostic monoclonal antibody in perennial alfalfa plants. *Biotechnol Bioeng* 1999; 64:135–143.
39. Loomis WD. Removal of phenolic compounds during the isolation of plant enzymes. *Meth Enzymol* 1974; 31:555–563.
40. Cunha T and Aires-Barros R. Large-scale extraction of proteins. *Mol Biotechnol* 2002; 20:29–40.
41. Kula MR, Kroner KH, and Hustedt H. Purification of enzymes by liquid–liquid extraction. *Adv Biochem Eng* 1982; 24:73–118.
42. Smith PK, Krohn RI, Hermanson GT, Mallia AK, Gartner FH, Provenzano MD, Fujimoto EK, Goeke NM, and Olsen BJ, Klenk DC. Measurement of protein using bicinchoninic acid. *Anal Biochem* 1985; 150:76–85.
43. Bradford MM. A rapid and sensitive method for the quantitation of microgram quantities of protein utilizing the principle of protein-dye binding. *Anal Biochem* 1976; 72:248–254.
44. Englard S and Seifter S. Precipitation techniques. *Meth Enzymol* 1990; 182:285–300.
45. Kumar A, Galaev IY, and Mattiasson B. Precipitation of proteins: Nonspecific and specific. In: Hatti-Kaul R and Mattiasson B. Eds. *Isolation and Purification of Proteins*. New York: Marcel Dekker, Inc., 2003, pp. 225–275.

46. Niederauer MQ and Glatz CE. Selective precipitation. In: Fiechter A. Ed. *Advances in Biochemical Engineering/Biotechnology*, Vol. 47. New York: Springer-Verlag, 1992, pp. 159–188.
47. Zaman F, Kusnadi AR, and Glatz CE. Strategies for recombinant protein recovery from canola by precipitation. *Biotechnol Prog* 1999; 15:488–492.
48. Sternberg M and Hershberger D. Separation of proteins with polyacrylic acids. *Biochim Biophys Acta* 1974; 342:195–206.
49. Hill RD and Zadow JG. Recovery of whey proteins from precipitated complexes of carboxymethyl cellulose and protein. *J Dairy Res* 1978; 45:77–83.
50. Fisher RR and Glatz CE. Polyelectrolyte precipitation of proteins: The effects of reactor conditions. *Biotechnol Bioeng* 1988; 32:777–785.
51. Zhang C, Lillie R, Cotter J, and Vaughan D. Lysozyme purification from tobacco extract by polyelectrolyte precipitation. *J Chromatogr A* 2005; 1069: 107–112.
52. Rawel HM, Kroll J, and Rohn S. Reactions of phenolic substances with lysozyme — physicochemical characterisation and proteolytic digestion of the derivatives. *Food Chem* 2000; 72:59–71.
53. Jerala R, Kroon-Zitka L, and Turk V. Improved expression and evaluation of polyethyleneimine precipitation in isolation of recombinant cysteine proteinase inhibitor Stefin B. *Protein Expr Purif* 1994; 5:65–69.
54. Davies DT, Holt C, and Christie WW. The composition of milk. In: Mepham TB. Ed. *Biochemistry of Lactation*. New York: Elsevier, 1983, pp. 72–117.
55. Gordon WG and Kalan EB. Proteins of milk. In: Wong NP. Ed. *Fundamentals of Dairy Chemistry*. New York: Van Nostrand Reihold Co., 1988, pp. 87–124.
56. Jenness R. Interspecies comparison of milk proteins. In: Fox PF. Ed. *Developments in Dairy Chemistry*. Vol. 1. New York: Applied Science Publishers, 1982, pp. 87–114.
57. Velander WH, Johnson JL, Page RL, Russell CG, Subramanian A, Wilkins TD, Gwazdauskas FC, Pittius C, and Drohan WN. High-level expression of a heterologous protein in the milk of transgenic swine using the cDNA encoding human protein C. *Proc Natl Acad Sci USA* 1992; 89:12003–12007.
58. Drohan WN, Wilkins TD, Latimer E, Zhou D, Velander W, Lee TK, and Lubon H. A scalable method for the purification of recombinant human Protein C from the milk of transgenic swine. In: Galindo E and Ramirez OT. Eds. *Advances in Bioprocess Engineering*. Netherlands: Kluwer Academic Publishers, 1994, pp. 501–507.
59. Morcol T, He Q, and Bell SJ. Model process for removal of caseins from milk of transgenic animals. *Biotechnol Prog* 2001; 17:577–582.
60. Cole KD, Lee TK, and Lubon H. Aqueous two-phase partitioning of milk proteins. Application to human protein C secreted in pig milk. *Appl Biochem Biotechnol* 1997; 67:97–112.
61. Capezio L, Romanini D, Pico GA, and Nerli B. Partition of whey milk proteins in aqueous two-phase systems of polyethylene glycol-phosphate as a starting point to isolate proteins expressed in transgenic milk. *J Chromatogr B, Anal Technol Biomed Life Sci* 2005; 819:25–31.

62. Degener A, Belew M, and Velander WH. Zn(2+)-selective purification of recombinant proteins from the milk of transgenic animals. *J Chromatogr A* 1998; 799:125–137.
63. Ozyurt S, Kirdar B, and Ulgen KO. Recovery of antithrombin III from milk by expanded bed chromatography. *J Chromatogr A* 2002; 944:203–210.
64. Gardner TC. Delipidation Treatments for Large-Scale Protein Purification Processing. M.S. thesis. Virginia Tech, Blacksburg, VA, 1998.
65. Parker MH, Birck-Wilson E, Allard G, Masiello N, Day M, Murphy KP, Paragas V, Silver S, and Moody MD. Purification and characterization of a recombinant version of human alpha-fetoprotein expressed in the milk of transgenic goats. *Protein Expr Purif* 2004; 38:177–183.
66. Luque S, Mallubhotla H, Gehlert G, Kuriyel R, Dzengeleski S, Pearl S, and Belfort G. A new coiled hollow-fiber module design for enhanced microfiltration performance in biotechnology. *Biotechnol Bioeng* 1999; 65:247–257.
67. Baruah GL and Belfort G. A predictive aggregate transport model for microfiltration of combined macromolecular solutions and poly-disperse suspensions: Model development. *Biotechnol Prog* 2003; 19:1524–1532.
68. Baruah GL, Couto D, and Belfort G. A predictive aggregate transport model for microfiltration of combined macromolecular solutions and poly-disperse suspensions: Testing model with transgenic goat milk. *Biotechnol Prog* 2003; 19:1533–1540.
69. Baruah GL and Belfort G. Optimized recovery of monoclonal antibodies from transgenic goat milk by microfiltration. *Biotechnol Bioeng* 2004; 87:274–285.
70. Ziomek CA. Minimization of viral contamination in human pharmaceuticals produced in the milk of transgenic goats. *Dev Biol Stand* 1996; 88:265–268.
71. Levy JH, Weisinger A, Ziomek CA, and Echelard Y. Recombinant antithrombin: production and role in cardiovascular disorder. *Semin Thromb Hemost* 2001; 27:405–416.
72. Horowitz B, Bonomo R, Prince AM, Chin SN, Brotman B, Shulman RW. Solvent/detergent-treated plasma: A virus-inactivated substitute for fresh frozen plasma. *Blood* 1992; 79:826–831.
73. Drohan WN, Miekka SI, Griko YV, Forng RY, Stafford RE, Hill CR, Mann DM, and Burgess WH. Gamma irradiation of intravenous immunoglobulin. *Dev Biol (Basel)* 2004; 118:133–138.
74. Van Holten RW, Ciavarella D, Oulundsen G, Harmon F, and Riester S. Incorporation of an additional viral-clearance step into a human immunoglobulin manufacturing process. *Vox Sang* 2002; 83:227–233.
75. Lindsay M, Gil GC, Cadiz A, Velander WH, Zhang C, and Van Cott KE. Purification of recombinant DNA-derived factor IX produced in transgenic pig milk and fractionation of active and inactive subpopulations. *J Chromatogr A* 2004; 1026:149–157.
76. Edmunds T, Van Patten SM, Pollock J, Hanson E, Bernasconi R, Higgins E, Manavalan P, Ziomek C, Meade H, McPherson JM, and Cole ES. Transgenically produced human antithrombin: Structural and functional comparison to human plasma-derived antithrombin. *Blood* 1998; 91:4561–4571.

77. van Berkel PH, Welling MM, Geerts M, van Veen HA, Ravensbergen B, Salaheddine M, Pauwels EK, Pieper F, Nuijens JH, and Nibbering PH. Large scale production of recombinant human lactoferrin in the milk of transgenic cows. *Nat Biotechnol* 2002; 20:484–487.
78. Pedersen L, Mollerup J, Hansen E, and Jungbauer A. Whey proteins as a model system for chromatographic separation of proteins. *J Chromatogr B* 2003; 790:161–173.
79. Van Cott KE, Williams B, Velander WH, Gwazdauskas F, Lee T, Lubon H, and Drohan WN. Affinity purification of biologically active and inactive forms of recombinant human protein C produced in porcine mammary gland. *J Mol Recognit* 1996; 9:407–414.
80. Koles K, van Berkel PH, Mannesse ML, Zoetemelk R, Vliegenthart JF, and Kamerling JP. Influence of lactation parameters on the N-glycosylation of recombinant human C1 inhibitor isolated from the milk of transgenic rabbits. *Glycobiology* 2004; 14:979–986.
81. Zhou Q, Kyazike J, Echelard Y, Meade HM, Higgins E, Cole ES, and Edmunds T. Effect of genetic background on glycosylation heterogeneity in human antithrombin produced in the mammary gland of transgenic goats. *J Biotechnol* 2005; 117:57–72.

13 Analytical Strategy for Biopharmaceutical Development

Drew N. Kelner and Mahesh K. Bhalgat

CONTENTS

13.1 INTRODUCTION

The development of biopharmaceuticals is a complex process that requires significant resource commitments with respect to personnel, time, and money. Given the tremendous cost involved, a well-defined strategy for product development is essential to enable efficient process development and subsequent introduction of the biopharmaceutical into clinical trials. In addition to a strategy for development of the cell culture, purification, formulation, and fill/finish of the drug product, a long-range plan for analytical development and testing that both supports the process development effort as well as provides the analytical methods required for product characterization, release and stability testing is required. This chapter will highlight the critical features of an analytical strategy for biopharmaceutical development and provide guidance towards monitoring the safety, purity, and efficacy of new biotechnology drugs. Please note, however, that the testing methods discussed in this chapter should not be regarded as fixed or all-inclusive.

13.2 REGULATORY GUIDANCE

One of the critical questions faced by the analytical organization is the strategy of how deeply to pursue product characterization at the stage of initial clinical trials (IND in the United States) relative to the expectations for extensive product characterization and knowledge of the impact of the process on product quality at the commercial license application (BLA in the United States). There are published regulatory guidelines that provide a framework upon which the strategy for protein product quality testing and analytical characterization can be constructed. The U.S. Code of Federal Regulations (CFR) provides general guidelines for the IND and BLA stages of product development in Title 21, Sections 312.23 (IND) and 610.10 to 610.14 (BLA). These guidelines discuss product characteristics such as identity, purity, quantity (strength), potency, and safety but do not delineate specific analytical tests. These regulations do recognize that final specifications for the drug substance and drug product are not expected at the IND stage, since modifications in the method of preparation of the drug substance and drug product are likely during the course of development. The expectation is explicitly stated that supplements to the CMC section of the IND need to be submitted during scale-up and progression through the various stages of clinical testing, but no specifics are provided with respect to the analytical requirements for these supplement submissions. It should be noted that according to 21 CFR 312.23(a)(7), the only basis for a clinical hold based on the CMC section is an identified safety concern or lack of sufficient data to evaluate safety. While application of this guidance to the IND characterization strategy is open to debate due to the difficulties inherent in defining what constitutes evidence of a clinical safety concern, the regulations clearly provide some opportunities to streamline the characterization data relative to that included in a BLA submission. In recent years the expectations at the IND phase have been dynamic, as regulatory expectations in the United States, Europe, and other jurisdictions have heightened in recent years. This creates some strategic difficulties in the continuing effort to meet worldwide regulatory requirements while advancing products to the clinic as efficiently as possible.

Other guidelines published by FDA do address characterization of biotechnology products at various stages of product development. However, in some cases these guidelines are quite dated, such as a 1985 Points to Consider document on biotechnology drugs [1], which does specify commonly used methods of characterization (in 1985) such as HPLC, sequencing, peptide mapping, electrophoretic techniques, bioassays and circular dichroism. A 1995 Guidance for Industry [2] refers to 21CFR Section 312.23 without providing significant additional details with respect to specific analytical tests. A more complete description of the analytical requirements at the IND phase is available for monoclonal antibodies in the form of a 1997 Points to Consider document [3], which specifies physicochemical tests for demonstration of structural integrity, such as SDS-PAGE, IEF, HPLC, and mass spectrometry. In addition, this document discusses assays demonstrating the specificity of the antibody for its target antigen, potency assays, and methods for characterization of impurities. Since this PTC document is several years old, and the pace of technological advance in analytical biochemistry is rapid, we should consider newer technologies as potential substitutes in the design of the analytical strategy, such as using CE-based IEF (cIEF) rather than conventional slab gel-based methods.

The guidance documents published by the International Conference on Harmonization (ICH) provide specific information on analytical testing of biotechnology products. While information is presented in the guidance document on specification setting [4], these guidelines are intended for the new marketing application stage rather than the IND stage. Nevertheless, this document provides an excellent framework for development of the analytical strategy for biotechnology products. The overall guiding principle stated focuses on the critical importance of a defined analytical strategy, since it is recognized that characterization of the protein by physicochemical, biological and immunochemical means, coupled with characterization of the impurities present in the product, are required for specification setting. In addition, it is noted that acceptance criteria need to be justified based on the data from the preclinical and clinical lots, stability data, and relevant development data. The guidelines recognize that biotechnology proteins are produced in living cells and are therefore structurally heterogeneous, such that characterization of the inherent structural heterogeneity and demonstration of its consistency in the preclinical and clinical lots is an important feature of the characterization strategy. In addition, it is noted that heterogeneity can be introduced by the manufacturing process as well as upon storage, and that this heterogeneity needs to be evaluated with respect to its potential impact on the safety and efficacy of the product.

The ICH Specifications guideline provides significant detail with respect to analytical testing of biotechnology products. Table 13.1 summarizes the basic principles stipulated in the ICH guidelines, focused on the key product attributes (identity, purity, quantity, potency, physicochemical properties, and safety). The principles discussed apply specifically to biotechnology products at the new marketing application (BLA) stage, focusing on analytical requirements that need to be met to develop product specifications. This guideline provides an excellent framework for the analytical strategy at the BLA stage but, as mentioned above, there is no clear guidance for the IND stage.

The specific analytical testing conducted to ensure a safe, pure, and efficacious product as a result of a well-controlled process depends on the specific product and is usually discussed with the appropriate government regulatory agency on a case-by-case basis.

13.3 OVERVIEW

Once a decision has been made to advance a molecule from research into process development with the intention of introducing the candidate into human clinical trials, an intensive process development effort targeted at definition of a cell culture and purification process will generally be initiated. This effort requires analytical support from the outset in order to determine if the conditions under evaluation are suitable for preparation of the product for use in humans. It should be noted that it is difficult to state unequivocally the quality targets that must be met for clinical trials, since the extent of product purity is dependent on the intended use of the product. For example, products that are intended for repeated administration and use at high dosage levels will likely require higher purity than those intended for less frequent dosing and lower dose levels. In addition, since biopharmaceuticals are produced in living cells, they are, by their nature, heterogeneous, and therefore some variability from lot to lot is not only anticipated but is also acceptable from the perspective of regulatory authorities, as long as the safety and efficacy impact of the product heterogeneity is well understood by the manufacturer, and suitable process controls are in place to demonstrate the capability to produce material that does not stray beyond the bounds of clinical experience.

TABLE 13.1
ICH Guidelines for Biotechnology Product Characterization (Basic Principles)

Attribute	Characteristics	Comments
Identity	• Qualitative • Specific • Physiochemical, biological and immunochemical methods • Multiple tests may be required	• MAb: FDA PTC recommends IEF and subclass-specific ELISA • Western blot, CE, HPLC, peptide mapping methods often used
Process-related impurities	• Host cell protein (HCP) • Host cell DNA • Process chemicals • Chromatographic leachables	• Immunoassay commonly used for HCP • Clearance studies may be used to eliminate need for lot-to-lot testing at BLA (except HCP)
Product-related impurities	• Degradation products • Molecular variants with properties different than that of desired product • Truncated forms, chemically degraded forms, and aggregates commonly found	• Structural heterogeneity inherent and expected • Pattern of heterogeneity should be characterized • Only variants outside of range of clinical product considered as impurities (whether result from production process or storage) • Consistency in heterogeneity pattern reduces need to demonstrate safety and efficacy of individual forms • See guideline for methods
Quantity	• Protein content defined by physicochemical method	• Determine extinction coefficient (by AAA) if UV spectroscopy used for protein content
Potency	• Animal-based, cell-based or, biochemical assay • MAb: Ligand binding may be acceptable potency assay	• Potency assay for drug product permits alternative surrogate assay for drug substance (i.e., physicochemical methods) • MAb: binding affinity, avidity and epitope mapping expected
Physicochemical properties	• Primary structure • Physical properties • Higher-order structure	• Confirm sequence using peptide mapping, sequencing, and mass spectrometry

(Continued)

TABLE 13.1
Continued

Attribute	Characteristics	Comments
	• Molecular weight/size • Isoform/charge pattern	• Disulfide bond pattern and free sulfhydryls should be characterized • Glycoproteins: characterize sugar content, glycan structure/sequence, glycosylation sites • SE-HPLC, SDS-PAGE, mass spectrometry, CE-SDS • IEF, cIEF, IEX-HPLC
Safety	• Microbial load • Sterility • Pyrogen testing • Turbidity • Mycoplasma and adventitious viruses	• Bioburden testing • Required for final dosage form • LAL • Visual and instrumental analysis • Cell culture fluid tests

A number of workhorse methods are generally required for support of intensive process development (as described below). This work is best achieved by developing high throughput assays, such as by use of automated immunoassay systems for immunochemical methods such as determination of host cell protein impurities, as well as by use of HPLC and CE-based methods. Obviously the analytical methods used for development of specific unit operations should be focused on evaluation of the intended purpose of the unit operation. For example, an ion exchange chromatography step targeting host cell protein and DNA removal should be evaluated for its capability to clear these process-related impurities, and it may be advisable to also evaluate the charge heterogeneity profile of the product before and after the step to determine its impact, if any, on product charge distribution.

Effective support of process development also requires production of a well-characterized reference standard, which is generally derived from a high purity research lot or from material produced early in development that is believed to be of sufficient purity and sufficiently representative of the desired product to be reflective of the material targeted for the clinic. There is obviously an inherent problem here with respect to timing, as the best reference standard would be material produced late in process development when the cell culture, process, purification process, and formulation are locked down, but the need for a reference standard to support development of the process and release the first scaled-up lots precludes the luxury of waiting until late in development for production of the reference standard. Clearly, the first reference standard can be considered as an interim standard until such time that a more representative lot can be substituted following crossover testing to ensure that the early reference standard is truly reflective of the desired product.

During the development of the process, the analytical function, in collaboration with the Product Quality Organization, needs to define the test panels that will be implemented for in-process, product release, and stability testing of the product such that appropriate analytical methods can be developed and qualified prior to release of the GMP material destined for the clinic. In addition, the analytical group will be engaged in product characterization activities throughout process development. In addition to characterizing the reference standard, analysis of the product to characterize product-related variants such as isoforms and glycoforms is required to gain an understanding of the molecular heterogeneity of the product with the goal, as stated above, of demonstrating whether isolated fractions of the variants (if available) have the potential to impact the safety and efficacy of the product. For example, glycoprotein preparations expressed in mammalian cells will generally contain multiple glycoforms, including nonglycosylated forms or incompletely glycosylated forms that may have altered specific activity, solubility and stability. In addition, since the glycosylation pattern can impact pharmacokinetic parameters, the pharmacokinetic properties of isolated glycoforms may need to be evaluated in animal models. Finally, an effort is generally made during process development to evaluate the instability mechanism(s) of the product, focusing on the identification of major degradation products so that the stability test panel for the product can target the known degradation mechanisms using assays that have been shown to be capable of detecting the degradation products. The characterization work and the work targeted at understanding the degradation mechanisms may involve not only biochemical methods but also biophysical methods that evaluate the higher-order structure of the product under various conditions using techniques such as circular dichroism and fluorescence spectroscopy as well as analysis of aggregation and self-association behavior using analytical ultracentrifugation and laser light scattering techniques.

13.4 SETTING SPECIFICATIONS

Guidance for specification requirements of biotechnology products comes primarily from ICH Q6B (Q6B Specifications: Test Procedures and Acceptance Criteria for Biotechnological/Biological Products. *Federal Register* 64, August 18, 1999:44928), which defines specifications as a list of tests, analytical procedures or methods, and appropriate acceptance criteria that specify numerical limits, ranges, or other criteria for results. These specifications establish a set of criteria to which a drug substance and drug product should conform to be acceptable for its intended clinical use. Specifications constitute the critical quality standards proposed and justified by a manufacturer and approved by regulatory authorities. They are designed and selected as one element of an overall manufacturing control strategy that includes a validated manufacturing process and raw material, in-process, and stability testing.

13.5 ANALYTICAL TESTING STRATEGIES

Each assay used for evaluation of process consistency and product quality should have a clearly defined purpose targeted at one or more of the product quality attributes: potency, quantity, identity, purity, and safety. Since one test can cover more than one quality attribute, it is conceivable that test panels can be streamlined by designing some tests to cover multiple quality attributes. For example, SDS-PAGE with quantitative laser densitometry of Coomassie Blue-stained bands can be used both as an identity and a purity assay. Similarly, N-terminal sequencing provides data on both purity and product identity, since N-termini that differ from the encoded N-terminal residue can be considered as product-related impurities. While tests can be streamlined in this manner, it is important to recognize that more than one test may be required to cover each quality attribute. This strategy, which is known as orthogonal testing, ensures that quality attributes are thoroughly evaluated. A given analytical method has the potential to be applied to several points in the process, including evaluation of in-process samples, product release (specification testing) and stability testing.

When designing the stability test panel, it is critical to target known degradation mechanisms based on product development experience as well as forced degradation experiments using such parameters as high temperature, altered pH, or product shear.

When designing in-process tests, it is critical that the tests be simple, fast assays that will give accurate measurements of the quality of the process. However, in addition to process monitoring using analytical methods, process validation studies must be performed prior to product registration to validate the removal of key impurities of concern such as whole cells, DNA, or antibiotics. Steps in the process whose purpose is to remove putative exogenous viruses that must be added and validated might contaminate the product.

Figure 13.1 shows the steps followed during the manufacture of a typical biopharmaceutical product along with the attributes tested during the manufacturing process as well as on the final product. Table 13.2 lists the analytical techniques typically used for testing these attributes along with relevant practical considerations.

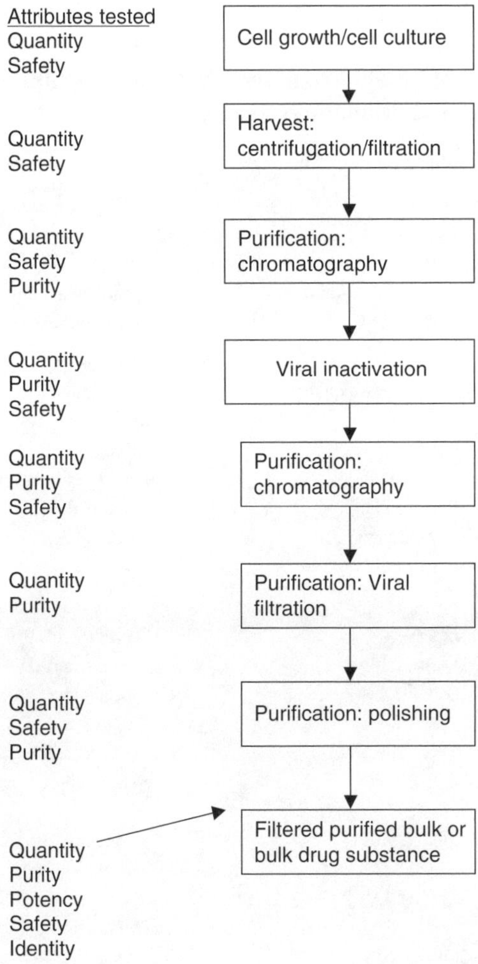

FIGURE 13.1 Typical biopharmaceutical production scheme with attributes tested at different steps identified on the left.

13.6 QUALITY ATTRIBUTES

13.6.1 Potency

Methods that monitor the potency of the drug substance and final product are essential for process control and reproducible assessment of efficacy. Potency assays evaluate the biological response elicited by the active ingredient.

TABLE 13.2
Methods and Techniques Used in Testing Product Attributes and Practical Considerations in Method/Technique Selection

Attribute	Methodology/Techniques	Practical Considerations
Quantity (Strength)	A280	Not suitable for crude samples
	ELISA	Labor intensive
	HPLC:	
	Protein A affinity	For monoclonal antibodies
	RP-HPLC	For microbial expression
Identity	Western blotting	Labor intensive
	N-terminal sequencing	Low throughput
	MS-MS	Requires significant expertise
	CE methods	High throughput
	HPLC methods	Simple, high throughput
	Peptide mapping	Complex method
	Immunochemical methods	"Dipstick" technology attractive
Purity		
Process-related	ELISA	Host cell proteins, Protein A
	Western blot	Supplement to ELISA for HCP
	QPCR DNA	High throughput, accurate, host cell-DNA specific
	Threshold DNA	Not host cell-DNA specific
	ICP, AA	Chromatographic leachables
Product-related		
Mass:	SDS-PAGE	Simple method for multimers and clips
	CE-SDS	High resolution and quantitative
	SE-HPLC	aggregates; clipped forms detected under denaturing conditions
	Protein MS	High resolution MW profiling
	AUC	Hydrodynamic properties in formulation buffer
	DLS and PCS	Particle-size distribution
Charge:	IEF gels	Simple method
	CIEF, CZE	Quantitative, higher resolution
Other variants:	Peptide mapping	Postsynthetic modifications, deamidation, oxidation
	RP-HPLC	Oxidation, clips (small proteins)
Safety:	Microbial load	Microbiological analysis
	LAL	Endotoxin analysis
	HIAC	Particle counting
	Microbiological assays	Adventitious virus, mycoplasma

TABLE 13.2
Continued

Attribute	Methodology/Techniques	Practical Considerations
Potency:	Cell-based bioassays	Required for agonists
	In vitro binding assay	May be suitable for antagonists
	Enzyme activity assay	Suitable for enzymes, cofactors and enzyme inhibitors
	Biacore assay	May be suitable for antagonists

Generally a cell-based potency assay will be required to directly monitor the biological response elicited by the product. For example, cytokines like interleukin-2 can be tested for potency using cells in culture that have been exposed to the cytokine in a T-cell proliferation assay. While cell-based potency assays provide definitive evidence for the bioactivity of the molecule, they are generally low throughput methods due to the labor-intensive nature of *in vitro* bioassays. In addition, due to the reliance of the assay on biological systems, these methods have relatively low precision compared to biochemical assays such as HPLC.

In some instances, it may be feasible to use an *in vitro* binding or activity assay to assess product potency. For example, it may not be necessary to directly measure the biological response in the case of enzymes, enzyme inhibitors, enzyme co-factors, or antagonists, since the biological activity is directly dependent on *in vitro* binding (receptor or ligand for an antagonist) or activity (enzyme, enzyme inhibitor, enzyme co-factor). In the case of agonists, a cell-based bioassay that measures the biological response is required because binding is a necessary but not sufficient condition for bioactivity, since agonist activity involves a signaling pathway following binding to elicit a biological response. Nevertheless, *in vitro* binding assays can be used as surrogate potency assays for agonists, not for product release but rather for use in support of process development, where high throughput may be required to assess the impact of process modifications on product quality. Similarly, biosensor data, such as that provided by Biacore® technology, can be used for bioactivity assessment of antagonists or as a surrogate potency assay for agonists to support process development.

13.6.2 Quantity

For the assessment of product quantity (strength), it is essential that a well-characterized reference standard be made available relatively early in product development (as described above). The concentration value assigned to the standard must be defined by a method capable of providing absolute quantitation, such as amino acid analysis. Once a reference standard is available, this standard can be used for assessment of product strength. The simplest means of strength determination is to use ultraviolet spectroscopy at 280 nm based on the determined extinction coefficient of the protein. The extinction coefficient is generally calculated theoretically, then verified by amino acid analysis of the initial reference standard. In cases where the samples contain interfering substances, such as albumin, immunochemical methods such as Enzyme-linked immunosorbent assay (ELISA) may be required. In some cases, a high throughput chromatographic assay using HPLC may be desirable, either because the spectroscopic measurement is subject to interference or because the HPLC-based method is more robust than a spectroscopic method for given sample types. For example, the concentration of monoclonal antibody product in both crude in-process as well as more highly purified downstream samples can be determined by HPLC using a Protein A column. For other types of protein samples, it may be necessary to evaluate crude upstream samples using immunochemical methods such as ELISA. It should be noted that it may be possible to measure the concentration of product in crude samples using methods such as reversed phase HPLC, particularly in the case of systems where high expression levels are achieved in cell culture, such as the case of microbial fermentation.

13.6.3 Product Identity

For demonstration of product identity, western blotting has long been widely used and accepted. This electrophoretic method, which is somewhat labor intensive, can be subject to nonspecific cross-reactivity. Another long-established method for demonstration of product identity is N-terminal sequencing, which is a low throughput method due to the long run times required for sequencing sufficient cycles to provide definitive proof of product identity (generally, 15 to 20 sequencing cycles).

As technology has advanced, new candidate methods for the determination of product identity have become available. A particularly sophisticated method involves tandem sequencing by MS–MS, which is a rapid method that provides high definitive sequence data. This method requires significant expertise and expensive equipment and therefore may not be a suitable candidate for use in a quality control laboratory. Alternative methods that have been used recently for identity testing include chromatographic methods and CE-based methods that provide a definitive fingerprint pattern, such as reversed phase HPLC, ion exchange HPLC, capillary isoelectric focusing (cIEF) and capillary zone electrophoresis (CZE). While peptide mapping provides a highly definitive molecular fingerprint, the complexity of the method, coupled with its capability to detect minor variants that can complicate interpretation of the results with respect to demonstration of product identity, suggest that this method may be too complex for routine use in a QC laboratory for identity testing. A recently emerging and attractive option is to develop rapid immunochemical methods using "dipstick" technology that use a specific immobilized binding partner, such as the antigen for an antibody or receptor for a receptor antagonist.

13.6.4 Product Purity

The purity of a biological product is defined as the measurement of the active drug substance in relation to the total substances (not including additives) present in the final product [5]. In addition to the biologically active ingredient and desirable additives, the product may contain impurities and contaminants. Impurities are defined as all process related nonadventitious substances present that are not considered to be the active material, additives, or excipients. Examples of common impurities in recombinant DNA-derived biologicals are presented in Table 13.3 together with a list of the analytical methods typically used for their determination. Impurities can be divided into two main categories: process- and product-related impurities. A good discussion on the former category which includes host cell proteins, host cell DNA, certain reagents used during bioprocessing, and chromatographic leachables (e.g., Protein A derived from the chromatographic resin used for purification of monoclonal antibodies) was the subject of a recent Well Characterized Biotechnology Pharmaceutical (WCBP) Chemistry, Manufacturing, and Controls (CMC) Strategy Forum [6]. Product-related impurities include aggregates (both soluble and insoluble), charge variants due to deamidation and other chemical processes, oxidation products, and N- and C-terminally truncated and internally clipped forms. While excipient levels need to be controlled, these additives are not considered as impurities, since they are deliberately included for control of pH, osmolality, conductivity, and for enhancing product stability in the final dosage form.

13.7 PROCESS-RELATED IMPURITIES

13.7.1 Host Cell Proteins

Manufacturing processes are designed to remove host cell protein impurities and minimize their levels for safety reasons. The levels of these impurities must be significantly reduced during processing to ensure that these potentially antigenic impurities are eliminated, or reduced to levels that will not elicit an immune response. The elicitation of an immune response can function as an

TABLE 13.3
Common Impurities of rDNA-Derived Protein Pharmaceuticals

Impurities	Detection Method
Endotoxin	LAL,[a] rabbit pyrogen
Host cell proteins	Western Blots, Immunoassays
Other protein impurities (media)	SDS-PAGE[b], HPLC[c], Immunoassays
DNA	DNA hybridization, Total DNA byThreshold, qPCR[d]
Protein mutants	Peptide mapping
Aggregates	SEC[e], Light scattering,
Oxidized methionines	Amino acid analysis, Peptide mapping, Edman degradation analysis
Proteolytic clips	IEF[f], SDS-PAGE (reduced), HPLC, Edman degradation analysis
Deamidation	IEF (standard comparison), HPLC
Monoclonal antibodies	SDS-PAGE, immunoassays
Amino acid substitutions	Amino acid analysis, Peptide mapping
Viruses (endogenous)	CPE[g], HAd[h], Electron microscopy, Reverse transcriptase activity

[a] Limulus amebocyte lysate.
[b] Sodium dodecyl sulphate polyacrylamide gel electrophoresis.
[c] High-performance liquid chromatography.
[d] Quantitative Polymerase Chain Reaction.
[e] Size exclusion chromatography.
[f] Isoelectric focusing.
[g] Cytopathic effect.
[h] Haemadsorption.

Source: Adapted from Garnick RL, *J. Pharm. Biomed. Anal.*, 1989; 7(2):255–266.

adjuvant, which can result in an antibody response to the product itself. This response can either be specific to the drug product, resulting in the inability of the product to hit its target, or if the drug product is similar to an already existing protein in the body, cross-reactivity that could result in antibodies to self-antigens. Depending on the antibodies elicited, the physiologic result may vary from negligible (transient effect) to severe (e.g., prolonged immune response to self-antigens and life-threatenting anaphylactic shock). It is also possible that the host cell protein may elicit a physiological response *in vivo* if present at high enough concentrations (if the host cell protein sequence is sufficiently homologous to that of the homologous protein in the recipient). The type of production cell line used will impact the specific concerns for removal of host cell protein impurities. Microbially derived products are more likely to act as an adjuvant due to the antigenic nature of microbial proteins, whereas mammalian-derived host cell proteins are more of a concern for reactivity to self antigens, and, in addition, may be sufficiently homologous to human proteins to elicit physiological responses.

The considerations that should be evaluated for monitoring HCP are the technologies and reagents available, whether the product is derived from mammalian or microbial cells, and dosing strength and frequency. The type of technology employed to monitor clearance of host cell proteins and the associated reagents are critical to providing adequate understanding and verification of host cell protein clearance throughout the process. Current technology is predominantly ELISA-based, although substrates and detection methods may vary. SDS/PAGE and Western blot techniques should also be used as supplements to ELISA-based results. Production of reagents for host cell protein impurities can be done in-house, although reagents are also available commercially. The benefit of producing reagents in-house is greater control over reagent production, with the goal of driving the immunization procedures such that the greatest number of epitopes are targeted. Commercial antibodies will most likely have reduced epitope coverage for the host cell line being used, and can therefore result in a greater chance of a host cell protein impurity evading detection, and therefore being present in the final material.

There are several approaches taken within the industry with respect to the reagents used for ELISA testing. One approach is to use process-specific reagents, and the second is to use a generic assay, based on platform methodologies, for all products produced in the same cell line. The benefit of the generic approach is that results for all products can be directly compared to each other.

Monitoring of host cell proteins should ideally be performed throughout the process. Evaluation of buffers used in the process, and the subsequent impact on quantitation of the host cell proteins, should also be evaluated, due to the potential inhibitory properties of various buffers. A generic understanding of potential inhibitors can be evaluated by testing the ability to quantitate host cell protein standards that have been subjected to various conditions and reagents, such as heat, extreme pH changes, salt, alcohol, etc. Inhibition of immunoassay quantitation can have an impact on downstream clearance measurements, and thereby understanding the impact of the process buffer for each unit operation on the ability to quantitate host cell proteins, especially by immunoassay, is critical.

Explicit regulatory guidance on appropriate tests and levels of these impurities is lacking, and therefore evaluation of multiple parameters is important to ensure product success and patient safety. Setting of the appropriate limits for drug substance should be determined by process consistency, an understanding of assay sensitivity, and evaluation of the presence of any particular host cell protein that might skew results by providing either a very high antigenicity relative to other antigens or, on the other hand, very low antigenicity in the animal such that high levels of the antigen may be present and undetected in the final product. Limits for clinical trials should be tight enough to ensure patient safety. For commercial products, there should ideally be a good understanding, and enough time for development of reagents, that will ensure the values obtained for the assay are appropriate and verifiable with orthogonal methods to ensure adequate coverage and quantitation. Upon obtaining data from a large number of runs ($n > 30$), a tighter specification can be set after knowledge of process performance is obtained.

At the IND stage, generic host cell protein assays are generally applied to determine residual host cell protein levels. While a process specific assay has been developed in some cases within the industry for BLA filings [7], this is not a requirement if it can be shown that the generic assay is suitable for detecting the host cell proteins present in the purified material, and if the method has appropriate performance characteristics to support commercial production. While a target of 100 ppm is suitable at the IND stage, this value is provided as guidance for process development only, such that failure to meet this target should not impede progress toward the Phase I/II clinical trials, as long as this issue is discussed at the appropriate time with the relevant regulatory authorities. Whether the target should be tightened at the BLA stage should be evaluated on a case-by-case basis. This assessment should take into account the protein dosage, frequency of administration and levels present in product administered during early clinical trials. For example, chronic administration of a high protein dosage may warrant a lower host cell protein limit for the commercial product due to the risks of chronic exposure to relatively high amounts of host cell protein. Once again, consultation with the regulatory authorities is recommended in cases where the host cell protein level can be considered a potential safety issue.

13.7.2 Host Cell DNA

Host cell DNA can be quantitated by a variety of methods. Historically there has been a heavy reliance on hybridization methods such as the dot blot and slot blot. These labor-intensive methods have relatively low throughput and have relatively high failure rates, and are rapidly being supplanted by alternative technologies, including total threshold DNA analysis, spectrophotometric methods such as binding to Pico Green and quantitative polymerase chain reaction (qPCR). It should be noted that while the hybridization method is specific for host cell DNA, the threshold and Pico Green techniques measure total DNA, including that introduced from nonhost sources, such as the HySoy medium used for product fermentation in mammalian systems. In this respect, the qPCR method is a more suitable substitute for hybridization methods, since the PCR-based method is specific for host cell DNA. Current industry standards are focused on the clearance of host cell DNA, such that measurement of total DNA is not required at either the FIH or BLA stage. Finally, we should target a DNA level of $\leq$10 ng/dose, the expectation currently specified by the World Health Organization.

13.7.3 Other Impurities

Chromatographic leachables such as Protein A, derived from the affinity resin used for monoclonal antibody purification, are generally measured by immunochemical methods. A number of ELISA-based methods are currently available for this determination, including kit-based methods. It should be noted that the best practice is to use an assay targeting the specific protein A type used in the antibody purification process, as a variety of antibody purification resins are currently available that use Protein A derived from natural or recombinant sources. For nonantibody products, chromatographic leachables such as heavy metals from IMAC columns can be measured using atomic absorption or ion-coupled plasma (ICP) spectroscopy methods.

13.8 PRODUCT-RELATED IMPURITIES

The regulatory authorities recognize that proteins produced in living systems are structurally heterogeneous, and that numerous modifications, including glysosylated forms and other modified forms, are often found as fully active components of purified product preparations. The aforementioned molecular heterogeneity needs to be evaluated with respect to the potential impact on the safety and efficacy of the product. This characterization effort includes an evaluation of the pattern of product heterogeneity relative to that seen in the material used for preclinical and clinical studies. In addition to the natural heterogeneity present in biotechnology protein products, structural variants can occur during manufacture and storage of the drug substance and drug product. According to ICH guidelines, the manufacturer defines the range of heterogeneity of the active ingredient, such that only variants that differ from those found in the clinical product are considered as product-related impurities, whether these variants are derived from production or storage. The analytical methods described below are typically used to evaluate the molecular heterogeneity of protein products resulting from biosynthesis, manufacturing, and storage. It should be noted that during product development the chromatographic methods used for biochemical characterization of the product can be used to support the development of unit operations and should also be applied to degraded samples generated upon storage or under forced degradation conditions to determine the utility of the methods as stability indicating assays. While full characterization of degraded forms is not required at the IND stage, this data should be included in the BLA filing, with a statement in the IND noting that characterization of degraded forms is ongoing.

13.9 MASS/SIZE DISTRIBUTION

In addition to the expected mass of the active ingredient, proteins have the potential to undergo physical stresses that result in aggregated forms due to covalent and noncovalent interactions as well as truncation of the amino acid sequence, which may be chemically derived or catalyzed by trace levels of protease activity. A number of analytical methods are used to evaluate the molecular weight distribution of the product. Denaturing methods such as SDS-PAGE can determine the presence of dimeric and higher multimeric covalent forms as well as low molecular weight (clipped) forms. Detection limits are about 200 ng/band for Coomassie stained gels and approximately 10- to 100-fold higher sensitivity for silver stained gels. The emerging technology of capillary electrophoresis (CE) has significant potential to supplant conventional slab gels. Current data has already shown that CE-SDS provides a high resolution, quantitative method that has the potential to provide information on structural heterogeneity that cannot be resolved on conventional slab gels. We anticipate that the CE-based methods will likely supplant conventional slab gel-based SDS-PAGE due to the higher resolution, improved reproducibility, and easier quantitation inherent in the CE technology.

Size exclusion chromatography using high performance liquid chromatography (SE-HPLC) under nondenaturing conditions is often used for both aggregate detection (both covalent and noncovalent) and detection of low molecular weight impurities. The use of online multi-angle light scattering (MALS) detection can significantly enhance the sensitivity of detection of high molecular weight multimeric species and can, when coupled to refractive index and UV detectors, provide data on the absolute molecular weight distribution. MALS detection is recommended due to the heightened sensitivity of the potential physiological consequences of aggregates in biopharmaceuticals (including potential immunogenicity). In some cases size exclusion HPLC under denaturing conditions can be used for characterization of low molecular weight (clipped) forms of the product.

Protein mass spectrometry can be used for high sensitivity, high resolution molecular weight profiling, though it should be noted that the experimental conditions used for protein MS typically disrupt noncovalent aggregates. An alternative method for size profiling under nondenaturing conditions is by use of analytical ultracentrifugation, which provides data on the hydrodynamic properties of the protein sample. Analytical ultracentrifugation-velocity sedimentation (AUC-VS) is an emerging technique that can, with recent advances in software, offer similar resolution to SE-HPLC. The main advantage of AUC-VS is that samples can be analyzed directly in the formulation buffer, without subjecting the sample to dilution or solid phase interactions as experienced in SE-HPLC. The main disadvantage of AUC-VS is that artifacts can skew the results. Qualitative agreement between AUC-VS and SE-HPLC is an excellent way of demonstrating the specificity and recovery of SE-HPLC.

Finally, soluble and insoluble aggregates may be present at trace levels; such particulates can be characterized using a variety of techniques, including visual assessment and turbidity measurements by visible spectroscopy as well as dedicated particle-size distribution instrumentation such as dynamic light scattering (DLS) and photon correlation spectroscopy (PCS). The lower resolution methods have historically been sufficient at the IND stage, but recently regulatory scrutiny has intensified in this area, and instrumental methods are being applied, at least in some cases, at the IND stage. Instrumental particle-size data should be available by the BLA stage.

13.10 CHARGE VARIANTS

Charge heterogeneity can be introduced into protein preparations via glycosylation (e.g., terminal sialic acids), deamidation, variable levels of heavy chain C-terminal lysine in monoclonal antibodies, N- or C-terminal truncations of other proteins (such as C-terminal arginine heterogeneity in erythropoietins), and via other mechanisms. Such charge heterogeneity can be characterized using IEF gels, capillary IEF methods, capillary zone electrophoresis (CZE) and ion exchange chromatography on HPLC. At the BLA stage, the charge heterogeneity analysis should be quantitative with established specifications, whereas reporting quantitative values is sufficient at the IND stage such that process history can be established for the molecule.

It should be noted that protein deamidation, which proceeds through a relatively unstable succinimide ring intermediate, can result in the formation of both aspartic acid and isoaspartic acid. These two reaction products have the same charge and often cannot be differentiated using HPLC analysis of the intact protein, such that peptide mapping methods are generally required for characterization of aspartate isomerization. This structural feature can be evaluated at the BLA submission stage using one of a number of potential strategies. One commonly used approach is digestion with Asp-N, since the enzyme is capable of clipping on the N-terminal side of aspartic acid, while isoaspartic acid is refractory to cleavage by the enzyme [8].

An alternative approach uses cation exchange chromatographic analysis of S-adenosyl-L-homocysteine, the reaction product formed in stoichiometric quantities by the incubation of isoaspartic acid-containing polypeptides with isoaspartyl methyltransferase [9].

While the succinimide ring intermediate that results from deamidation is relatively unstable in aqueous systems, these forms are sometimes detected in peptide mapping (using LC-MS) as degradation products in biopharmaceutical preparations.

13.11 OXIDIZED AND CLIPPED FORMS

Reversed phase HPLC (RP-HPLC) is a method commonly used for detection of clipped forms and other variants, such as oxidized forms of the protein. In some cases, such as monoclonal antibodies and highly glycosylated proteins such as the erythropoietins, it may not be feasible to use chromatographic analysis of the intact protein for detection of oxidized and possibly clipped forms. In such cases, peptide mapping may be the only viable alternative for detection of product-related variants such as oxidized forms of the protein.

13.12 SAFETY TESTING

The goal of safety testing is to ensure that the drug product can be safely administered without causing an overt toxic effect or an immunological reaction. Routine analytical safety assessment can be differentiated from the clinical safety assessment that can only be obtained in human clinical trials. Routine safety testing involves assurance that the product does not contain microbial contaminants, pyrogens or substances that render the dosage form in a turbid state. Tests commonly used include microbiological evaluation for microbial load (also known as bioburden), testing of cell culture fluid for mycoplasma and adventitious viruses, sterility testing of the final packaged product, and pyrogen testing such as endotoxin analysis using the limulus amoebocyte lysate (LAL) inhibition test. For assessment of product turbidity the visual appearance is assessed by visual inspection and the analysis of levels of subvisible particles is carried out using instrumental methods such as particle counting (HIAC). Finally, product antigenicity is an important consideration for biopharmaceuticals which can, in some instances, elicit an immune response. While clinical assessment of the immunological safety can only be obtained in clinical trials, it is critical to conduct animal experiments as a preliminary test of product immunogenicity. This preclinical testing is complicated by the fact that the administration of any human protein in an animal has the capacity to elicit an immune response. Nevertheless, model systems have been developed to assess the immunogenicity of human therapeutics in animal models [10]. Viral testing is a critical component of safety testing, but has not been covered here since it is the subject of another chapter in this book and has been the subject of several publications [11–13].

13.13 PRODUCT CHARACTERIZATION

Extensive product characterization beyond that carried out for routine lot release is required for regulatory filings of biopharmaceuticals. This characterization work involves structural analysis at both the primary and higher-order levels of structure. While primary structural analysis relies heavily on widely used and familiar methods such as N-terminal sequencing and peptide mapping, the advent of modern methods of mass spectrometry has added powerful new tools that permit rapid and detailed structural analysis. The two major modes of protein mass spectrometry, electrospray ionization mass spectrometry (ESI-MS) and matrix assisted laser desorption-time of flight mass spectrometry (MALDI-TOF MS) have been extensively reviewed in the literature. Recently a new hybrid technology that uses an electrospray source with time of flight detection, quadrupole time of flight mass spectrometry (Q-TOF), has advanced protein mass spectrometry to new levels of mass precision while allowing on-the-fly mass determination of peptides and peptide sequencing.

13.14 BIOPHYSICAL ANALYSIS

Evaluation of the secondary and tertiary structure of the protein preparation is a component of the physicochemical analysis of protein products. Typically, biophysical methods are carried out both to characterize the product and also to provide data on the stability of the protein under defined processing conditions, such as ranges of pH, salt concentration and temperature. This data is used to support both purification process and formulation development.

During development of the IND-enabling process, biophysical analyses using circular dichroism (CD), fluorescence spectroscopy, and Fourier-transform infrared spectroscopy (FT-IR) are typically performed to support process and formulation development. Fluorescence measurements are generally carried out based on both the intrinsic fluorescence of the protein as well as in the presence of compounds that probe surface hydrophobicity using compounds such as ANS (extrinsic fluorescence). For the IND filing, data demonstrating intact secondary and tertiary structure using a single technique for each parameter is recommended. In most cases, FT-IR and near-ultraviolet CD spectroscopy are satisfactory methods for demonstrating intact secondary and tertiary structure, respectively. The rationale for not including additional biophysical data at the IND stage is based on two factors (i) the desire to maintain a concise format in the product characterization section of the IND, and (ii) the inherent risk in providing more extensive data that requires comparability assessment when material from the BLA-enabling process is introduced into clinical trials. For the BLA filing, additional biophysical data, including that obtained using orthogonal techniques such as far-UV CD and intrinsic fluorescence spectroscopy for corroboration of the secondary and tertiary structural analysis, respectively, can be presented. In addition, the melting temperature should be obtained using differential scanning calorimetry (DSC) and, if feasible, isolated product variants should be analyzed using biophysical methods to determine if their properties vary significantly from those of the major form of the protein. Finally, forced degradation samples should be subjected to biophysical analysis to determine if such samples have biophysical properties that differ from those of the native protein. The final decision on which data to include in the BLA submission requires consideration on a case-by-case basis.

13.15 PROCESS ANALYTICAL TECHNOLOGIES

Controlling the quality of biological pharmaceuticals requires attention to in-process control. This offers an economical and efficient way of assessing the quality of the end product. As stated by CDER [14], the goal of PAT is to understand and control the manufacturing process, which is consistent with the current drug quality system: quality cannot be tested into products; it should be built-in or should be by design. In fact, building quality into the product from the beginning is as important as carrying out tests on the final product.

For successful implementation of PAT, it is important to understand that the goal of PAT is not simply to introduce online monitoring equipment but to introduce a complete system and approach to analyzing and controlling manufacturing through timely measurements of critical quality and performance attributes. Manufacturers are sensitive to the needs of industry in this regard and are providing comprehensive solutions, such as the one offered by Siemens. The comprehensive solutions are geared towards providing tools for

- Multivariate data acquisition and analysis tools
- Modern process analyzers or process analytical chemistry tools
- Process and endpoint monitoring and control tools
- Continuous improvement and knowledge management tools

In addition, more specific techniques are being evaluated for their use as PAT tools. Near Infra Red Chemical Imaging (NIR-CI) is an example of a noninvasive method for rapid analysis [15].

13.16 ANALYTICAL TESTING OF BIOGENERIC PHARMACEUTICALS

Current US regulations require the filing of a new biologics license application for products made with recombinant DNA technology, even if the product is considered to be identical in structure to a naturally occurring substance or a previously approved product produced in a conventional way (FDA Talk Paper, January 7, 1983. Regulating Recombinant DNA products). The rationale for this regulation is that the inherent heterogeneity of biopharmaceuticals dictates that the product structure is process-dependent and therefore the manufacture of an existing biopharmaceutical by a new manufacturer results, in essence, in a new product that requires clinical evaluation. More recent commentary and responses to questions raised by the FDA on biogeneric pharmaceuticals or follow-on protein therapeutics can be helpful in determining the areas that still need to be addressed [17,18]. Needless to say, this is an area where evolution is evident and readers are encouraged to stay tuned.

REFERENCES

1. Office of Biologics Research and Review, center for drugs and biologics, FDI. Points to consider in the production and testing of new drugs and biologicals produced by recombinant DNA technology. April 10, 1985.
2. Food and Drug Administration [Docket No. 95D090164] FDA Guidance Document Concerning Use of Pilot Manufacturing Facilities for the Development and Manufacture of Biological Products. July 1995.
3. Food and Drug Administration [Docket No. 94D-0259] Points to Consider in the Manufacture and Testing of Monoclonal Antibody Products for Human Use. February 1997.
4. ICH Harmonized Tripartitie Guideline. Q6B Specifications: Test Procedures and Acceptance Criteria for Biotechnological/Biological Products. Federal Register 64, August 18, 1999: 44928; www.ich.org/MediaServer.jser?@_ID=432&@_MODE=GLB.
5. American Society for Testing and Materials (ASTM), Draft Standard Guide for Determination of Purity, Impurities, and Contaminants in Biological Drug Products. Developed under ASTM Subcommittee E-48.01 (Materials for Biotechnology), Task Group .05 (R. L. Gamick, Chairman), Philadelphia, PA, February (1988).
6. Heather Simmerman and Raymond P. Donnelly Defining Your Product Profile and Maintaining Control Over It, Part 1 Session One of the WCBP CMC Strategy Forum, July 19–20, 2004 BioProcess International 2005; 6:32–37.
7. Eaton LC. Host cell contaminant protein assay development for recombinant bio-pharmaceuticals. *J. Chromatogr* 1995; 705:105–114.

8. Zhang W, Czupryn MJ, Boyle PT, and Amari J. Characterization of asparagine deamidation and aspartate isomerization in recombinant human interleukin-1. *Am J Pharm Res* 2002; 19(8):1223–1231.
9. Zhan W and Czupryn MJ. Analysis of isoaspartate in a recombinant monoclonal antibody and its change isoforms. *J Pharm Biomed Anal* 2003; 30:1479–1490.
10. Schellekens H. Bioequivalence and the immunogenicity of biopharmaceuticals. *Nat Rev Drug Discov.* 2002; 1(6):457–462.
11. Brown F, Griffiths E, Horaud F, and Petricciani JC (Eds): *Safety of biological products prepared from mammalian cell culture. Quality of biotechnology products: Viral safety evaluation of biotechnology products derived from cell lines of human or animal origin (ICH Harmonised Tripartite Guideline).* Dev Biol Stand. Basel, Karger, 1998; 93:177–201.
12. Viral safety. Chapter 14 of this volume.
13. Valera CR, Chen JW, and Xu Y. Application of Multivirus Spike Approach for Viral Clearance Evaluation. *Biotechnol Bioeng* 2003; 84:714–722.
14. http://www.fda.gov/cder/OPS/PAT.htm.
15. Lewis EN, Schoppelri J, and Lee E. Molecular spectroscopy workbench – Near-infrared chemical imaging and the PAT initiative. *Spectroscopy*, 2004; 19(4):26–36.
16. Garnick RL. Safety aspects in the quality control of recombinant products from mammalian cell culture. *J Pharm Biomed Anal* 1989; 7:255–266.
17. Food and Drug Adminstration [Docket No. 2004N-0355] Scientific Considerations Related to Developing Biotechnology Products. October 2004 (Weblink: http://www.fda.gov/ohrms/dockets/dockets/04n0355/04N-0355-EC9.html.
18. Food and Drug Adminstration [Docket No. 2004N-0355] Scientific Considerations Related to Developing Biotechnology Products. August 2004 (weblink:http://www.fda.gov/ohrms/dockets/dockets/04n0355/04N-0355-EC-9-Attach-1.pdf).

14 Evaluation of Viral Clearance in Purification Processes

Amitava Kundu and Karl Reindel

CONTENTS

14.1 INTRODUCTION

The use of continuous cell lines of mammalian origin in the manufacture of recombinant proteins and monoclonal antibodies introduces the potential of viral contamination in the purified product. Although to date no biotech products have been implicated in the transmission of infectious viruses, there are some recent documented instances of adventitious viral contamination of cell culture based products [1,2]. In each of these cases, contamination was most likely thought to be from an adventitious source, such as the medium or the serum used in the cell culture process. Additionally, rodent cell lines that are very commonly used in the manufacturing of monoclonals are known to express multiple copies of endogenous retroviral genomes [3–5], although none of these endogenous retroviruses or retrovirus-like particles (RVLP) have been shown to be infectious to humans. Nevertheless, the regulatory agencies worldwide require the quantification of the retrovirus titers in the cell culture harvests and the validated clearance of these retrovirus-like particles in the downstream purification steps using a specific model virus such as xenotropic murine retrovirus (X-MuLV), to provide assurance that the purified drug product is free of these endogenous contaminants [6–8].

RVLPs have been detected in both CHO and hybridoma cell lines. Published literature has shown the presence of RVLPs in CHO cell lines without any evidence of infectivity. Usually two types of particles (A- and C-type) have been observed using transmission electron microscopy (TEM). The A-type particles are located within the cytoplasm, often in association with centrioles and C-type particles. Little is known about the C-type particles; however their intracellular location and their low level of budding suggest that they are analogous to intracisternal A-type (IAPs) found in other rodent cell lines which are intracellular, lack a retroviral envelope and are noninfectious [9]. An explanation for the lack of infectivity of CHO C-type RVLPs comes from experiments carried out at Genentech using a recombinant CHO cell line [10]. Typically, between 10^3 and 10^6 RVLPs/ml are present in unprocessed bulks from industrial cell culture processes. Although the majority of C-type RVLPs produced by hybridoma cells appear to be noninfective (in S^+L^- assays), approximately one in a million hybridoma C-type RVLPs have the ability to replicate in S^+L^- cells [9]. However, the ability of these C-type RVLPs to replicate in human cells has been difficult to prove. C-type RVLPs are more abundant as a rule in hybridomas than those from CHO cells with levels typically ranging from 10^6 to 10^8 RVLPs/ml in the unprocessed bulk from industrial cell culture processes [9].

The possibility of viral contamination in the final product can arise either from the original source of the cell lines or from adventitious introduction of virus during the production processes. Viruses can be introduced into the master cell bank (MCB) by several routes such as (a) derivation of cell lines from infected animals; (b) use of virus to establish the cell line; (c) use of contaminated biological reagents such as animal serum components; (d) contamination during cell handling. Adventitious viruses can be introduced during production by several routes including, but not limited to, the following: (a) the use of contaminated biological reagents such as animal serum components; (b) the use of a virus for the induction of expression of specific genes encoding a desired protein; (c) the use of a contaminated reagent, such as a monoclonal antibody affinity column; (d) contamination during medium handling, (e) introduction of viruses from the manufacturing personnel due to a noncompliance of the current good manufacturing practices (cGMPs).

Direct testing of the final drug product for the absence of virus cannot ensure that the product is free from viral contamination. Direct methods are often designed to detect known specific contaminants and thus the testing methodologies may fail to pick up the presence of other unknown or unsuspected virus contaminant. Secondly, the methods developed may be so specific that they may fail to pick up variants of known potential contaminants as was seen in the initial hepatitis C screening kits [11]. The third limitation on direct testing methods is the inherent inability of these methods to detect very low levels of viruses. The ability to detect low concentrations of virus is also limited by statistical sampling considerations. As a result, the final product may contain virus that may escape detection by the direct testing methods.

To minimize the presence of viral contaminants in the final product, three complementary approaches are widely used by the manufacturers of biotechnology products. These are (a) selecting and testing cell lines and raw materials for the absence of undesirable viruses which may be infectious and pathogenic for humans; (b) assessing the capacity of the production processes to clear infectious viruses; and (c) testing the products at appropriate steps of production for the absence of contaminating infectious viruses. This article discusses in detail the design and requirements of the second approach, namely the evaluation of the production processes to clear adventitious viruses and retroviruses or retroviral-like particles that are commonly detected in the unprocessed bulk of cell cultures using hybridomas or murine cell lines.

14.2 HEALTH RISK FROM VIRUS CONTAMINATION

From a theoretical point of view, viruses of nonhuman hosts are less of a threat than viruses specific for humans, and the greater the evolutionary distance of a host species from humans, the lesser of a threat a virus specific for that species will be to man. However, the species barrier is not perfect and infection of humans can occur with what are normally considered to be viruses of animal origin — zoonoses. There are well-recognized zoonoses (such as rabies) and, clearly, if an animal species which is the host of known zoonotic agents is involved in the production of a final drug, then these zoonotic agents will certainly be of concern. In addition, it is also possible that a benign infection in one species can result in a fatal infection in another species. Simian B virus, which causes a benign herpes virus infection of monkeys, is potentially fatal for humans; similarly the Hantaan family of rodent viruses causes a nonapparent infection in the host species but a serious disease in humans — haemorraghic fever. Other nonhuman viruses which are known to cause diseases in human are Ebola (monkey), Lassa fever (rodent), Lymphocytic Choriomengitis Virus (rodent), equine morbilli (equine, bat) and many anthropod viruses such as West Nile Fever, Japanese encephalitis, and Western equine encephalitis [12].

14.3 RATIONALE AND ACTION PLAN FOR VIRAL CLEARANCE STUDIES

The ICH Q5A publication [6] clearly describes the different scenarios that can potentially occur in the manufacturing of biopharmaceuticals and recommends an appropriate action plan to provide assurance that the drug product is free of any viral contamination. These cases are described below.

Case A: Where no virus, virus-like particle or retrovirus-like particle has been demonstrated in the cells or the unprocessed bulk, virus removal and inactivation studies should be performed with nonspecific model viruses.

Case B: Where only a rodent retrovirus (or a retrovirus-like particle which is believed to be nonpathogenic) is present, evaluation should be done using a specific model virus such as a murine leukemia virus. Purified bulks should be tested using suitable methods having high specificity and sensitivity for the detection of the virus in question. For marketing authorization, data from at least three lots of purified bulk at commercial scale should be provided. Cell lines such as CHO, C127, BHK, and murine hybridoma cell lines have frequently been used as substrates for production with no reported safety problems related to viral contamination of the products. For these cell lines for which the endogenous particles have been extensively characterized and adequate clearance has been demonstrated, it is not necessary to assay for the presence of the noninfectious particles in the purified bulk.

Case C: When the cells or unprocessed bulk are known to contain a virus other than a rodent retrovirus for which there is no evidence of capacity for infecting humans, virus removal and inactivation evaluation studies should include the identified virus, if possible. In situations where it is not possible to use the identified virus, relevant or specific model viruses should be used to demonstrate acceptable clearance. Purified bulks should be tested using suitable methods having high specificity and sensitivity for the detection of the virus in question. For the purpose of marketing authorization, data from at least three lots of purified bulk manufactured at commercial scale should be provided.

Case D: Where a known human pathogen is identified, the product may be acceptable only under exceptional circumstances. In this instance, it is highly recommended that the identified virus be used for virus removal and inactivation evaluation studies. The process has to be shown to inactivate and remove the identified virus in the evaluation studies. Purified bulks should be tested using suitable methods having high specificity and sensitivity for the identified virus. For the purpose of marketing authorization, data from at least three lots of purified bulk manufactured at commercial scale should be provided.

Case E: When a virus, which cannot be classified by the existing methodologies is detected in the cells or unprocessed bulk, the product is considered unacceptable since the virus may prove to be pathogenic.

In all cases, characterization of clearance using nonspecific model viruses should be performed. It may be noted that the most common situations are cases A and B as described below. Usually cells or unprocessed bulk systems contaminated with a virus other than a rodent retrovirus are normally not used. However, when there are convincing and well-justified reasons for drug production using a cell line from cases C, D, or E as described below, these reasons should be discussed with the regulatory authorities. With cases C, D, or E, it is extremely important to have validated effective steps to inactivate/remove the virus in question from the manufacturing process.

14.4 CHOICE OF VIRUSES IN THE VIRAL CLEARANCE STUDIES

The viruses that are used in the clearance studies fall primarily into three categories, relevant viruses, specific model and nonspecific model viruses. Relevant viruses are viruses that are either (a) the identified viruses, or (b) of the same species as the viruses that are known, or likely to contaminate the cell substrate or any other reagents or materials used in the production process. A specific model virus is (a) closely related to the known or suspected virus (same genus or family), and (b) having similar physical and chemical properties to those of the observed or suspected virus. A nonspecific model virus is one that is used for the characterization of viral clearance of the process when the purpose is to evaluate the capacity of the manufacturing process to remove and inactivate viruses in general, that is, to characterize the robustness of the purification process.

The choice of viruses in the evaluation studies depends on the virus or virus-like particles that have been identified in the cells and unprocessed bulk and also on representing a wide spectrum of physicochemical properties in order to test the ability of the purification process to clear any adventitious viral contamination. Additionally, the stage of development of a product in part dictates the number of viruses used in the study. For example, for products derived from murine cell lines at the Phase I clinical trial or IND submission stage, it is usually sufficient to evaluate the clearance of murine retroviruses using a specific model virus such as xenotropic murine leukemia virus (X-MuLV). However, for licensure, it is mandatory to include at least a panel of four viruses, spanning a spectrum of widely different physicochemical properties. The use of model viruses is a very important concept in these studies and should therefore be properly selected to provide a claim that the purification process is sufficiently robust to assure that the final drug product is free from adventitious viral contamination. If a process study has demonstrated good clearance of viruses representing different virus groups and characteristics, then there is a high degree of assurance that any adventitious virus contamination, if unintentionally introduced into the system, would be cleared by the production process. The same arguments also apply to the risk posed by unknown viruses. Examples of model viruses representing a wide range of physicochemical properties that can potentially be used in viral clearance studies are presented in Table 14.1.

In addition to the above considerations, the other points that need to be considered in the selection of viruses are (a) viruses that can be grown to high titers are desirable, although this may not always be possible; (b) there should be a reliable and a sensitive assay for the detection of the viruses used at every stage of the scale-down manufacturing process; (c) consideration should be given to the health hazard which certain viruses may pose to the personnel performing the clearance studies.

While the viruses in Table 14.1 have been used in viral clearance studies, for practical reasons, it is not necessary to test all types of viruses in an evaluation study. In addition to specific model viruses that resemble closely the virus or retrovirus detected in the cells and unprocessed bulk, the selection of other viruses should give preference to viruses that display a significant resistance to physical and chemical treatments. Table 14.2 provides an example of a panel of viruses that can be used to validate a purification process for a product derived from murine hybridoma cell line.

In a mouse or hamster cell line, murine retroviruses detected in the cells or unprocessed bulk are one of the main virus groups of concern, since a very small proportion of these retroviruses have been shown to be capable of replicating, although none of these have been shown to be capable of infecting human cell lines. Therefore a model for a murine retrovirus must always be included in any study and murine leukemia virus (MuLV) is most commonly used as a model specific virus. Pseudorabies virus is a model for a herpes

TABLE 14.1
Example of Viruses Used in Viral Clearance Studies

Virus	Family	Genome	Enveloped	Size (nm)	Shape	Resistance
MuLV	Retro	RNA	Yes	80–110	Spherical	Low
Parainfluenza virus	Paramyxo	RNA	Yes	100–200^{+}	Pleo/spherical	Low
Sindbis virus	Toga	RNA	Yes	60–70	Spherical	Low
BVDV	Flavi	RNA	Yes	50–70	Pleo/spherical	Low
Pseudorabies virus	Herpes	DNA	Yes	150–200	Spherical	Medium
Poliovirus sabin type I	Picorna	RNA	No	25–30	Icosahedral	Medium
Encephalomyo-carditis virus	Picorna	RNA	No	25–30	Icosahedral	Medium
Reovirus 3	Reo	RNA	No	60–80	Spherical	High
SV40	Papova	DNA	No	40–50	Icosahedral	Very high
Parvoviruses	Parvo	DNA	No	18–24	Icosahedral	Very high

Source: Reproduced from ICH Topic Q5A: Viral Safety Evaluation of Biotechnology Products Derived from Cell Lines of Human or Animal Origin, 1997. With Permission.

TABLE 14.2
A Panel of Viruses Used for Virus Validation Studies

Virus	Genome	Enveloped	Size (nm)	Resistance
Murine leukemia virus (MuLV)	ss-RNA	Yes	80–120	Low
Pseudorabies virus (PRV)	ds-DNA	Yes	150–200	Low to medium
Reovirus 3 (Reo 3)	ds-RNA	No	60–80	high
Minute virus of mice (MVM)	ss-DNA	No	18–25	Very high

virus, and like retroviruses can establish latent infections within cells and thus escape detection. Reovirus 3 is often used in the studies as it is zoonotic and infects a wide variety of cell lines from different species. Minute virus of mice (MVM), a parvovirus, is the final virus selected in this study as it is a small, highly resistant virus that severely challenges the capacity of the production process to clear viruses. In addition, MVM has been implicated in a few instances in the contamination of production runs of Chinese Hamster Ovary (CHO) cell line derived products [1]. A model virus selection as presented in

Table 14.2 thus covers not only specific viruses or virus groups of concern, but also selects viruses that have the following characteristics: (a) DNA and RNA genomes with single and double stranded, (b) lipid-enveloped and nonenveloped, (c) large to small sizes, (d) low to an high resistance to physicochemical reagents.

14.5 SELECTION OF STEPS TO BE EVALUATED IN VIRAL CLEARANCE STUDIES

The ultimate objective of viral clearance studies is to demonstrate that the purification process is capable of eliminating substantially more virus that what may be potentially present in the unprocessed bulk. In addition, the purification process should have the capacity to clear any adventitious or unknown viral contaminants. In light of this, several steps in the process are usually studied independently by deliberately spiking viruses and measuring the clearance by estimating the virus titers in the load and the product by infectivity or other appropriate assays. The log clearance from each of the steps is then added together to give an overall log clearance of the purification process. When adding the log clearance from the process steps, one should consider log clearance from only the orthogonal steps (steps that inactivate/remove viruses by orthogonal mechanisms) that provide $>1\ \log_{10}$ of viral clearance. Although a purification process may consists of several steps, not all the steps need to be evaluated for virus clearance.

Several factors influence the decision of which steps should be studied when performing virus clearance studies. An important criterion in the selection of a step is to incorporate the ones that are deemed to be robust. The definition of robust in this case is a step that (a) can be scaled down accurately and (b) will reproducibly and effectively remove or inactivate a wide variety of potential viral contaminants [11]. Steps such as pH inactivation, solvent/detergent inactivation, and nanofiltration fall into this category. Steps such as precipitation, centrifugation, and other types of filtration are difficult to scale-down and are thus viewed on a case-by-case basis. Column chromatography steps such as hydrophobic interaction and ion exchange lies somewhere between the two extremes. Another consideration is whether or not a particular step will provide any significant level of virus clearance. Information available from the regulatory agencies and the large database that exists with contract virus testing laboratories can be very useful in identifying the process steps. Including at least one step that inactivates viruses such as a low pH step, another step that removes viruses based on a size-based mechanism such as nanofiltration, and other steps that remove viruses based on a binding (or nonbinding) mechanism such as an affinity chromatography step is highly desirable from a regulatory standpoint. Typically, in a monoclonal antibody purification process, proteinA, anion exchange and a cation exchange chromatography, nanofiltration, and a low pH step are evaluated for viral clearance.

14.6 SCALE DOWN OF MANUFACTURING PROCESS STEPS

Virus clearance studies are always performed with scale-down models of the manufacturing process steps. It is not feasible to perform viral clearance studies at the manufacturing scale because it would be inappropriate to introduce infectious virus into a cGMP manufacturing facility. Also, the volumes of virus needed to achieve a satisfactory spiking level at the manufacturing scale would be impractical and prohibitively expensive. Thus, in order for the virus clearance studies to be extrapolated to the manufacturing scale, it is imperative that the scale-down model is a true representation of the full-scale manufacturing process. The following paragraphs describes briefly the strategy for scaling down chromatography, nanofiltration and low pH inactivation steps.

Although scaling down a chromatography step is relatively straightforward, attention needs to be paid to the details of the scale-down process. While the column diameters between the manufacturing and the scale-down process may differ by 100-fold, yet the column heights should be the same at the two scales. Maintaining the same linear velocity at the two scales will ensure the same contact time. Additionally, the column volumes for each of the buffers used should be the same across the two scales. The process step at the small scale is loaded within the range observed in the manufacturing scale, using a typical load concentration from an earlier processing step. Having adjusted the process input parameters, the first step to ensure that the small-scale process is representative of the manufacturing scale is to compare the chromatograms with regards to pH, UV, and conductivity profiles. In addition, some product quality attributes such as percentage of monomer, percentage of aggregate, and level of host cell proteins and other product and process-related impurities as deemed appropriate for this step are measured and compared to the values obtained at the manufacturing-scale.

The primary purpose of a nanofiltration step is to provide viral clearance by size exclusion. For the purification of a monoclonal derived from cell culture, typically nanofiltration with very small pores (20 nm or less) is performed to provide clearance of enveloped and nonenveloped small viruses. Although the filtration area used in a manufacturing-scale process may be as much as 4000-fold or more than that used in a scale-down process, yet the volumetric loading measured in volume of load material filtered per unit surface area of the filter should be kept constant across the two scales. A typical load material from an earlier processing step obtained from manufacturing should be used for the scale-down studies. Since this step is usually carried out under constant pressure, the inlet pressure should be matched for the two scales and similar flush volumes in terms of liters of buffer flushed to recover any protein held up in the pores per unit surface area should be matched. An important output parameter is the average volumetric flux as a function of volumetric capacity, obtained during the course of the filtration, and this should have a similar trend for the two scales. The yield and resultant product pool concentration for the scale-down model should be within the ranges observed at the manufacturing scale.

The inactivation of enveloped viruses by low pH is a very common step in a monoclonal antibody purification process. Typically, low pH inactivation step is performed after the protein A affinity chromatography step. The product from the protein A affinity column is titrated to a pH of 3.8 or lower and incubated for a period of 15 to 60 min, depending on the monoclonal antibody that is purified. This brief exposure to low pH effectively inactivates most lipid enveloped viruses. Following this incubation, the protein solution is then titrated upwards to a pH of 5.0 or higher to prepare for the next step. While the inactivation of viruses is favored by low pH, there is an inherent risk of aggregating the target protein at these low pH conditions. Moreover, a strongly acidic solution that is used as a titrating solution for the low pH step can cause localized low pH conditions, if not adequately mixed. This could potentially cause aggregation of the protein solution. While the scale-down of a low pH inactivation step is relatively straightforward, care must be taken to ensure adequate mixing at this step during the addition of the acid to titrate to a low pH such that no undue aggregation of the target protein results across this step. Thus, upon completion of this low pH step and subsequent neutralization, the aggregate content of the protein during scale-down studies should be within the range observed during manufacturing.

Once the model has been established, typically the scale-down model is run in duplicate to ensure that the performance of the scale-down model is similar to the manufacturing-scale process. Under certain circumstances, there might be certain limitations in accurately scaling down the manufacturing-scale process. In these situations, the deviation should be noted and the implications of this deviation on the viral clearance results should be explained. In addition to scaling down the process steps, it is also important to use buffers and loads for each step that are representative of the manufacturing process. As a standard practice, the scale-down model verification experiments are done with buffers and process intermediates obtained from the full scale-manufacturing process.

In addition to scaling down the manufacturing process steps using representative process intermediates, it is also necessary to demonstrate the validity of the scale-down model with the virus spike in the load for each step, as the presence of virus spike has been shown in some cases to have a dramatic impact on the performance of the process step [11], in particular the yield across the bind and elute process steps. Many virus preparations, especially for enveloped viruses, include high concentrations of protein (including serum proteins), lipids, nucleic acid, and in some cases, phenol red as a pH indicator, all of which could detrimentally impact the performance of the scale-down model such that it no longer compares to the performance of the manufacturing-scale process. Thus, whenever possible, it is important to study the impact of the virus spiking on the process steps by performing mock spiking experiments with the medium in which the viruses are stored in advance of the actual studies. Results from these studies can then be used to decide on the ratio of the spike to the volume of the load material used at each process step. In general, the volume of the spike should not exceed more than 10% of the volume of the load material, to ensure that the composition and nature of the starting material is not significantly altered as compared to the load at that stage of the manufacturing process. In practice, virus-spiking studies will always be a compromise trying to add as much virus as possible in order to potentially claim the maximum virus clearance without negatively altering the performance of the process step.

14.7 ESTIMATION OF VIRUS TITERS

The quantitation of infectious virus particles in process samples for virus clearance studies is done primarily using either (a) a cytopathic effect (CPE) assay or (b) a plaque (or focus) forming assay. In a plaque assay, the virus titer is determined by dividing the number of plaques by the total volume of the original sample tested. This method is quantitative as it counts the number of plaques as a function of the virus dose. This method of computation is an averaging procedure that gives equal weight to equal volumes of the original suspension at different dilutions. The virus titer is normally expressed as a logarithmic value with a 95% confidence interval.

The second method used to quantitate infectious virus is based on the cytopathic effect. This is also known as the Tissue Culture Infectious Dose at 50% infectivity ($TCID_{50}$). This method is used to quantitate viruses that do not form plaques, but cause a change in the cellular morphology. This assay however is a quantal assay in the sense that the wells are scored either positively or negatively for the presence of infectious viruses in samples diluted to the end-point. The dilution of the sample needed to infect 50% of the culture wells is determined and used to calculate the virus titers. The accuracy of this assay depends on how reproducibly and reliably the infection rate at each dilution is determined. As a result, a large number of replicates at each serial dilution are done to ensure an accurate titer determination. For the calculation of the titers, either the Spearman–Kaerber method or the modified Spearman–Kaerber method is used. The virus titer using either of the methods is normally expressed as a logarithmic value with a 95% confidence interval. A detailed description of the method of virus determination using either the plaque or the $TCID_{50}$ assay can be found in this article [13].

14.8 CYTOTOXICITY AND VIRAL INTERFERENCE TESTING

Prior to performing virus-spiking studies, it is essential to perform cytotoxicity and virus interference studies on the process intermediates. This is a regulatory requirement because samples generated during an actual spiking study may cause significant problems in the titration of the virus thereby obtaining an accurate estimation of the virus present in the sample. These problems may arise from the cytotoxicity of the samples. Cytotoxicity assays are performed to demonstrate whether process intermediates are toxic to the indicator cell lines used in the virus titration assays. This can be determined by incubation of the nonvirus containing process intermediate on each of the indicator cell lines and assessing whether this causes any change in the cell morphology. In addition to being cytotoxic, the process sample might also interfere with the ability of the virus to infect the indicator cell lines or prevent detection of the appropriate virus-induced cytopathic effect. This is termed as viral interference. These studies are done by first exposing the indicator cell lines to the samples being tested and then infecting the cells with a known amount of the virus. By comparing the virus titer obtained in the treated versus the untreated control cells (cells with and without the exposure to the process samples), the degree of interference can be assessed. Interference cannot be measured from cytotoxicity and it is possible that samples that show little or no cytotoxicity can show significant interference.

Cytotoxicity and viral interference studies are usually done with multiple serial dilutions of the process samples until no interference or cytotoxicity is observed with two successive dilutions. These studies are typically done using a 96 well format microtiter plate (12 rows of 8 wells each), very similar to the setup of an ELISA assay. Typically, a small volume such as 50 μl of the process sample (either diluted or nondiluted) is pipetted into each of the wells for assessing cytotoxicity as well as viral interference. The least dilution (or no dilution) of the process sample at which neither viral interference nor cytotoxicity is observed is used to estimate virus titers in the various fractions (i.e., load, wash, elute etc.) of the unit operations that are evaluated during the virus-spiking studies as described in Section 14.9. Without the data from interference studies, the log clearance values obtained in the virus-spiking study for a particular process step may be either under- or overestimated, depending on whether the load or the product sample from that step is interfering or not.

Although cytotoxicity and interference can usually be eliminated by dilution of the samples, it is important to note that the minimum possible dilution of the sample should be performed in order to maximize the possibility of obtaining the best possible clearance data. This is especially important in the case of process steps where no virus is detected in the product and hence a theoretical titer must be assigned based on a Poisson distribution (see Appendix 14.A1).

An example of the results from cytotoxicity and viral interference studies and the minimum possible dilution of the process samples for a purification scheme comprising of three process steps (1 through 3) is presented in Table 14.3. The minimum dilution chosen is the one where no cytotoxicity or

TABLE 14.3
Cytotoxicity, Viral Interference and Minimum Valid Dilutions for Process Intermediates

Process Intermediate	Dilution	Cytotoxicity Observed	Viral Interference $Log_{10}(TCID_{50})$/ml	Minimum Dilution
Negative control	None	—	N/A	
Positive control	None	N/A	8.05 ± 0.32	
Certified titer	None	N/A	7.58 ± 0.28	
Load for process Step #1	Undiluted	—	7.93 ± 0.24	
	1:3	—	7.80 ± 0.35	None
	1:10	—	7.85± 0.43	
Product for process Step #1	Undiluted	—	7.68 ± 0.40	
	1:3	—	8.05± 0.36	None
	1:10	—	8.05 ± 0.35	
Load for process Step #2	Undiluted	+	TOX	
	1:3	—	TOX	1:10
	1:10	—	7.18 ± 0.36	
Product for process Step #2	Undiluted	—	8.05 ± 0.32	
	1:3	—	8.18 ± 0.44	None
	1:10	—	7.68 ± 0.24	
Load for process Step #3	Undiluted	—	8.05 ± 0.32	
	1:3	—	7.68 ± 0.40	None
	1:10	—	7.68 ± 0.40	
Product for process Step #3	Undiluted	+	TOX	
	1:3	—	7.18 ± 0.40	1:3
	1:10	—	7.93 ± 0.40	

− = No cytotoxicity is observed as per 8 wells.
+ = Cytotoxicity is observed as per 8 wells.

viral interference is observed. This table needs to be constructed for each of the viruses used in the evaluation study.

14.9 DESIGN OF VIRUS-SPIKING STUDIES

With the qualification of the scale-down models completed along with the cytotoxicity and viral interference experiments, the next step is the spiking of the process samples with concentrated virus preparations to predetermined levels and then measuring for the presence of viruses in the process solutions using appropriate infectivity assays. For each process step assessed, the possible mechanism of loss of infectivity should be described with regard to whether it is due to inactivation or removal. The number and nature of samples assayed for each process step depends on the type of step being studied. For the inactivation studies such as low pH inactivation of enveloped viruses, it is mandatory to take samples at different timepoints throughout the duration of the study as specified in the manufacturing process, such that an inactivation curve can be constructed. This is important as virus inactivation is not a simple, first-order reaction and is usually more complex, with a first phase 1 and a slow phase 2. It is highly recommended that the inactivation studies include at least one timepoint less than the minimum exposure time and greater than zero, in addition to the minimum exposure time. On the other hand, for the chromatography steps, in addition to the load and product fractions, flowthrough, wash, and regeneration samples are also assayed to understand the partitioning of the virus in the different fractions. However, for the filtration steps such as nanofiltration, only the product and the filtrate samples are usually assayed to determine the viral clearance. It is also important to note that during evaluation of nanofiltration steps for viral clearance, a prefiltration step on the spiked load needs to be incorporated prior to applying the load on the nanofilter. This prefiltration step is necessary for the removal of aggregated virus particles in the spiked load, if any, such that the nanofilter is challenged to remove only the nonaggregated virus particles.

Appropriate hold control samples at the process temperature (e.g., the hold control sample should be stored at 2 to 8 C only if the process step is carried out at 2 to 8 C) should also be included for each of the process steps studied, to ensure that the virus spiked into the process sample does not lose infectivity during the course of the study. Samples from the spiking studies should be titrated immediately upon collection. If this is not possible, and it is necessary to freeze samples prior to titration, then appropriate controls with the stock virus solution should be carried out.

14.10 CALCULATION OF LOG REDUCTION FACTORS IN A VIRAL CLEARANCE STUDY

The following sections provide examples of virus clearance studies performed for a typical monoclonal antibody purification process and shows how the log reduction values for each process step and the manufacturing process as a whole is estimated. Since the virus titers are normally expressed with 95% confidence intervals, the same should be done when reporting viral clearance for each process step and the production process as a whole.

In the example shown in Table 14.4, the log reduction value (LRV) for the chromatography step is calculated as the difference between the viral load in the spiked load and the product pool which is $Log_{10}(7.66 \pm 0.24) - Log_{10}(4.32 \pm 0.40) = Log_{10}(3.34 \pm 0.47)$. In this specific example, the 95% confidence interval of the LRV is calculated as the square root of the sum of the squares of the confidence intervals of the load and product fractions. Also, it may be noted that hold control titer was within the expected titer range of the spiked load, which indicated that there was no significant decrease in the virus titer over the time course of the study. On the other hand, if the hold control titer was not within the expected titer range, then the LRV value has to be calculated based on the hold control value instead of the spiked load. In this specific example, it would have been $Log_{10}(7.46 \pm 0.37) - Log_{10}(4.32 \pm 0.40) = Log_{10}(3.14 \pm 0.54)$.

Table 14.5 is an example of a viral clearance study of a purification process for an antibody expressed in a murine hybridoma cell line.

In this specific example, the total log_{10} clearance of the purification process as a whole for each virus is calculated by adding up the log_{10} clearances from each of the steps while the 95% confidence interval of the overall purification process is calculated as the square root of the sum of the

TABLE 14.4
Viral Clearance Calculations for a Chromatography Step

Process Sample	Titer ± 95% CI (Log_{10} $TCID_{50}$/ml)	Volume (ml)	Viral Load (Log_{10} $TCID_{50}$/ml)
Spiked load	5.68 ± 0.24	95.2	7.66 ± 0.24
Hold control	5.48 ± 0.37	95.2	7.46 ± 0.37
Flowthrough + wash	2.27 ± 0.40	233.7	4.64 ± 0.40
Product pool	2.68 ± 0.40	43.6	4.32 ± 0.40
Column strip	2.55 ± 0.49	73	4.89 ± 0.49

CI: Confidence interval.

TABLE 14.5
Summary of Viral Clearance for an Antibody Purification Process

	$\log_{10}$ (Clearance + 95% CI)	
Process Step	**Specific Model Virus**	**Nonspecific Model Virus**
Chromatography Step #1	N/A	1.80 ± 0.44
Low pH inactivation Step	$>5.71 \pm 0.28$	N/A
Chromatography Step #2	3.87 ± 0.51	4.00 ± 0.47
Chromatography Step #3	3.34 ± 0.47	0.48 ± 0.56
Viral filtration	$>5.63 \pm 0.43$	$> 4.80 \pm 0.40$
Total $\log_{10}$ clearance	$>18.55 \pm 0.86$	$>10.60 \pm 0.94$

squares of the confidence intervals of each of the process steps. It is important to note here that the log reduction of 0.48 ± 0.56 for the nonspecific model virus for chromatography step #3 is not included in the calculation for the overall log clearance for the nonspecific model virus as this LRV is $<1 \log_{10}$.

14.11 ASSESSMENT OF THE SAFETY FACTOR IN THE FINAL DRUG PRODUCT

Having obtained the overall log reduction value for the entire purification process, it is important to put this number in the context of risk assessment of the final drug product. This assessment is performed following the recommendations of the regulatory guidelines [6–8]. According to these guidelines, the level of clearance demonstrated should be substantially in excess of the potential virus load in one dose of the final product, as calculated from the endogenous virus particle count obtained by transmission electron microscopy (TEM) of at least three lots of unprocessed bulk at the manufacturing-scale. An example of such a calculation with the assumptions is provided below:

Assumptions

1. Number of viral particles estimated in the unprocessed bulk by TEM: 10^8/ml
2. Calculated viral clearance for the specific model virus (model for the endogenous viral particles): $>18.55 \pm 0.86$
3. Volume of unprocessed bulk required to make a dose of product: 2000 ml

Calculation of estimated particle per dose and the safety factor

1. Number of viral particles in a dose, if there was no clearance: $2000 \times 10^8/\text{ml} = 2 \times 10^{11}$ or 11.3 $\log_{10}$
2. Thus safety factor is ($>18.55\log_{10}$)–($11.3\log_{10}$) => $7.25\log_{10}$ or in other words there is less than one virus particle per 17.8 million doses ($10\hat{\ }7.25 = 17.8 \times 10^6$)

This calculation is relevant only to those viruses for which an estimate of the starting numbers is available, as in the case of endogenous retroviruses or retrovirus-like particles. For other nonspecific model viruses for which viral clearance studies are performed, but no such estimate exists in the starting material, there is no requirement for a specific clearance value, although the expectation is that there are at least a couple of process steps where a significant LRV is obtained consistently.

14.12 QUANTITATIVE POLYMERASE CHAIN REACTION ASSAY FOR VIRUS QUANTITATION

While the cell-based infectivity assays as described in Section 14.7 are viewed as the gold standard for the estimation of virus titers in viral clearance studies, quantitative polymerase chain reaction (Q-PCR) is rapidly gaining acceptance as an alternative and complementary method for estimation of virus particles in virus clearance studies. A number of reports, primarily from the work done collaboratively between Genentech and the Division of Monoclonal Antibodies, Center for Biologics Evaluation and Research of FDA have been published [14–17] that employs a real time Q-PCR method for the quantification of a range of model viruses that are commonly used for virus validation studies. These studies have shown comparable log reduction values across orthogonal process chromatography and nanofiltration steps for typical model viruses.

Real time Q-PCR is based on the 5′–3′ exonuclease activity of TaqDNA polymerase and the amount of virus is determined by quantifying viral genomic DNA or RNA using an appropriate detection system. As in traditional PCR, Q-PCR incorporates primers that amplify target-specific regions of nucleic acids but unlike traditional PCR, also employs a fluorogenic probe that is labeled with a fluorescent reporter dye at the 5′ end and a quencher dye at the 3′ end. This probe anneals to the region between the primer sets. When the probe is intact, the proximity of the reporter dye to the quencher dye results in the suppression or quenching of the fluorescence. However, during amplification, the TaqDNA polymerase cleaves the probe, resulting in the release of the reporter dye and a concomitant increase of fluorescence of the reporter dye that is directly proportional to the amount of PCR product accumulated. Thus, there is an increase of the fluorescence with the number of PCR cycles. The PCR cycle during which the system begins to detect the fluorescence is defined as the threshold cycle (C_T). The more target DNA/RNA present in a test sample at the outset, the earlier is the threshold cycle reached. A standard curve is used to quantify the amount of DNA/RNA in the test sample. This standard curve is generated using a serial dilution of a known concentration of a standard DNA/RNA. This standard DNA/RNA should have identical primer/probe binding sequences and amplification efficiencies as the target DNA/RNA. The C_T value obtained for each standard DNA/RNA dilution is plotted against the $\log_{10}$ of the corresponding standard DNA/RNA concentration and a straight line fit is obtained for the standard curve using a linear regression analysis. The target DNA/RNA concentration in the test sample is then determined from this standard curve using the experimentally obtained C_T value of the test sample.

Typically, the linear range of the standard curve is over 100,000-fold or 5-logs and hence reliable quantitation is possible over this wide range. This method is highly sensitive with a detection limit of approximately one virus particle per reaction. In quantitative terms, it has a limit of detection of 0.6 fg of DNA per microliter of test sample. This may translate to a 100-fold higher sensitivity as compared to a $TCID_{50}$ assay [15]. This higher sensitivity of the Q-PCR assay can be potentially useful in claiming a higher log reduction value in cases where the virus is cleared to nondetectable levels. Secondly, it has a much higher sample throughput. A cell-based infectivity assay is very labor and time intensive requiring 7 to 14 days to get an output reading. On the other hand, the Q-PCR method can provide a result easily within a day on multiple samples. Q-PCR assays are also much more cost effective and easier to perform, as it does not involve the expensive reagents needed to grow and keep the cells alive. Additionally, a Q-PCR assay can potentially have less interference from the presence of buffer components, salts, and protein concentrations most likely due to the extraction process, which employs an efficient wash step to remove possible interfering components. Another important advantage that relates to virus clearance studies is the ability to use this assay to quantitate the removal of pH labile enveloped viruses such as X-MuLV across the Protein A chromatography step, which is commonly used as an initial capture step in a monoclonal antibody purification process. Typically, Protein A chromatography employs a low pH elution buffer to elute the bound antibody. Such low pH elution buffers can partially inactivate enveloped viruses and thus it is not possible to quantitate the physical removal from the inactivation of the viruses across the Protein A chromatography step using an infectivity based assay. Since Q-PCR quantitates both infectious as well as noninfectious virus particles, it can be used to quantitate the number of residual virus particles in the protein A product. Thus, the reduction factor solely due to the physical removal of the virus particles across this step can be estimated and hence claimed in the overall virus clearance calculations. However, the one disadvantage of using Q-PCR is the inability to quantitate the enveloped virus clearance for virus inactivation steps such as low pH, which is commonly used as an orthogonal step to nanofiltration and chromatography for virus clearance. This limitation stems from the fact that Q-PCR does not rely on infectivity for quantitation, whereas low pH steps only impacts the infectivity of the virus. Thus the total number of virus particles (infectious and noninfectious) is unaltered across the low pH step.

Although, Q-PCR has been employed successfully to estimate virus clearance across nanofiltration step, it might be important to remember that virus stocks used for spiking studies sometimes can contain a certain portion of free DNA fragments not associated with the intact virion, but are nevertheless large enough to be detected by the Q-PCR assay. These free fragments can potentially pass to the filtrate side with the product and result in false positives.

14.13 IDENTIFICATION OF WORST-CASE SITUATIONS

While process evaluation studies can be conducted at process extremes to test the robustness of a process, it is not feasible to conduct viral clearance studies at process extremes, as these studies are very expensive and time consuming. Instead, it may be prudent to perform virus validation studies under worst-case conditions, if such conditions can be properly identified. The identification of such conditions however hinges on the understanding of the factors that influence the mechanism of clearance.

In the case of pH inactivation studies, high protein concentrations may have a protective effect on the virus inactivation. Thus a combination of high protein concentration, high pH and low exposure time (all within process ranges) will result in the worst case. In the case of solvent–detergent inactivation, the combination of lowest solvent–detergent concentration, low exposure time and low exposure temperature (all within process ranges) results in the worst case. In case of nanofiltration, combination of process conditions that result in the greatest degree of decay in the volumetric flux relative to the initial flux will most likely be the worst case, for viruses where the pore size of the filter is similar to that of the virus. For a chromatography step operated in a flowthrough mode, such as an anion exchange step, usually the higher loading represents a worst case. However, for binding chromatography steps, the lowest protein concentration during binding could present the virus with more sites to bind onto the column and thus co-elute with the product.

14.14 COLUMN SANITIZATION AND REUSE OF CHROMATOGRAPHY RESINS

While the nanofiltration step is done with disposable filters and are thus used only once, the same cannot be said of the chromatography resins. Chromatography resins such as recombinant protein A affinity resins are prohibitively expensive to be used only once in a manufacturing process. Moreover, if a resin is used only once, all the process-scale chromatography columns will need to be packed and tested for each lot processed, which would make the manufacturing process very inefficient. Thus, it is not uncommon to reuse chromatography resins for as many as 100 cycles. However, multiple uses of the chromatography resins poses a safety risk because declining performance of the media may lead to diminished virus removal capabilities. Virus clearance studies thus need to be performed both on new resins and on resins at the end of their production lifetime. Additionally, viruses may be bound onto the resin, and if the resin is not properly sanitized or regenerated to either inactivate or remove the virus, buildup of viruses can occur, which can potentially contaminate the next or several later batches of the product purified with the reused resin. There is documented evidence that viruses can bind onto chromatography matrices and survive several cycles of purification before eluting with the product [11]. In an experiment conducted by Pharmacia, IgG was fractionated using a three-column step purification process. The product from the intermediate Q-Sepharose chromatography step was processed on the final CM Sepharose step by dividing the product from the Q step into three different fractions. The CM Sepharose column performed three cycles of purification without any regeneration or sanitization in-between. To study the effect of potential virus carryover, virus was spiked only onto the load of the first purification cycle. The load material for the second and third purification cycles were not spiked. The IgG product fractions from the first and third cycles were then assayed for the presence of infectious virus. No virus was detected in the product of the first cycle, indicating complete clearance of the virus from the load fraction. However, significant amounts of infectious virus was detected in the product from the third purification cycle, suggesting that the virus spiked during the first purification cycle must have bound onto the chromatography resin and eluted during the subsequent purification cycle.

The previous example underlines the importance of designing effective sanitization steps in between cycles to avoid carryover of virus particles from one cycle to the next. Such a sanitization regime needs to be validated by performing solution spiking studies in the sanitization buffer for a duration equivalent to that exposed in the production columns for that process step. In addition, small-scale reuse studies with spiking the virus in only the first load and not the second, but sanitizing the column in-between runs in a manner similar to the manufacturing process should be performed. If the sanitization regime is effective in inactivating and removing the bound virus, then no virus should be detected in the product from the second cycle.

As an alternative to performing small-scale studies using new and aged resins, a second approach is to perform virus removal validation studies on the new resin only, and then monitor during production, chromatography performance attributes such as product step yield or product impurity levels that might decay prior to virus LRV [18]. This approach requires the identification of such a performance attribute, but obviates the need for measuring virus LRV by used media. Reuse studies on Protein A media have been successfully used as a model to prove the validity of this alternative approach. However, with this approach the solution spiking studies as well as the small scale reuse studies for determining virus carryover still needs to be done.

14.15 LIMITATIONS OF VIRAL CLEARANCE STUDIES

Although viral clearance studies play an important role in assuring the safety of the drug product from a viral contamination perspective, these studies by themselves are by no means a guarantee of the safety of the final product. It is extremely important to remember that there are a number of factors in the design and execution of these studies that can lead to an incorrect estimation of the overall viral clearance.

Virus clearance studies are done in scale-down models and even with accurate scale-down, there is no guarantee that the virus clearance will be identical at the two scales. The viruses that are used in these studies are produced in cell culture that may differ from the native virus in their susceptibility to inactivation and removal. Small variations in the process at the manufacturing-scale might impact the virus clearance, and thus studies should be performed under worst case conditions, if these can be clearly identified. Overall virus clearance is obtained by summing up the clearance of apparently orthogonal steps, but this might lead to an overestimate if the clearance actually occurs through a similar mechanism in seemingly independent processing steps. While the clearance mechanism is relatively well defined in the low pH inactivation and nanofiltration steps, the removal of viruses across chromatography steps occurs through a complex combination of hydrophobic and ionic mechanisms and thus there may be some overlap between orthogonal chromatography steps. Due to the nature of the evaluation studies, the process steps are loaded with extremely large amounts of viruses, which for the most part is extremely different from the real life situation, where virus contamination, if present, will only be at a much lower level. The clearance values thus estimated from an overloaded case may not match with the actual clearance that may be obtained from a much lower level of contamination. Last, but not the least, the expression of reduction factors as logarithmic reductions in titer implies that while residual virus infectivity may be greatly reduced, it will never be reduced to zero.

14.16 RE-EVALUATION OF VIRAL CLEARANCE

The impact of a process change on viral clearance need to be reassessed, if there is a change, major or minor, to the manufacturing process. If the change is deemed to have an impact on the viral clearance, then the viral clearance needs to be reevaluated as needed to determine the impact of the manufacturing changes on the safety of the drug product. It is possible that some changes in cell culture conditions can lead to a significant change in the number of endogenous retroviral particles in the unprocessed bulk that could then have an impact on the overall safety of the drug product. Also, changes in process steps, such as an introduction of a new step, or deletion of an existing step, or substitution of an existing step with a new one, might have an impact on the viral clearance.

14.17 BRACKETED GENERIC APPROACH TO VIRUS CLEARANCE STUDIES

The purification schemes for monoclonal antibodies tend to be similar across different antibody subclasses and certainly within the same subclass. The arrangement of the capture and the polishing chromatography steps and the virus inactivation and filtration steps are also somewhat similar with minor variations in buffer conditions such as conductivity and pH and protein concentration. Thus, it is theoretically possible to bracket the virus clearance of process steps with respect to the operating parameters (this is also known as matrix approach) for a range of commonly used model viruses using one antibody and then extrapolate the results to another antibody with a similar sequence of downstream process steps. This concept of extrapolation from one antibody to another antibody is termed as generic approach. Although this bracketed generic approach is yet to be officially accepted by the regulatory authorities, a few publications and presentations based on this concept have been presented to the biotech community. A synopsis of those is presented in this article.

Low pH inactivation is very commonly used as a virus inactivation step in a monoclonal antibody purification process. Since this step has been reliably shown to inactivate $>4 \log_{10}$ of large enveloped viruses such as X-MuLV in quite a few commercial purification processes, the concept of a bracketed generic virus clearance was investigated [19,20]. Based on the data, the authors proposed that a bracketed generic clearance log reduction value of 4.6 $\log_{10}$ of X-MuLV can be applied to monoclonal antibodies that have a purification process that include a low pH step meeting the following criteria:

1. pH of incubation is $\leq$3.8
2. Incubation time is $\geq$30 min
3. Incubation temperature is $\geq$14°C
4. Buffer system is citrate or acetate
5. Total protein concentration is $<$40 mg/ml
6. Sodium chloride concentration is $\leq$500 m*M*
7. pI of the monoclonal is between 3 and 9
8. Low pH incubation step is performed on a cell-free harvest intermediate after the initial capture step of the recovery process
9. Product is not a retrovirus targeted monoclonal antibody

In another publication [16], the authors have published the results of a bracketed generic clearance of SV40, a nonenveloped model virus across a Q-Sepharose Fast Flow (QSFF) chromatography step. The clearance of SV40 as a function of several key process parameters was experimentally investigated. The authors proposed that a bracketed generic clearance of 4.7 $\log_{10}$ of SV40 can be applied to monoclonal antibodies that have a flowthrough QSFF purification step meeting the following criteria:

1. Flow rate of 76 though 600 cm/h
2. Bed height of at least 11 cm
3. Load capacity of $<$250 mg IgG/ml resin
4. pH of the equilibration and load in the range of 7.0 through 8.5
5. Conductivity of the equilibration and load in the range of 25 through 100 m*M* NaCl

If officially accepted by the regulatory authorities, this approach has the potential to beneficially impact companies that have a pipeline of several antibodies with similar purification process. The information could be used to make decisions during process development of new products, to support postapproval changes without additional viral clearance studies, to resolve manufacturing deviations and finally for IND and BLA submissions. Furthermore, this approach can afford considerable flexibility to academic IND sponsors with limited resources that can be focused on other areas of research than on viral clearance studies.

14.18 MULTIVIRUS-SPIKING APPROACH FOR VIRUS CLEARANCE STUDIES

Virus clearance studies are typically performed by spiking one model virus at a time into the load of an unit operation step and then evaluating the clearance of the virus across this step. However, the specificity of the Q-PCR method, that is the ability to quantify multiple types of target sequences associated with different viruses in a single sample without cross interference, opens up the possibility of spiking multiple viruses in the load sample of an unit operation step. This multivirus-spiking approach has been shown to work experimentally for a protein A affinity and an anion exchange chromatography (QSFF) step for three model viruses, X-MuLV, MMV and SV40 [17]. As all the model viruses are spiked simultaneously and evaluated for their clearance in a single experiment, this approach can provide significant time, manpower, and cost savings as compared to the traditional approach of evaluating virus clearance using one virus at a time using infectivity assays.

Comparability of the multivirus spike approach to the single virus spike approach, both using Q-PCR methods was established using a set of well-defined criteria. These are clearly outlined in the reference cited in the previous paragraph. The first criterion required that the chromatograms of the single and multivirus spike runs are equivalent to the chromatograms generated during manufacturing. This comparison was based on the UV_{280}, pH and conductivity curves of the chromatogram. The second criterion required that the protein recovery achieved during single and multivirus spike be within the acceptable range established during manufacturing. The third criterion required that the virus clearance obtained for the multivirus spike approach should be within 1 $\log_{10}$ of the single virus spike data. As shown in Table 14.6, the virus clearance data obtained with single and multivirus spike approach are very comparable to each other.

14.19 VIRUS CLEARANCE ACROSS MEMBRANE ADSORBERS

Although anion exchange chromatography using columns packed with positively charged chromatography beads remains the most widely used approach, membrane adsorbers possessing a charged functionality in the pores are being evaluated as an alternative to column chromatography for commercially viable biotechnology processes [21–25]. While the limitations of pore diffusion and pressure drop across packed beds limit the flow rates that can be used in this step at manufacturing-scale, membrane adsorbers by virtue of convective mass transfer (as opposed to diffusive mass transfer), can be operated at very high flow rates without any appreciable loss in binding capacity for trace levels of impurities and contaminants. Additionally, there is minimal pressure drop at high flow rates. The other advantages of membrane chromatography lies in the disposable nature of the modules, which eliminates the need to clean and reuse the modules unlike column chromatography, which needs to be regenerated and sanitized after each use. Furthermore, since the volumes of the membrane adsorbers are much smaller than the corresponding chromatography columns, a significant savings in buffer volume can be realized using membrane adsorbers at manufacturing-scale.

An important feature of anion exchange chromatography is its ability to remove viruses. In fact, in a monoclonal antibody purification process it is usually one of the robust steps for viral clearance. Thus, it is important that any potential replacement has to match the viral clearance capabilities of the anion exchange column chromatography step. To this end, several studies have been done to evaluate the clearance of model viruses across membrane adsorbers. In fact, the Food and Drug Administration has already approved a biotechnology product made using a membrane adsorber as one of the process steps [21]. In a study conducted at Amgen (Thousand Oaks, CA), a *Q* membrane adsorber was evaluated against an existing *Q* chromatography step. A preliminary viral clearance study using two model viruses, MuLV (murine leukemia virus) and MMV (minute mice virus) showed >5 $\log_{10}$ of clearance, very comparable to the existing column chromatography step. In another collaborative study

TABLE 14.6
Comparison of Single and Multiple Virus-Spiking Studies

		Viral Clearance Factor ($\log_{10}$)		
Process Step	**Spike Type**	**X-MuLV**	**MMV**	**SV40**
Protein A	Single	2.6	1.9	1.8
	Multiple	3.0	2.1	2.2
QSFF (100 mg/ml load)	Single	>6.2	>5.4	>5.3
	Multiple	>6.1	>6.0	>5.4
QSFF (250 mg/ml load)	Single	>6.2	>5.4	>5.3
	Multiple	>6.1	>6.0	>5.4

Source: Reproduced from Valera C, Chen J, and Xu Y. *Biotechnol Bioeng* 2003; 84: 714–722. With permission.

between Abgenix (Fremont, CA) and Sartorius, >5.5 $\log_{10}$ of MuLV and PRV (pseudorabies virus), >6.7 $\log_{10}$ of MMV, and >7.3 $\log_{10}$ of Reo-3 (Reovirus) was obtained across a Q membrane adsorption process step. In a study conducted by Millipore [22], >5.5 $\log_{10}$ of clearance of MuLV, SV40 (simian virus), and MMV have been shown for a membrane adsorption process step. These studies show that the membrane adsorbers are capable of providing acceptable clearance of the model viruses commonly used to validate a purification process.

14.20 CONCLUSIONS

Viral clearance studies of the purification process is one of the complementary arms of the three pronged approach that is undertaken by the manufacturers of biopharmaceuticals to minimize virus contamination. The other approaches to ensure viral safety of the final product are, (a) selecting and testing cell lines and raw materials for the absence of undesirable viruses which may be infectious and pathogenic for humans and (b) testing the products at appropriate steps of production for the absence of contaminating infectious viruses. Although no approach by itself can assure that the drug product is free of virus contamination, the combination of all the three complementary approaches greatly minimizes the probability of a virus contamination.

To ensure the validity of the viral clearance studies, it is imperative that the scale-down studies represent the manufacturing process as closely as possible. The qualification of the scale-down models along with the use of representative buffers and load materials for each of the process steps is critical towards establishing the equivalence of the scale-down models to the manufacturing-scale process. Additionally, the choice of the specific and the nonspecific model viruses in the evaluation studies is of paramount importance toward establishing that the purification process is capable of removing the endogenous retroviral contaminants and any other adventitious viral contamination from unknown sources. Finally, to ensure a scientifically sound viral clearance package, it is important to follow a rational approach, all of which are clearly explained in the various sections of this article.

APPENDIX: USE OF THE POISSON DISTRIBUTION TO DETERMINE VIRUS TITERS

An understanding of the use of Poisson distribution is useful when trying to design viral clearance studies for steps that usually clear viruses to the limit of detection, such as a low pH inactivation step and a nanofiltration step. When a sample contains a very low concentration of virus, there is a discrete possibility that if only a small fraction of the sample is tested for virus, that fraction will test negative due to the random distribution of the virus in the total sample. The probability p that the sample analyzed does not contain infectious virus is expressed by $p = ((V - v)/V)y$, where V is the total volume of the container, v is the total volume of the fraction tested and y is the absolute number of infectious viruses randomly distributed in the sample. If V is sufficiently relative to v (i.e., $V \gg v$), then the Poisson distribution approximates to the following equation:

$$p = e^{-cv} \quad \text{or} \quad c = [-\ln(p)]/v \tag{14.1}$$

where c is the concentration of the infectious virus and v is the volume of the fraction tested. The amount of virus which would have to be present in the total sample in order to achieve a positive result with 95% confidence ($p = .05$) is calculated as:

$$c = -[\ln(0.05)]/v = 3/v \tag{14.2}$$

The following example shows the difference between the estimated residual viral infectivity for a 400 μl sample vs. a 4000 μl sample tested for infectivity using a $TCID_{50}$ assay. In each of the cases, no infectivity was detected in any of the wells for either dilutions. Thus, the theoretical amount of virus present in the entire sample was estimated using a Poisson distribution.

In the case of data presented in Table 14.7, using Equation 14.2, the infectivity is calculated as <0.83 $\log_{10}$ $TCID_{50}$/ml. However, if the total number of wells tested at no dilution was increased to 80 and the sample still tested negative in each of the 80 wells (data presented in Table 14.8), with each well having an inoculum volume of 50 μl, then using Equation 14.2, the infectivity would be calculated as <−0.13 $\log_{10}$ $TCID_{50}$/ml, roughly a decrease of about a log unit in infectivity. This would translate to an increase of approximately one log

TABLE 14.7
Determination of Infectivity Using a $TCID_{50}$ Assay with 400 μl Sample Volume

$\log_{10}$ Dilution Factor	Number of CPE Positive Wells	Total Number of Wells
0.0	0	8
1.0	0	8
2.0	0	8
3.0	0	8

Inoculum volume in each well: 50 μl.

TABLE 14.8
Determination of Infectivity Using a $TCID_{50}$ Assay with 4000 μl Sample Volume

Log_{10} Dilution Factor	Number of CPE Positive Wells	Total Number of Wells
0.0	0	80
1.0	0	80

Inoculum volume in each well: 50 μl.

of clearance across this step. This example thus underlines the advantage of assaying higher volumes of the sample when no virus is detected in a volume of a sample tested at the lowest possible dilution.

DISCLAIMER

The statements in this chapter reflect the professional views of the authors and are not necessarily the official practices and positions of PDL BioPharma.

REFERENCES

1. Garnick RL. Experiences with viral contamination in cell culture. In: Brown F, Lubiniecki AS, Eds. *Viral Safety and Evaluation of Viral Clearance from Biopharmaceutical Products*, Vol. 88. Basel: Karger, 1996; pp. 49–56.
2. Rabenau H, Ohlinger V, and Anderson J. Contamination of genetically engineered CHO-cells by epizootic haemorrhagic disease virus (EHDV). *Biologicals* 1993; 21: 207–214.
3. Anderson KP, Lie YS, and Low ML. Defective endogenous retrovirus-like sequences and particles of Chinese Hamster Ovary Cells. *Dev Biol Stand* 1990; 75: 123–132.
4. Bartal AH, Feit C, and Erlandson RA. Detection of retroviral particles in hybridoma secreting monoclonal antibodies. *Med Microbiol Immunol* 1986; 174: 325–332.
5. Lie YS, Penuel EM, and Low ML. Chinese Hamster ovary cells contain transcriptionally active full-length type C proviruses. *J Virol* 1994; 68: 7840–7849.
6. ICH Topic Q5A: *Viral Safety Evaluation of Biotechnology Products Derived from Cell Lines of Human or Animal Origin*, 1997.
7. Center for Biologics Evaluation and Research: *Points to Consider in the Manufacture and Testing of Monoclonal Antibody Products*, 1997.

8. The European Agency for the Evaluation of the Medicinal Products: *Note for Guidance on Virus Validation Studies: The Design, Contribution and Interpretation of Studies Validating the Inactivation and Removal of Viruses*, 1996.
9. Adamson SR. Experiences of virus, retrovirus and retrovirus-like particles in Chinese hamster ovary (CHO) and hybridoma cells used for production of protein therapeutics. In: Brown F, Griffiths E, Horaud F, Petricciani JC, Eds. *Safety of Biological Products Prepared from Mammalian Cell Culture*. Basel Karger, 1998; 93:89–96.
10. Markus-Sekura CJ, Klutch M, Lundquist M, and Dunlap R. Increased expression of CHO cell endogenous retrovirus-like particles detected by electron microscopy after treatment with viral inducting agents or cytokines. *In Vitro Toxicol* 1991; 4: 13–25.
11. Darling A. Validation of biopharmaceutical purification processes for virus clearance evaluation. *Mol Biotechnol* 2002; 21: 57–83.
12. Robertson JS. Viruses and assuring viral safety. In: Brown F and Lubiniecki AS, Eds. *Process Validation for Manufacturing of Biologics and Biotechnology Products*. Basel: Karger, 2003; 113: 73–77.
13. Darling A. Virus assay methods: Accuracy and validation. *Biologicals* 1998; 26: 105–110.
14. Xu Y and Brorson K. An overview of quantitative PCR assays for biologicals: quality and safety evaluation. In: Brown F and Lubiniecki AS, Eds. *Process Validation for Manufacturing of Biologics and Biotechnology Products*. Basel: Karger, 2003; 113: 89–98.
15. Shi L, Chen Q, Norling L, Lau A, Krejci S, and Xu Y. Real time quantitative PCR as a method to evaluate xenotropic murine leukemia virus removal during pharmaceutical protein purification. *Biotechnol Bioeng* 2004; 87: 884–896.
16. Curtis S, Lee K, Blank G, Brorson K, and Xu Y. Generic/Matrix evaluation of SV40 clearance by anion exchange chromatography in flow-through mode. *Biotechnol Bioeng* 2003; 84: 179–186.
17. Valera C, Chen J, and Xu Y. Application of multivirus spike approach for viral clearance evaluation. *Biotechnol Bioeng* 2003; 84: 714–722.
18. Brorson K, Brown J, Hamilton E, and Stein KE. Identification of protein A media performance attributes that can be monitored as surrogates for retrovirus clearance during extended re-use. *J Chromatogr A* 2003; 989: 155–163.
19. Xu Y and Brorson K. Matrix Approach. Presented at the First International Symposium on Virus Safety; January 17, 2005, Kitasato Institute, Tokyo, Japan.
20. Brorson K, Krejci S, Lee K, Hamilton E, Stein K, and Xu Y. Bracketed generic inactivation of retroviruses by low pH treatment for monoclonal antibodies and recombinant proteins. *Biotechnol Bioeng* 2003; 82: 321–329.
21. Galliher P and Fowler E. Validation of impurity removal by the Campath-1H biomanufacturing process. Presented at the IBC's Biopharmaceutical Production Week, Paradise Point, San Diego, CA, November 12, 2001.
22. Dermawan S, Zhou J, Solamo F, Hong T, and Tressel T. Process development of a flow-through anion exchange membrane chromatography in protein

purification. Presented at the ACS National Meeting, San Diego, CA, March 13, 2005.

23. Phillips M and Lutz H. Membrane adsorber technology for trace impurity removal applications. Presented at the ACS National Meeting, Boston, MA, March 23, 2003.
24. Gosh R. Protein separation using membrane chromatography, opportunities and challenges. *J. Chromatogr* 2002; 952: 13–27.
25. Gottschalk U, Fischer-Fruehholz S, and Reif O. Membrane adsorbers, a cutting edge process technology. *Bioprocess Int* 2004; 2: 56–65.

15 Advances in Viral Clearance

Kurt Brorson

CONTENTS

15.1 INTRODUCTION

The viral safety of biotechnology products has traditionally been a key concern both of regulators and industry.[1–6] It has also proven to be a stumbling block for early product development, particularly for inexperienced sponsors such as small start-up firms and academic investigators.

15.1.1 The Critical Path Initiative

The revolution in biotechnology has raised new hope for the prevention, treatment, and cure of serious illnesses. However, FDA is aware of a growing concern that many of the new basic science discoveries, like sequencing of the human genome, have not translated into more effective, affordable, and safe medical products. For example, the number of new Biologics License Applications (BLAs) has declined from 33 in 1997 to 14 in 2003. At the same time, in some estimates the cost of a single drug development has soared from $1.1 billion in 1995 to $1.7 billion in 2002.[7] The current development path is becoming increasingly challenging, inefficient, and costly. Improvements in the science supporting medical product development are urgently needed to translate the tremendous advances in the basic sciences into useful products.

In 2004, FDA proposed the critical path initiative to develop a new product development toolkit (e.g., animal or computer-based predictive models, biomarkers for safety and effectiveness, new clinical evaluation techniques, better manufacturing technologies) to improve the predictability and efficiency of the development process spanning laboratory concept to commercial product (http://www.fda.gov/oc/initiatives/criticalpath). Three key areas were identified:

- *Assessing safety*: developing tools and standards for both preclinical and clinical stages of development.
- *Demonstrating medical utility*: developing tools for evaluating efficacy, better clinical trial designs, and efficacy standards.
- *Industrialization*: developing a better toolkit and standards for designing high-quality products and developing mass production capacity.

Regarding industrialization, FDA occasionally perceives a hesitance on the part of industry to introduce state-of-the-art science and technology because of concern about regulatory impact. This hesitance has led to retention of outdated technologies associated with high in-process inventories, long development times, low factory utilization rates, significant product wastage, compliance problems; all of which drive up costs and decrease productivity.

Modern manufacturing technologies are needed to improve efficiency and increase flexibility while maintaining high-quality standards. Further research (academic and industrial) and data sharing (FDA, industry and academia) are necessary to make these efficiencies a reality.

Virus safety is one area of industrialization where improvements and streamlining are feasible.[3] Initiatives have been started for standardization of viral clearance studies and unit operations, new technologies to detect viruses or clear viruses with improved robustness, understanding and improving robustness in currently implemented bioprocessing steps, controlling viruses in unconventional source materials, and adoption of risk assessment and mitigation. All of these will lower economic barriers for early stage development while providing for greater assurance of viral safety.

15.1.2 Critical vs. Noncritical Operating Parameters

A clear understanding of critical operating parameters is required for design of a successful viral safety strategy. Operational parameters are process inputs that are directly controlled. Typically, these parameters are physical or chemical (e.g., temperature, process time, column flow rate, column wash volume, reagent concentration, or buffer pH). Performance parameters are process outputs that may be monitored to ensure or confirm acceptable process performance; in the context of this discussion the $\log_{10}$ removal capacity of one or more viruses is the relevant performance parameter.

Defining critical operation parameters is inherently subjective; therefore, the draft PDA process validation technical report (TR42; www.pda.org/PDF/PubsCatalpdf),[8] currently near completion, will suggest that the term critical operational parameter be reserved for a limited subset of parameters that significantly affect critical product quality attributes when varied outside a meaningful and narrow (or difficult to control) operational range. By contrast, noncritical operational parameters are all other parameters considered outside this definition. It is important to note that distinctions between critical vs. noncritical are not always obvious. There must be strong scientific justification behind the designation of which parameters are or are not critical. Justification can be acquired through small-scale studies, manufacturing experience and consultation of the peer-reviewed scientific literature. The field of bioprocessing science is active and fluid; advances in science might change a particular unit operation parameter's designation over time. Parameters deemed noncritical for one performance attribute, like step yield, may be critical for other aspects of unit operation performance, like viral clearance. It is even possible for some parameters to be more critical for clearance of one virus than another.[9,10] All of this must be understood in a scientific and mechanistic context.

Another consideration for understanding critical operating parameters is the position of the operating set point and range relative to its edge of failure. Processes are run at set points, and the operating range relative to the acceptable limit that could determine the criticality of that parameter. If the operating range is relatively close to the acceptable limit, this would place product quality or process in jeopardy when relatively minor excursions occur. If the acceptable limit is wide relative to the operating range, then the parameter can be categorized as noncritical. There are also instances, however, when the operating range is the acceptable limit; in this case edge of failure would be relative to the set point rather than the operating range.

15.1.3 Robustness Concept

Unit operation robustness, the reliability of a unit operation and insensitivity to minor process variations, is critical for bioprocessing. Unit operations clear/inactivate viruses by specific mechanisms and can be characterized for robustness, based on an understanding of critical and noncritical process variables.

Certain unit operations were reported anecdotally to be highly robust for virus removal[1] and their mechanism of virus removal or inactivation is known. Robustness of some of these operations has been verified experimentally and described in the scientific literature. Clearance was found to be dependent on a few critical unit operation parameters but relatively less sensitive to others, so long as these parameters remained within acceptable defined manufacturing ranges. Establishing the mechanism of clearance by these unit operations enhances the confidence that changing parameters unrelated to their mechanism of clearance will not impact LRV. For example, low-pH inactivation of murine retroviruses was, as expected for a chemical reaction, highly dependent on time, temperature, and pH but relatively independent of the type of model protein or salt content in the matrix.[11]

15.1.4 Streamlined Approaches

By focusing on mechanisms and critical variables of unit operations, it is possible to adopt streamlined approaches to virus removal validation, particularly in the clinical phase of product development.[1]

Streamlined approaches include generic validation, bracketing, and the combination of the two approaches. Generic validation was proposed in FDA's monoclonal antibodies points to consider document (1997): "A Generic Clearance Study is one in which virus removal and inactivation is demonstrated for several steps in the purification process of a model antibody. These data may then be extrapolated to other antibodies following the same purification and virus removal/inactivation scheme as the model antibody." To remove an early barrier to product development, FDA has accepted generic virus removal/inactivation data to support IND use of monoclonal antibody products in cases where sufficient justification has been provided. FDA compares the unit operations of the model and new products and an assessment is made whether critical operating parameters are identical. Generally, these critical parameters are the same or similar to those matched in scaled-down validation studies of large-scale operations, and should focus on those that are mechanistically most likely to impact clearance. When the products move to advanced stages of development or on to marketing applications, a reevaluation takes place concerning the need for product specific validation data.

FDA's monoclonal antibodies points to consider document also describes the streamlining approach of bracketing: "In some cases, sponsor may demonstrate virus removal/inactivation for a particular module at two different values of a given parameter (e.g., ionic strength, dwell time, temperature) and use any values of that parameter falling within this range." This approach is similar to the design space concept introduced by the draft ICH Q8 document (available at www.ich.org).[12]

It is also feasible to validate viral clearance at worst-case set points and then justify operation at a set point within the acceptable side of the worst-case set point. For example, validating clearance by a flow through anion exchange column packed at the minimum bed height likely to be used in operation would be worst-case because chromatography performance, contact time, and resolution increases with bed height. Thus, one could validate virus removal at a minimum bed height and then operate at a higher bed height. Justification of this approach requires a detailed understanding of the mechanism of viral clearance by the unit operation.

Combining generic validation and bracketing imparts more flexibility by defining design spaces where robust viral clearance may be assured on a product-independent basis. The definition of these design spaces should be supported by a mechanistic understanding of the viral clearance step, and data to show that product-specific impacts are unlikely. If widely adopted, predefined design spaces can afford considerable flexibility to IND sponsors. This flexibility will be particularly important for small firms or academic sponsors with limited resources that can be more productively applied to other aspects of product development.

15.1.5 Virus Spike Quality

The quality of virus spikes can impact the results of a validation study. For example, use of aggregated virus in a filtration study can artificially increase the clearance capacity of the filter that is measured in the small-scale validation study. Other impurities, such as DNA or extraneous proteins from the virus spike, can clog virus filters, resulting in an underestimate of their performance capacity. Buffers or other components can also change the pH or conductivity of a process sample after spiking, resulting in impaired inactivation by low pH steps or altered clearance by ion exchange chromatography. It is always prudent to assess and minimize the impact of the virus spike on the performance of scaled-down models to ensure that they are truly representative of the large-scale commercial unit operation.

15.2 VIRUS FILTRATION

Virus filtration is a common safety step in biopharmaceutical and plasma derived product manufacture.[13] Virus filters are cast polymeric membranes with a complex internal porous structure. Solutions pass through this pore network while viruses and other particles carried along by the fluid may be retained on the membrane surface or penetrate some distance into the membrane structure prior to entrapment. For virus filters, critical parameters include transmembrane pressure (if constant pressure), feed stock composition (buffer and protein), flux (if constant flux), total throughput or percentage of flow decay. Note that both pressure and flux should be monitored for filters run under either constant flow or constant pressure modes. In constant flow mode, the flux would be a critical operating parameter while pressure would be a critical performance attribute. In constant pressure mode, the opposite is true.

15.2.1 FILTER CLASSES

Virus filters target two broad classes of viruses: large viruses (e.g., retroviruses) and small viruses (parvoviruses, others). Some large virus filters are also effective at clearing medium size viruses.

Retention that occurs because virus particles are too large to pass through a pore is referred to as sieving or size exclusion. This mechanism, assuming uniformity in pore-size and integrity of the filter unit, should be robust over operating conditions within those recommended by the filter vendor. In a recent study of three brands (Millipore Viresolve NFR, Pall Ultipor DV50 and Asahi-Kasai Planova 35N), large virus filters were shown to be remarkably efficient at removing a 64 to 82 nm sized bacteriophage virus, up to 8 to 9 $\log_{10}$ when run under standardized conditions.[14] This bacteriophage is slightly smaller than retroviruses, the virus targeted by these filters. In a separate study of Viresolve NFR filters, virus removal was found to be highly dependent on the size of the model virus, as expected for size-based sieving, but was less dependent on buffer composition, process time and pressure, or by membrane lot and model protein.[15]

For small virus filters, a more challenging technology, one additional filtration performance parameter with a clear impact on virus removal is pressure-adjusted flow rate, a parameter not normally controlled in validation studies.[16] When flow rates decline past 75% after extended processing at constant pressure, the $\log_{10}$ reduction values (LRVs) also decreased, perhaps due to changes in effective pore-size due to fouling. In commercial bioprocessing, flow declines of this magnitude are rarely achieved, but this finding argues that this performance parameter should be monitored in commercial processing and validation studies. In validation studies, the extent of flow decay is probably a more meaningful end-point definition than volumetric throughput per filter surface area for some filters. Thus, given a choice between the two endpoint definitions, matching the flow-rate performance attribute between the large-scale commercial operation and the small-scale models for those filters should be a higher priority than matching volumetric throughput.

Based on the correlation of filtration robustness on particle size (for non-fueled filters), it can be argued that validation data for a size-exclusion-based small virus filter can be generated with one small parvovirus or even a small bacteriophage like ϕX174 and then applied to large viruses. This argument would assume that no model virus: process protein interactions occur during the validation studies and that the model virus remains monodispersed during the duration of the study.

15.2.2 Filter Ratings

Based on the targeted class and filter capabilities, filter vendors have developed ratings to allow for selecting the appropriate virus filter for an application. However, retention ratings vary considerably among filter manufacturers. In some cases, manufacturers have assigned a rating associated with a particular type of virus (e.g., parvovirus or retrovirus). In other cases, a virus size rating has been assigned based on either the retention of a particular model (e.g., bacteriophage) at a given LRV or an average pore-size rating is established from a mathematical model for the permeability. In yet other cases, a molecular weight retention or passage rating has been assigned.

CDER and PDA are working on developing uniform ratings for virus filters to enhance clarity for end users. So far, this effort has led to a rating system for large virus filters (published as part of PDA's technical report on virus filtration TR41; www.pda.org/pubs/publications/publications_search.asp) based on a 6 $\log_{10}$ retention of a 64 to 82 nm bacteriophage PR772.[13,14,17] Large virus filters from three vendors have been demonstrated by an independent lab to remove this level or more of PR772 under defined conditions. Development of ratings for small virus filters by the CDER/PDA committee is ongoing, but may prove to be more technically challenging due to the impact of flow decay on filter efficiency.

15.3 PHYSICO-CHEMICAL INACTIVATION

15.3.1 Solvent/Detergent Treatment

Solvent/detergent (SD) treatment, initially developed by the New York Blood Center, is a widely used safety measure for plasma-derived products, as well as a subset of monoclonal antibodies.[6] SD treatment is believed to inactivate lipid enveloped viruses by dissolving their membranes. The concentration of the solvent and detergent are the most critical parameters; the New York Blood Center procedure stipulates treatment with 1% tri-(*n*-butyl) phosphate (TNBP, the solvent) and 1% triton X-100 (the detergent) for 4 h at 30°C. When the proper SD concentration is set, the dissolution reaction is very rapid (complete within minutes). Studies have shown that the reaction is more robust to changes in temperatures and protein concentrations, as long as they are maintained in ranges that are acceptable in a commercial manufacturing environment.[18,19] Nonetheless, time, temperature, and completeness of mixing are typically controlled within the range recommended by the New York Blood Center.

15.3.2 Low-pH Inactivation

Low-pH inactivation is a common step in mAb purification processes employed after protein A affinity chromatography. It is particularly advantageous because protein A columns eluate at low pH anyway, so extending the acid incubation of the eluate for one to two hours does not significantly impact bioprocessing logistics or the mAb integrity. Low-pH incubation disrupts the capsid and membrane structure of X-MuLV and other retroviruses in a manner sufficiently extensive to destroy infectivity; this disruption can be visualized by electron microscopy as membrane and capsid blurring and virus aggregation.[11] Because the damaged particles still remain after treatment, genomic RNA and RT activity can be detected following low pH exposure.

The operating parameters typically controlled during low-pH inactivation are pH, time and temperature of incubation, salt content, protein concentration, aggregates, impurities, model protein pI and sequence, and buffer composition. However, low-pH incubation inactivates murine retroviruses by driving pH-dependent chemical reactions such as irreversible conformational changes on viral surface proteins. The chemical reactions are dependent on time, temperature, and buffer pH. Unrelated factors such as concentrations of salts or nonviral proteins would have at most secondary effects on reaction efficiency; perhaps by complexing and shielding the viral capsid or envelope from the destructive effects of H^+ ions. A recent matrix study has confirmed that pH, time, and temperature are the most critical operating parameters for low-pH incubation, while the other parameters have, at most, secondary effects.[11] For four model proteins, a robust inactivation of X-MuLV occurs within 30-min at pH 3.8 when the temperature, buffer conditions, protein, and NaCl concentration is within ranges typical of commercial bioprocessing.

15.3.3 Heat Treatment

Heat treatment is regarded to be a reliable measure to inactivate viruses in final dosage forms. For example, 21 CFR 640.81 (e) & (f) mandates heating of final containers of Albumin (Human) as a viral safety measure within 24 h after completion of filling. Heat treatment is conducted so that the solution is heated continuously for 10 to 11 h, at an attained temperature of 60 ± 0.5°C. Either 0.08 ± 0.016 mmol sodium caprylate, or 0.08 ± 0.016 mmol sodium acetyltryptophanate plus 0.08 ± 0.016 mmol sodium caprylate per gram of protein is added as a stabilizer(s) to prevent protein degradation. The mechanism of virus inactivation by heat treatment is believed to be irreversible denaturation of heat sensitive viral proteins, such as reverse transcriptase in retroviruses. Although used rarely and mostly at early IND stages, this step is also feasible for recombinant DNA and monoclonal antibody products, as long as the heat does not impact the protein drug substance quality attributes.

15.4 CHROMATOGRAPHY

Chromatography steps are introduced into a bioprocess scheme to capture a protein API or remove non-API impurities; many concomitantly clear viruses. Different columns clear virus by different mechanisms. Depending on the chromatographic step and the model virus, 4 to 6 $\log_{10}$ LRVs are achievable, but LRVs can be less for some viruses and some columns. If viral clearance can be mechanistically explained, selection of critical parameters to maintain for adequate clearance can be justified. For a column, these can include column bed height, linear flow rate, flow rate to bed-volume ratio (contact time), buffer, pH, temperature, concentration of protein, impurities, salt, and product.

15.4.1 Protein A

Protein A media specifically binds antibodies (Abs), while viruses are partitioned from Ab intermediates by flowing uninhibited through the column.[20,21] Some amount of nonspecific sticking occurs during loading, so trace amounts of virus can be dislodged with the product by the pH change during elution. Critical operating parameters in this instance would include protein load, buffer composition, flow rate, and bed height. The impact of these factors is likely to be complex, for example, a low mAb load may allow more nonspecific binding sites to remain on the column after loading, allowing for more, not less, nonspecific carry-over of virus. Similarly, a higher bed height would allow for more nonspecific binding sites. These effects may need to be assessed for individual process fluids.

15.4.2 Ion Exchange Chromatography

Ion exchange (IEX) unit operations are believed to remove viruses from in-process intermediates by ionic binding.[9] Experience from the gene therapy and vaccine fields has shown that viruses can be partitioned from protein contaminants based on charge difference. Thus, it can be surmised that a flow through anion exchange unit operation conducted in neutral, low conductivity buffers removes negatively charged viruses from positively charged mAbs by binding them with high avidity while the mAb flows through.

Matrix studies defining critical operating parameters for robust SV40 and X-MuLV removal by Q-anion exchange chromatography have been performed. The predominant factors impacting SV40 clearance were pH and conductivity, suggesting that this virus bound the medium with behavior similar to proteins.[9] Factors like bed height, flow rate (contact time), and protein load content had negligible effect on SV40 LRV in the ranges studied, which extended beyond the typical range in a manufacturing environment.

In contrast, X-MuLV clearance can be less robust to extremely short contact times, especially in combination with high load density and impurity content.[10] The overall impact of pH and conductivity on LRV was more subtle than for SV40 and mostly in the extreme case of high protein load density. This is in contrast to protein A, where a low protein load would be predicted to be the worst-case. X-MuLV is an enveloped virus, predicted to have more extensive surface heterogeneity. Heterogeneous surface charges may explain the more complex chromatographic behavior of X-MuLV relative to SV40. However, based on these two studies, a design space for efficient removal of both viruses was defined for flow through anion exchange chromatography.

15.4.3 Media Age

Regulators are also concerned about the robustness of virus removal by chromatography after extensive cleaning and reuse (cycling) of resins. The theoretical concern is that resin degradation or fouling over time might impair viral clearance.[2] Recent studies with protein A chromatography and anion exchange chromatography, however, found that viral clearance was remarkably stable after extensive resin cycling. These studies identified useful surrogate performance attributes that changed prior to or simultaneously with reduced viral clearance; for protein A chromatography, decreases in step yield and antibody breakthrough in the flow through appeared to be the most sensitive indicators of degraded column performance.[20] For AEX columns, increases in band spreading and in back pressure or the appearance of impurities in the process fluid was indicative of the end of their effective functional lifetime — and loss of their ability to clear viruses.[22] Overall, rigorous scientific investigations demonstrated that viral clearance achieved by chromatography unit operations was quite robust (e.g., protein A chromatography and AEX chromatography).

15.5 NEW REMOVAL/INACTIVATION TECHNOLOGY

Emerging technology promises to complement the currently available methods for virus clearance. Some emerging technologies are nearing commercialization while others require additional development. Ion exchange membrane adsorbers have ligand:virus-binding properties similar to those of AEX chromatography, but membranes possess certain practical advantages. For example, ligand:target binding to membranes is largely kinetic and not limited by pore diffusion; thus, membranes allow very high flow rates, short processing times, and low pressure drops. Membranes are disposable and generally require less floor space and specialized equipment than do columns, while their performance validation is simplified because post-use cleaning is not necessary. Ion exchange membranes have already been used successfully to bind and then release virus particles in vaccine production;[23] removing viruses from process intermediates should be even simpler, since the particles are discarded with the disposable adsorber and not recovered. Thus, it is desirable to develop membranes incorporating ligands and with operating conditions that favor tight interactions between viruses and membranes, because the binding need not be reversible.

Broad-spectrum pulsed light inactivates a variety of mammalian viruses, but robustness (e.g., interference by high protein concentrations) must be addressed before this technology can be recommended for widespread use in bioprocessing or treatment of raw materials.[24] Virus:product partitioning by flocculation followed by microfiltration[25] and micelle-based extraction[26] may also become a realistic method at some point; however, significant technical issues, such as maximizing product recovery and improving virus partitioning robustness, must be resolved before these methods become successfully commercialized.

15.6 NEW VIRUS DETECTION METHODS — Q-PCR

With the advent of fluorogenic 5′-nuclease-based quantitative PCR (Q-PCR), a new series of assays became available to measure virus titers in process and validation samples by quantifying components of the viruses such as genomic nucleic acids or enzymes like reverse transcriptase, that are more easily measured than infectivity.[27] These assays offer increased precision and sensitivity over standard infectivity assays. They directly quantify total particle counts rather than complete infectious virions. Q-PCR assays lend themselves to measuring clearance by unit operations that remove viruses (e.g., chromatography and filtration) but not those that inactivate them (e.g., solvent/detergent or low pH); infectivity assays can still be used to measure clearance by these unit operations.

One particular advantage of Q-PCR assays is that independent assays can be used to quantify more than one virus type in a single preparation; thus, clearance studies can be performed by spiking three or more viruses into feedstock for a single column or filter and the clearance of each virus measured simultaneously in separate assays.[28] Q-PCR assays can also be used to quantify viral clearance of process-scale steps instead of scale-down models. For example, a type C particle specific Q-PCR or Q-PERT assay can be used to track endogenous virus loads before and after an initial capture step.[21] These approaches should streamline validation time and costs.

To perform the Q-PCR assays, viral nucleic acids from process or validation samples are extracted using standard molecular biology kits. Often, a nuclease step precedes viral capsid destruction to eliminate interference by free nucleic acids. The nuclease step can be critical for interpretation of validation studies; for example, free nucleic acids can pass through a virus filter while virus particles are retained, leading to wild underestimates of clearance. After the nucleic acids are extracted, they are used directly in the Q-PCR assay (DNA viruses) or they are reverse transcribed into cDNA using conventional molecular techniques (RNA viruses).

A variation on this technology, Q-PERT, can be used to quantify retroviruses.[21] In this three step assay, reverse transcriptase in the retroviruses creates a copy of an irrelevant template RNA from a bacteriophage (MS2). The level of cDNA generated during this step will depend on the amount of RT (and retroviruses) in the test article. This assay has been proposed to quantify endogenous retrovirus levels in cell culture harvests and perhaps retrovirus-like particle production by insect cells.[29]

ACKNOWLEDGMENTS

I thank Michael Hanson, Scott Lute, Dr. Joe Kutza and Dr. Patrick Swann (CDER/FDA) for careful review of this chapter.

REFERENCES

1. Points to Consider in the Manufacture and Testing of Monoclonal Antibody Products for Human Use; Docket No. 94D-0259; Food and Drug Administration: Rockville MD, 1997.
2. Guidance on Viral Safety Evaluation of Biotechnology Products Derived From Cell Lines of Human or Animal Origin, Q5A; International Conference on Harmonization of Technical Requirements for Registration of Pharmaceuticals for Human Use: Geneva Switzerland, 1998.

3. Brorson, K., Norling, L., Hamilton, E., Lute, S., Lee, K., Curtis, S., and Xu, Y., Current and future approaches to ensure the viral safety of biopharmaceuticals. *Dev. Biol.* (Basel), 2004, 118, 17–29.
4. Farshid, M., Taffs, R., Scott, D., Asher, D., and Brorson, K., The clearance of viruses and transmissible spongiform encephalopathy agents from biologicals. *Curr. Opin. Biotechnol.*, 2005, 16, 1–7.
5. Roberts, P., Virus safety of plasma products. *Rev. Med. Virol.*, 1996, 6, 25–38.
6. Horowitz, B. and Ben-Hur, E., Efforts in minimizing risk of viral transmission through viral inactivation. *Ann. Med.*, 2000, 32, 475–484.
7. Gilbert, J., Henske, P., and Singh, A., Rebuilding Big Pharma's Business Model; Windhover Information: November 2003.
8. PDA process validation task force: Process validation of protein manufacturing (TR42); PDA: Bethesda MD. PDA *J. Pharm. Sci. Technol.*, 2005, 59, 5–9.
9. Curtis, S., Lee, K., Blank, G.S., Brorson, K., and Xu, Y., Generic/matrix evaluation of SV40 clearance by anion exchange chromatography in flow-through mode. *Biotechnol. Bioeng.*, 2003, 84, 179–186.
10. Chen, Q., Bracketed Matrix Evaluation of X-MuLV and SV40 Clearance by Anion Exchange Chromatography in Flow-Through Mode. Presented at BioProduction 2005, Amsterdam, October 25–27, 2005.
11. Brorson, K., Krejci, S., Lee, K., Hamilton, E., Stein, K., and Xu, Y., Bracketed generic inactivation of rodent retroviruses by low pH treatment for monoclonal antibodies and recombinant proteins. *Biotechnol. Bioeng.*, 2003, 82, 321–329.
12. Pharmaceutical Development Q8 (draft); International Conference on Harmonization of Technical Requirements for Registration of Pharmaceuticals for Human Use: Geneva Switzerland, 2004.
13. PDA virus filter task force: Virus Filtration (TR41); PDA: Bethesda MD, 2005.
14. Brorson, K., Sofer, G., Robertson, G., Lute, S., Martin, J., Aranha, H., Haque, M., Satoh, S., Yoshinari, K., Moroe, I., Morgan, M., Yamaguchi, F., Carter, J., Krishnan, M., Stefanyk, J., Etzel, M., Riorden, W., Korneyeva, M., Sundaram, S., Wilkommen, H., and Wojciechowski, P., Large pore size virus test method recommended by the PDA virus filter task force. *PDA J. Pharm. Sci. Technol.*, 2005, 59, 177–186.
15. Brough, H., Antoniou, C., Carter, J., Jakubik, J., Xu, Y., and Lutz, H., Performance of a novel viresolve NFR virus filter. *Biotechnol. Prog.*, 2002, 18, 782–795.
16. Bolton, G., Cabatingan, M., Rubino, M., Lute, S., Brorson, K., and Bailey, M., Normal flow virus filtration — detection and assessment of endpoint in Bioprocessing. *Biotechnol. Appl. Biochem.*, 2005, 42, 133–142.
17. Lute, S., Aranha, H., Tremblay, D., Liang, D., Ackermann, H.W., Chu, B., Moineau, S., and Brorson, K., Characterization of coliphage PR772 and evaluation of its use for virus filter performance testing. *Appl. Environ. Microbiol.*, 2004, 70, 4864–4871.
18. Biesert, L. and Suhartono, H., Solvent/detergent treatment of human plasma — a very robust method for virus inactivation. Validated virus safety of OCTAPLAS. *Vox Sang*, 1998, 74, 207–212.

19. Horowitz, B., Lazo, A., Grossberg, H., Page, G., Lippin, A., and Swan, G., Virus inactivation by solvent/detergent treatment and the manufacture of SD-plasma. *Vox Sang*, 1998, 74, 203–206.
20. Brorson, K., Brown, J., Hamilton, E., and Stein, K.E., Identification of protein A media performance attributes that can be monitored as surrogates for retrovirus clearance during extended re-use. *J. Chromatogr. A*, 2003, 989, 155–163.
21. Brorson, K., Swann, P.G., Lizzio, E., Maudru, T., Peden, K., and Stein, K.E., Use of a quantitative product-enhanced reverse transcriptase assay to monitor retrovirus levels in mAb cell-culture and downstream processing. *Biotechnol. Prog.*, 2001, 17, 188–196.
22. Norling, L., Lute, S., Emery, R., Khuu, W., Voisard, M., Xu, Y., Chen, Q., Blank, G.S., and Brorson, K., Impact of multiple re-use of anion-exchange chromatography media on virus removal. *J. Chromatogr. A*, 2005, 1069, 79–89.
23. Specht, R., Han, B., Wickramasinghe, S.R., Carlson, J.O., Czermak, P., Wolf, A., and Reif, O.W., Densonucleosis virus purification by ion exchange membranes. *Biotechnol. Bioeng.*, 2004, 88, 465–473.
24. Roberts, P. and Hope, A., Virus inactivation by high intensity broad spectrum pulsed light. *J. Virol. Meth.*, 2003, 110, 61–65.
25. Akeprathumchai, S., Han, B., Wickramasinghe, S.R., Carlson, J.O., Czermak, P., and Preibeta, K., Murine leukemia virus clearance by flocculation and microfiltration. *Biotechnol. Bioeng.*, 2004, 88, 880–889.
26. Kamei, D.T., King, J.A., Wang, D.I., and Blankschtein, D., Separating lysozyme from bacteriophage P22 in two-phase aqueous micellar systems. *Biotechnol. Bioeng.*, 2002, 80, 233–236.
27. Xu, Y. and Brorson, K., An overview of quantitative PCR assays for biologicals: quality and safety evaluation. *Dev. Biol.* (Basel), 2003, 113, 89–98.
28. Valera, C.R., Chen, J.W., and Xu, Y., Application of multivirus spike approach for viral clearance evaluation. *Biotechnol. Bioeng.*, 2003, 84, 714–722.
29. Brorson, K., Xu, Y., Swann, P.G., Hamilton, E., Mustafa, M., de Wit, C., Norling, L.A., and Stein, K.E., Evaluation of a quantitative product-enhanced reverse transcriptase assay to monitor retrovirus in mAb cell-culture. *Biologicals*, 2002, 30, 15–26.

16 Protein A Affinity Chromatography for Capture and Purification of Monoclonal Antibodies and Fc-Fusion Proteins: Practical Considerations for Process Development

Sanchayita Ghose, Thomas McNerney, and Brian Hubbard

CONTENTS

16.1 INTRODUCTION

Monoclonal antibodies and Fc-fusion proteins form the largest and most rapidly expanding category of biopharmaceuticals today with annual sales exceeding \$8 billion and applications across a wide range of diseases [1]. With the growth of this class of biomolecules, significant attention is now being focused on reducing manufacturing costs and streamlining process development activities to enable the rapid progression of these therapeutic candidates through clinical development. Protein A affinity chromatography has come to be used as the industry-wide standard for capture and purification of antibodies and Fc-fusion proteins. This chapter will provide a brief introduction to antibodies and Fc-fusion proteins, describe the basics of Protein A chromatography and discuss practical considerations for the development of this process step in an industrial context.

16.2 MONOCLONAL ANTIBODIES AND Fc-FUSION PROTEINS

An antibody (or immunoglobulin) is a protein synthesized by an animal in response to the presence of a foreign substance (antigen). The antibody has specific affinity for the foreign material that elicited its synthesis. The binding site on the antigen is referred to as the epitope. Antibodies are attractive tools to develop therapeutics because of multiple applications for which they can be employed *in vivo*, all related to their ability to bind specifically to a target. Some of these applications include, (i) blocking a cellular receptor to prevent interaction with its ligand, (ii) transferring a signal to a cell by binding to a specific receptor, (iii) activating the immune system to destroy a specific cell type by binding to a receptor found primarily on that cell type, and (iv) additional functions can be coupled to an antibody including conjugation with a toxin to kill a specific cell type, using targeted radioactivity to deliver a dose of radiation to tumors or coupling an enzyme to an antibody to convert a harmless prodrug to a toxic compound only at the target site. Therapeutic antibodies are usually monoclonal (i.e., they are mass produced from a single-cloned cell line) and can recognize only one antigen.

This therapeutic modality has finally started to meet the promise of being magic bullets against disease that had been predicted by scientists over a decade ago [2]. This is in sharp contrast to the mood in the 1980s when a number of murine monoclonal antibodies failed in clinical trials due to human immune response (referred to as HAMA) against those early murine-derived molecules [3]. The emergence of antibodies as an attractive therapy is the result of the evolution of monoclonal antibody technology over the last two decades (such as CDR grafting, molecular phase display, and transgenic animals) from 100% mouse protein through chimeric and humanized proteins to fully human antibodies [4,5]. This has transformed the stagnant state of antibody therapeutics in early 1980s to a scenario where antibodies have started to dominate the therapeutic landscape. Currently, not only are there 18 FDA approved antibodies on the market (listed in Table 16.1), but also, a majority of drugs from biotech companies that are currently in clinical or preclinical investigation are monoclonal antibodies.

Till date, all recombinant antibodies developed for therapeutic applications have been of the IgG class [3,6,7] because they have the highest serum half-life compared to other classes (IgA, IgM, IgE, IgD). In addition, a considerable degree of versatility is also inherent in the different IgG subclasses (IgG1–4), which have different abilities to trigger effector cascades and therefore can be selected accordingly to fulfill different therapeutic requirements.

The basic structure of an IgG molecule is composed of two heavy chains (H) and two light chains (L) joined together by covalent and noncovalent association to adopt an overall conformation that resembles the letter Y (Figure 16.1). Each heavy chain is composed of three constant domains (C_H1, C_H2, C_H3) and one variable domain (V_H) while each light chain is composed of one constant domain (C_L) and one variable domain (V_L). The stem of the Y, which was obtained by cleavage using the enzyme papain by Rodney Porter et al. [8] and subsequently crystallized, is called the Fc fragment (fragment crystallizable). The remaining molecule, that is, each arm of the Y, was shown to bind antigen in a manner equivalent to the original antibody and was named as the Fab (Fragment antigen binding) fragment. Thus, the Fc fragment consists of two carboxy-terminal domains of each heavy chain (C_H2 and C_H3 domains) while each Fab fragment is formed by the amino-terminal domain of the H-chain (V_H, C_H1) as well as the two domains of the light chain (V_L, C_L) (Figure 16.1). Furthermore, the Fc and the Fab units of the intact antibody are joined by a flexible polypeptide region called the hinge region that allows facile variation in the angle between the Fab units. This kind of mobility is called segmental flexibility which can enhance the formation of an antigen–antibody complex.

TABLE 16.1
Monoclonal Antibodies Approved by the FDA

Trade Name	Target	Indication	Company	Year of Approval	Antibody Type
Orthoclone OKT3	CD3	Acute kidney transplant rejection	Ortho Biotech	1986	Murine
ReoPro	Platelet GP IIb/IIIa	Prevention of blood clot	Centocor	1994	Murine
Rituxan	CD20	Non-Hodgkin's lymphoma	Genentech/ Biogen-IDEC	1997	Chimeric
Panorex	17A-1	Colorectal cancer	GlaxoSmithKline	1995	Murine
Zenapax	IL2Rα (CD25)	Acute kidney transplant rejection	Hoffman-LaRoche	1997	Humanized
Simulect	IL2R	Prophylaxis of acute organ rejection in allogenic renal transplantation	Novartis	1998	Chimeric
Synagis	RSV	Respiratory synctial virus	Medimmune	1998	Humanized
Remicade	TNFα	Rheumatoid arthritis	Centocor	1998	Chimeric
Herceptin	Her2/neu/ ErB2	Metastatic breast cancer	Genentech	1998	Humanized
Mylotarg	CD33	Acute myelogenous lymphoma	Wyeth-Ayerst	2000	Humanized
Campath-1H	CD52	B-cell chronic lymphocytic leukemia	Millenium/ ILEX	2001	Humanized
Zevalin	CD20	Non-Hodgkin's lymphoma	Biogen IDEC	2002	Murine
Humira	TNFα	Rheumatoid arthritis	Abbott	2002	Human
Bexxar	CD20	Non-Hodgkin's lymphoma	Corixa/GSK	2003	Murine
Xolair	IgE	Allergy	Genentech/ Novartis	2003	Humanized
Erbitux	EGFR/ Her1	Colon cancer	Imclone/ BMS/Merck	2004	Humanized
Avastin	VEGF	Metastatic colon cancer	Genentech	2004	Humanized
Raptiva	CD11a	Psoriasis	Genentech/Xoma	2004	Humanized
Tysabri[a]	α4-Integren	Multiple sclerosis	Biogen/Idec	2004	Humanized

[a]Withdrawn in 2005 due to toxicity issues in some patients.

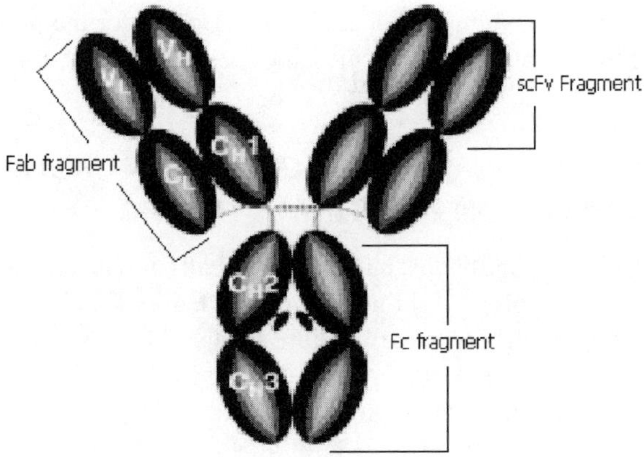

FIGURE 16.1 Structure of an IgG antibody.

TABLE 16.2
Fc-Fusion Proteins Approved by the FDA

Trade Name	Indication	Company	Year of Approval	Fusion Type
Enbrel	Rheumatoid arthritis, psoriasis, ankylosing spondylitis	Amgen	1998	Soluble TNFα receptor fused to IgG1 Fc
Amevive	Psoriasis	Biogen-Idec	2003	Extra cellular portion of leukocyte function antigen-3 (LFA-3) fused to IgG1 Fc

On the other hand, Fc-fusion proteins consist of the constant regions of antibodies (immunoglobulins) fused to an unrelated protein or protein fragment. Such constructs have become popular laboratory tools for the study of protein function since the Fc moiety assures them a longer *in vivo* half life [9]. These molecules can have a range of functions depending on the nature of their fusion partner. Two Fc-fusion proteins have been approved till date (Table 16.2) and a number of this class of molecules are undergoing clinical development. Structurally, the common motifs that Fc-fusion proteins share are the C_H2 and C_H3 domains of antibody heavy chains. Fc-fusion proteins are typically dimeric as the antibody heavy chains are held together by disulfide bonds. Even though biologically quite different from antibodies, the Fc tag imparts these molecules with a strong affinity towards Protein A, enabling the use of a very similar purification scheme as monoclonal antibodies. Hence, for the purpose of this chapter, both these classes of molecules have been considered together.

16.3 PURIFICATION OF MONOCLONAL ANTIBODIES AND Fc-FUSION PROTEINS

Till date, monoclonal antibodies have been typically produced by mammalian cell culture to ensure proper folding and glycosylation. Efficient recovery and purification of antibodies from cell culture media is a critical part of minimizing manufacturing costs [10]. Figure 16.2 shows the typical breakdown of costs associated with an antibody production process [11]. As can be seen from Figure 16.2, a significant percentage (~30 to 40%) of the total manufacturing cost of therapeutic antibodies is incurred during purification. In fact, with significant improvements in cell culture titers (>2 to 3 g/l), downstream purification has the potential of becoming the bottleneck in antibody drug production. Thus the continued commercial success of these biomolecules hinges on the rapid and successful development of economic, robust, and efficient downstream operations.

Chromatography, by virtue of its high resolving power has made itself indispensable for downstream purification of biomolecules [12]. Various combinations of chromatographic steps have been employed for the purification of monoclonal antibodies [13,14]. Over the years, Protein A affinity chromatography has come to be used as the industry-wide standard for direct capture and purification of monoclonal antibodies and Fc-fusion proteins owing to the high degree of selectivity it offers [15]. The highly specific binding between the Fc-region of an antibody and Protein A leads to widespread use of Protein A chromatography as the capture step in the process and gives a large purification factor starting directly from complex solutions such as clarified cell culture harvest media. It can remove >99.5% of product impurities in a single step with

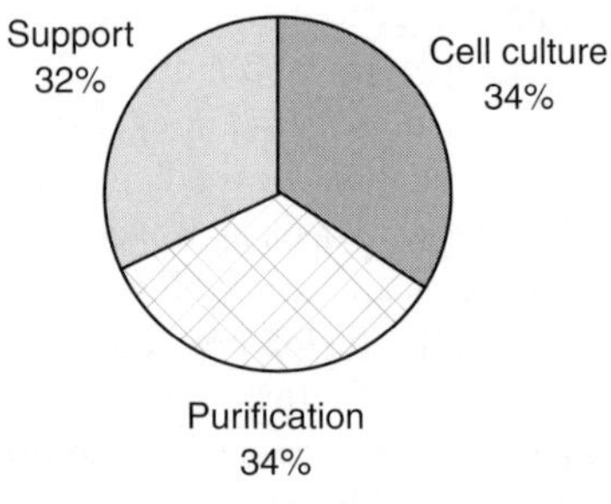

FIGURE 16.2 Representative distributions of costs in an antibody production process.

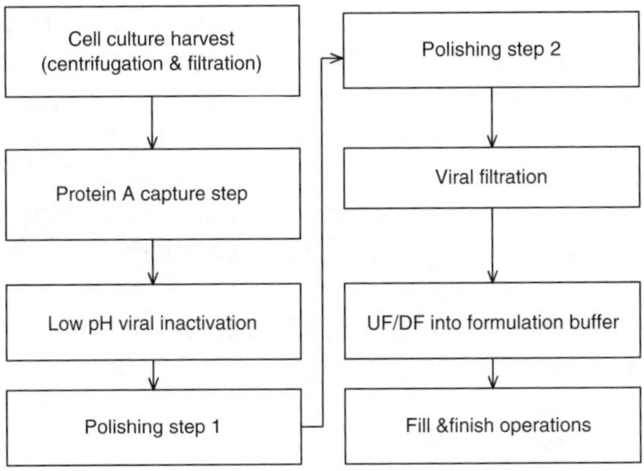

FIGURE 16.3 Generic purification process for antibodies and Fc-fusion proteins.

high yields and often minimal method development [16]. The high degree of purification from this process step helps make the entire downstream process very robust, since in general only trace contaminants need to be removed after this unit operation (high molecular weight aggregates, residual host cell proteins and leached Protein A). Usually only one to two chromatographic steps are required following the Protein A capture step in these processes [17]. This has helped companies move towards platform processes for antibody purification [18,19] with important implications for time to market and process harmonization for multiproduct manufacturing. Figure 16.3 shows a flow sheet for a typical generic template that is used for purification of antibodies and Fc-fusion proteins. Apart from the capture and polishing steps, there are also unit operations dedicated specifically for viral reduction. This chapter focuses on the Protein A capture step while the polishing and viral reduction steps is dealt with in other chapters.

16.4 PROTEIN A AFFINITY CHROMATOGRAPHY

Staphylococcal Protein A, or SpA, is a type I membrane protein from the bacterium *Staphylococcus aureus*. SpA has high specificity for the Fc region of antibodies which has led to its widespread use as a powerful affinity ligand for several immunological and purification applications. Protein A is a ~42 kDa protein consisting of a single polypeptide chain. The chain is made up of five homologous IgG binding domains followed by a C-terminal region necessary for cell wall attachment [15]. The IgG-binding domains are designated as E, D, A, B, C in the order from the N-terminus and are named in the order of their discovery) and share 65 to 90% amino acid sequence identity [20–22].

The Z-domain is a 58 amino acid synthetic analogue of the B-domain which has been very well characterized and extensively studied in the literature [16].

High selectivity and good physiochemical stability has made Protein A the preferred generic ligand for affinity purification of antibodies and molecules tagged with an antibody Fc-region. Since its discovery in 1972 [23,24], Protein A chromatography has become the workhorse of antibody purification and has received growing attention as the importance of therapeutic antibodies in the biotech industry has kept increasing.

The interaction between IgG and Protein A has been studied in detail [25,26] by x-ray crystallography of the complex between a human Fc fragment and a 58 amino acid fragment spanning the B-domain of Protein A. The three-dimensional (3D) structure of the complex revealed two antiparallel α-helices on domain B interacting with both the C_H2 and C_H3 domains of Fc region. The interaction has been shown to primarily consist of hydrophobic interactions along with some hydrogen bonding and two salt bridges [27]. Eleven residues of the Protein A domain and nine residues of Fc were suggested to be involved in binding [26,28]. The primary binding site for Protein A on the Fc region is at the juncture of C_H2 and C_H3 domains. Experimental data indicates that induced fit occurs, explaining the harsh conditions required for elution [15].

Elution off a Protein A column is typically achieved by lowering the pH of the mobile phase. Studies have revealed that a highly conserved histidyl residue is present in the center of the Protein A binding site of IgG [29]. This residue aligns facing a complimentary and similarly conserved histidine residue on Protein A itself [22,30]. At alkaline or neutral pH, these residues are uncharged and there are no restrictions on interfacial contact. In fact, the hydrophobic character of the uncharged immidazole rings contribute to net hydrophobicity at the interface, strengthening the association [15]. At low pHs, the complementary histidine groups take on a positive charge resulting in electrostatic repulsion between the two proteins and a concomitant reduction in the hydrophobic contact area between them. This electrostatic repulsion is strong enough to elute the antibody off the Protein A column. Several attempts have been described in the literature to avoid the low pH elution by employing weakly hydrophobic competitors such as glycyl-tyrosine, ethylene glycol or by using chaotropic salts but all of these methods have met with limited success [15,31]. Moreover, antibodies can potentially become irreversibly denatured in high concentrations of organic solvents or chaotropes disqualifying them for preparative applications.

Apart from the classical binding site, some immunoglobulins have been shown to have an alternate binding site for Protein A on their heavy chain variable domain [32–35]. In particular, IgMs as well as some IgGs and IgAs that contain heavy chains from the human V_H3 gene family have been shown to exhibit this behavior [36,37]. The heavy chain variable domains of antibodies can be classified into six distinct subfamilies (V_H1 to V_H6) on the basis of DNA sequence homology [36,38,39]. Nearly half of the human V_H germline genes belong to the V_H3 subfamily [40,41]. There has been work in the literature on variable region interactions with protein A in free solution focussing on identifying the binding site and establishing the structural basis for these interactions [34,42–44].

The affinity of Protein A for immunoglobulins varies with species and subclass [29,45]. Human IgGs are bound with very strong affinities, except for IgG3, which is very weakly bound [46]. Some classes of murine antibodies have much lower affinities; however their binding can be enhanced by using high concentrations of kosmotropic salts, glycine and/or lower temperatures [15,47,48]. Most of the current antibody drug candidates are humanized or human monoclonal IgGs 1, 2, and 4, produced in Chinese Hampster Ovary (CHO) cells. IgG3s are not selected as therapeutic candidates due to their short half-life [7]. Hence Protein A can be conveniently used to directly capture antibodies from cell culture fluid under physiological conditions.

16.4.1 Protein A Chromatographic Stationary Phases

There are a wide variety of commercially available Protein A resins. These vary with respect to the source of the Protein A ligand (natural wild type vs. recombinant), coupling chemistry or bead characteristics (e.g., backbone matrix, particle size of the bead, and pore-size distribution). Recombinant Protein A lacks the cell wall associated region of natural Protein A, however the antibody binding is indistinguishable [15]. The C-terminal region of the recombinant molecule might be altered to facilitate its purification itself. Moreover, it might incorporate features (e.g., C-terminal cysteine or polylysyl sequences) to facilitate coupling of the ligand to the stationary phase.

Hahn et al. [49] have recently compared a large number of Protein A resins with respect to their transport characteristics and equilibrium binding capacities using polyclonal human IgG as the feed material. Some of the commercially available Protein A resins are listed in Table 16.3. The two leading manufacturers of Protein A chromatographic resins are Amersham Biosciences (now a division of GE Healthcare) and Millipore Corp. While Millipore has adopted the controlled pore glass (CPG) matrix for their resins, Amersham employs agarose with varying degrees of cross-linking as their backbone of choice. Stationary phase backbone is an important factor to be kept in mind even for an affinity resin such as Protein A because nonspecific interactions can occur with the backbone leading to variations in the Protein A eluate purities with respect to host cell protein levels. CPG is quite hydrophobic as compared to agarose and thus exhibits significantly higher levels of nonspecific interactions. Several wash steps have been developed for Prosep A resins to specifically address this issue.

TABLE 16.3
Commercially Available Protein A Resins For Preparative Chromatography

Resin Name	Vendor	Source of Protein A	Backbone Matrix	Particle Diameter (μm)
nProtein A Sepharose FF	GE Healthcare	Natural (coupled by cyanogen bromide activation)	4% cross-linked agarose	45–165
rProtein A Sepharose FF	GE Healthcare	Recombinant (epoxy activation; thioether coupling)	4% cross-linked agarose	45–165
rmp Protein A Sepharose FF	GE Healthcare	Recombinant (multipoint attachment by reductive amidation)	4% cross-linked agarose	45–165
MabSelect	GE Healthcare	Recombinant (Epoxy activation)	Highly cross-linked agarose	40–130 (average ~85)
MAbXtra	GE Healthcare	Recombinant (Epoxy activation)	Highly cross-linked agarose	Average ~75
MAbSelect SuRe	GE Healthcare	Recombinant; alkali stabilized (Epoxy activation)	Highly cross-linked agarose	Average ~85
ProSep-vA High Capacity	Millipore	Natural	Controlled pore glass (1000 Å pore size)	75–125
ProSep-rA High Capacity	Millipore	Recombinant	Controlled pore glass (1000 Å pore size)	75–125
ProSep-vA Ultra	Millipore	Natural	Controlled pore glass (700 Å pore size)	75–125
IPA-500	Repligen Corp	Natural	cross-linked agarose	90
Protein A Ceramic HyperD	Ciphergen	Recombinant	Polyacrylamide gel in ceramic macrobead	50
Poros 50 A High Capacity	Applied Biosystems	Recombinant	Polystyrene Divinyl benzene	50

The first commercial Protein A resin from was natural Protein A Sepharose FF which involved multipoint attachment of the Protein A ligand to the base matrix. Subsequently, GE Healthcare introduced the recombinant version of this resin in which the ligand was attached only at a single point. This can give the recombinant resin a higher binding capacity due to greater flexibility of the attached ligand [15]. Single-point attachments however can potentially lead to higher ligand leaching. Amersham Biosciences also has a resin available in which the recombinant ligand is bound to the base matrix by multipoint attachment. These resins are based on 4% cross-linked agarose backbone and have flow rate limitations and problems associated with compressibility of agarose. MAbSelect was introduced by Amersham in 2000 to address this shortcoming in the Sepharose FF line of products. A higher degree of cross-linking was employed in this bead making it more rigid [50]. This resin is increasingly being adopted for newer products. Recently, GE Healthcare also launched another Protein A resin called MAbXtra® based on the same backbone chemistry as MAbSelect®, but with a wider pore size to improve mass transport, and thereby dynamic binding capacity [51]. The decrease in surface area due to the larger pores was compensated by an increase in ligand density. SuRe® is yet another resin launched in 2005 by the same manufacturer which has the same backbone as MAbSelect but a genetically modified Protein A ligand to help withstand alkaline conditions.

Despite increased nonspecific binding of contaminants to the backbone, the Prosep A resins have the advantage of better pressure-flow characteristics due to the rigid CPG backbone. The Prosep A resin comes in two pore sizes: 700 and 1000 Å. The smaller pore size was recently introduced to give a larger surface area and thus increase binding capacity. The decreased pore size may however lead to increased mass transfer limitations for larger molecules [52]. For the most part, Millipore has adopted natural Protein A as their ligand of choice. Recently the production method for this ligand was modified to eliminate animal-derived raw materials and this product series has been termed vegan even though for chromatographic purposes the resin is identical to its earlier version [53].

Junbauer and Hahn [54] have summarized the equilibrium saturation capacity, equilibrium dissociation constant and effective diffusion coefficients for the most commonly used Protein A media. The authors mention that agarose-based media possess higher dynamic binding capacities offset by increased mass transfer resistance. In contrast, the Prosep A media showed lower binding capacities with improved mass transport properties. Typically in the industry, choice of the Protein A resin is application-specific and depends on the best compromise between capacity, product purity, and flow characteristics (which contribute to throughput). The relevance of some of these factors is explained in greater detail in the subsequent section.

16.4.2 Practical Considerations for Developing a Protein A Step in an Industrial Process

The basic protocol of Protein A chromatography is relatively straightforward: bind at neutral pH and elute at acidic pH. Even an unoptimized Protein A step can yield a highly purified antibody. The ease and simplicity of methods development on this mode of chromatography has been a key reason for its widespread adoption for monoclonal antibody and Fc fusion protein purification all the way from the molecular biology laboratory to large-scale production processes. Despite these significant advantages, the use of Protein A chromatography for process-scale purification does involve several critical challenges that are described below.

16.4.2.1 Binding Capacity and Process Throughput

The primary disadvantage of Protein A affinity chromatography is the high cost of the resin. Figure 16.4 shows a comparison of the costs for some widely used resins on a per L basis. As can be seen from the figure, Protein A media are almost an order of magnitude more expensive as compared to traditional chromatographic media. The high cost coupled with the large production quantities for antibodies have caused the binding capacity on Protein A to become a key parameter that has significant influence on process economics in industrial purification processes. Given this fact, resin manufacturers are

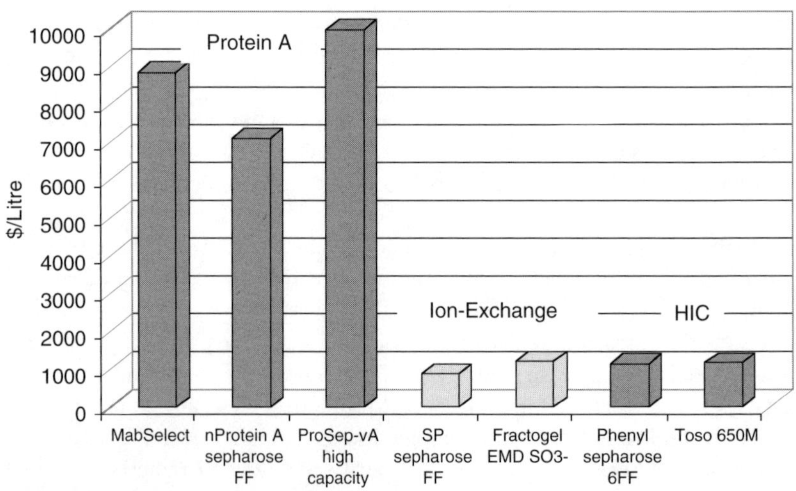

FIGURE 16.4 Typical costs of chromatographic resins.

continuously introducing new versions of Protein A media aimed at providing better binding capacity. Some recent examples include the introduction of MAbXtra (over MabSelect) from Amersham Pharmacia Biotech and the introduction of ProsepA media from Millipore in a smaller pore size. These changes to resin morphology also have their limits. For example, decreasing pore size will increase surface area for binding, but will ultimately lead to increased mass transfer resistance [52] and therefore low dynamic capacity at higher flow rates.

Another important consideration in using Protein A chromatography for bioprocessing applications is production rate or throughput [55]. Since Protein A is used as the capture step, the harvested and clarified cell culture fluid is directly loaded on to a Protein A column. Unlike the polishing steps, the load to this step is usually very large and dilute in antibody concentration. Figure 16.5 shows a typical chromatogram for a Protein A process step. As can be seen from the figure, the load time is a significant proportion of the total process time. This makes throughput a particularly critical parameter for this mode of chromatography. Moreover, since the media is very expensive, rather than using a large column to process a batch of antibody in a single cycle, typical bioprocess applications run a smaller column for several cycles to purify a single batch. This reduces the risk of capital loss if the column is compromised during operation and also brings the column diameter into a practical range. Cycling increases the total purification time and thereby decreases the production rate. Thus, processing time can be an important factor in Protein A step development.

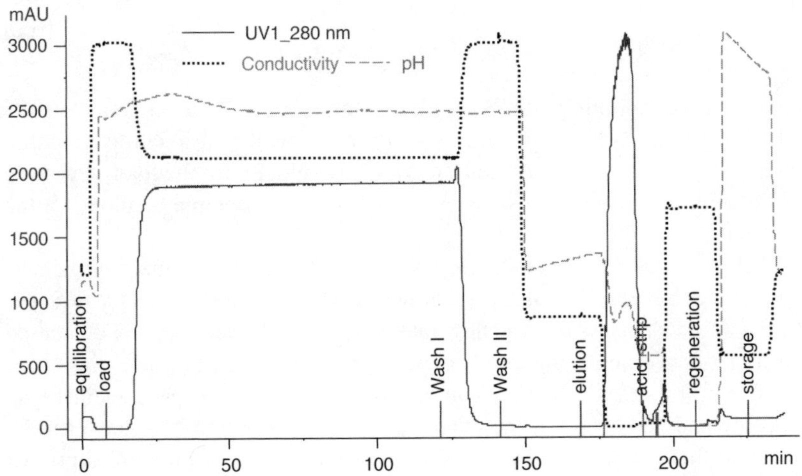

FIGURE 16.5 Typical chromatogram for a Protein A process.

Fahrner et al. [56] discuss the importance of considering the optimal flow rate on Protein A and suggest that higher flow rates will reduce process time without significantly affecting process capacity. Processing time has been mentioned to be critical to process development for three reasons [57]. First, if purification is the limiting factor in a production facility, then a direct improvement in process time will increase throughput. Often, Protein A is the first capture step and is often the rate-limiting step due to the large volume loads and low protein concentrations associated with the process. Second, the product stability in the harvested cell culture fluid can limit allowable hold and processing times. Third, cell culture fluid is a rich medium that can promote an increase in bioburden. Minimizing processing time can help to decrease bioburden contamination. The authors even suggest using a resin with a slightly lower binding capacity but better flow characteristics to enable a decrease in the overall processing time.

Productivity or volumetric production rate can be defined as the mass of protein purified in one cycle divided by the processing time taken, divided by the column volume to make it independent of scale [55]

$$\begin{aligned} \text{Productivity } (P) &= \left(\frac{\text{mass of product}}{\text{column volume}}\right) \Big/ (\text{time}) \\ &= \left(\frac{VQ_\text{d}}{V}\right) \Big/ \left(\frac{VQ_\text{d}}{C_0 u_\text{L} A} + \frac{NV}{u_\text{N} A}\right) \end{aligned} \tag{16.1}$$

Rearranging terms, Equation 16.1 reduces to

$$P = \frac{1}{(L(1/C_0 u_\text{L}) + (N/Q_\text{d} u_\text{N}))} \tag{16.2}$$

where V is the column volume; A is the column cross-sectional area; L is the column length; Q_d is the binding capacity taken as the surrogate for column loading; C_0 is the load concentration; u_L is the velocity for the load step; u_n is the velocity for the nonload steps; N is the number of column volumes for the nonload steps.

Fahrner et al. [55] have developed a methodology for comparing the performance of several Protein A resins with respect to their productivity under different flow rates. As loading flow rate was increased, bed capacity decreased but its influence on throughput was offset by a reduction in processing time. The overall process throughput can thus be plotted in a 3D space against load flow rate and bed height as the other two axes. Shukla et al. [58] have further extended this comparison by including a consideration of pressure drop that will limit the maximum attainable flow rate at a given bed diameter and column height and also considered the impact of column cycling. Figure 16.6 shows

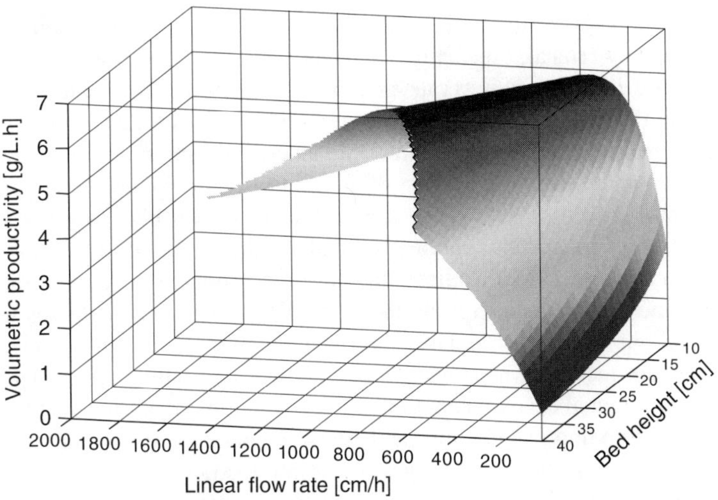

FIGURE 16.6 Plot of productivity vs. bed height and loading flow rate.

a 3D plot of productivity vs. bed height and linear flow rate for a fixed bed diameter. As can be seen from the figure, at a given bed height productivity first increases and then decreases with increasing flow rate. Very high flow rates at high bed heights are not possible due to flow rate limitations and are removed from the plotted surface. The optimal operating regime is defined by several combinations of bed height and linear flow rate. Due to the wall effect, column pressure drops depend on the column diameter in addition to bed height. Thus, another important attribute of Protein A resins to kept in mind is their pressure-flow behavior. Bead particle diameter, uniformity of particle size-distribution, resin compressibility can play a dominant role in determining the pressure-flow behavior.

To increase throughput compared to the traditional packed bed mode of operation, technologies such as simulated moving-bed chromatography (which increases throughput by moving into a continuous operation mode from a batch operation) has also been evaluated for the Protein A affinity step [59,60]. Additionally, expanded-bed chromatography (which enables elimination of the cell harvest step and may allow increased throughput due to higher flow rates during column loading) has also been investigated with Protein A media [61,62]. However, both these technologies have met with limited success and have not yet been employed in any commercial therapeutic antibody manufacturing process.

It is to be remembered that throughput is not the sole consideration while selecting a Protein A resin for a downstream process. Resin choice is quite frequently made earlier in clinical development and is dominated by product purity considerations. In addition, due to the high cost of the resin, binding capacity can be equally critical, especially for an early-stage process. Operating at the point of highest productivity alone might not always be prudent because it might mean that the resin-binding capacity is not utilized to its fullest extent. While using a very slow loading flow rate would give the maximum resin utilization, such an approach is obviously not desirable from a production time point of view. Thus, the two objectives of high productivity and high capacity are at odds with each other. Conventionally, a rule of thumb in industrial Protein A operation is to use a single loading flow rate that results in a column residence time of 5 to 6 min for loading. In fact, the best strategy in many cases would be to maximize loading capacity while making sure productivity is adequate and does not make downstream purification the rate-limiting step in the process. An additional development tool can be to employ a dual-flow rate loading strategy to help improve binding capacity while maintaining high productivity [63]. Intuitively, in the initial stages of column loading when all the binding sites are available, one should be able to flow at a faster flow rate and save on processing time. Once all the readily accessible sites are blocked, a slower flow rate could then be used to enable the protein to diffuse into all the pores and bind to the less readily accessible sites. Using an appropriate chromatographic model and equilibrium, and transport parameters, the operating conditions for such a strategy can easily be optimized for a given resin and column dimensions. Moreover, the authors have proposed the use of a weighted combination of capacity and productivity (as the objective function) for simultaneous optimization of both throughput and capacity. This can be a useful design tool and can give the user discretion over what combination of binding capacity and throughput would be the best process-fit based on facility scheduling or economic constraints [63].

16.4.2.2 Elution Conditions

As mentioned earlier, low pH is the most commonly used method for eluting Protein A chromatographic columns. However, several proteins are known to unfold and tend to form aggregates under low pH conditions [64,65]. Aggregation phenomena observed during Protein A elution can be categorized as shown in Figure 16.7 [67]. It can be due to (i) soluble high molecular weight generation as determined by analytical size exclusion chromatography, (ii) visible turbidity due to insoluble particle formation which can be either the antibody or contaminant proteins, and (iii) combination of (i) and (ii). Shukla et al. [66] have proposed several strategies to address some of the above-mentioned problems. Stabilizing additives such as salts, urea, and amino acids can be added to stabilize the product as it elutes off the Protein A column. Lowering the operating temperature and slowing down the kinetics of aggregation can be a viable

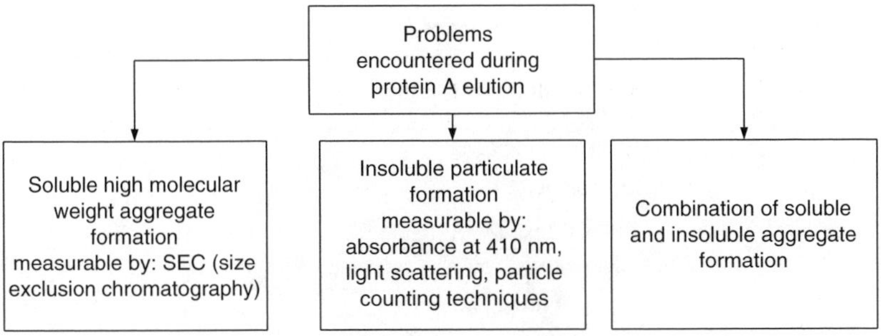

FIGURE 16.7 Aggregation and precipitation phenomena observed during protein A chromatography.

strategy if the antibody is prone to soluble aggregate formation. If the turbidity observed in the elution pool is predominantly due to precipitation of contaminating proteins, pretreating the cell culture fluid to remove impurities (by use of additional depth filter) can be another option. Finally, manipulating the pH transition between wash and elution by controlling the buffering species and its strength has also been shown to be an effective strategy. The appropriate elution condition that needs to be chosen is very product-specific and will depend on the problem at hand. Thus even though Protein A chromatography lends itself to the possibility of generic processing conditions, complete templating of all parameters is not possible even for this process step, as shown in Figure 16.8 [19]. Determining appropriate elution conditions is one of the areas that will require significant process development effort.

16.4.2.3 Wash for Impurity Removal

Typical operating conditions for Protein A comprise of column equilibration with a buffer at neutral pH, direct loading of cell culture fluid, a wash with the equilibration buffer to remove unbound contaminants followed by elution. Often, an intermediate wash is also included in-between the equilibration wash and elution steps and can help to serve various purposes. Despite the high specificity of Protein A, host cell protein contaminants are still present as varying levels in the elution pool. Even though the residual levels of contaminants can be cleared in the subsequent polishing steps, it is desirable to minimize the impurity level in the Protein A step itself to increase the overall robustness of the process. An intermediate wash step can help to reduce impurity levels and even help to minimize turbidity in the elution pool, if the turbidity is due to precipitation of contaminants. Resins with a hydrophobic backbone (such as Prosep A) have a higher level of contaminants in the elution pool due to

Parameter
Resin
Residence time during loading
Resin load capacity
Bed height
Operating temperature
Equilibration/post-load wash buffer
Equilibration buffer volume
Post load wash volume
Wash II buffer
Wash II buffer volume
Elution buffer
Elution buffer pH
Strip buffer
Strip buffer volume
Flush buffer
Flush buffer volume
Regeneration buffer
Regeneration buffer volume
Storage buffer
Storage buffer volume

FIGURE 16.8 Process parameters for a Protein A chromatographic step. Light: development required; dark: predetermined condition.

nonspecific interactions of the contaminants with the backbone. In such cases disruption of this nonspecific interaction by employing hydrophobic electrolytes such as tetramethyl ammonium chloride (TMAC) or a combination of detergent and salt can be beneficial [67]. For agarose-based resins, which have minimal nonspecific interactions, sometimes an intermediate pH wash is used to minimize co-elution of contaminants. It is to be noted that if the intermediate wash contains additives that are not desirable in the elution pool, a pre-elution wash might be required to prevent mixing of the intermediate wash buffer constituent with the eluate.

16.4.2.4 Protein A Leaching

Protein A leaching is yet another problem associated with the use of this mode of chromatography. This is a serious concern for the drug industry because

Protein A is known to cause immunogenic responses in humans and has been proven toxic in a number of clinical trials [68]. One should pay special attention during development of the polishing steps to reduce leached Protein A to safe and acceptable levels. Leaching can occur by three different pathways: breakdown of the support matrix, breakdown of the immobilization linkage, and proteolytic cleavage of the interdomain sequences of Protein A [15]. To avoid the first two, it is recommended to select a resin from an established manufacturer with a history of good ligand stability. In industrial processes, Protein A leaching is primarily a result of proteolytic degradation due to proteases that can be present in the cell culture fluid. Thus, addition of chelators such as EDTA to cell culture harvest (to inhibit metallo-proteases) or holding the cell culture load at lower temperatures can also help to reduce Protein A leaching. Storage conditions for resin storage can also have an influence on the level of leached Protein A. Storing the Protein A resin under slightly acidic conditions has also been seen to be beneficial in minimizing proteolytic degradation during storage.

16.4.2.5 Resin Lifetime

Efficient cleaning of Protein A resin is crucial because the high cost of this resin makes extensive cycling of Protein A columns imperative. Sodium hydroxide solutions are commonly used for cleaning and sanitizing chromatographic systems in a GMP manufacturing environment [69]. Even though Protein A is physiochemically stable under strong acidic conditions, it cannot withstand strong alkaline conditions. Exposure of Protein A resins to sodium hydroxide solutions has been shown to cause a decrease in its binding capacity [70], thus preventing the use of high concentrations of NaOH for cleaning and regenerating the resin. Currently Protein A columns are cleaned using sequential washes with a strip and regeneration solution. Typically, strip refers to cleaning of the resin by a mechanism similar to its elution (low pH in this case) and thus low concentration of acids (such as 1 *M* acetic acid or 0.1 *M* phosphoric acid) is used as strip solutions. Regeneration solutions clean the resin by a complimentary mechanism — mildly alkaline solutions (e.g., 50 m*M* NaOH) or denaturing conditions (e.g., urea, guanidine) are commonly used regeneration agents [71,72]. Cleaning of the column with the harsher regeneration solution might not be required after every cycle. Typically, in the industry, a Protein A column is cleaned by the strip solution after every cycle and regenerated once every manufacturing campaign prior to column storage.

To overcome the limitation of alkaline instability, GE Healthcare has very recently introduced a new version of Protein A resin (called SuRe®) that is expected to be resistant to alkaline conditions such as 0.1 *M* NaOH. To achieve this, a number of asparagine (the most alkali-sensitive amino acid) residues in the Z-domain of Protein A were replaced with other amino acids using protein engineering techniques and a new ligand was composed as a tetramer of four identical modified Z-domains [73]. Use of 0.1 to 0.5 *M* NaOH as been recommended for cleaning and sanitization of this resin [74]. The replacement of expensive solutions like urea and guanidine by NaOH is expected to result in significant raw material cost savings. However, this product is relatively new and its application in large-scale processes remains to be seen.

In a commercial antibody process, the lifetime of the large protein A column is validated for at least 100 cycles so that the high cost of the resin is amortized over several production batches. Millipore and GE Healthcare have demonstrated the use of their leading Protein A resins for over 300 cycles using appropriate cleaning regimen [72,75]. However, since resin lifetime is a strong function of the nature of the feed load material and the product, a validated resin lifetime study is required for a commercial process. Such lifetime studies are carried out using a qualified scale-down model using representative feed load after the Protein A process conditions are completely defined. The number of cycles for which the resin is validated is case-specific and is chosen based on economic considerations as well as performance of the column over extensive cycling.

16.4.3 Process Flow Sheet

Figure 16.9 shows a typical process flow sheet for a Protein A step along with the specific purpose of some of the segments. The rationale for the development of certain segments in the Protein A step has already been mentioned in Section 16.4.2 and this section outlines some other general guidelines. Equilibration of the column is usually done under neutral conditions (~pH 6.0 to 7.5) and the commonly used buffer systems are tris, phosphate, or citrate. The equilibration buffer also contains moderate concentration of salt to minimize nonspecific electrostatic interactions. Commonly used buffer systems during elution are citrate and acetate which have good buffering capacity in the lower pH range. It is desirable to avoid the use of halide containing salts (such as NaCl) in the buffers at lower pHs to avoid corrosion of stainless steel tanks. Finally, if an acidic strip is followed by a sodium hydroxide regeneration solution, it is recommended to flush the column (usually with the equilibration buffer) to prevent acid/base reactions inside the column.

16.5 CONCLUSIONS

Protein A affinity chromatography continues to be the state-of-the-art technique for capture and purification of antibodies and Fc-fusion proteins. It forms the

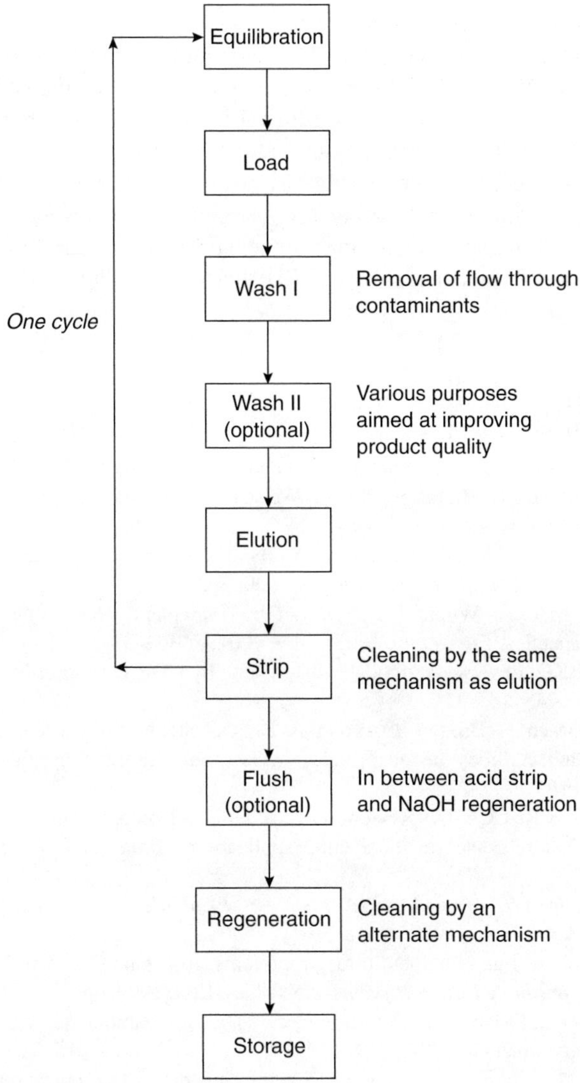

FIGURE 16.9 Typical flowsheet for a protein A process.

foundation for generic process development of these biomolecules, which is a key strategic initiative for the biotech industry today. The yield for this process step is very high (>90%) and a high degree of purification factor (>2 logs) can easily be obtained. Several attempts have been made to purify antibodies in three chromatographic steps without recourse to a Protein A step [16,76] but such nonaffinity purification schemes require significantly greater methods development time and resources and are potentially less robust, which is not desirable in a timeline driven development scenario. Even though Protein A chromatography is relatively straightforward to develop for laboratory-scale separations, a number of areas require special consideration and development time when production scale separations are being designed. These include stabilization of the product during low pH elution, optimization of binding capacity, and throughput, improving product purity through the development of specific wash steps and increasing resin lifetime through improved regeneration conditions.

REFERENCES

1. Walsh G. Modern antibody based therapeutics. *Biopharm* 2004; Dec.: 18–25.
2. Cochlovius B, Braunagel M, and Welschof M. Therapeutics antibodies. *Mod. Drug Discov.* 2003; 6: 33–38.
3. Stockwin LH and Holmes S. Antibodies as therapeutic agents: Vive la renaissance! *Expert Opin. Biol. Ther.* 2003; 3: 1133–1152.
4. Djik MA van, Winkel J, and GJ van de. Human antibodies as next generation therapeutics. *Curr. Opin. Chem. Biol.* 2001; 5: 368–374.
5. Lo BKC. Review: Antibody humanization by CDR grafting. *Meth. Mol. Biol.* 2004; 248: 135–159.
6. Rohrbach P, Broders O, Toleikis L, and Stefan D. Therapeutic antibodies and antibody fusion proteins. *Biotechnol. Genet. Eng. Rev.* 2003; 20: 137–163.
7. Jefferis R. CCE IX: Review. Glycosylation of recombinant antibody therapeutics: relevance to therapeutic applications. *Biotechnol. Prog.* 2005; 21: 11–16.
8. Goding JW. *Monoclonal Antibodies: Principles and Practice*. London: Academic Press, 1983.
9. Aruffo A. Immunoglobulin fusion proteins. In: Chamow S and Ashkenazi A, Eds. *Antibody Fusion Proteins*. NY: Wiley-Liss, 1999, pp. 221–241.
10. Keller K, Friedmann T, and Boxman A. The bioseparation needs for tomorrow. *Trends Biotechnol.* 2001; 19: 438–441.
11. Myers J. Economic consideration in the development of downstream steps for large-scale commercial biopharmaceutical processes: Production and Economics of Biopharmaceuticals conference, San Diego, November, 2000.
12. Wheelright SM. *Protein Purification: Design and Scale-up of Downstream Processing*. Munich (Germany): Hanser, 1991.
13. Naveh D and Siegel RC. Large-scale downstream processing of monoclonal antibodies. *Bioseparation* 1991; 1: 351–366.
14. Ostlund C. Large-scale purification of monoclonal antibodies. *Trends Biotechnol.* 1996; 166: 288–293.

15. Gagnon P. Protein A affinity chromatography. In: *Purification Tools for Monoclonal Antibodies*. Tucson, AZ: Validated Biosystems, 1996, pp. 155–158.
16. Follman D and Fahrner RL. Factorial screening of antibody purification processes using three chromatographic steps without Protein A. *J. Chromatogr. A* 2004; 1024: 79–85.
17. Fahrner RL, Knudsen H, Basey C, Galan W, Feuerhelm D, Vanderlaan M, and Blank G. Industrial purification of pharmaceutical antibodies: Development, operation and validation of chromatography processes. *Biotechnol. Genet. Eng. Rev.* 2001; 18: 301–327.
18. Blank G. The future of process development for monoclonal antibody manufacturing. IBC Conference on Antibody Production and Downstream Processing, San Diego, CA, March, 2005.
19. Hubbard BH and Shukla AA. Platform approaches to monoclonal antibody purification. IBC Conference on Antibody Production and Downstream Processing, San Diego, CA, March, 2005.
20. Hjelm H, Sjodahl J, and Sjoquist J. Immunologically active and structurally similar fragments of protein-A from *Staphylococcus aureus*. *Eur. J. Biochem* 1975; 57: 395–403.
21. Sjoquist J. Structure and immunology of protein A. *Contrib. Microbiol. Immunol.* 1972; 1: 83–92.
22. Moks T, Abrahmsen L, Nilsson B, Hellman U, Sjoquist J, and Uhlen M. Staphylococcal protein-A consists of five IgG-binding domains. *Eur. J. Biochem* 1986; 156: 637–643.
23. Hjelm H, Hjelm K, and Sjoquist J. Protein A from *Staphylococcus aureus*. XXIII. Its isolation by affinity chromatography and its use as an immunosorbent for isolation of immunoglobulins. *FEBS Lett.* 1972; 28: 73–76.
24. Kronvall G. Surface component in group A, C, and G streptococci with nonimmune reactivity for immunoglobulin G. *J. Immunol.* 1973; 111: 1401–1406.
25. Diesenhofer J, Jones TA, Huber R, Sjodahl J, and Sjoquist J. Crystallization, crystal structure analysis and atomic model of the complex formed by a human Fc fragment and fragment B of protein A from *Staphylococcus aureus*. Hoppe–Seyler's Z. *Physiol. Chem.* 1978; 359: 975–985.
26. Diesenhofer J. Crystallographic refinement and atomic models of a human Fc fragment and its complex with fragment B of protein A from *Staphylococcus aureus* at 2.9- and 2.8-Å resolution. *Biochemistry* 1981; 20: 2361–2370.
27. Li R, Dowd V, Stewart D, Burton SJ, and Lowe CR. Design, synthesis and application of a Protein A mimetic. *Nat. Biotechnol.* 1998; 16: 190–195.
28. Cedergren L, Andersson R, Jansson B, Uhlen M, and Nilsson, B. Mutational analysis of the interaction between staphylococcal protein A and human IgG_1. *Protein Eng.* 1993; 6: 441–448.
29. Burton DR. Immunoglobulin G: functional sites. *Mol. Immunol.* 1985; 22: 161–206.
30. Lindmark R, Movitz J, and Sjoquist J. Extracellular protein A from a methicillin-resistant strain of *Staphylococcus aureus*. *Eur. J. Biochem.* 1977; 74: 623–628.

31. Bywater R, Eriksson GB, and Ottosson T. Desorption of immunoglobulins from Protein A-Sepharose CL-4B under mild conditions. *J. Immunol. Meth.* 1983; 64: 1–6.
32. Inganas M. Comparison of mechanisms of interaction between protein A from *Staphylococcus aureus* and human monoclonal IgG, IgA and IgM in relation to the classical Fcγ and the alternative F(ab$'$)$_2$ protein A interactions. *Scand. J. Immunol.* 1981; 13: 343–352.
33. Vidal MA and Conde FP. Alternative mechanism of Protein A — immunoglobulin interaction: The VH associated reactivity of a monoclonal human IgM. *J. Immunol.* 1985; 135: 1232–1238.
34. Akerstrom A, Nilson Bo HK, Hoogenboom HR, and Bjorck L. On the interaction between single chain Fv antibodies and bacterial immunoglobulin-binding proteins. *J. Immunol. Meth.* 1994; 177: 151–163.
35. Starovasnik MA, O'Connell MP, Fairbrother WJ, and Kelley RF. Antibody variable region binding by Staphylococcal Protein A: Thermodynamic analysis and location of the Fv binding site on E-domain. *Protein Sci.* 1999; 8: 1423–1431.
36. Sasso EH, Silverman GJ, and Mannik M. Human IgA and IgG F(ab$'$)$_2$ that bind to staphylococcal protein A belong to the VHIII subgroup. *J. Immunol.* 1991; 147: 1877–1883.
37. Sasso EH, Silverman GJ, and Mannik M. Human IgM molecules that bind staphylococcal protein A contain VHIII H chains. *J. Immunol.* 1989; 142: 2778–2783.
38. Tomlinson IM, Walter G, Marks JD, Llewelyn MB, and Winter G. The repertoire of human germline V_H sequences reveals about fifty groups of V_H segments with different hypervariable loops. *J. Mol. Biol.* 1997; 227: 776–798.
39. Ignatovich O, Tomlinson IM, Jones PT, and Winter G. The creation of diversity in the human Immunoglobulin V_λ repertoire. *J. Mol. Biol.* 1997; 268: 69–77.
40. Walter MA and Cox DW. Analysis of genetic variation reveals human immunoglobulin VH-region gene organization. *Am J. Hum. Genet.* 1988; 42: 446–451.
41. Schroeder HW, Hillson JL, and Perlmutter RM. Structure and evolution of mammalian VH families. *Int. Immunol.* 1990; 2: 41–50.
42. Hillson JL, Karr NS, Opplinger IR, Mannik M, and Sasso EH. The structural basis of germline-encoded V_H3 immunoglobulin binding to Staphylococcal Protein A. *J. Exp. Med.* 1993; 178: 331–336.
43. Potter KN, Li Y, Mageed RA, Jefferis R, and Capra JD. Anti-idiotypic antibody D12 and superantigen SPA both interact with human V_H3-encoded antibodies on the external face of the heavy chain involving FR1, CDR2 and FR3. *Mol. Immunol.* 1998; 35: 1179–1187.
44. Potter KN, Li Y, and Capra JD. Staphylococcal Protein A simulataneously interacts with Framework region 1, Complementarity-Determining Region 2, and Framework Region 3 on human V_H3-encoded Igs. *J. Immunol.* 1998; 157: 2982–2988.

45. Huse K, Bohme HJ, and Scholz GH. Purification of antibodies by affinity chromatography. *J. Biochem. Biophys. Meth.* 2002; 51: 217–231.
46. Langone JJ. Protein-A of *Staphylococcus aureus* and related immunoglobulin receptors produced by streptococci and Pneumonococci. *Adv. Immunol.* 1982; 32: 157–252.
47. Schuler G and Reinacher M. Development and optimization of a single-step procedure using protein A affinity chromatography to isolate murine IgG1 monoclonal antibodies from hybridoma supernatants. *J. Chromatogr.* 1991; 587: 61–70.
48. Tu YY, Primus FJ, and Goldenberg DM. Temperature affects binding of murine monoclonal IgG antibodies to protein A. *J. Immunol. Meth.* 1988; 109: 43–47.
49. Hahn R, Schlegel R, and Jungbauer A. Comparison of Protein-A affinity sorbents. *J. Chromatogr. A* 2003; 790: 35–51.
50. GE technical note on "Chemical and Functional Stability of a new Protein A Media with high Dynamic Binding Capacity Based on Novel, Highly Rigid Agarose Beads" http://www1.amershambiosciences.com/aptrix/upp00919.nsf/(FileDownload)?OpenAgent&docid=D1A8F2CE4D763D46C1256EB400417E1F&file = 18114904AA.pdf
51. Malmquist G, Lacki KM, Ljunglof A, and Johansson H. Engineering a new generation of affinity media for antibody purification: Presentation at Prep conference, Washington D.C., May, 2003.
52. McCue JT, Kemp G, Low D, and Quinones-Garcia I. Evaluation of Protein-A chromatography media. *J. Chromatogr. A* 2003; 989: 139–153.
53. Lebreton B, Lazzarechi K, McDonald P, and O'Leary R. Performance evaluation of Prosep Va: an "animal-free" protein A resin. Recovery of Biological Products XI, Banff, Canada, September 2003.
54. Jungbauer A and Hahn R. Engineering protein A affinity chromatography. *Curr. Opin. Drug. Discov. Develop.* 2004; 7: 248–256.
55. Fahrner RL, Whitney DH, Vanderlaan M, and Blank GS. Performance comparison of Protein-A affinity chromatography sorbents for purifying recombinant monoclonal antibodies. *Biotechnol. Appl. Biochem.* 1999; 30: 121–128.
56. Fahrner RL, Iyer HV, and Blank GS. The optimal flow rate and column length for maximum production rate of protein A affinity chromatography. *Bioprocess Eng.* 1999; 21: 287–292.
57. Iyer H, Henderson F, Cunningham E, Webb J, Hanson J, Bork C, and Conley L. Considerations during development of a Protein-A based antibody purification process. *BioPharm* 2002; January: 14–20.
58. Shukla AA, Hinckley PJ, and Hubbard B. Productivity comparisons on Protein-A affinity chromatography. ACS National Meeting, Boston, MA, August, 2002.
59. Gottschlich N and Kasche V. Purification of monoclonal antibodies by simulated moving bed chromatography. *J. Chromatogr. A* 1997; 765: 201–206.
60. Thommes J, Conley L, Pieracci J, and Sonnenfeld A. Protein A affinity simulated moving bed chromatography for continuous monoclonal antibody purification. Recovery of Biological Products XI, Banff, Canada, September: 2003.

61. Fahrner RL, Blank G, and Zapata G. Expanded bed protein A affinity chromatography of a human monoclonal antibody: process development, operation and comparison with a packed bed method. *J. Biotechnol.* 1999; 75: 273–280.
62. Blank GS, Zapata G, Fahrner R, Milton M, Yedinak C, Knudsen H, and Schmelzer C. Expanded bed adsorption in the purification of monoclonal antibodies: a comparison of process alternatives. *Bioseparation* 2001; 10: 65–71.
63. Ghose S, Nagrath D, Brooks C, Hubbard B, and Cramer SM. Use and optimization of a dual flow rate loading strategy to maximize throughput in Protein A affinity chromatography. *Biotechnol. Prog.* 2004; 20: 830–840.
64. Chen T. Formulation concerns of protein drugs. *Drug. Develop. Ind. Pharm.* 1992; 18: 1311–1354.
65. Krishnamurthy R and Manning MC. The stability factor: importance in formulation development. *Curr. Pharm. Biotechnol.* 2002; 3: 361–371.
66. Shukla AA, Hinckley PJ, Gupta P, Yigzaw Y, and Hubbard B. Strategies to address aggregation during Protein-A chromatography. *BioProcess International* 2005; May: pp. 36–45.
67. Reifsnyder DH, Olson CV, Etcheverry T, and Prashad H, Builder SE. Purification of insulin-like growth factor-I and related proteins using underivatized silica. *J. Chromatogr. A* 1996; 753: 73–80.
68. Terman DS and Bertram JH. Antitumor effects of immobilized protein A and staphylococcal products: linkage between toxicity and efficacy, and identification of potential tumoricidal reagents. *Eur. J. Cancer Clin. Oncol.* 1985; 21: 1115–1122.
69. Sofer GK, and Nystrom LE. Process hygiene. In: *Process Chromatography. A Practical Guide*. New York: Academic Press, 1989, pp. 93–105.
70. Tejeda-Mansir A, Espinoza R, Montesinos RM, and Guzman R. Modelling regeneration effects on protein-A affinity chromatography. *Bioprocess Eng.* 1997; 17: 39–44.
71. Millipore technical note # TB1021EN00 on "Cleaning ProSep®-A Protein A Affinity Chromatography Media", http://www.millipore.com/publications.nsf/docs/66lllf (accessed September 2005).
72. GE Heathcare technical note No. 18117765 on "The use of NaOH for CIP of rProtein A media: a 300 cycle study". http://www.bioprocess. amershambiosciences.com/Applic/upp00738.nsf/va_temp/ABA5D988F09C9CA6C1256 D97005AAF4F/$file/18117764AA.pdf (accessed September 2005).
73. GE Heathcare technical note No. 11-0011-64 AA on "MabSelect SuRe — studies on ligand toxicity, leakage, removal of leached ligand, and sanitization". http://www1.amershambiosciences.com/aptrix/upp00919.nsf/(FileDownload)?OpenAgent&docid = 6127EE90B62C21F4C1256F6A000DC19C&file = 11001164AA.pdf (accessed September 2005).
74. GE Heathcare technical note No. 11-0011-65 AA on "MabSelect SuRe: alkali-stabilized Protein A derived medium for capture of monoclonal antibodies". http://www1.amershambiosciences.com/aptrix/upp00919.nsf/(FileDownload)?

OpenAgent&docid = BD6F2CD251195D7BC1256F6A000DC219&file = 11001165AA.pdf (accessed September 2005).

75. Millipore technical note No. TB1650EN00 on "ProSep-vA Ultra Media 300 Cycle Lifetime Studies". http://www.millipore.com/publications.nsf/docs/tb1650en00 (accessed September 2005).
76. Duffy SA, Moellering B, Prior G, Doyle K, and Prior C. Recovery of therapeutic grade antibodies: protein A and ion-exchange chromatography. *Biopharm* 1989; June: 34–47.

17 Polishing Methods for Monoclonal IgG Purification

Pete Gagnon

CONTENTS

17.1 INTRODUCTION

Protein A affinity chromatography has become well established as the preferred capture step for purification of human monoclonal IgG for *in vivo* applications, so much so that it has become largely regarded as generic. This has created the desire for an equally generic block of polishing methods to complete the purification. This chapter will discuss the suitability of anion exchange, cation exchange, hydrophobic interaction, and ceramic hydroxyapatite (CHT™ Bio-Rad) chromatography as candidates for this application. The strengths and limitations of each method will be discussed individually, and the merit of combining defined subsets with protein A toward the possibility of a generic overall IgG purification scheme will be considered.

17.2 ANION EXCHANGE CHROMATOGRAPHY

Anion exchange is nearly as universal as protein A in the purification of monoclonal IgG. It is employed frequently as the last chromatography step because of its ability to scavenge endotoxins that may have entered the process via contaminated manufacturing materials or inappropriate sample handling. Its proven ability to reduce other key contaminant classes such as nucleotide, virus, and leached protein A — in addition to its ability to remove host cell proteins (HCPs) — make it an even stronger candidate [1]. The fact that IgG is usually soluble under binding conditions adds to its utility.

Anion exchange is frequently applied in a format referred to as flow-through mode. Buffer conditions are set so that the antibody passes through the column while strongly electronegative contaminants are captured. The majority of established applications are still performed on conventional ion exchangers such as Sepharose Fast Flow Q™ (GE Healthcare) but charged membrane filtration (Pall Corporation, Sartorius) is becoming increasingly popular. Flow-through applications are effective for DNA, retrovirus, and endotoxin reduction but sacrifice much of the exchanger's ability to reduce leached protein A and HCP [1,2].

Effective reduction of protein A contamination requires that the antibody be captured and eluted in a gradient. Protein A has an isoelectric point of 4.9 to 5.1, making it much more electronegative than most IgGs [1,3]. At pH 8.0 to 8.5 it tends to elute at about 0.30 *M* NaCl, depending on the choice of exchanger and buffers. IgGs, with their typically alkaline pIs, usually elute at a third that salt concentration or less. This suggests that quantitative removal of protein A should be a simple matter, but overlooks the fact that leached protein A is affinity-complexed to the IgG. This creates hybrid molecules with intermediate charge characteristics. Leached protein A is seldom intact; it dominantly exists in the form of individual IgG-binding domains of about 6.6 kDa, with lesser subpopulations of two or three domains [1,4]. Each IgG molecule has two primary fragment crystallization (Fc) protein A binding sites, and two secondary fragment antibody (FAb) sites. The combination of multiple binding sites and multiple degraded forms gives rise to a spectrum of charge-hybrids. IgG tends to dominate their retention characteristics, with the result that they co-elute to varying degrees with the product. The higher the mass proportion of protein A in a given hybrid, the later it elutes relative to the uncomplexed IgG peak [1,5].

Anion exchange is generally regarded as a high capacity technique, but the alkaline pI of most IgG monoclonals limits the ability to exploit it. An operating pH of 8.0 to 8.2 is most common, with some processes employing values as high as 8.5. Under these conditions, IgG binding capacity on conventional anion exchangers is seldom higher than 10 mg/ml and often half that or less. Higher pH can increase IgG capacity but most process developers avoid this because of the elevated risk of deamidation and proteolysis. USP buffers such as phosphate are sometimes used when anion exchange is the last purification step. Phosphate concentration is often reduced to 0.02 *M* to minimize interference with IgG binding but capacity is still significantly less than for buffers such as Tris, and the lack of pH control manifests itself as process variation. One way to compensate for both problems is to employ anion exchangers with very high capacities, like tentacle TMAE™ (E. Merck), UNOsphere Q™ (Bio-Rad), or Q Sepharose XL™ (GE Healthcare). Rather than exploiting their full capacity potential, you can raise buffer concentration sufficiently to ensure adequate pH control, and still have much more capacity than is possible with conventional exchangers. The elevated charge densities of these media exert greater constraints on buffer selection, and development is required to minimize equilibration volumes and maintain pH control in conjunction with salt-elution steps, but the combination of capacity and process control they offer is compelling.

17.3 CATION EXCHANGE CHROMATOGRAPHY

Thanks to the same alkaline pIs that limit capacity on anion exchangers, IgG capacity on cation exchangers is typically high. Capacities of 10 to 20 mg/ml are common even for conventional exchangers, and capacities more than twice are achievable for cation exchange analogs of the media mentioned above. Average mass removal of HCP is modestly better than anion exchange. Cation exchange is generally more effective than anion exchange for reduction of leached protein A. Protein A barely binds under the range of conditions commonly used for IgG purification [1]. As a result, the relative charge differentiation between IgG and IgG-protein A complexes is maximized.

By the same token, it might be expected that cation exchange would offer outstanding removal of DNA and endotoxin, but it is usually inferior to anion exchange. This is an artifact of the buffer conditions under which sample is applied. Operating pH is typically in the range of 4.5 to 5.5 to enhance the positive charge on the antibody in order to maximize binding capacity to the cation exchanger. This has the inadvertent result of also maximizing IgG's charge complementarity to strongly electronegative contaminants. Both DNA and endotoxin are polyphosphorylated, making them liquid-phase cation exchangers. As the equilibrated sample awaits loading on the column, stable charge complexes form between IgG and DNA, or IgG and endotoxin. Some of these complexes survive elution, carrying the contaminants along with the product. A secondary liability of these associations is that DNA fragments or endotoxins occlude positive charge sites on the IgG that would normally contribute to binding the cation exchanger. This creates subpopulations of more weakly bound IgG that broaden the elution peak and contribute to product losses [1].

Another limitation of cation exchange is that most IgGs are partially insoluble under the low pH, low conductivity conditions required to achieve high-binding capacity. Off-line sample equilibration is impossible because the IgG forms aggregates and precipitates progressively with time. Loading antibodies in this state inevitably reduces capacity and causes significant product losses, not to mention column fouling. Increasing backpressure during loading and recovery $<80\%$ are warning signs. These limitations can be avoided through loading the sample by in-line dilution. IgG at the target pH, but with minimal salt to maintain solubility, is loaded through one pump. Diluting buffer is loaded simultaneously through another pump at a proportion sufficient to achieve binding conductivity. The two streams meet at the mixer and pass to the column. The precolumn contact time of the antibody to desolubilizing conditions is limited to seconds. This technique has the additional advantage of better reproducibility than direct loading: over the time course of direct sample loading, the latter part of the load has more opportunity to aggregate or precipitate, which makes sample composition an uncontrolled process variable. With online dilution, precolumn antibody contact time with desolubilizing conditions is constant regardless of the time required to load the total sample volume. This technique requires experimentation to identify the minimum salt concentration to maintain antibody solubility in the preload sample and determine the most appropriate in-line dilution factor. Low dilution factors require less buffer and minimize process time. High dilution factors support higher binding capacities, which result in the product being eluted at a higher protein concentration.

Acetate buffer at pH 4.0 to 4.5 is most commonly used for IgG purification but it is not ideal. IgGs have a progressive tendency toward permanent denaturation below pH 4.5. This is typically manifested as aggregation in the purified antibody population. In addition, the low pH attacks 316L stainless buffer vessels, columns, and chromatography systems. This problem can be ameliorated but not eliminated. Acetate can be replaced with MES, which is zwitterionic and does not contribute to conductivity. It is thus able to achieve binding capacity at pH 5.5 equivalent to what acetate achieves at pH 4.5. Even at pH 5.5, corrosion remains a concern with high-salt elution buffers, especially halide salts such as NaCl. Acetate is one option, especially if acetate is the primary buffering system, but citrate may be generally preferable. Its molar conductivity is similar to NaCl and its strong buffering capacity from pH 4.5 to 7.0 allow it to enhance pH control in conjunction with elution steps. Another option is to elute the IgG with a pH step. Selectivity is different from salt elution, which may be either favorable or unfavorable in terms of purification, but it minimizes corrosivity. It can also be advantageous if the subsequent chromatography step requires sample application at low conductivity. Despite precautions to minimize the corrosive effects of low pH buffers it is prudent to store them in plastic or polymer-lined vessels. If not, their metal leaching characteristics and the secondary effects of leached metals on the purification process and product must be thoroughly characterized.

17.4 HYDROPHOBIC INTERACTION CHROMATOGRAPHY

Hydrophobic interaction chromatography (HIC) is on par with ion exchange in its ability to remove HCP, but with its own distinct selectivity. Under the high salt conditions used to achieve binding, the charge complexes that plague ion exchangers are completely dissociated. DNA is unretained. Endotoxins in high salt solutions tend to self-associate into secondary structures with their hydrophobic lipid-A components internalized [6]. Their retention behavior is dependent on the choice of HIC media. On weakly hydrophobic media, they are unretained [1]. On strongly hydrophobic media, some of these structures dissociate and their constituents bind. Clearance is good in either case since IgG typically elutes earlier than bound endotoxin. Removal of leached protein A is typically inferior to ion exchangers. Protein A is more hydrophobic than most IgGs but not dramatically so. The differential in retention behavior of protein-A complexed IgG vs. clean IgG is therefore relatively modest. 50% leachate reduction can require sacrificing 50% of the product, although it can be better with weakly hydrophobic antibodies [7]. HIC is also gaining a reputation for effective removal of IgG aggregates, but under inappropriate conditions it can create them as well.

Hydrophobic interaction chromatography is the most inherently compromised of the primary fractionation methods. Weakly hydrophobic media offer product recoveries up to 90% with low risk of denaturation but require very high concentrations of binding salts to achieve good capacities. The required salt levels are typically well into the range where IgG begins to precipitate, making direct sample application impossible; in-line dilution is almost always necessary [8]. Stronger hydrophobic media support good capacities at relatively low salt concentrations and can often be loaded directly, but recoveries are lower and the risk of denaturation is higher. Recovery can often be improved to 80 to 90% by the inclusion of 1 to 2 *M* urea or an organic solvent in the elution buffer [1]. This has the additional benefit of eluting the IgG in a sharper, more concentrated peak. The best of these enhancers have no effect on subsequent purification methods, but they involve other compromises. Urea carries a risk of carbamylating the product. This is easily controlled but requires validation. 10% ethylene glycol is very effective but is human-toxic and likewise requires careful validation. Propylene glycol avoids the toxicity issue but has not been well characterized for this application. Folklore suggests that water is an effective eluent but IgGs are inherently unstable in water, compounding their vulnerability to denaturation in the presence of a potentially denaturing surface such as a hydrophobic solid phase. Uncontrolled pH drift elevates that risk.

One effective strategy is to employ the most hydrophobic medium that does not cause excessive losses or denaturation [9]. This will generally support the best capacity at the lowest salt concentration, and likewise elute the antibody at the lowest possible salt concentration. Phenyl media are the most widely used in IgG purification, and a good place to start. If recovery is good and there are no signs of aggregation, try butyl. If recovery is poor or the antibody shows an increased tendency toward aggregation, try weaker media such as PPG (Tosoh). Very weakly hydrophobic media (Ether, Tosoh) are useful for highly labile antibodies such as IgMs but seldom necessary for IgGs.

A reliable rule of thumb concerning capacity has yet to emerge, but if you are obtaining <10 mg IgG per ml of gel, you are probably not realizing the full benefits of the technique. Dynamic binding capacities of 15 to 20 mg/ml are the norm and capacities approaching 50 mg/ml are occasionally observed [1,10]. Such capacities however are not the result of process development skill. An apparent correlation exists between antibody hydrophobicity and dynamic capacity. If you encounter a strong HIC binder, exploit it. If your capacity runs to the low side, try higher concentrations of binding salt and console yourself with the fact that a less hydrophobic antibody is likely to exhibit better solubility and stability in final formulation.

Ammonium sulfate has been the binding salt of choice in the literature but bears two major liabilities (1) At alkaline pH it releases ammonia gas. Volatilization of the gas can destabilize buffer pH while soluble ammonia can cause alkaline hydrolysis of the product. (2) The nutritive value of ammonia nitrogen is a concern for municipal waste water authorities, and hence more expensive to dispose of. Potassium phosphate avoids these functional limitations and gives nearly identical molar selectivity, but substitutes phosphate for nitrogen: an equivalent disposal liability. Sodium sulfate avoids this problem and is a more effective binding promoter but its solubility is limited to about 1.1 *M*. Sodium citrate lacks all these limitations but is more viscous and exerts strong buffering capacity. Sodium chloride is neutral but more corrosive than nonhalide salts. Although its molar effectivity is lower than traditional HIC salts, its lower molecular weight compensates somewhat, so that the mass concentration required to achieve comparable capacity is only modestly greater. All of these salts mediate unique selectivites that may either benefit or not benefit the purification.

17.5 CERAMIC HYDROXYAPATITE

Ceramic hydroxyapatite (CHT) is second only to protein A in its ability to remove HCP from monoclonal IgG preparations. Contaminating proteins mostly elute earlier than IgG. DNA typically elutes after IgG, due to the strong affinity of its phosphoryl moieties for the calcium sites. Protein A–IgG complexes usually elute after IgG, often supporting better removal efficiency than ion exchangers [11]. Aggregates elute after monomer and their clearance is generally superior to other adsorptive methods, at far higher sample loads than can be tolerated by size exclusion [12,13].

Ceramic hydroxyapatite (type I, 40, Bio-Rad) easily supports linear flow rates greater than 600 cm/h at low backpressures [14]. It lacks the physical resilience of polymeric supports but it can reliably support more than 50 cycles in large-scale industrial columns with the proper choice of buffers and appropriate care and maintenance. CHT scavenges contaminating metal ions. This has two very different ramifications: On the positive side, it strips contaminating metal ions that are nonspecifically bound to proteins. Such contaminants create uncontrolled charge heterogeneity, which manifests as reduced separation performance and amplified process variability, for both HIC and ion exchange [1]. On the negative side, metals scavenged from process buffers discolor the media. To the extent that discoloration may reflect a change in its chemical composition, validation is required to show that fractionation performance is unreduced. The simplest way to minimize discoloration is to prefilter all process buffers off-line through 80 μm CHT type 1. This media supports very high flow rates at low backpressure and can be run in tandem with buffer microfiltration.

IgG capacity on CHT (type I, 40 μm) can range from 25 to 60 mg/ml, depending on the antibody and buffer composition. Achieving the maximum particularly requires control of phosphate concentration in the equilibration buffer and sample. 1 m*M* phosphate can reduce capacity by 15% and 5 m*M* can reduce it by more than half [14]. Some antibodies may support capacities >40 mg/ml, even in 5 to 10 m*M* phosphate buffers, but the relationship between capacity and phosphate concentration persists. Neutral salts have a more modest effect: 50 m*M* NaCl reduces capacity about 20%; 150 m*M* about 35%, making the method fairly tolerant of salt-containing samples [14]. Dynamic capacity reaches its maximum for essentially all antibodies at about pH 6.5 [14]. This suggests that MES should be well suited for pH control, but it has been demonstrated to degrade CHT within as few as 10 runs (L. Cummings, personal communication, 2005). This may be a general phenomenon with nonphosphate buffers, which would suggest a strategy of buffering with phosphate, except for the capacity issue [14]. If adequate capacity cannot be achieved with phosphate buffer, use MES for pH control and augment with a minimum of added phosphate to stabilize the CHT. Chelating agents and pH values below 6.5 destabilize CHT. It is also important to avoid anhydrous phosphates. The heating process used to eliminate hydration water induces formation of polyphosphates that interact strongly with CHT-calcium, alter media selectivity, and reduce protein-binding capacity.

Elution has traditionally been performed with phosphate gradients. This works well for removing the bulk of HCP and also supports effective removal of aggregates and protein, but for many antibodies a sodium chloride gradient (0.0 to 1.0 *M*) superimposed on a constant phosphate buffering environment

(5 to 10 m*M*) supports better fractionation. There are preliminary indications that sodium chloride gradients become more effective in loose correlation with increasing antibody pI, but it seems advisable at present to explore both options. Sodium chloride gradients also improve DNA removal.

17.6 INTEGRATED PURIFICATION SCHEMES

Given that protein A is the first step and anion exchange the last, the remaining questions are (1) how many additional steps, (2) which ones, (3) in what order, are required to complete the process. The current industry norm is to have one additional fractionation step in the process. This may not accommodate all antibodies. Particularly if the anion exchange step is applied in flow-through mode, the compromise to HCP and leached protein A removal efficiency may require a fourth step. Media manufacturers continue to dangle the prospect of a 2-step capture/polish process before process developers. HCP, DNA, and endotoxin removal often give enticing results, but the real challenge lies with adequate removal of leached protein A and virus. Three steps is the practical minimum with current feedstreams and purification tools.

The question of which additional step is a matter of complementarity. There is no way to predict which combination will prove most effective. At the same time, the necessity to identify the best process is absolute. Antibodies are typically administered at high dosages, some in the range of grams per patient per year. This increases the risk that trace contaminants — possibly accumulating over the course of prolonged therapy — may mediate adverse reactions. Since the goal is to identify which combination offers the best overall complementarity to protein A, the only realistic option is to screen them in the context of integrated procedures. Table 17.1 to Table 17.3 provide a series of three-step protocols that can be conducted at any scale to provide sufficient material for complete evaluation. Process order has been organized to avoid the need for intermediate sample preparation steps like buffer exchange, concentration, or dialfiltration. For any of these processes that give adequate results, consider converting the anion exchange step to a flow-through format. This in turn raises the issue of process order; if you obtain better continuity by making the anion exchange as the second step instead of last, go with the flow.

If none of these options prove satisfactory, then a wide range of alternatives exists for further investigation. One is to add another fractionation step. This is probably the shortest path to having a clinical-ready process in hand but it may not be the best solution in the long run. Every step involves special sample loading requirements and may complicate integration of the overall process. Every step requires hardware, column media, buffers, and salts that add expense to the manufacturing process. Every step requires validation. Every step involves product losses. For all these reasons, it is worthwhile to evaluate

TABLE 17.1
Purification Feasibility Screening by Protein A Affinity, Cation Exchange, and Anion Exchange Chromatography

Protein A elution buffer: Any, with conductivity as low as possible. 1 ml column
Cation exchange, high capacity S (see Table 17.4), 1 ml column, bind and elute mode
Sample preparation: titrate sample to running pH

Buffers:

A: 0.05 *M* MES, pH 5.5
B: A + 1.0 *M* NaCl, pH 5.5

Fractionation:

Flow rate: per gel manufacturer's recommendation
Equilibrate: buffer A until pH of column effluent matches buffer A
Load sample: volume equivalent to 10–20 mg IgG per ml of gel, load by in-line dilution 1 part sample, 9 parts buffer A
Wash: 5 CV buffer A
Elute: 15 CV linear gradient to 30% buffer B
Strip: 5 CV buffer B

Comments: The biggest challenge with this prep is having enough salt in the protein A elution buffer to maintain antibody solubility without having an excess that will inflate the dilution factor required to permit the antibody to bind the cation exchanger. The in-line dilution factor given above will accommodate most antibodies. Half that dilution factor will serve for many. Greater dilution will be required by some. If 20 mg binds well, try 40. If 40 binds, increase it again

Anion exchange, high capacity Q (see Table 17.4) 1 ml column, flow-through mode
Sample preparation: titrate pH to 7.5, dilute with buffer A until conductivity is 12 mS
Buffers:
A: 0.02–0.05 *M* Tris, plus NaCl to yield conductivity of 12 mS, pH 7.5
B: 0.02–0.05 *M* Tris, plus 1 *M* NaCl, pH 7.5

Fractionation:
Flow rate: per gel manufacturer's recommendation
Equilibrate: buffer A until pH and conductivity of effluent equals buffer A
Load sample
Wash, until UV absorbance returns to baseline
Strip: 5 CV buffer B

Comments: These conditions are adapted from Reference 2, which describes effective removal of retrovirus from a variety of Q exchangers. Most antibodies will flow through. Where you put this step in the process sequence is optional. If the sample conductivity from the previous step is high, the process may flow better if the Q step directly follows protein A. However, if the antibody requires more than 0.1 *M* NaCl to elute from S, it may be possible to raise the operating pH of the cation exchange step significantly, maybe convert to a pH elution, either of which will drop the conductivity into a range that permits efficient placement of the Q step last

TABLE 17.2
Purification Feasibility Screening by Protein A Affinity, Hydrophobic Interaction, and Anion Exchange Chromatography

Protein A elution buffer: any. 1 ml column
HIC, Phenyl (see Table 17.4) 1 ml column, bind and elute mode
Sample preparation: titrate pH to 7.0. Immediately prior to sample application, dilute 1:1 with 2× concentration of buffer A or gradually dissolve dry ammonium sulfate in sample to a final concentration of 1 *M*. The former is less likely to create aggregates; the latter minimizes sample dilution

Buffers:
A: 0.05 *M* $NaPO_4$, 1.0 *M* ammonium sulfate, 7.0
B: 0.05 *M* Na phosphate, 2.0 *M* urea, pH 7.0

Fractionation: Flow rate: per gel manufacturer's recommendation
Equilibrate: 5 CV buffer A
Load sample: volume equivalent to 10–20 mg IgG per ml of gel
Wash: 5 CV buffer A
Elute: 15 CV linear gradient to buffer B
Strip: 5CV buffer B

Comments: One of the advantages of this prep is that it is tolerant to any conductivity required to elute the antibody effectively from protein A and keep it soluble during low pH viral inactivation. Dilution with 2 × buffer concentrate is the least potentially troublesome option for sample equilibration. Precipitation of IgG will be apparent at the solid–liquid interface of dissolving ammonium sulfate crystals. This will mostly disappear after the salt is completely in solution. Resist the temptation to filter out the haze since the salt will cause massive losses through antibody adsorption to the filter membrane. You can load the sample as long as it does not contain obvious precipitates. Strong HIC columns spontaneously dissociate weak aggregates. Some antibodies will remain soluble in ammonium sulfate concentrations >1 *M*, some won't. For those that do, this may support higher capacities without resorting to in-line dilution — so long as the binding capacity of the media is not limiting

Anion exchange, 1 ml column, flow-through mode, as described in Table 17.1

Comments: Most antibodies will elute from phenyl at a low enough salt concentration to make dilution feasible as a sample preparation method for the Q step. For those that elute at high salt concentrations, it may be more practical to perform the Q step directly following the protein A step

three-step processes with different chromatography media (Table 17.4), buffers, and gradient conditions before committing to a four-step process.

17.7 VIRUS INACTIVATION AND FILTRATION

It has become common practice to conduct a low pH viral inactivation step immediately following elution from protein A. The IgG is already at low pH

TABLE 17.3
Purification Feasibility Screening by Protein A Affinity, Hydroxyapatite, and Anion Exchange Chromatography

Protein A elution buffer: no citrate or chelating agents. 1 ml column

Anion exchange, 1 ml column, flow-through mode, as described in Table 17.1

Comments: The conductivity at which IgG elutes from CHT, whether in a phosphate gradient or NaCl, is likely to require so much dilution to prepare for Q that it will usually be more practical to put the Q step immediately after protein A

Hydroxyapatite, CHT type I 40, 1μml column, bind and elute mode

Sample preparation: titrate pH to 6.5

Buffers:

A: 5 m*M* $NaPO_4$, pH 6.5

B: A + 1 *M* NaCl

C: 0.5 *M* $NaPO_4$, pH 6.5

Fractionation: Flow rate 300–600 cm/h

Equilibrate: buffer A until column effluent is pH 6.5

Load sample: volume equivalent to 10–20 mg IgG per ml of gel

Wash: 5 CV buffer A

Elute: 20 CV linear gradient to 100% buffer B

Strip: 5 CV 100% buffer C

Comments: If the antibody elutes within the NaCl gradient, reduce the gradient slope to optimize fractionation of aggregates, leached protein A, and other contaminants. If the antibody fails to elute within the confines of the NaCl gradient, increase the phosphate concentration by 5 m*M*. If the NaCl gradient does not support the hoped for removal of aggregates, substitute a phosphate gradient (0–60% C, 20 CV; strip with 5 CV 100% C). Optimize slope for aggregate removal

5 m*M* phosphate may not be the best choice for manufacturing use. If buffering capacity is inadequate, augment with MES. This also provides the opportunity to reduce the phosphate concentration even lower, which will increase antibody binding capacity, but be prepared to retain at least 2 m*M* phosphate to maintain stability of the CHT

and it is a simple matter to hold it there for a specified interval prior to the next process step. Virus removal filtration is likewise easily integrated. Placing it prior to the last binding step in the process supports reconcentration of the product following dilution from the viral filtration step, however it can go elsewhere according to the demands of overall process flow. You can reasonably expect 3 to 5 logs of virus reduction for each chromatography step, and up to 5 logs each for the inactivation and filtration steps. Following the final fractionation step, a combined dialfiltration and membrane concentration step is commonly employed to put the product in its final form.

TABLE 17.4
Process Chromatography Media for Polishing Monoclonal IgG

Media	Manufacturer	Comments
Anion exchange:		
UNOsphere Q	Bio-Rad	High capacity
Fractogel EMD TMAE	E. Merck	High capacity
Q Sepharose Fast Flow	GE Healthcare	Low capacity
Q Sepharose XL	GE Healthcare	High capacity
Mustang Q	Pall Corporation	Flow-through
Toyopearl Super Q-650	Tosoh Bioscience	High capacity
Cation exchange:		
UNOsphere S	Bio-Rad	High capacity
Fractogel EMD SO3	E. Merck	High capacity
S Sepharose Fast Flow	GE Healthcare	Low capacity
S Sepharose XL	GE Healthcare	High capacity
Toyopearl S-650	Tosoh Bioscience	Low capacity
Hydrophobic interaction:		
Fractogel EMD Propyl	E. Merck	Moderately hydrophobic
Fractogel EMD Phenyl	E. Merck	Strongly hydrophobic
Phenyl Sepharose Fast Flow (hi sub)	GE Healthcare	Very strongly hydrophobic
Phenyl Sepharose HP	GE Healthcare	Strongly hydrophobic
Toyopearl Ether-650	Tosoh Bioscience	Weakly hydrophobic
Toyopearl PPG-600	Tosoh Bioscience	Moderately hydrophobic
Toyopearl Phenyl-650	Tosoh Bioscience	Strongly hydrophobic
Toyopearl Butyl-600	Tosoh Bioscience	Very strongly hydrophobic
Ceramic hydroxyapatite:		
CHT type I, 40 μm	Bio-Rad	

17.8 FUTURE TRENDS

The recent introduction by the U.S. Food and Drug Administration of its Process Analytical Technology (PAT) initiative has fundamentally altered the regulatory landscape for process developers and manufacturers [15]. In broad brush, this initiative gives manufacturers the latitude to alter approved processes in order to incorporate advances in manufacturing technology, with the idea that using the latest technology will help ensure that patients receive the safest and most effective product possible. The caveat is that any such changes be supported by the most advanced analytical methods to document that proposed changes reduce risk to the patient population. This includes not only end-product testing but also analytical methodologies that improve the quality of in-process monitoring and control. The benefit for manufacturers is the potential to reduce overall process costs. The net result for process developers is that process development never stops. This compounds the burden already created by the surge of new product applications and makes it more essential than ever that developers have an efficient platform for evaluating purification options.

Two areas of technical interest are the emerging generation of high capacity chromatography media and adsorptive filtration media, both addressed previously in this chapter. High capacity ion exchangers are already available from many suppliers and do in fact have the capability to improve process control by virtue of being able to support acceptable capacities under conditions that support more effective process control. Their ability to support good capacities under less extreme buffer conditions, and thereby minimize potential damage to the product, likewise supports PAT. The fact that they support better process economy is incidental to regulators but an important motivation for manufacturers. These new products seem likely to quickly displace conventional ion exchangers and it is to be hoped that their performance features will be extended to other selectivities as well.

The replacement of chromatography with filtration remains a tantalizing possibility, especially from the perspective of rapid processing of large process volumes, and filtration media have advanced substantially for more than two decades since filtration proponents have been promoting this idea [16]. Anion exchange membranes have become competitive for some flow-through applications, but the capabilities of chromatography media have not stood still, and there remains a limitation in the ability of skids to deliver the volumetric flow rates to exploit the potential of membrane-based fractionation. Molecular shear may compromise throughput even after the flow limitations have been addressed. In addition, many of the sample limitations that affect chromatography afflict membrane separations at least as severely, and will not be solved by a simple change of physical format; one example: preadsorption precipitation of IgG under cation exchange binding conditions.

A third area to watch concerns new adsorptive selectivities, whether chromatography-based or otherwise. Central among these are capture technologies that may offer alternatives to protein A. Most of the current and emerging alternatives are affinity-based. Beyond the obvious process economy issues, the key issue is toxicology. Protein A is well known to be a potent immunotoxin but more than 200 published studies have thoroughly characterized its toxicology profile, and all the approved products purified with it have proven to be clinically safe [1]. Potential alternatives, including ligands derived from a protein A, require thorough toxicological evaluation in order to be considered, and this compounds the validation burden. It may prove worth the cost but it represents a far more demanding task than simply switching from one protein A product to another.

In addition to protein A replacements, mixed-mode adsorption mechanisms represent real promise for antibody purification. As discussed above, there are serious compromises associated with both cation exchange and HIC. If mixed-mode selectivities can provide competitive process capabilities in conjunction with good process control, they can be very valuable. The ability of CHT to remove aggregates and leached protein A has proven the feasibility of this rationale. Hydrophobic charge induction™ (Pall Corporation) and other possibilities may likewise prove to be legitimate process options in the coming years.

17.9 CONCLUSIONS

A single generic polishing method for IgG purification is not a practical reality. The purity requirements are too stringent and IgGs are too diverse in their chromatographic behavior to be accommodated by such a scheme. They do however exhibit sufficient similarity to permit effective application of a limited set of purification options under a defined range of conditions. Each of the templates suggested here is likely to give promising results, but one will usually emerge as being more suitable for a given antibody, or better suited to your established conventions in material handling. After selecting one of these templates or another of your choice, evaluate media options according to your preferences. As you identify the most productive media and conditions at each step, refine the loading conditions, establish capacities, optimize the fractionation, and proceed with the other tasks of process optimization.

REFERENCES

1. Gagnon P, Purification Tools for Monoclonal Antibodies, Tucson, Validated Biosystems, 1996.
2. Curtis S, Lee K, Brorson K, and Xu Y, Generic/matrix evaluation of SV40 clearance by anion exchange in flow-through mode, *Biotechnol. Bioeng.*, 2003, 84, 179–186.
3. Sjoquist J, Meloun B, and Hjelm H, *Staphylococcus aureaus* after digestion with lysostaphin, *Eur. J. Biochem.*, 1972, 29, 572–578.
4. Diesenhofer J, Jones TA, Huber R, Sjodahl J, and Sjoquist J, Crystallization, crystal structure analysis and atomic model of the complex formed by a human Fc fragment and fragment B of protein A from *Staphylococcus aureaus*, *Hoppe Seylar's Z. Physiol. Biochem.*, 1978, 359, 975–985.
5. Bloom J, Wong MF, and Mitra G, Detection and reduction of protein A contamination in immobilized protein A purified monoclonal antibody preparations, *J. Immunol. Meth.*, 1989, 117, 83–89.
6. Weary M and Pearson F III, A manufacturer's guide to depyrogenation, *BioPharm. Manuf.*, 1988, 1, 22–29.

7. Shadle PJ, Erickson JC, Scott RG, and Smith TM, Antibody Purification, U.S. Patent 5,429,746, 1995.
8. Gagnon P, Grund E, and Lindback T, Large scale process development for hydrophobic interaction chromatography, Part 1, Gel selection and development of binding conditions, *BioPharm*, 1995, 8, 21–27.
9. Gagnon P and Grund E, 1996, Large scale process development for hydrophobic interaction chromatography, Part 3, Factors affecting capacity determination, *BioPharm*, 1996, 9, 34–39.
10. Mitoma S, Development of a New Toyopearl® Hydrophobic Interaction Chromatographic Resin for Protein Purification, 4th HIC/RPC Bioseparation Conference, Palm Harbor, FL, February 15–19, 2004.
11. Franklin S, Protein A removal from IgG on CHT Ceramic Hydroxyapatite support, Bio-Rad Technical note, 2849, 2004.
12. Josic D, Loster K, Kuhl R, Noll F, and Reusch J, Purification of monoclonal antibodies by hydroxyapatite HPLC and size exclusion HPLC, *Hoppe Seylar's Z. Physiol. Biochem.*, 1991, 372, 149–156.
13. Franklin S, Removal of aggregate from an IgG4 product using CHT ceramic hydroxyapatite, Bio-Rad Technical note 2940, 2003.
14. Gagnon P, An enigma unmasked; how hydyroxyapatite works and how to make it work for you, http://www.validated.com/revalbio/pdffiles/hxyapt.pdf, (accessed July 1998).
15. U.S. Food and Drug Administration, Guidance for Industry, PAT — A framework for innovative pharmaceutical development, manufacturing, and quality assurance, http://www.fda.gov/cvm/guidance/published.html (accessed September 2004).
16. Gottschalk U, Downstream processing of monoclonal antibodies: from high dilution to high purity, *BioPharm Int.*, 2005, 18, 42–58.

18 Making Changes to a Biopharmaceutical Manufacturing Process during Development and Commercial Manufacturing: The REMICADE® Story

Peter W. Wojciechowski, Hendrik I. Smit, Michele M. Myers, Paul J. Voronko, Timothy Laverty, R. Andrew Ramelmeier, and Richard C. Siegel

CONTENTS

18.1 INTRODUCTION

The lifecycle for a biopharmaceutical manufacturing process begins with pre-clinical process development, followed by process scale-up and improvements through manufacturing campaigns for human clinical trial materials leading up to the licensed, commercial manufacturing process. After licensure, process changes may be required to accommodate further scale-up and improvements as well as transfer to new or additional manufacturing sites. The drug substance must be demonstrated to be comparable to provide an assurance that the safety and efficacy of the drug used by a patient is the same as that used in clinical trials and for which licensure was granted.

While the drug substance itself must remain unchanged, process changes are unavoidable due to the changing availability of equipment and raw materials, the need to scale-up production, technology transfer, process improvements, and changing regulatory requirements. This is the story of the lifecycle for the REMICADE® manufacturing process from early process development in the early 1990s to its current large-scale commercial manufacturing at multiple sites.

18.2 INFLIXIMAB STRUCTURE, FUNCTION, AND FORMULATION

The REMICADE® brand of infliximab was the first drug product in the class of tumor necrosis factor α (TNF-α) inhibitors approved for use in humans. It is a lyophilized formulation of the drug substance, infliximab, and is approved for the treatments of autoimmune disorders including rheumatoid arthritis, Crohn's disease, ankylosing spondylitis, and psoriatic arthritis.

The infliximab drug substance is a purified, recombinant DNA-derived, chimeric human–mouse IgG monoclonal antibody (MAb) which binds to and neutralizes human TNFα with high affinity ($K_a = 1 \times 10^{10}\ M^{-1}$). The cA2 IgG contains murine heavy (H) and light (L) chain variable regions (V_H and V_L, respectively) derived from the murine anti-TNF-α MAb, A2, ligated to genomic human heavy and light chain constant regions (Figure 18.1). The infliximab drug substance is manufactured using a recombinant murine myeloma cell line transfected with expression plasmids encoding the H and L chains.

The cA2 IgG molecule contains 1328 amino acids and consists of 2 identical H chains and 2 identical L chains which associate by noncovalent H–H and H–L interactions and covalent H–H and H–L disulfide bonds. The oligosaccharide structures result in five IgG glycoforms containing 0 to 4 galactose residues distributed between the 2 N-linked biantennary oligosaccharides structures

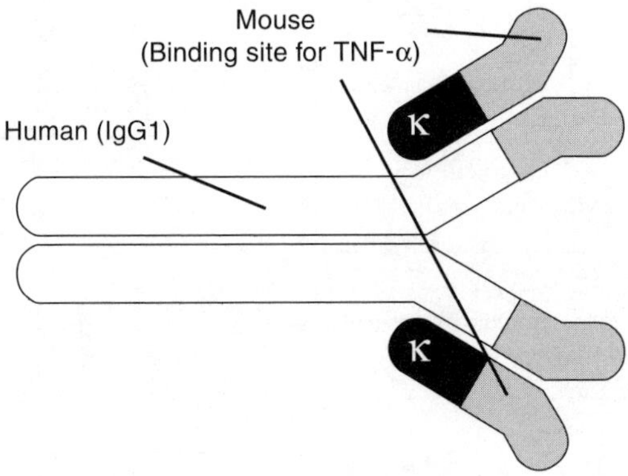

FIGURE 18.1 Structure of Infliximab (REMICADE drug substance). The molecule is a chimeric (mouse/human) monoclonal antibody that binds to TNF-α with high specificity and high avidity.

located on Asn-300 of each H chain. The molecular mass range for the cA2 IgG glycoforms is 148,770.5 to 149,419.1 Da.

18.3 INFLIXIMAB MANUFACTURING PROCESS OVERVIEW

The infliximab drug substance is manufactured by continuous perfusion cell culture. The expansion of the antibody secreting cells and production of the chimeric monoclonal antibody occur in the first two manufacturing stages: preculture and expansion (Stage 1) and large-scale cell culture production by continuous perfusion (Stage 2). REMICADE is purified and formulated to pre-formulated bulk (PFB) from cell supernatant (harvest) in Stages 3 through 9 of the manufacturing process as shown in Figure 18.2.

Purification of infliximab begins with the filtration of clarified harvest material and the purification of cA2 IgG by Protein A affinity chromatography. It is during this purification step that the vast majority of impurities including viruses, media components, and host cell species are removed. The purified material in the eluted product stream is frozen and stored prior to pooling for the subsequent downstream purification steps. In Stage 4, Protein A purified monoclonal antibody is thawed at room temperature, pooled, pH adjusted, and filtered in preparation for Solvent/Detergent (S/D) viral inactivation, the first dedicated viral clearance step in Stage 5a. Cation exchange chromatography at Stage 5b is designed to remove the S/D reagents as flow through to waste while the product remains bound to the column.

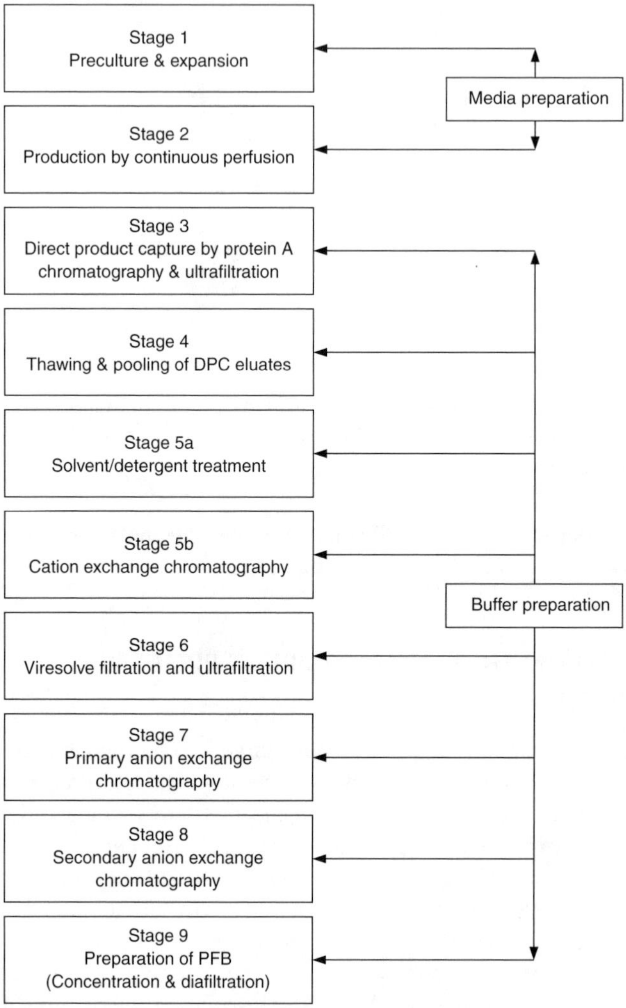

FIGURE 18.2 An overview of the REMICADE production process. Each stage of the production process is shown. Stages 1 through 3 are considered part of the upstream production process. Stages 4 through 9 are considered part of the downstream production process.

In Stage 6, the eluted product from Stage 5b is diluted and undergoes a second, dedicated viral clearance step by tangential flow filtration (TFF). The resulting product stream is then concentrated by ultrafiltration.

In Stages 7 and 8, the product is further polished by two anion exchange chromatography steps. The first is performed in flow-through mode and the second in a capture and elution mode of operation. Viruses and DNA are removed by the first step while viruses, DNA, antibody aggregates, and residual media components are removed by the final chromatography step.

In the final stage, Stage 9, the product is concentrated and diafiltered into a buffer containing stabilizers to prepare the preformulated bulk drug substance which is frozen for storage and shipping to the drug product manufacturing site.

18.4 PROCESS CHANGES AND PRODUCT COMPARABILITY DURING DEVELOPMENT

Clinical development of REMICADE followed a path common to new drugs and involved four main steps that affected the speed of drug development. First, toxicology studies were performed in animal models to test safety prior to use in humans. Second were safety studies in humans, also known as Phase-1 clinical trials to provide the initial proof of concept. Third was the preliminary efficacy and dose ranging in humans during Phase-2 clinical trials. The fourth and pivotal stage was a Phase-3 efficacy study in a larger population of human patients. These data laid the foundation for the BLA (Biologic License Application) submission to the US FDA. While products with fast-track indications (such as oncology treatments) may involve pivotal trials at Phase-2 or -2b, from which the BLA submission can be made, REMICADE followed the more traditional path of conducting all three clinical trial phases.

To meet the needs of the clinic, a manufacturing process must be developed for the drug substance (bulk drug) and the drug product (formulation and filling into a primary container and delivery device). As clinical development is the critical path to the advancement of a product candidate, process development is focused initially on delivering drug product for preclinical toxicology studies and Phase-1 trials. The process at this point may not yet be designed for commercial manufacturing therefore, process improvements are usually required after Phase-1. To ensure that the changes have no impact on product quality and efficacy, rigorous comparability studies need to be conducted.

As a company gains experience with drug development and manufacturing technology platforms, they are able to design initial processes that are better suited to manufacturing, thus minimizing process changes and the associated risks for demonstrating product comparability. In the event that product comparability cannot be achieved or significant improvements are desired, a bridging study may be needed to assure the safety of the new formulation prior to continuing pivotal clinical trials. The cost/benefit of product improvements vs. licensure delays often represent a business challenge.

Figure 18.3 shows the progression of process changes for infliximab through the stages of clinical development. The clinical and regulatory milestones are

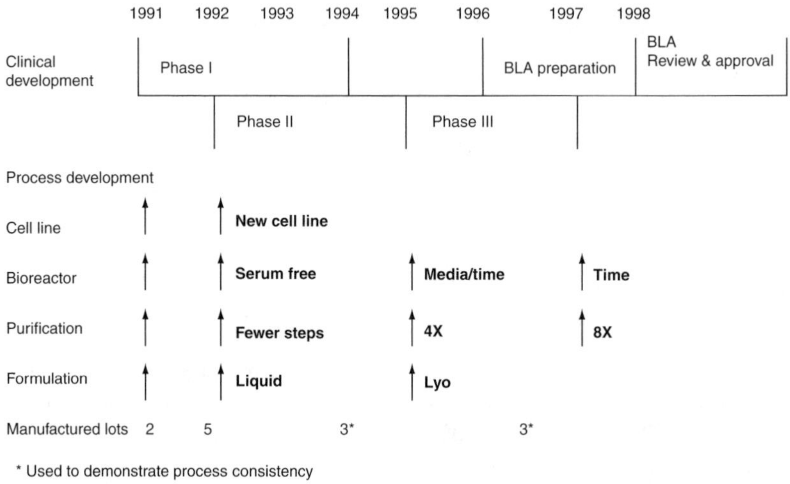

FIGURE 18.3 A timeline showing the elements of the manufacturing process that were changed during clinical and early commercial development of REMICADE and the infliximab drug substance.

highlighted at the top of the chart. Process changes in all aspects of the process including cell line, bioreactor, purification, and formulation were required as the product progressed toward commercial development. The Phase-1 process was designed primarily to make material for preclinical toxicology studies and Phase-1 human clinical trials. Clinical development was considered to be the critical business path, as infliximab was given fast-track orphan designation for Crohn's disease. Manufacturing and process development timelines were chosen to fit within those for clinical development.

To increase the productivity of the bioreactor step, several process changes were made. Through clonal selection, the productivity of the cell line was improved greater than tenfold. The final clone produced approximately 92 μg/mL in T-flasks, while the original clone produced <9 μg/mL in T-flasks. The media was improved to increase productivity another twofold and reduce the need for animal-derived components. The bioreactor duration was extended from 45 to 50 days of continuous perfusion to 60 days. Although the titers tend to decrease after reaching a peak during the perfusion run (Figure 18.4), extending the run improves the overall productivity by reducing downtime due to turnaround and maintaining a higher average cell density and titer.

The downstream purification process also posed significant development challenges. The initial Phase-1 process was comprised of several UF/DF and other steps, which were removed for the Phase-2 and -3 processes. To increase

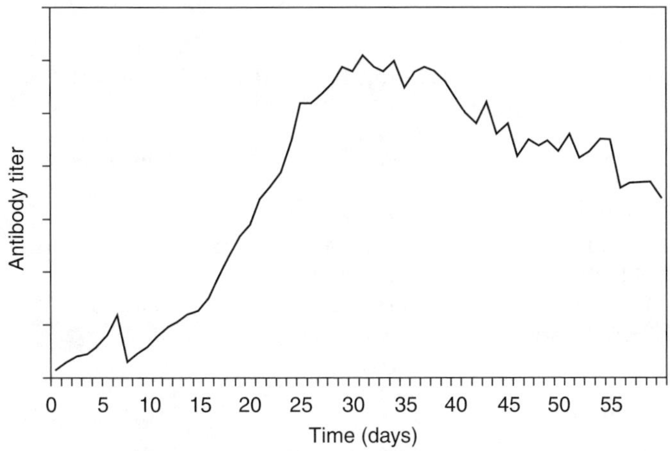

FIGURE 18.4 Antibody titer as a function of time in an infliximab perfusion bioreactor.

process throughput and resin utilization, the loading on the ion exchange resins was increased by 3- to 4-fold. To sharpen the separation and improve the purity of the product, high-performance resins and smaller beads were introduced. Two dedicated virus reduction steps with independent mechanisms of action (virus inactivation by S/D and removal by virus filtration) were introduced. In order to prepare for commercial manufacturing, the batch size was increased fourfold over that used at Phase 3 and 8-fold over that for the Phase-2 process.

The formulation of REMICADE required changes to develop the final version of the commercial drug product. For Phases 1 and 2, slightly different liquid formulations were developed but to ensure an adequate shelf life for commercial inventory and distribution a lyophilized formulation was developed, first for a 250 mg dose at Phase-3 and then later for a 100 mg dose.

Product comparability was established for infliximab at every stage of clinical development shown in Figure 18.3. In 1992, changes to the Phase-1 process were needed to meet clinical needs and to progress towards a commercially viable process. These changes for Phase-2 manufacturing included:

- New cell line with higher productivity
- New media to replace serum with serum-free components
- Purification steps to achieve a higher purity
- Improved liquid formulation

Analytical tests were selected to demonstrate comparability of identity, purity, primary structure, and bioactivity (Table 18.1). All of the test results

TABLE 18.1
Analytical Tests That May Be Used to Establish Product Comparability

Protein	**Aggregates**	**Activity**
A280	SEC HPLC	*In vivo* bioassay
A280/A260	Western blot	Cellular bioassay
SDS-PAGE (CB)	Analytical Ultracentrifuge	Radioimmunoassay
SDS-PAGE (silver)	Light scattering	E LISA
Western blot	Visible particles	**General**
Peptide mapping	Subvisible particles	pH
tryptic	**Carbohydrate**	Appearance
Lys-C	Sialic acid	Excipients
rp HP LC	Neutral sugar content	Bioburden/sterilty
N-terminal sequence	N-linked oligosaccharides	**Purity**
C-terminal sequence	Monosaccharides	SDS-PAGE
Isoelectric focusing	Isoelectric focusing	Endotoxins
Amino acid composition	Capillary electrophoresis	Bioburden
Circular dichroism	% NGNA	Host DNA
Methionine oxidation		Host cell protein
Deamidation		Leachates
Fluorescence Spec		
Mass Spec		

conformed to the known reference standard and the expected specifications. Despite a change in cell line and media, no significant change was measured, including the oligosaccharide structure. In 1995 changes were made in order to meet Phase-3 clinical needs and to continue to improve the productivity, quality and control of the commercial process. Several changes were made for the Phase-3 process, as follows:

- Improved media composition
- Increased bioreactor run time
- Fourfold scale up of the purification process including high performance resins
- Formulation change from liquid to lyophilized

In 1996 and 1997, additional changes were made after Phase-3 and consistency lots. A relatively minor change was made to the infliximab formulation post-Phase-3, decreasing the container size from 250 to 100 mg. Comparability was demonstrated by testing the clinical lots side-by-side using WEHI bioassay, pharmacokinetics AUC, and assays which probe the mechanism of action (such as inhibition of TNF-α receptors p55 and 75-sf2, neutralization of TNF-α, and complement-mediated lysis of a transgenic cell expressing membrane-bound TNF-α).

As the market projections were being estimated for REMICADE, it became apparent that the Phase-3 process would not supply the commercial needs. The initial process validation included both a 1 $\times$ and 4 $\times$ purification process (Stages 4 to 9 in Figure 18.2). To expand the scale of the purification process further to 8 $\times$, the regulatory agencies required Centocor to verify comparability for additional consistency lots.

18.5 PROCESS CHANGES AND PRODUCT COMPARABILITY FOR COMMERCIAL MANUFACTURING

Process validation of the first commercial REMICADE manufacturing process in Leiden, The Netherlands was completed with the successful execution of five consistency batches. These were used to demonstrate reliability of the process and comparability of the product to that used in clinical trials. The results of process validation were used to support the initial licensure of REMICADE around the world. Not long after process validation was complete, post-approval process changes were pursued to further enhance the reliability of supply while continuing to ensure product quality. One example of these post-approval change projects, an expansion of the DPC stage, is described in detail here.

The Direct Product Capture (DPC, Stage 3) step produces a highly purified, stable process intermediate, which can be stored in a concentrated, frozen state. In 2001, two 60 cm diameter Protein A columns were employed in the Leiden manufacturing plant in a unit operation that is schematically represented in Figure 18.5. The processing of bioreactor harvests by DPC was the rate-limiting step for the production of REMICADE and the production requirements with respect to DPC could be met only at high utilization rates of both 60 cm chromatography systems.

Several scenarios identified to increase the DPC throughput were evaluated and prioritized with respect to quality and regulatory requirements, capital investments, operating cost impact, production capacity, implementation lead-time, and possible risks to all of the above.

The improvement scenario selected was an essentially linear scale-up of the DPC step that minimized the risk of significant changes to process performance or characteristics of the purified intermediate product. Scalable process parameters related to the chromatography step included the resin properties, linear flow rates, load ratio, feedstock composition, column packing, and bed

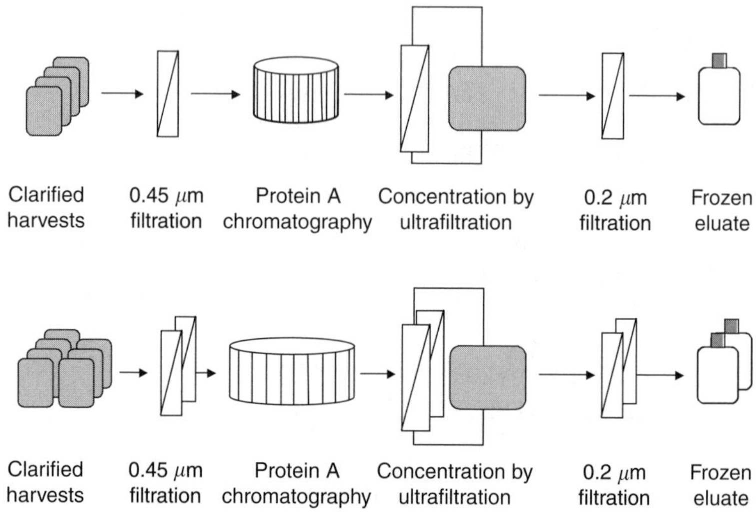

FIGURE 18.5 An overview of the Leiden Direct Product Capture stage as it was operated with a 60 cm column (top) and after scale-up to an 80 cm column (bottom).

height. In order to scale-up the process, all of the parameters mentioned were maintained constant and column diameter was increased from 60 to 80 cm. The linear scalability of the Protein A chromatography step had been previously established in laboratory-scale experiments using columns with diameters of 2.6 and 5.0 cm and at the 60 cm manufacturing scale. Verification of comparable product quality and process performance was performed at the 80 cm scale in manufacturing.

A linear scale-up of the concentration step performed by ultrafiltration following Protein A chromatography was achieved by maintaining operating parameters including the type of membrane, its nominal pore size, flow geometry, channel height, and channel length. Important fluid dynamic parameters were maintained constant including feed and filtrate volume per membrane area, flux rates, feed, retentate, and filtrate pressures, and the concentration factor. Scale-up was achieved by a doubling of the membrane area to accommodate the 1.8-fold increase in the size of the chromatography column volume.

A schematic overview of the post-approval DPC-process is outlined in Figure 18.5. Agreements were reached with all internal stakeholders including Quality Assurance and Regulatory Affairs to perform the following validation work to support the change prior to implementation:

- Installation Qualification (IQ) of the larger chromatography columns
- IQ of the larger UF systems

- Five validation DPC runs in which all routine in-process tests and additional impurity tests were performed
- Cleaning validation of the larger chromatography column for 3 runs
- Cleaning validation of the larger UF system for 3 runs
- Additional stability testing was performed on affected batches

The following submissions in support of the described process changes were issued and approved:

- USA: Prior Approval Supplement
- EU: Type I variation
- Canada: Notifiable Change

These submissions contained information including:

- Description of the proposed change
- Purpose of the proposed change
- Summary of the validation work performed
- In-process test results of five consecutive DPC runs at enlarged scale
- Comparability of impurity clearance before and after the change
- Comparability of chromatographic profiles before and after the change
- Batch release data of the pre-formulated bulk (PFB) lot that contained material from the first five DPC runs performed at manufacturing scale

The implementation of the process change was governed by internal change control procedures to ensure comparability to the process defined in prior submissions. The internal change control procedure consists of initiation, review, revision, approval, and amendment (if required) of change request forms with associated implementation plans. After approval by the regulatory authorities, the process change was executed by completion of the implementation plans. Scale-up of the DPC process stage was fully implemented within 17 months and resulted in a significant throughput increase.

18.5.1 Scale-Up and Post-Approval Changes

Soon after approval of the Leiden manufacturing facility, it became apparent that demand for REMICADE would outpace production capacity. Plans to scale-up and add a second manufacturing site in Malvern, Pennsylvania were initiated in 1997. The Malvern manufacturing process consists of the same nine approved stages used to produce the drug substance in Leiden, shown in

FIGURE 18.6 1000 L perfusion bioreactor and spin filter for manufacturing of infliximab at the Malvern manufacturing site.

Figure 18.1. The Malvern facility is a twofold scale-up of the Leiden process. Changes were implemented in the manufacturing process to accommodate the capabilities of the new facility. The Malvern facility is largely hard-piped and has more automation than in Leiden. An example of this type of difference is the two different virus filtration skids employed in the facilities. A 1000 L bioreactor and a virus filtration skid employed at the Malvern manufacturing site are shown in Figure 18.6 and Figure 18.7. Additional selected changes to the process are summarized in Table 18.2.

All process changes were in place during comparability lot manufacture and were therefore included in process validation studies. Manufacturing of four consecutive comparability lots (one more than our minimum requirement of three lots) to validate the Malvern manufacturing process was initiated in October 2000. In-process controls and specifications employed in the Malvern manufacturing facility are identical to those used in the Leiden manufacturing facility. In addition to meeting all in-process specifications, clearance of impurities, host cell proteins, and host cell DNA were measured throughout the purification process. Process validation studies were conducted to demonstrate comparability of the process in the new facility with the Leiden manufacturing facility, as shown in Table 18.3.

To further demonstrate comparability, the PFB manufactured in the Malvern facility was subjected to additional characterization beyond routine release tests. The results of the routine release testing for four consecutive lots of PFB produced in Malvern manufacturing were compared to the results from the release tests for three Leiden PFB lots prepared in the same time frame (2001)

FIGURE 18.7 Virus filtration skid for processing of infliximab at the Malvern manufacturing site.

and three Leiden PFB process validation lots, manufactured between 1997 and 1998. It was important to demonstrate comparability to current lots being manufactured in Malvern and Leiden as well as to historical control materials that were the basis of the original process validation. All ten lots were subjected to concurrent analysis by WEHI, SDS-PAGE, IEF, and GF-HPLC assays in order to compare the lots directly (side-by-side analysis) and to minimize run-to-run variability. The PFB lots were also analyzed using additional characterization testing to demonstrate that primary structure (as determined by N-terminal sequence analysis and peptide mapping), secondary structure (as determined by circular dichroism analysis), post-translational modifications (as determined by C-terminal lysine content measurement, oligosaccharide mapping, and mass spectrometry), and hydrodynamic properties (as determined by sedimentation velocity analytical ultracentrifugation) were comparable. In addition to biochemical characterization of the product, the levels of residual impurities in the PFB prepared in the new manufacturing facility were shown to be comparable to those prepared in the previously validated facility.

18.6 REGULATORY STRATEGIES TO SUPPORT PROCESS CHANGES

Centocor, Inc. was granted FDA approval for its biological license application (BLA) for the manufacture of REMICADE (infliximab or cA2 IgG) in 1998.

TABLE 18.2
Select Process Changes Implemented in the New REMICADE Production Facility

Process Step	Leiden Process	Malvern Process	Rationale for Change
Stage 1: Preculture and expansion	A single WCB vial is used per batch	Mulitple WCB vials are used per batch	Additional WCB vials are required to increase the number of cells to support bioreactor scale-up without adding additional generations
	50 L perfusion seed bioreactor is used	100 L perfusion seed bioreactor	Direct twofold scale-up in volume
Stage 2: Production by continuous perfusion	A 500 L perfusion bioreactor is used	A 1000 L perfusion bioreactor is used	Direct twofold scale-up in volume
	An internal spin filter is used for cell retention	An external spin filter is used for cell retention	Direct twofold scale-up in filter surface area
	Supernatant is stored at 8 to 14°C after clarification	Supernatant is stored unclarified at 2 to 8°C	Site differences and supported by additional process validation
Stage 3: Direct product capture by Protein A chromatography	One size Protein A column (with a 14 cm bed height)	Two different sizes of Protein A column (with 8 and 14 cm bed heights)	New site practice supported by additional process validation
Stage 4: Thawing and pooling of DPC Eluates	4 × batch size	8 × batch size	Direct twofold scale-up
Stage 5: Solvent/detergent treatment and cation exchange chromatography	A 100 cm column (12 cm bed height)	A 140 cm column (same bed height)	Direct twofold scale-up
Stage 6: Viresolve filtration and ultrafiltration	Batch processed in two parts	Batch processed in one part	To improve process efficiency
	Membrane surface areas of 1 × for virus filtration and 1 × for ultrafiltration	Membrane surface areas of 4 × for virus filtration and 4 × for ultrafiltration	Direct fourfold scale-up to accommodate larger batch size and processing in one part
Stage 7: Primary Anion exchange chromatography	A 100 cm column	A 120 cm column (higher bed height)	Twofold scale-up of column volume Bed height increase supported by additional process validations
Stage 8: Secondary anion exchange chromatography	A 100 cm column (17 cm bed height)	A 140 cm column (same bed height)	Direct twofold scale-up
Stage 9: Preparation of PFB	1 × membrane area	2 × membrane area	Direct twofold scale-up

TABLE 18.3
Process Validation Studies Supporting the New REMICADE Production Facility

Process Step	Process Validation Requirements
Stage 1: Preculture and expansion	Viability at thaw and accumulated generations must be comparable to the validated process during cell expansion
	Accumulated generations must be comparable to the validated process in the seed bioreactor
Stage 2: Production by continuous perfusion	Number of generations, peak-specific productivity, virus expression, and product quality produced throughout cell culture must be comparable to the validated process
	Product quality throughout storage must be comparable to the validated process
Stage 3: Direct product capture by protein A chromatography and ultrafiltration	Product purity, identity, potency, and stability must be comparable to the validated process
	Virus removal must be comparable to the validated process
Stage 7: Primary anion exchange chromatography	Clearance of DNA and virus must be comparable to the validated process

The following year, EMEA approved the REMICADE marketing authorization application (MAA). The original REMICADE BLA and MAA included a single drug substance manufacturing facility, Centocor, B.V., in Leiden, The Netherlands, and a single drug product manufacturing facility, Parkedale, in Rochester, Michigan.

In 2001, Centocor qualified the manufacturing facility in Malvern, Pennsylvania. The Malvern manufacturing facility was approved by the FDA in 2002, and by the EMEA in 2003. From 2002 to the present, Centocor has increased capacity at the Malvern manufacturing facility by introducing changes including a more efficient DPC ultrafiltration process and modifications to the design and operation of the 1000 L continuous perfusion bioreactor.

Concurrently, Centocor continued to increase REMICADE production at its Leiden manufacturing facility through process optimization and facility expansion. In 2001, Centocor initiated a two-phase facility extension for the introduction of additional nonproduct processing areas and the qualification of downstream manufacturing areas. FDA approval for both phases was received in early 2003. In 2002, Centocor increased the efficiency of the DPC process (Stage 3) through a linear scale-up of the cA2 IgG Protein A purification and concentration steps. From 2003 to the present, Centocor has continued to increase manufacturing capacity at the Leiden facility, installing additional manufacturing equipment such as 500 L production bioreactors and 50 L seed bioreactors.

In order to increase lyophilization capacity, a total of five drug product manufacturing facilities were filed to the BLA and MAA. The addition of five manufacturing facilities and continual process optimization has required major submissions using the best available regulatory strategies and early and ongoing communications with the health authorities. As a result, Centocor has accumulated a record of first-round review approvals. Out of approximately 50 major, REMICADE-related submissions to FDA — that is, Prior Approval and CBE-30 supplements — all but one were approved during the first review round.

To maintain a steady supply of REMICADE drug product and meet market demand to ensure adequate supply for existing patients, Centocor has made use of the Comparability Protocol (CP) as a regulatory filing strategy. The CP allows for downgrades to a lesser reporting category. To date, seven CP have been submitted to the FDA for building expansions and increased lyophilization capacity, and all but one resulted in submission downgrades.

Centocor has also pursued global market approvals, registering and launching REMICADE in 54 countries as well as the EU, Canada, and Japan. Whether in the US or abroad, Centocor's regulatory expertise and submission strategies have proven beneficial, keeping manufacturing continuous while ensuring regulatory compliance.

19 Linear Scale-Up of Ultrafiltration of High Viscosity Process Streams

Christopher Daniels, Mark Perreault, Brian Gierl, P.K. Yegneswaran, Marshall G. Gayton, David Serway, Ann L. Lee, John Rozembersky, and Narahari S. Pujar

CONTENTS

19.1 SUMMARY

Scale-up of ultrafiltration processes is generally considered straightforward due to the availability of membrane cassettes that are linearly scalable. The fluid dynamics in each membrane channel is meant to be preserved upon scale-up, which is achieved by geometric similarity of the membrane cassettes at the two scales, and appropriate scale-up of fluid volumes and flow rates. While this is true for most ultrafiltration processes, such simple rules are not adequate in the case of highly viscous process streams and/or at high flow rates due to the greater demand placed on pressure drops in the system. This chapter illustrates some of the issues that may be encountered in such a scale-up, using a case study of ultrafiltration of a bacterial polysaccharide exhibiting such characteristics.

During a manufacturing-scale ultrafiltration run of bacterial polysaccharide ultrafiltration, higher feed pressures were observed than those seen at small-scale. High feed pressures are not uncommon during ultrafiltration of polysaccharide solutions, due to their high viscosity caused by their high molecular weight and charge. However, this lack of scalability in differential pressure, defined as feed pressure minus the retentate pressure, was particularly striking, considering that linear scale-up principles were utilized to design process conditions at large-scale. An investigation was conducted to determine the root cause of the higher differential pressure at full-scale. The investigation addressed the following aspects: membrane cassette construction, membrane cassette compression at large-scale, and hydrodynamics in the large-scale ultrafiltration skid hardware. The largest contributor to the high pressures was found to be the skid hardware, consisting of the stainless steel membrane holder and the associated piping. It was also found that variations in membrane cassette construction and compression had much smaller contributions, with the contributions accentuated due to the high viscosity of the process stream studied. This case study provides important insights into design and scale-up of ultrafiltration processes of high viscosity streams.

19.2 ULTRAFILTRATION OF POLYSACCHARIDE PROCESS STREAMS AND LINEAR SCALE-UP

Ultrafiltration of bioprocess solutions is frequently performed as a recovery step to reduce batch volume and exchange buffer, while providing clearance of small-molecular weight impurities. This has been employed in the recovery [1–4] and purification of polysaccharides [5,6]. However, the ultrafiltration of polysaccharides is complicated by the high viscosity of these process streams, which is exacerbated as the stream is concentrated. Increased viscosity and non-Newtonian behavior, such as the shear-thinning characteristics of these polysaccharides has a fundamental impact on mass transfer parameters [7–9], and novel modules have shown improvement over standard modules for polysaccharide solutions [10]. Pumping and mixing characteristics must be carefully considered when designing systems and developing operating parameters for these process streams.

The concept of linear scale-up is frequently employed in the scale-up of ultrafiltration processes [11]. Linear scale-up ensures equivalent ultrafiltration performance upon scale-up by (1) choosing the large-scale membrane area based on equal volume processed per area of membrane (also termed membrane loading) to that at small-scale, (2) appropriate membrane cassette design, and (3) choice of hydrodynamic conditions. The choice of the appropriate area based on a constant membrane loading sets-up the correct trajectory for scale-up. This is supplemented by a membrane cassette design where the flow path length in the membrane cassette is kept constant upon scale-up, while increasing the total number of channels to achieve the desired filter area for full-scale operation. This is a feature of commonly available cassettes from membrane manufactures. Finally, the hydrodynamics within each of the membrane channels is kept constant, by keeping the specific flows (flow/membrane area) and pressures constant. Maintaining all three aspects of scale-up described above is important for maintaining equivalence in all subprocesses in a membrane process — nonspecific adsorption, mass transfer, fouling, liquid and solute fluxes, leading to equivalence in measurable process parameters such as step time, concentration factor, and impurity removal. Needless to say, other operational factors not mentioned here (e.g., ramp-up time to achieve steady state, recirculation time prior to permeation) also need to be uniform during scale-up. Appropriate scale-up is immediately evidenced in the similarity of the flows and pressures at any given time that are not explicitly controlled (i.e., pressure in a flow rate-controlled operation or flow rate in a pressure-controlled operation).

The above discussion can be illustrated using equations governing ultrafiltration [12,13]:

$$J_{\mathrm{v}} = L_{\mathrm{m}}(\Delta P_{\mathrm{TM}} - \sigma_0 \Delta\Pi_{\mathrm{TM}}); \quad \text{with } \Delta\Pi_{\mathrm{TM}} = f(C_{\mathrm{w}}) - f(C_{\mathrm{p}}) \qquad (19.1)$$

and

$$J_{\mathrm{v}} = k \ln \frac{C_{\mathrm{w}} - C_{\mathrm{p}}}{C_{\mathrm{b}} - C_{\mathrm{p}}} \qquad (19.2)$$

where J_{v} is the permeate flux, L_{m} is the membrane permeability, ΔP_{TM} is the transmembrane pressure, $\Delta\Pi_{\mathrm{TM}}$ is the transmembrane osmotic pressure, σ_0 is the osmotic reflection coefficient, k is the mass transfer coefficient, and C is the solute concentration at the wall (subscript w), bulk (subscript b), and permeate (subscript p). The osmotic pressure is a function of the local solute concentration. For a fully retained solute, $C_{\mathrm{p}} = 0$ and $\sigma_0 = 1$. The linear scale-up principles outlined previously ensure that the parameters in Equation 19.1

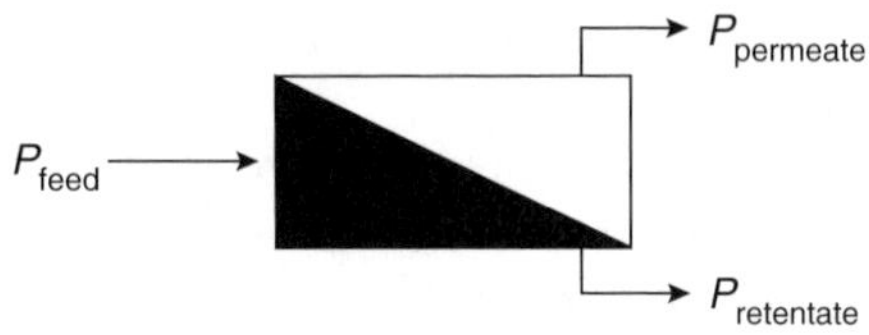

FIGURE 19.1 Definition of pressures during ultrafiltration.

and Equation 19.2 are the same at different scales, resulting in equivalent performance. On the other hand, changes to any of the controlled parameters has the potential to change the other parameters to satisfy Equation 19.1 and Equation 19.2, resulting in a potential for altered performance. For example, changes to system hydrodynamics upon scale-up can change the transmembrane pressure drop and the mass transfer coefficient, leading to changes in other (not controlled) parameters and ultimately the ultrafiltration performance. Also, it should be noted that Equation 19.1 and Equation 19.2 represent the osmotic pressure model; other competing models provide similar conclusions.

This chapter provides a case study of a highly viscous polysaccharide process stream, where conventional scale-up principles were not adequate. The data and analysis in this study is centered on the measured differential pressure defined as difference between feed and retentate pressures, as shown in Figure 19.1.

$$\Delta P_{\text{observed}} = P_{\text{feed}} - P_{\text{retentate}} \tag{19.3}$$

The differential pressure at a given cross-flow rate provides a measure of the hydraulic resistance in the retentate flow path, and allows for comparison across scales even when small differences in retentate pressures are present.

The observed differential pressure at any scale can be divided into two contributions (Equation 19.4):

$$\Delta P_{\text{observed}} = \Delta P_{\text{cassette}} + \Delta P_{\text{hardware}} \tag{19.4}$$

where $\Delta P_{\text{cassette}}$ is the contribution due to the membrane cassette and $\Delta P_{\text{hardware}}$ is the contribution due to the membrane holder, and the piping leading to it from the pressure guage where the above pressures are measured (typically close to the membrane holder). For appropriate scale-up, the minimum principle is that $\Delta P_{\text{cassette}}$ has to be constant across scales. It is also important that even $\Delta P_{\text{hardware}}$ be constant across scales. The latter ensures that measured pressure drops, which include this component, can be used an indication of process performance and scale-up during routine operation. Furthermore, differences in $\Delta P_{\text{hardware}}$ can lead to differences in available operating ranges. Typically

$\Delta P_{\text{hardware}}$ is small, and hence the focus of the scale-up is the membrane cassette.

$\Delta P_{\text{hardware}}$ can be further broken down to the individual areas of pressure loss in the hardware [11]. In general, $\Delta P_{\text{hardware}}$ can be related to total flow rate, Q using a power-law equation for a Newtonian fluid [11]:

$$\Delta P_{\text{hardware}} = K_{\text{hardware}} Q^a \tag{19.5}$$

where a is an exponent required to address turbulence in these areas. The pressure drop in the membrane cassette, $\Delta P_{\text{cassette}}$, is also composed of multiple contributions, each one of which could be represented with equations similar to Equation 19.5. The most important among them is the flow in the membrane channel. Simplistically, flow through a rectangular membrane channel can be described using the concept of hydraulic diameter (denoted as D_{h}). For a channel of rectangular cross section of width w and height h:

$$D_{\text{h}} = \frac{4 \times \text{Flow area}}{\text{Perimeter}} = \frac{4wh}{(2w + 2h)} \tag{19.6}$$

In order to evaluate the pressure drop in a membrane channel, we can use the hydraulic diameter in the expression for friction factor for fully developed laminar flow between two parallel plates (true since $w \gg h$) for a Newtonian fluid:

$$f = \frac{24}{\text{Re}}, \quad \text{where Re} = \frac{D_{\text{h}} V \rho}{\eta} \tag{19.7}$$

where V is the velocity in the channel and η and ρ are the viscosity and density of the liquid, respectively. Upon substitution of this friction factor into standard pressure drop expression [14], this leads to

$$\Delta P_{\text{channel}} = 12 \frac{\eta Q L}{N w h^3} \tag{19.8}$$

where L is the length of the channel, Q is the volumetric flow rate and N is the number of membrane channels. The height of the channel is determined by a polymer spacer that separates the retentate side of a membrane from the screen, as shown schematically in Figure 19.2. A lot of simplifications are inherent in Equation 19.8 and need to be justified, but will not be attempted here. For example, these membranes contain a floating screen, intended to generate additional turbulence in the channel, to maximize mass transfer. In Equation 19.8, the screen is assumed to be a solid channel separator. The other important assumption in Equation 19.7 is that the flow is assumed to be

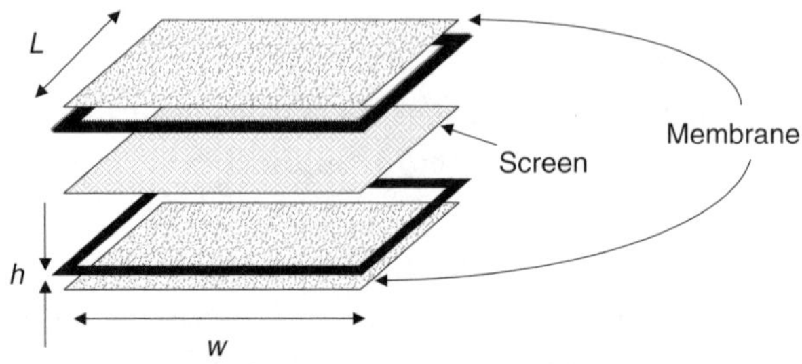

FIGURE 19.2 A membrane channel.

laminar, which may not be valid due to the turbulence-promoting screen. While rigorous pressure drop calculations would require a correlation generated from independent experimental data, Equation 19.8 provides insights on the impact of key membrane cassette dimensions. For example, the pressure drop is inversely proportional to h^3, and hence small changes in channel height can have a large effect on pressure drop, especially for high-viscosity fluids.

For non-Newtonian flow, such as the shear-thinning polysaccharide process streams, the expression for pressure drop is more complicated. For a power-law fluid, where the fluid viscosity η is given by:

$$\eta = K\dot{\gamma}^{n-1} \tag{19.9}$$

where $\dot{\gamma}$ is the shear rate, and K and n are constants, with $n < 1$ for a shear-thinning liquid. The pressure drop is given by [14]:

$$\Delta P_{\text{channel}} = 2\frac{LK}{h}\left(\frac{4n+2}{n}\right)^n\left(\frac{Q}{Nwh^2}\right)^n \tag{19.10}$$

Equation 19.10 reduces to Equation 19.8 for the case of a Newtonian liquid where $n = 1$.

19.3 MATERIALS AND METHODS

Ultrafiltration at laboratory-scale was performed using Maximate™ membrane cassettes from Pall Corporation (Northborough, MA and East Hills, NY). Each cassette consists of 14 parallel flow channels stacked horizontally, to give the total 2 ft^2 of membrane area. Large-scale filtration was performed using Pall Corporation Maxisette™ membrane cassettes, which consist of 15 sections that are each roughly equivalent to one Maximate cassette (2 ft^2), for a total of 30 ft^2 per cassette. The number of parallel flow channels is increased to 13 to

TABLE 19.1
Dimensions and Characteristics of Pall Corporation Ultrafiltration Cassettes

	Small-Scale	Large-Scale
Cassette name	Maximate	Maxisette
Area per cassette (ft^2)	2	30
Number of individual channels	14	210
Type of spacer	Suspended screen	Suspended screen

compensate for a narrower width of the cassette. Cassette characteristics are summarized in Table 19.1.

At laboratory-scale, ultrafiltration was performed using a set-up consisting of the Maximate™ membrane holder, a Flowtech Labtop 250 rotary lobe pump (Atlanta, GA), a Yokogawa ADMAG CA magnetic flow meter (Newnan, GA), and pressure gauges for feed, retentate, and permeate flows located in close proximity to the membrane holder ports. A similar set-up was used at large-scale with an Alfa-Laval G & H 822 rotary lobe pump (Richmond, VA) albeit the skid was fully automated with clean-in-place (CIP) and steam-in-place (SIP) capability. Again, pressure monitoring gauges and transducers were located near the entrance of the membrane holder. Filtration experiments with these cassettes were also performed at Pall Corporation (Northborough, MA), with a manual skid. The cassettes are held within the membrane holder on the skid, with the torque uniformly applied to a defined specification, obtaining a uniform torque manually at lab-scale (4-bolt system at Merck, and 8-bolt system at Pall) or by a six-piston hydraulic membrane holder at large-scale.

The process fluid for ultrafiltration experiments was clarified bacterial fermentation broths of different serotypes of *Streptococcus pneumoniae* obtained from Merck & Co., Inc. (West Point, PA), water and carboxymethylcellulose (Sigma-Aldrich, St. Louis, MO) solutions to simulate the viscous process fluid.

19.4 RESULTS AND DISCUSSION

Figure 19.3 shows an illustration of the pressure observed during ultrafiltration of one of the *S. pneumoniae* clarified fermentation broths at large-scale using 8 Maxisette cassettes. Data are compared to a parallel lab-scale ultrafiltration performed with the same feed material with a single Maximate cassette. These experiments were conducted at a fixed cross flow rate of 2 LPM/ft^2,

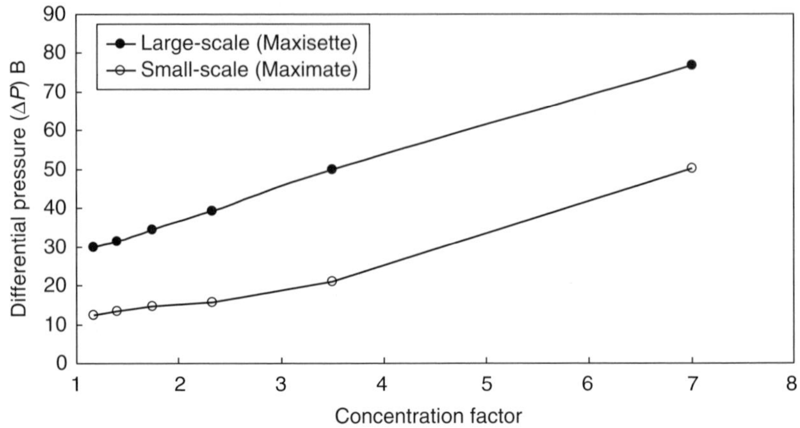

FIGURE 19.3 Differential pressure (feed pressure — retentate pressure) for a large-scale and a corresponding small-scale ultrafiltration with the same polysaccharide clarified fermentation broth feed stream.

and a relatively fixed retentate pressure, typically 5 to 10 psig. In this figure, differential pressure is plotted as a function of the concentration factor (inverse of volume reduction). With concentration the differential pressure increases, due to increasing viscosity caused by the increase in the polysaccharide concentration. The low-shear viscosity increases dramatically in this case from 2.4 to 41.2 cP over the course of concentration. The increase in differential pressure with concentration was observed at both scales. The striking feature, however, was that the large-scale run had a greater differential pressure at the beginning of the step by approximately 20 psi, and this persisted throughout the step.

The difference in differential pressure at the two scales under similar flow conditions indicated differences in the flow resistance at the two scales, despite the use of linear scale-up principles. The fact that the same feed stream was used at both scales indicated that it was inherent to differences in the membrane cassette and the hardware, and not to any differences in the process streams at the two scales. The pumping mechanism was determined not to alter the process stream in any meaningful way (data not shown), and hence the use of different pumps at the two scales was ruled-out as a potential cause for the difference. The higher differential pressure at large-scale was also seen with other bacterial fermentation broths. The relative constancy of the difference in the differential pressures is also striking, despite the large increase in viscosity with concentration. This is most likely related to the shear-thinning and drag-reducing nature of the process fluid, resulting in some element of self-correction to the higher-flow

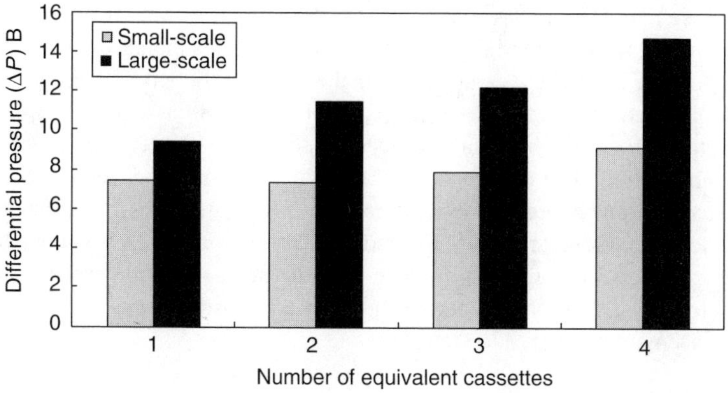

FIGURE 19.4 Differential pressure for a large-scale (Maxisette) and a corresponding small-scale (Maximate) ultrafiltration with water. Volumetric flow rate is increasing along the x-axis, but volumetric flow rate per membrane area is constant, to maintain a constant cross-flow rate of 2LPM/ft2.

resistance at large-scale. More work is needed to quantitatively understand this effect.

At appropriate flow rates, lack of scalability, represented by a lack of constancy of $\Delta P_{\text{observed}}$, could be a sign of poor scale-up of either the membrane or the hardware or both, per Equation 19.3. For example, $\Delta P_{\text{cassette}}$ will not be constant upon scale-up if the membrane cassette configuration is not scaled-up linearly, i.e., if the path length and channel height are not kept constant. While linear scale-up is usually explicitly accounted for in the design of the membrane cassette, the linear scale-up of the hardware is usually ignored, resulting in the potential for $\Delta P_{\text{hardware}}$ not being constant.

We evaluated both contributions and their change upon scale-up. The cassette pressure drop cannot be independently measured, since the contribution from the hardware always exists. Therefore, once the hardware pressure drop is independently characterized, the cassette scale-up can be appropriately estimated. In order to make the evaluation of the two components more straightforward, water was used as the fluid. Whereas water does not mimic the rheology of polysaccharide process streams, it provides an excellent tool for characterizing the intrinsic hydraulic resistance of the system, without the complication of non-Newtonian behavior.

Figure 19.4 shows the pressure drop in both the small- and large-scale systems, with increasing number of membrane cassettes. The cross-flow rate per unit area (henceforth termed cross-flow rate) is kept constant at 2 LPM/ft^2. Data in Figure 19.4 obtained with water mimics the data shown in Figure 19.3 with the polysaccharide process stream. The large-scale system offers more hydraulic resistance than the small-scale system for the same flow rate; furthermore, the difference in the differential pressures at the two scales increases as additional cassettes are used (i.e., at larger total flows). For the sake of brevity, data were obtained only up to 4 cassettes, even though 8 cassettes were used in the actual process. The data can be extrapolated relatively easily with this Newtonian fluid. The difference in differential pressures between one small-scale cassette and 4 large-scale cassettes is approximately 7 psi, and assuming a linear extrapolation, it may be expected to be approximately 14 psi with 8 large-scale cassettes. The higher difference with the polysaccharide process stream of 20 to 25 psi, is probably due to the higher viscosity of the polysaccharide process stream compared to water.

19.4.1 Contribution to Pressure due to Differences in System Hardware

In order to measure the hardware pressure drop, blank Maximate and Maxisette cassette frames, with the membrane, screen, and spacer materials removed, were tested with water at a variety of cross-flow rates. Data are shown in Figure 19.5. The x-axis in this figure is again expressed as the number of membrane cassettes, and represents an increase in the total flow rate with a fixed cross-flow rate per membrane area of 2 LPM/ft^2.

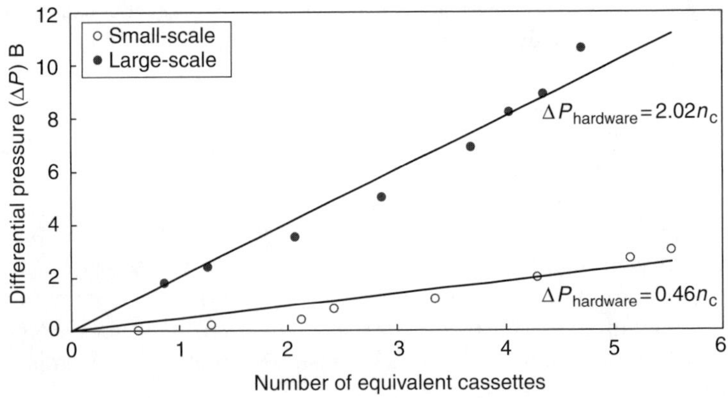

FIGURE 19.5 Differential pressure for large-scale (Maxisette) and a corresponding small-scale (Maximate) skid hardware with water. Volumetric flow rate is increasing along the x-axis with increasing number of cassette blanks (n_c), while keeping volumetric flow rate per membrane area is constant, to maintain a constant cross-flow rate of 2 LPM/ft^2.

Consistent with the data with membrane cassettes, the differential pressure in the large-scale system is higher than that in the small-scale system. Even in the case of scale-up from a single small-scale cassette to a single large-scale cassette, a difference in the differential pressure is seen. This difference becomes more severe as additional membrane area is added. Figure 19.5 illustrates an important problem upon a typical scale-up — from one small scale cassette to multiple large-scale cassettes — where the difference between the two differential pressures was as large as 10 psi. This can be expected to only increase with more viscous process streams.

The contribution from the hardware can now be subtracted from the total pressure drop in Figure 19.4. This is shown in Figure 19.6. We see that the pressure drop across the membrane cassette remains relatively constant as additional cassettes are added, implying good scalability of the membrane cassettes. Thus, the increase in total observed pressure drop in Figure 19.4 is due almost entirely to the hardware. While actual measurements (or calculations) were not made to estimate the different contributions within the hardware, one major fluid flow restriction occurs in the flow distribution manifold, which splits the retentate flow between the eight stacked cassettes and collects the permeate flow. Retentate recirculation occurs through 4 in. piping and is reduced to 0.5 in. ports within the membrane holder before entering the membrane flow channels, in addition to a number of right-angle bends within the membrane holder itself.

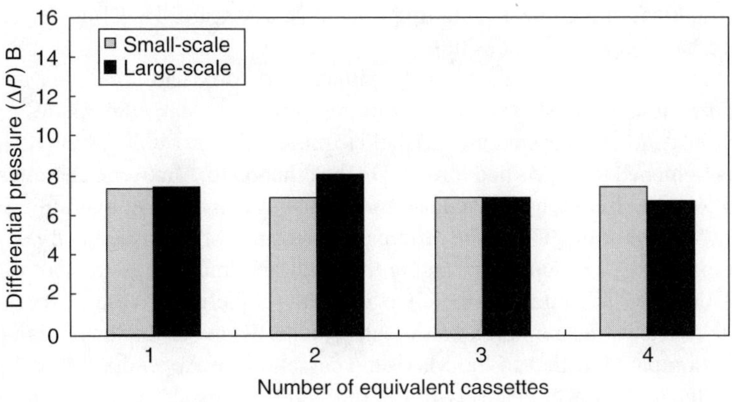

FIGURE 19.6 Differential pressure for large-scale (Maxisette) and small-scale (Maximate) ultrafiltration corrected for skid hardware pressure drop. This represents pressure drop due to the membrane cassette(s) alone.

19.4.2 Contribution to Pressure due to Membrane Cassette Construction

Even though the above set of experiments demonstrated that the membrane channels did not contribute to the lack of scalability, this contribution to the total pressure drop was explored in some detail. This is because small differences in membrane geometry can have a big impact on the total pressure drop, especially with viscous process streams.

Although membrane cassettes are typically scaled-up linearly by the membrane manufacturers, the geometry of the cassettes at different scales have to be carefully evaluated to assess the impact on linear scale-up. In the case of the Maximate and Maxisette cassettes, the length of the membrane channel is the same, but the channel width is lower in the case of the Maxisette cassettes. The widths of the Maximate and Maxisette cassettes are 36 mm and 27 mm, respectively. This 25% reduction in channel width is partially made up by increasing the number of membrane layers from 11 in the Maximate to 13 in the Maxisette cassette, resulting in the potential for only a ~11.5% increase in the differential pressure.

Another factor to be considered is that differences in the torquing of the membrane cassettes at the two scales has the potential to compress the membrane cassettes differently, resulting in differences in the channel height. This is due to the use of different torquing mechanisms at each scale — typically a torque wrench at small scale vs. an automated hydraulic torquing system at larger scales. Since channel height is the smallest dimension, subtle changes in this dimension can lead to relatively large changes in the membrane pressure drop, due to the inverse cubic relationship per Equation 19.8, leading to a breakdown of the linear scale-up principle at any scale. The effect is further exacerbated at higher viscosities.

Figure 19.7 shows data for Maximate and Maxisette cassettes at two compressions to illustrate the effect on the membrane cassette. As described previously, Maximate cassettes are held together using a 4-bolt system with the bolts tightened to a specified torque. On the other hand, Maxisette cassettes are held together by a 6-bolt dynamic auto-torque system at a set hydraulic pressure. As seen in the figure, the differential pressure for Maxisette increases by approximately 25% upon increasing the holder hydraulic pressure from 2000 to 2500 psig. The effect of the change in the torque in the case of the Maximate cassette was negligible, although it could be due to the lower clamping force compared to that in the Maxisette cassette (see Appendix A), and possibly due to the lower viscosity of the solution (i.e., lower pressure) studied in that system. Data at the higher viscosity at different compression torques was not obtained with the Maximate cassettes. It should also be noted that in some cases, increase in pressure due to increased torque could be caused by creeping

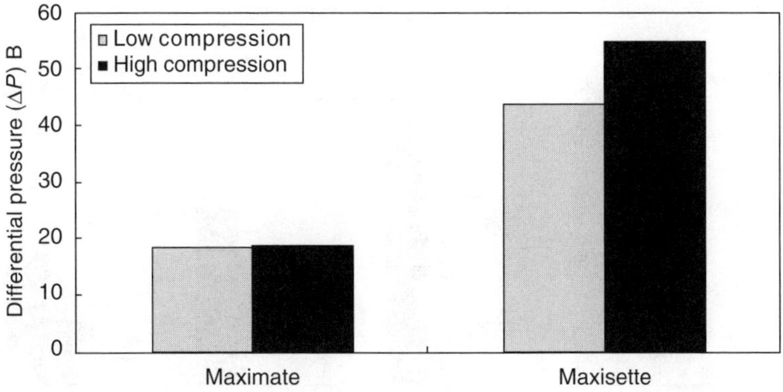

FIGURE 19.7 Differential pressure for membrane cassette made up of the standard spacer material at each of the two scales — small (Maximate) and large (Maxisette). Data are shown for two different torque levels. Low and high levels correspond to 65 and 80 in-lb for Maximate cassettes and 2000 and 2500 psi for Maxisette cassettes. Data on Maximate was obtained with a 10 cP CMC solution and data on Maxisette was obtained with a 250 cP CMC solution.

of sealing gasket into the flow ports; it was however determined not to be the case in this experiment.

In order to avoid any potential for cassette compression, new cassettes were constructed using a less compressible, high-density polyethylene (HDPE) spacer. Although the spacer is not the only component that can impact cassette compressibility (the potting polymer being the other), the spacer is the limiting determinant of channel height. Three different incompressible spacer thicknesses were evaluated — 7, 9, and 10 thousandths of an inch (mil), as well as the standard spacer. Data from pressure-drop studies with cassettes made with these spacers is shown in Figure 19.8. The 7 mil spacer provides pressure drops that are closest to the standard spacer. This spacer thickness was then chosen for construction of large-scale cassettes, to provide linear scalability from the Maximate cassettes with the standard spacer. The new spacer was then evaluated for compressibility at different compression pressure settings at large-scale and the data are shown in Figure 19.9. The data show an improved tolerance to changes in the compression at the Maxisette scale, a desired feature in order to maintain linear scale-up.

These data indicate that even in the case of linearly scaled cassettes, differences in the dynamic membrane channel geometry also have to be considered, especially for high viscosity streams. It should be noted however, per data in Figure 19.7, that the standard spacer is quite adequate for low-viscosity process solutions.

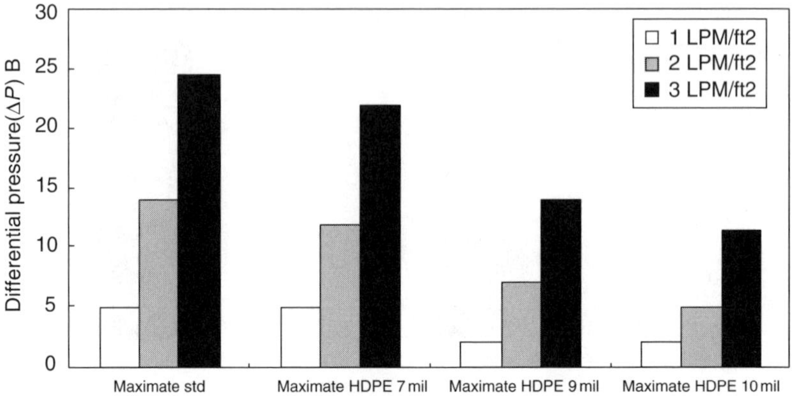

FIGURE 19.8 Differential pressure for Maximate cassettes made using different spacers at three different cross-flow rates with water.

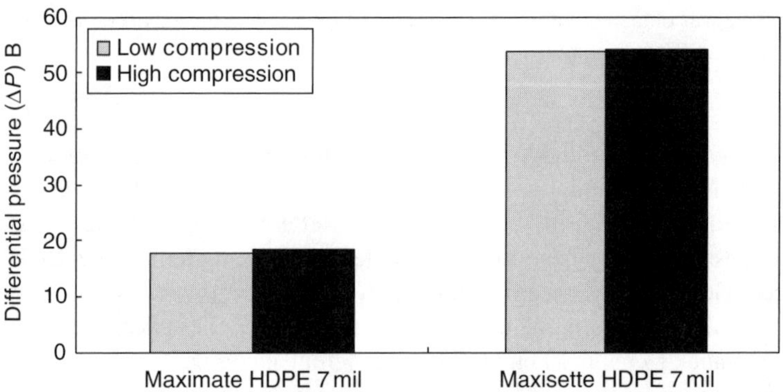

FIGURE 19.9 Differential pressure for membrane cassette made up of a HDPE spacer material. Data are shown for two different torque levels. Low and high levels correspond to 65 and 80 in-lb for Maximate cassettes and 2000 and 2500 psi for Maxisette cassettes. Data on Maximate was obtained with a 10 cP CMC solution and data on Maxisette was obtained with a 250 cP CMC solution.

This study did not cover pressure drops on the permeate side of the membrane. Permeate flows are generally small and do not have large pressure losses associated with them, however an analysis similar to the one described above could be done to discern the impact of any differences in geometries on the permeate side of the membrane cassette, which do exist in Maximate and Maxisette cassettes.

Finally, a statement on the impact of process performance due to the higher-differential pressure in the particular example described above is warranted.

No measurable impact on actual process performance other than higher pressures was observed, presumably due to constancy in the membrane cassette hydrodynamics. In some cases, the high pressures at large-scale resulted in the need to reduce the cross-flow rate so that the pressure rating on the system was not exceeded, although even in these instances, no other process impact was observed.

19.5 CONCLUSIONS

Linear scale-up of membrane systems are not as straightforward when high-viscous process solutions and large cross-flow rates are involved. While membrane cassettes are scaled linearly, system hardware is often overlooked and can be the cause for poor linear scale-up. The design of full-scale production equipment often involves necessary flow restrictions, bends in order to satisfy space and equipment requirements, as well as flow distribution in narrow channels in the membrane holder. Whereas the scale-up factor in the membrane area can exceed 100-fold, the scale-up factor in the piping cross-sectional area is usually much less (e.g., a 0.75 in. ID scaled up to a maximum of 4 in. ID leading to only a fivefold cross-sectional area scale-up). This can introduce a significant hardware pressure drop, which is exacerbated at higher-flow rates and viscosities.

While most membrane cassettes are scaled-up linearly, high viscosity and high flow rates expose any minor shortcomings in cassette construction and differences in compression. Stacked configurations of flat-sheet membranes result in bends and split fluid flows into a greater number of membrane channels, which can introduce appreciable pressure drops. Attention has to be paid in ensuring similar torque levels at different scales. A complicating factor in both the membrane and hardware pressure drop evaluation is that pressure gauges are not always located immediately next to the membrane housing, resulting in pressure readings that will differ slightly from the true pressure at the entrance and exit of the membrane channel.

A simple set of experiments using a Newtonian fluid such as water at varying flow rates with the particular membrane systems involved, would allow determination and comparison of the hydrodynamics in the systems, which should be kept constant upon scale-up. These could be supplemented with limited studies with a model non-Newtonian fluid. These types of studies can lead to a more accurate *a priori* prediction of scale-up performance in membrane systems with macromolecular process fluids. Such information is undoubtedly useful during skid design, construction, and start-up, leading to a reduction in unexpected scale-up outcomes.

APPENDIX: CONVERSION OF TORQUE AND PRESSURE TO CLAMPING FORCE

1. The Maximate cassettes are held together using a bolts at a specified torque. The torque can be converted to a clamping force using:

$$\text{Force/bolt} = \frac{\text{Torque(in.-lb)}}{\text{Friction factor}} \times \text{Bolt diameter} \tag{19.A1}$$

   Friction factor is assumed to be 0.21 [15]. The bolt diameter on the Maximate system is 0.375 in. At 80 in.-lbs, the total force on a Maximate by a 8-bolt hardware is 8127 in.-lbs.
   The clamping force is obtained by:

$$\text{Clamping force} = \frac{\text{Force/bolt} \times \text{No. of bolts}}{\text{Cassette perimeter}} \tag{19.A2}$$

   For the Maximate cassette perimeter of approximately 28 linear in., the total clamping force on the perimeter of the Maximate cassette is 290 lbf/linear perimeter inch.

2. The Maxisette cassette are held together by an auto-torque system at a specified pressure. The pressure can be converted to a clamping force using:

$$\begin{aligned}\text{Force/bolt} = {} & \text{Hydraulic pressure (psig)} \\ & \times \text{Hydraulic cylinder piston area (in}^2\text{)}\end{aligned} \tag{19.A3}$$

   The clamping force is then calculated using Equation 19.A2 with number of pistons substituted for number of bolts.

For a Maxisette with a 2500 psig and a piston area of 2.36 in^2, and 6 pistons, the clamping force on the perimeter of the cassette of 70 linear inches is 506 lbf/linear perimeter inch.

ACKNOWLEDGMENTS

The authors gratefully acknowledge Ralph Brooks, Jennifer Blatteau, Richard Saurman, Mary Kinsey, William Simmler, Pete Cottrell, and Richard Diliberto for their assistance with studies performed in large-scale equipment, and Russel Lander for useful discussions.

REFERENCES

1. Lo YM, Yang ST, and Min DB. Kinetic and feasibility studies of ultrafiltration of viscous xanthan gum fermentation broth. *Journal of Membrane Science* 1996; 117:237–249.
2. Bergmaier D, Lacroix C, Guadalupe Macedo M, and Champagne CP. New method for exopolysaccharide determination in culture broth using stirred ultrafiltration cells. *Applied Microbiology and Biotechnology* 2001; 57:401–406.
3. Brou A, Ding LH, and Jaffrin MY. Extraction and concentration of polysaccharides using a rotating disk filtration system. *Filtration* 2003; 3:162–168.
4. Brou A, Jaffrin MY, Ding LH, and Courtois J. Microfiltration and ultrafiltration of polysaccharides produced by fermentation using a rotating disk dynamic filtration system. *Biotechnology and Bioengineering* 2003; 82:429–437.
5. Gilbert FB, Poutrel, B, and Sutra, L. Purification of type 5 capsular polysaccharide from Staphylococcus aureus by a simple efficient method. *Journal of Microbiological Methods* 1994; 20:39–46.
6. Gonccalves, VMM, Takagi, M, Lima, RB, Massaldi, H, Giordano, RC, and Tanizaki, MM. Purification of capsular polysaccharide from *Streptococcus pneumoniae* serotype 23F by a procedure suitable for scale-up. *Biotechnology and Applied Biochemistry* 2003; 37:283–287.
7. Pritchard M, Howell JA, and Field RW. The ultrafiltration of viscous fluids. *Journal of Membrane Science* 1995; 102:223–235.
8. Charcosett C and Choplin L. Concentration by membrane ultrafiltration of a shear-thinning fluid. *Separation Science and Technology* 1995; 30:3649–3662.
9. Howell J, Field R, and Wu D. Ultrafiltration of high-viscosity solutions: theoretical developments and experimental findings. *Chemical Engineering Science* 1996; 51:1405–1415.
10. Gehlert G, Luque S, and Belfort G. Comparison of ultra- and microfiltration in the presence and absence of secondary flow with polysaccharides, proteins, and yeast suspensions. *Biotechnology Progress* 1998; 14:931–942.
11. Van Reis R, Goodrich EM, Yson CL, Frautschy LN, Dzengeleski S, and Lutz H. Linear scale ultrafiltration. *Biotechnology and Bioengineering* 1997; 55:737–746.
12. Zeman L and Zydney A. *Microfiltration and Ultrafiltration: Principles and Applications.* New York: Marcel Dekker, 1996.
13. Choe TB, Masse P, Verdier A, and Clifton MJ. Flux decline in batch ultrafiltration: concentration polarization and cake formation. *Journal of Membrane Science* 1986; 26:1–15.
14. Perry RH and Green DW. *Perry's Chemical Engineers' Handbook.* Section 5. 7th ed. New York: McGraw Hill, 1997.
15. Rozembersky J. Personl Communication.

20 A Membrane Chromatography Application: A Rapid, High Capacity Gene Therapy Vector Purification Tool

Ajay R. Lajmi, Robert Kutner, and Jakob Reiser

CONTENTS

20.1 INTRODUCTION

Gene therapy may be defined as the introduction of genetic material into cells for therapeutic purposes. The transfer of genetic material into the target cells or tissues is mediated by vectors. Such vector systems may involve viruses. These vectors are referred to as viral vectors. Alternatively, vectors may involve nonviral systems such as naked plasmid DNA. Gene therapy vectors may be administered either *in vivo* by direct infusion into the patient or *ex vivo* where the vector is introduced into target cells that are extracted from a patient. Ultimately, cells bearing a vector are infused back into the patient.

Among commonly used viral vectors are adenoviral (Ad) vectors, vectors based on adeno-associated virus (AAV), and retroviral vectors of which lentiviral (LV) vectors are a subclass. Over the past decade, despite a few setbacks considerable progress has been made in minimizing gene therapy side effects without compromising efficiency [1,2]. Although the majority of clinical trials are currently in Phases I and II where safety and feasibility are examined on a small group of patients, there have been notable clinical responses in more than fifteen of them that have advanced to Phase III clinical trials in the United States and elsewhere [3]. Gene therapy treatments have initially targeted inherited disorders but recently an increasing number of investigations have been focused on infectious diseases, cancer, neurodegenerative disorders, and organ transplant rejection.

As gene therapy vectors advance to late stages of clinical trials, efficient methods of chromatographic capture and purification will be necessary to scale-up to manufacturing scale. The purpose of this chapter is therefore, to present the current developments in the purification of four of the most popular and emerging gene therapy vectors namely, Ad, AAV, LV, and plasmid DNA, as well as to introduce recent developments in the application of membrane ion exchange chromatography technology for the purification of such vectors.

20.1.1 ADENOVIRAL VECTORS

Adenoviral viruses contain double-stranded DNA. The icosahedrally shaped virion is approximately 90 nm in diameter and is composed of a protein shell (capsid) that surrounds a DNA–protein core complex [4]. Ad vectors are popular as high vector titers and high levels of transgene expression can be achieved. However, loss of transgene expression and the ability of Ad vectors to elicit strong immune responses are major drawbacks. The development of helper-dependent Ad vectors that carry deletions in all viral genes has helped to decrease immunogenicity, prolong transgene expression and improve the prospects of Ad vectors for long-term gene therapy [5,6].

20.1.2 Vectors Based on Adeno-Associated Virus

Adeno-associated virus includes a group of small nonenveloped viruses with a diameter of 20 to 25 nm. They contain a single-stranded DNA genome encapsulated by a protein capsid. Vectors based on AAV are currently in various stages of clinical trials to treat hemophilia A, hemophilia B, cystic fibrosis, Parkinson's disease, AIDS, rheumatoid arthritis and hyperlipidemia. AAV vectors are nonimmunogenic and nonpathogenic to humans. There are well over 100 known serotypes of AAV. These vectors have a packaging capacity of up to 4.5 kb with the ability to infect a broad range of nondividing and dividing cells. The first generation AAV vectors required a helper virus such as Ad or herpesvirus but recent advances in identifying the required helper genes from Ad has resulted in helper-free AAV production systems [7–9]. Inefficient large-scale production is one of the major drawbacks of this vector system although progress has been made to increase vector production levels to 100 l volumes [10,11].

20.1.3 Lentiviral Vectors

Along with oncoretroviruses like Moloney Murine Leukemia virus and spumaviruses, lentiviruses are a subclass of retroviruses. These enveloped viruses contain RNA and range from 90 to 140 nm in diameter. The envelope proteins are responsible for specificity as well as for cell attachment. To increase the host-range of LV vectors, a heterologous envelope glycoprotein such as the vesicular stomatitis virus G (VSV-G) glycoprotein is used. This process is referred to as pseudotyping. LV vectors transduce nondividing cells as well as dividing cells and fully differentiated cells [12] with stable integration into the target cell genome. Some of the disadvantages of retroviral vectors include the potential of insertional mutagenesis that may result in oncogene activation and a general lack of high titer production methods [13,14].

20.1.4 Plasmid DNA

Plasmids used in gene therapy applications consist of covalently closed circular DNA molecules that range in size from 1 to 200 kb. Plasmids >15 kb in size are particularly susceptible to mechanical damage due to shear stress during isolation, purification, and handling, therefore precautions need to be taken when processing large plasmids [15]. Some of the advantages of plasmid-based gene therapy approaches include low immunogenicity and improved safety properties relative to virus-based approaches. Naked plasmid DNA can be used in vaccines. One such vaccine is currently being marketed for veterinary applications against West Nile virus. Additional plasmid vaccines to be used in human clinical applications are in either Phase I or II clinical trials [16]. Plasmids can be delivered into cells either chemically by complexing with agents such as liposomes or by using membrane-disrupting procedures such as electroporation.

20.2 RECENT DEVELOPMENTS IN VECTOR PURIFICATION

Adenoviral, AAV, and retroviral vectors are produced in mammalian cells. One way to release Ad and AAV vectors from cell pellets is by applying multiple freeze–thaw cycles. Retroviruses such as LV on the other hand are released into the supernatant. The viral vectors may be separated from the cellular debris by either centrifugation followed by filtration or by using a series of filters with decreasing porosity.

20.2.1 PURIFICATION OF AD VECTORS

A classical method of Ad vector purification has involved cesium chloride (CsCl) density gradient centrifugation. This process typically takes one to two days and generates vector stocks of variable quality. A major drawback of this method is its limited scalability making it unsuitable for large-scale vector production.

Although Ad have been traditionally isolated by CsCl density gradient ultracentrifugation, recently other purification methods based on ion exchange chromatography [17–19], size exclusion chromatography [20,21] and hydrophobic interaction chromatography [17] have been reported. A general Ad purification scheme shown in Figure 20.1 summarizes some of the key steps. Goerke and coworkers [22] have recently reported an Ad purification process that involves selective precipitation of host cell DNA as well as proteins with 2% domiphen bromide, a cationic detergent resulting in 3 log reduction of DNA with host cell protein levels of 15 $\mu g/10^{11}$ viral particles in yields of 58 to 86% as determined by anion exchange chromatography. Examples in published literature however, report either one- or two-step chromatography processes followed by ultrafiltration, diafiltration, and sterile filtration. As the

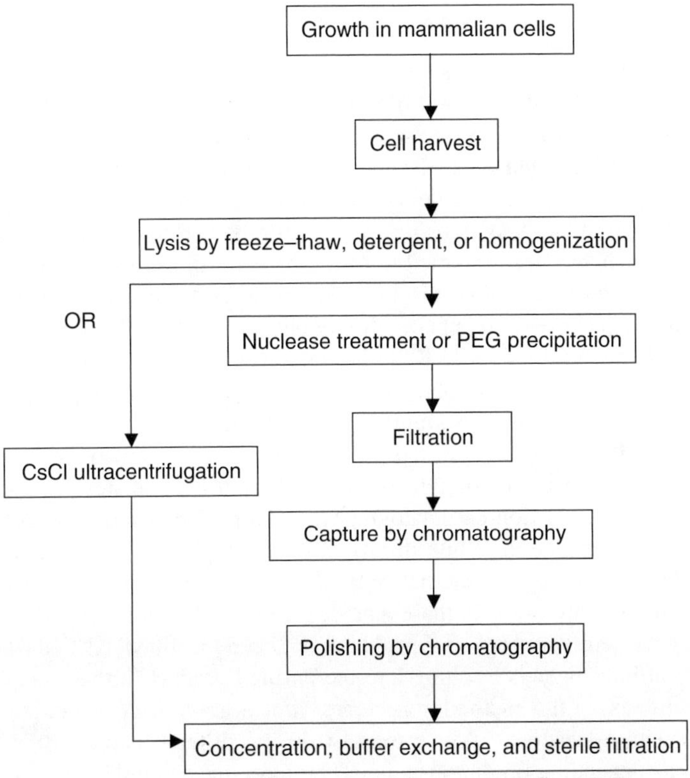

FIGURE 20.1 A general process flow chart for Ad and AAV purification.

Ad capsid protein composition changes with the serotype, the ionic charge on the Ad particle can be modulated in solution. Ion exchange chromatography can therefore be used for the capture step.

Huyghe and coworkers [17] reported on a two-column process to purify Ad vectors. Approximately 3×10^{12} Ad particles present in a crude cell lysate were captured on a 1.7 ml DEAE anion exchange column at a flow rate of 1 ml/min. A follow-up step included a zinc metal ion affinity column. The overall recovery was 32% and the purity of the virus preparation was good, as judged by protein gel analysis. Green and coworkers [23] purified Ad vectors using DEAE anion exchange column chromatography and reversed phase ion-pair chromatography. Kamen and Henry [24] as well as Arcand and coworkers [25] have developed a process that involved capture of Ad particles from 20 l of lysate on a DEAE anion exchange column resulting in 80% recovery. The first chromatography step was followed by a polishing step using gel filtration.

20.2.2 Purification of AAV Vectors

Cesium chloride density gradient centrifugation has been used successfully in the past for the purification of AAV-2 vectors resulting in virus particles of sufficient purity to generate x-ray diffraction quality crystals [26]. However, this approach is limited due to the poor scalability in the production of sufficient vector quantities to meet the demands of Phase III clinical trials and beyond.

Purification of AAV-5 using mucin affinity chromatography involving agarose beads has been reported to result in vector preparations whose purity was comparable to that of vectors purified by CsCl gradient centrifugation [27]. Several groups have used heparin affinity chromatography for the purification of AAV-2 and -3 vector stocks [28]. Zolotukhin et al. [29] purified AAV vectors by centrifugation using discontinuous iodixanol gradients followed by heparin affinity chromatography resulting in 30 to 60% overall recovery. Snyder and Flotte [30] described a three-step purification procedure involving capture of the vector on a heparin column followed by a purification step using phenyl hydrophobic interaction chromatography, and a final polishing step involving another heparin column. While affinity chromatography has been demonstrated in the literature as a good alternative for the purification of AAV vectors, it has several limitations. Some of these include the fact that affinity ligands for some of the AAV serotypes such as type 8 have not been identified. Also, the possibility of affinity ligands leaching into the purified product further complicates the usefulness of this method to generate clinical-grade AAV vector stocks.

As with Ad vectors, AAV vectors also display different capsid protein compositions depending on the serotype. Therefore, one can utilize ion exchange chromatography to exploit the differences in the ionic charge properties of these vectors in solution. Several AAV vector purification methods involving beaded ion exchange chromatography media have been reported. These included (a) a vector capture step using a heparin column followed by another purification step involving a PEI anion exchange column [31]; (b) a capture and purification step using either a PEI or Q anion exchange chromatography column [32]; (c) a two-step purification protocol involving a strong cation exchange chromatography resin followed by a strong anion exchange resin [33]; and (d) capture of the AAV vector by anion exchange chromatography using a strong anion exchange resin with subsequent polishing by gel filtration chromatography [34].

20.2.3 Purification of Lentiviral Vectors

Lentiviral vector production for large-scale *in vivo* applications that require high-titer stocks is challenging due to the lack of simple procedures capable of rapidly processing large volumes of cell culture supernatant. The traditional ultracentrifugation-based approaches are limited in terms of their capacity to handle large volumes, thus making this procedure extremely tedious. One problem with ultracentrifugation-based approaches is that cell-derived components are concentrated along with the vector particles leading to potential immune and inflammation responses [35].

Thus, chromatography-based approaches are needed in order to purify lentivirus vectors of contaminating host cell components. Methods based on anion exchange chromatography of HIV-1 vectors pseudotyped with VSV-G have been established [36,37]. Schauber et al. [38] described a similar procedure for HIV-1 vectors pseudotyped with the baculovirus GP64 glycoprotein. Yields and purity of the virus stocks resulting from these procedures were not reported, but these approaches may lead to vector stocks of improved purity, increased infectivity, and reduced toxicity. The purification of inactivated HIV-1 particles that involved a two-step TMAE and Q anion exchange chromatography procedure yielding virus preparations with >95% purity as judged by gel filtration chromatography analysis was also reported [39]. Size-exclusion chromatography has also been used to purify HIV-1 vectors albeit with a prior concentration of the cell culture supernatant by cross-flow filtration [3,40].

20.2.4 Purification of Plasmid DNA

Plasmid DNA is typically produced in bacterial cells such as *Escherichia coli*. Plasmid DNA can be released from bacterial cell pellets by alkaline cell lysis [42]. The lysate containing plasmid DNA is neutralized using potassium acetate resulting in the precipitation of proteins along with the cellular debris including bacterial DNA [42]. Addition of high concentrations of calcium chloride to the lysate prior to clarification has been reported to effectively precipitate most of the RNA contaminants [44]. Clarification of the plasmid DNA containing lysate can be achieved by centrifugation or by dead-end filtration consisting of two filters in series with decreasing porosity [43]. Plasmid DNAs from clarified cell lysates are then purified further by different techniques such as CsCl gradient ultracentrifugation in the presence of ethidium bromide, or ion exchange chromatography using beaded, or membrane media. Plasmid DNA purification methods involving anion exchange chromatography has been reported by Schleef [41] and by Prazeres and coworkers [45]. The former used 15% isopropanol in the wash and elution buffers. The high costs of the buffer and handling and disposal expenses could make such a process economically unfavorable during scale-up. A general purification scheme based on literature examples for plasmid DNA is summarized in Figure 20.2.

In most instances, a reduction in the RNA impurities is generally carried out prior to the plasmid capture by anion exchange chromatography. But this requires an additional diafiltration step before the chromatography step. Several modifications of the process shown in Figure 20.2 that are discussed below have

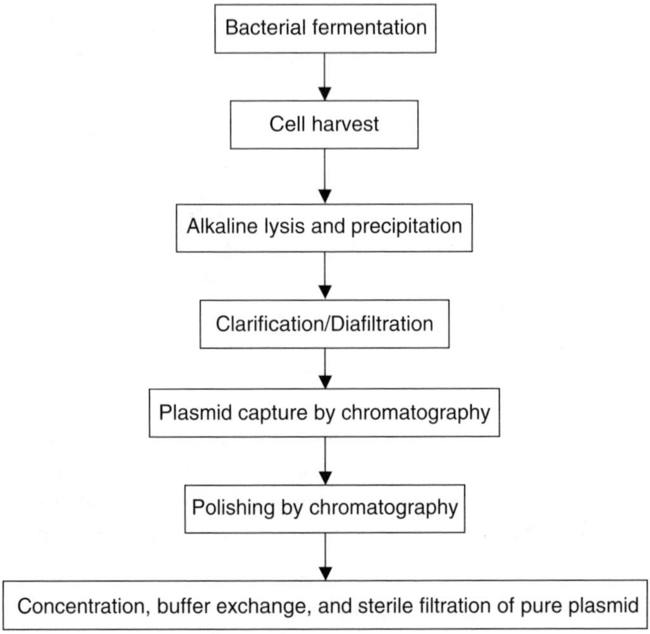

FIGURE 20.2 Flow chart for a generic process for isolation of plasmid DNA.

been explored recently in an attempt to address critical plasmid purification process issues such as plasmid capacity on chromatography media, plasmid purity with respect to RNA, and endotoxin contamination as well as plasmid recovery.

Levy and coworkers [46] used a similar but more elaborate plasmid purification process that involved RNase A digestion, partial purification of plasmid DNA by PEG precipitation, followed by filtration involving nitrocellulose filters before capture on anion exchange chromatography media, with further purification by another nitrocellulose filtration step and an anion exchange chromatography column. Eon-Duval and Burke [44] screened several anion exchange chromatography sorbents for plasmid polishing for trace RNA removal following primary purification using precipitation and tangential flow filtration (TFF). They found that for an approximately 5 kb plasmid, polishing on both Fractogel®* DEAE and POROS®** 50HQ resulted in >98.0% RNA

---

*Fractogel is a registered trademark of Merck KGaA Darmstadt.

**POROS is a registered trademark of Applied Biosystems.

removal with >94.0% plasmid recovery when the plasmid was loaded in 0.63 *M* NaCl and 0.72 *M* NaCl, respectively, in 50 m*M* phosphate buffer pH 7.0 at a flow rate of 150 cm/h.

Hydroxyapatite chromatography media have been shown to be useful for plasmid DNA purification [47]. Giovannini and Freitag [48] reported on the effects of the ratio of calcium and phosphorous in such media on the dynamic binding capacities of plasmid DNA. They concluded that the Hydroxyapatite media containing low calcium to phosphorous (C/P) ratios resulted in the best binding capacities (446 μg/g for a 4.7 kb plasmid and 59 μg for 11.4 kb plasmid). Sagar and coworkers [49] have shown that the dynamic binding capacity of some reversed phase beaded chromatography media can be increased by two- to threefold by adding up to 1 *M* NaCl to a plasmid feed stream.

Alternative protocols for the purification of plasmid DNA were also reported. Horn and coworkers [50] reported an overall yield of 50% following two successive PEG precipitations followed by a size-exclusion chromatography step. Lander and coworkers [51] demonstrated that cetyltrimethylammonium bromide (CTAB) selectively precipitated plasmid DNA from proteins, genomic DNA, RNA, and endotoxin. TFF has been shown to be effective in removing >99% of the contaminating RNAs from plasmid DNA after a precipitation step involving high concentrations of calcium chloride followed by centrifugation and micro-filtration [52]. Alternatively, by extending the alkaline lysis step in 0.2 *M* NaOH and 1% SDS from under 30 min to 24 h resulted in lower levels of RNA and endotoxin in a lysate containing a 10 kb plasmid while further reduction of these contaminants was accomplished by TTF [53]. Prazeres et al. [54] have recently published a detailed review of plasmid DNA purification by different chromatography techniques.

Thus, several methods exist for capture, purification, and polishing of gene therapy vectors that include both chromatographic as well as nonchromatographic techniques. All of the methods discussed above have their merits and drawbacks that impact the decision to adopt them for process development and ultimately transfer to manufacturing scale. The following sections will discuss membrane chromatography purification for capture of Ad, capture, and purification of AAV and for capture of LV.

20.3 MEMBRANE-BASED CHROMATOGRAPHY APPROACHES

Over the past decade several reviews have appeared in the literature that have chronicled the advances made in membrane chromatography [55–58]. Large biomolecules such as plasmid DNA and viral vectors cannot diffuse efficiently into the pores of beaded chromatography media and bind only to the surface of the particle due to pore-size limitations [59]. As a result, these chromatography media have low binding capacities for large biomolecules. A major challenge therefore, is to develop new capture and purification methods that would alleviate some of the downstream processing bottlenecks in an attempt to improve productivity of gene therapy purification.

Membrane chromatography media have large convective pores that provide fast mass transfer rates with low backpressure. The interaction between large biomolecules and the active sites in the convective pores is not diffusion-limited [60–62]. This results in high binding efficiencies at fast flow rates. As a consequence, one would need smaller volumes of chromatography media to process large quantities of gene therapy vectors for commercial production. This has a direct impact on process economics such as reduced raw material costs, reduced cycle time, and increased productivity. Therefore, membrane chromatography is ideally suited for efficient capture of such large biomolecules from a large feed-stock. Since these units are prepacked, they eliminate the need for column packing and packing validation. These units could either be incorporated in single-use applications or re-used. Since membrane chromatography was initially targeted towards contaminant removal applications, the designing of membrane housings to reduce hold-up volume, and hence dispersion was not a major concern. Low hold-up volume units that provide elution volumes comparable to column chromatography are now available.

While Grunwald and Shields [63], as well as Enders and coworkers [60] reported plasmid purification using adsorptive membranes, Zhang and coworkers [43] described a large-scale plasmid purification method involving strong anion exchange membrane capsules. Approximately 1.5 kg of frozen bacterial cell paste was processed using a 260 ml capsule. Compared to traditional bead-based media, the dynamic binding capacity of Mustang Q membranes for plasmid DNA was found to be 20 to 25 times greater, and the flow rate was 55 to 550 times greater than conventional beaded anion exchange chromatography media [60]. Alternative macroporous adsorbents such as monoliths have also been investigated for plasmid purification and revealed a dynamic binding capacity of approximately 9 mg/ml [64].

20.4 AD CAPTURE BY ANION EXCHANGE MEMBRANE CHROMATOGRAPHY

A rapid, simple, and scalable process was developed in our laboratory with a minimum number of sample handling steps for chromatographic capture

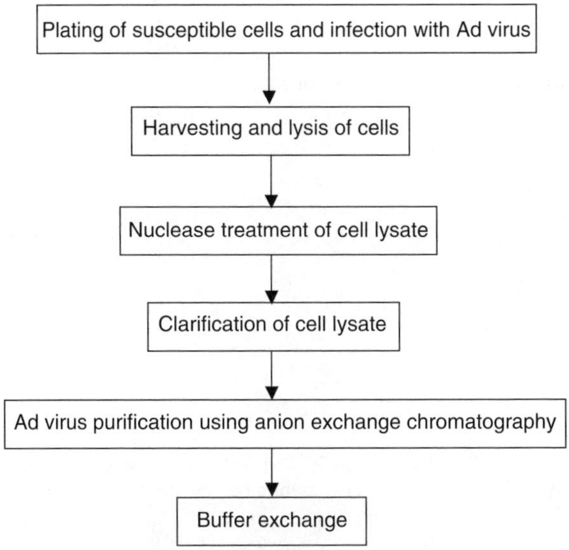

FIGURE 20.3 Flow diagram for capture of Ad from lysate.

of intact, infectious Ad viral particles using Mustang®* Q membranes (Figure 20.3). To measure the dynamic binding capacity of anion exchange membranes for Ad, CsCl gradient-purified Ad (1.24×10^{12} virus particles total) was loaded at various flow rates onto a Mustang Q Acrodisc^{+} with a membrane volume (MV) of 0.03 ml. Fractions were analyzed at 280 nm. Figure 20.4 shows a breakthrough curve by plotting the absorbance at 280 nm vs. time. Since the sample used was purified Ad with contaminating host cell proteins or nucleic acids below detection levels at this wavelength, an anion exchange HPLC assay provided a good correlation between the absorbance at 280 nm and the number of virus particles. Recently, Sweeney and Hennessey [65] reported a more accurate and robust spectrophotometric method for Ad particle quantitation. The breakthrough curve suggests that the dynamic binding capacity for Ad at 10% breakthrough at a flow rate of 3.0 ml/min or 100 MV/min was 1.9×10^{14} virus particles per ml of membrane. Up to 70% of the bound virus could be eluted following addition of 25 m*M* HEPES pH 7.4 buffer containing 1.0 *M* NaCl.

To measure the capacity of Mustang Q membranes for crude virus, DNase/RNase treated supernatant from a freeze–thaw lysate of Ad-infected 911 cells equivalent to ten 150 cm dishes were loaded onto a Mustang Q unit

*Mustang and Acrodisc are registered trademarks of Pall Corporation.

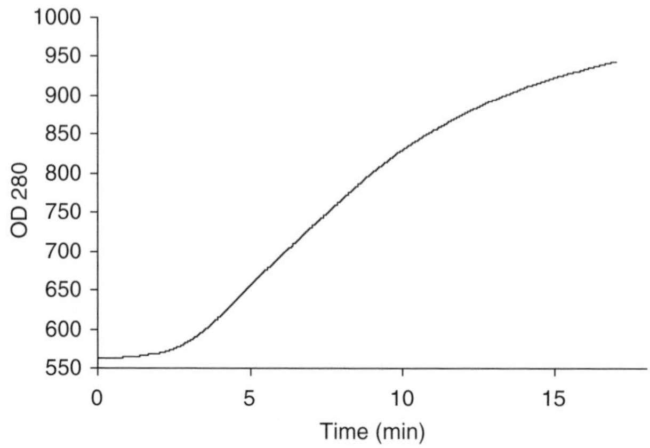

FIGURE 20.4 Dynamic binding capacity from breakthrough curve for the capture of CsCl-purified Ad on Mustang Q membranes. Purified Ad (1.24 × 10^{12} VP purified by CsCl centrifugation) were loaded onto a Mustang Q unit with a membrane volume of 0.03 ml using a buffer containing 25 m*M* HEPES, pH 7.4. The sample was loaded on the membranes using ÄKTA Purifier 100 system with Unicorn 3.2.1 software (Amersham Biosciences, Piscataway, NJ, USA). A flow rate of 3 ml per min was used. Virus breakthrough was monitored at 280 nm.

with a 0.03 ml MV in 25 m*M* HEPES pH 7.4 buffer containing of 0.2 *M* NaCl. Breakthrough was determined in the flow through fractions by real time PCR analysis. The results presented in Figure 20.5 show that the dynamic capacity for capture of Ad from crude cell lysates at 10% breakthrough was 4.9 × 10^{13} virus particles (VP) per ml of membrane at a flow rate of 100 MV/min. This has a profound influence on the process economics during scale-up. Evidently, one would require a much smaller anion exchange membrane chromatography device in order to capture Ad, compared to a conventional beaded chromatography media column at the manufacturing scale. For example, as titers of 10^{11} Ad particles per ml of cell culture media can now be routinely produced, a 1000 l batch that produces 10^{17} viral particles could be captured by a 2 l membrane chromatography unit in under an hour at a flow rate of 20 l/min. In contrast, based on the Ad capacities for anion exchange chromatography columns determined by Huyghe and coworkers [17], it would require a 100 l column to process that amount of Ad or one would have to perform 10 cycles on a 10 l column.

The results presented in Figure 20.6 show analytical size exclusion chromatograms of Ad purified from crude cell lysates by Mustang Q membrane chromatography or CsCl gradient centrifugation and following buffer exchange. The elution profiles were similar for both kinds of Ad preparations indicating

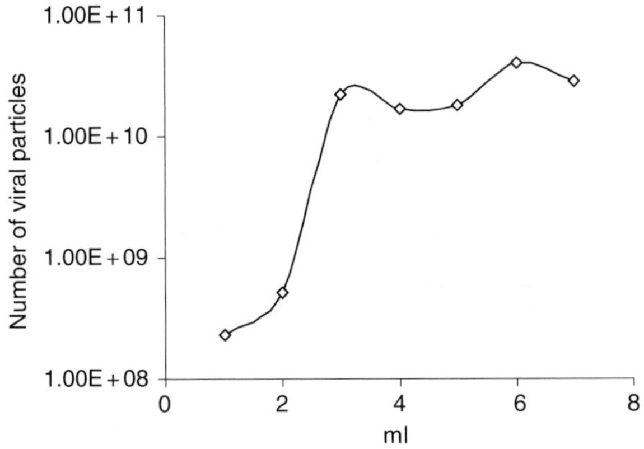

FIGURE 20.5 Dynamic binding capacity from breakthrough curve for the capture of Ad from lysate using real time PCR. Infected 911 cells equivalent to ten 15 cm dishes were subjected to 3 freeze–thaw cycles and the lysate centrifuged for 15 min at 5000*g* (4°C). The supernatant was incubated at room temperature for 30 min with 100 units DNase and 50 units RNase per ml of cell lysate followed by a filtration step using a PALL SuporCap™–50 Capsule (0.2 μm) to ensure that the lysate was free of particulate matter. The suspension was adjusted to a final concentration of 0.3 *M* NaCl. This crude lysate sample was applied directly onto a Mustang Q membrane equilibrated with 0.3 *M* NaCl in 25 m*M* HEPES, pH 7.4 to determine Ad breakthrough. Virus particles were detected by real-time PCR.

that Ad purified by strong anion exchange membranes was as clean as virus purified by standard CsCl centrifugation.

The attractive feature of the strong anion exchange membrane units is portrayed by the small amount of membrane volume needed for reliable capture of sizeable amounts of Ad from cell lysates. Also, purification of Ad vectors using membrane-based anion exchange chromatography is significantly faster and more cost-effective than the traditional CsCl protocol where an ultracentrifuge is needed as opposed to a syringe adaptable membrane chromatography unit. Also, protocols involving membrane-based anion exchange chromatography can easily be scaled up as in the plasmid DNA primary capture step [43].

One of the advantages of purification methods based on ion exchange chromatography compared to methods based on CsCl gradient centrifugation is the high ratio of the number of infectious viral particles compared to the total number of virus particles. Huyghe and coworkers [17] have reported ratios of 1:80 using DEAE anion exchange columns. Membrane anion exchange chromatography involving Mustang Q Acrodiscs on the other hand provided a ratio of

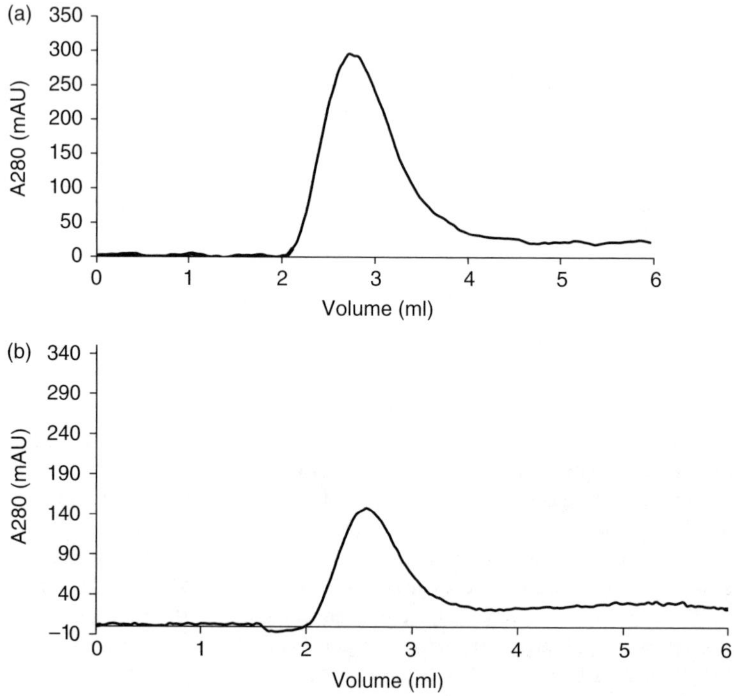

FIGURE 20.6 Analytical size exclusion chromatography of Ad purified by (A) CsCl gradient centrifugation and (B) Mustang Q anion exchange membrane chromatography. Ad particles purified by CsCl centrifugation (2.3×10^{12} VP total) or Mustang Q anion exchange chromatography (1.8×10^{11} VP total) were loaded onto a Amersham XK-16 column packed with Sepharose CL-4B (bed volume 2.0 ml) using a buffer containing 25 m*M* HEPES, pH 7.4. Virus elution was monitored at 280 nm.

1:9, (Table 20.1) indicating that this procedure was gentler on the virus than the procedures based on DEAE anion exchange columns.

A recent example showed rapid and efficient capture of Ad35 vector from a Benzonase and Triton X-100 treated 20 l cell culture supernatant containing 4×10^{15} VP on a 260 ml membrane volume Mustang Q capsule with 10-fold reduction in host cell proteins and 60 to 70% Ad35 recovery in one hour processing time [66]. Aggregation of Ad through association with host cell DNA during the purification process is of major concern as it impacts meeting regulatory guidelines for DNA levels in Ad dosage form. Konz et al. [67] have developed an Ad purification process that involves addition of polysorbate-80 throughout the process as well as spiking with 1 *M* sodium chloride at two intermediate steps in order to dissociate the DNA/Ad complex.

TABLE 20.1
Specific Infectivity of Ad Purified by CsCl and rAd Purified by MUSTANG Q Membrane Chromatography

	VP/ml	Plaque Forming Units/ml (PFU)	Ratio VP:PFU
CsCl	2.3×10^{13}	6.0×10^{11}	38:1
Mustang Q membrane	1.8×10^{12}	1.0×10^{11}	9:1

PFU/ml and VP/ml were determined as described (http://www.medschool.lsuhsc.edu/reiser/).

20.5 AAV PURIFICATION BY MEMBRANE ION EXCHANGE CHROMATOGRAPHY

Davidoff and coworkers [68] reported the first chromatographic purification method for AAV-8 vector particles based on membrane ion exchange chromatography involving Mustang Q units that generated vector stocks with >90% purity. The average yield of purified AAV-8 from five different vector preparations was 41% with an average dynamic binding capacity of approximately 1^{13} VP/ml of membrane. Electron microscopy of these purified stocks revealed typical icosohedral virions with <10% empty particles. The method took <5 h to process and it represents a significant advance over CsCl density gradient centrifugation-based techniques that are currently used for the purification of AAV-8 vector systems and will likely facilitate the transition of the AAV-8 vector system to the clinic.

Bataille et al. [69] showed scalability of membrane anion exchange chromatography for capture and purification of AAV2/1 from baculovirus insect cell culture medium. They observed a twofold reduction in host cell proteins as well as approximately 2-log reduction in DNA contamination with 70% AAV recovery of infectious titer units in the elution pool.

20.6 LENTIVIRAL VECTOR CAPTURE BY ANION EXCHANGE MEMBRANE CHROMATOGRAPHY

For the purification of LV vectors, Marino and coworkers [70] reported a preconcentration step that involved PEG precipitation of LV vector particles before capture on a small Mustang Q strong anion exchange membrane chromatography unit. Slepushkin and coworkers [40] showed that a VSV-G pseudotyped

HIV-1 based lentiviral vectors could be directly captured and eluted from clarified cell culture supernatants on a 60 ml membrane bed volume Mustang Q strong anion exchange membrane chromatography capsule without any preconcentration step. The vector was purified some 1000-fold as determined by a p24 ELISA assay with a 30% recovery of infectious particles. The purity and recovery of the vectors was similar to those purified using a size exclusion chromatography step. This latter method was used to purify the LV vectors that are currently being used in clinical trials [71]. The low recovery of infectious particles could be due to a combination of factors including variability in the infectivity assay and the presence of high salt in elution buffer that may compromise the integrity of the lentivirus envelope that is responsible for cellular attachment and fusion. In order to improve the infectious particle recovery during chromatographic purification of enveloped virus vectors, it may be beneficial to include 0.2 to 0.4% glycerol or 5 to 10% sucrose in the loading and elution buffers.

20.7 GENERAL MEMBRANE CHROMATOGRAPHY PURIFICATION OPTIMIZATION STRATEGIES FOR VIRAL VECTORS

20.7.1 VIRAL VECTOR CAPTURE

Optimization of the primary capture step from clarified cell culture lysate or supernatant should be performed on a scaled-down unit that is suitable for scouting the effect of loading pH, sodium chloride concentration, or conductivity in the load, wash and elution on product purity as well as yield. As the flow rate, in principle, has little effect on the dynamic binding capacity, a reasonable starting point is 10 to 20 MV/min.

One of the first parameters to optimize is the loading pH at which the vector binds to the membrane. If the isoelectric point (pI) of the target vector's surface proteins is known then one may use the strategy shown in Figure 20.7. In many cases this is rarely true. Hence, an optimum binding pH may be determined by testing loading in low conductivity (<8 mS/cm) at several different pH values. The operating pH range for most viral vectors is narrow as highly acidic and basic conditions adversely affect integrity of viral vectors. Therefore, bind- and elute conditions may be tested in the range between pH 6 and 9 on both anion exchange as well as cation exchange membranes. However, optimization of loading pH with Ad vectors is less complicated as they are known to bind to anion exchange chromatography media at pH > 6.5. Most viral vectors elute between 0.3 and 0.5 *M* NaCl with the exception of some lentiviral vectors that elute in a broad range between 0.5 and 1.5 *M* NaCl.

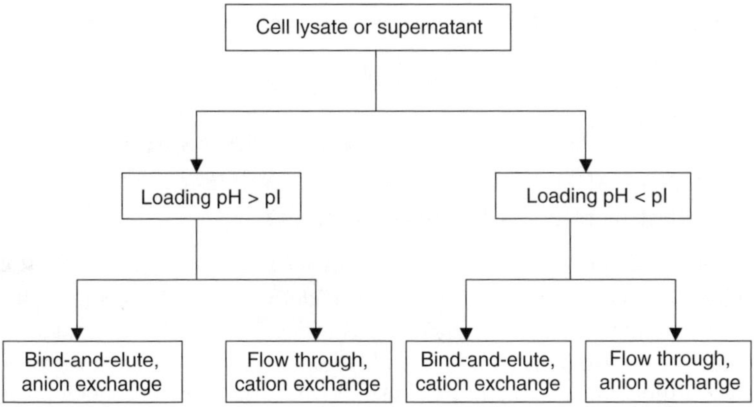

FIGURE 20.7 Flow chart for selecting ion exchange chemistry in viral vector purification.

20.7.2 Dynamic Binding Capacity

Accurate measurement of dynamic binding capacity of the target vector in the host cell lysate or supernatant would be an unrealistic expectation, given the complexity of the sample composition that may include host cell proteins and nucleic acids that interfere with the binding isotherm. However, an estimation of the dynamic binding capacity of the ion exchange membrane for the target vector on a scale-down unit would provide a reasonable control during scale-up. This may be accomplished by collecting flow through fractions during loading at 10 to 20 MV/min until the membrane is saturated. The dynamic binding capacity can then be calculated from a curve by plotting the number of viral particles versus the cell lysate or supernatant volume as the number of viral vector particles at 10% of saturation. Saturation may be defined as the number of viral particles that is at least 80% of that in the cell lysate or supernatant. Since virus vectors have a protein shell, online spectrophotometric monitoring at 280 nm wavelength is obscured by the host cell proteins and nucleic acids. A plaque assay or a real-time PCR assay would serve the purpose of measuring the number of viral particles in the flow through fractions. Several such binding isotherm experiments may be necessary in order to determine the dynamic binding capacity.

Once the bind and elute parameters are optimized together with the knowledge of an approximate dynamic binding capacity under the loading conditions that were tested during the process optimization, scale-up can be performed linearly with respect to the membrane volume. For example, if 5×10^{12} VP were bound to a 0.35 ml membrane unit from a cell lysate or supernatant at 3.5 ml/min, then 5×10^{13} VP could be bound on a 10 ml membrane unit at

100 ml/min or 1.3×10^{15} VP could be bound on a 260 ml unit at 2.6 l/min flow rate.

20.8 DOWNSTREAM PROCESSING AND PLASMID DNA CAPTURE BY ANION-EXCHANGE MEMBRANE CHROMATOGRAPHY

In addition to the target plasmid DNA, bacterial cell lysates contain RNA, cellular DNA, cellular proteins as well as endotoxins. While cationic cellular proteins will not bind to anion exchange chromatography media, separation of anionic proteins from the target plasmid DNA can be accomplished with a high salt wash following capture of the plasmid DNA. Endotoxins contain polyanionic and hydrophobic regions [72]. Hence, a successful strategy would entail an integrated multistep approach for endotoxin reduction to acceptable regulatory levels in the final formulation. Since both DNA and RNA are polyanionic, their separation based on anion exchange chromatography is particularly challenging. The complexity of separating plasmid DNA from RNA and genomic DNA requires utilization of several relevant purification technologies. Thus, as with endotoxins, effective removal of RNA and genomic DNA must be built into the overall downstream purification operation.

Large-scale production of plasmid DNA for clinical trials is commonly performed using bacterial cell pellets produced in large fermentors. This ultimately results in large volumes of bacterial cell lysates containing dilute quantities of plasmid DNA requiring efficient capture of the target molecule from a dilute feed stream at fast flow rates. In membrane chromatography, the mass transfer rates of large biomolecules such as plasmid DNA are not diffusion-limited due to the convective pores. Hence plasmid DNA can be efficiently captured from a large cell lysate pool at 10- to 20-fold faster flow rates and 10-fold higher dynamic binding capacities compared to beaded chromatography media [43]. This results in smaller chromatography units, shorter cycle time, higher productivity, and less raw material usage. Zhang and coworkers [43] reported on the capture of 1.52 g of a 4.5 kb plasmid DNA from 71 l of bacterial cell lysate on a 260 ml bed volume anion exchange membrane capsule with a yield of 94.7%. Although the eluted plasmid pool contained approximately 10% RNA contamination, a subsequent orthogonal chromatography step such as hydrophobic interaction chromatography (HIC) could be used to further purify the product.

20.9 CONCLUSIONS

Purification of large biomolecules including gene therapy vectors present significant challenges when conventional chromatographic resins are used due to their low binding capacities and limited flow rates. The examples presented in this chapter indicate that because membrane ion exchange chromatography media have convective pores they offer binding capacities that are up to an order of magnitude higher than those observed with conventional chromatography resins. Also membrane-based approaches have much higher flow rates, an economic factor compared to column chromatography. This makes the membrane chromatography-based approach practical and economical, attractive alternatives for the capture of viral vectors and plasmid DNAs from cell lysates or cell supernatants. Since the membrane chromatography media discussed in this chapter are scalable, they may be used in both disposable as well as reusable settings in a manufacturing process.

Although membrane chromatography offers clear advantages over conventional chromatography media, the widespread usage of membrane chromatography is in its infancy but gaining recognition as a very useful purification tool. However, membrane chromatography alone cannot achieve the purity and safety required for *in vivo* clinical studies. It is more than likely that a combination of membrane and conventional chromatography as well as in some cases, size-based separations like TFF will produce gene therapy vectors of the required purity and safety levels for clinical evaluation studies.

ABBREVIATIONS

AAV	adeno-associated virus
Ad	adenovirus
AIDS	acquired immunodeficiency syndrome
CsCl	cesium chloride
DEAE	diethylaminoethyl
DNA	deoxyribonucleic acid
ELISA	enzyme-linked immuno sorbent assay
g	gram
HSV	herpes simplex virus
kb	kilobase
l	liters
LV	lentivirus
ml	milliliters
MoMLV	Moloney Murine Leukemia viruses
mS	milli Siemens
MV	membrane volume
nm	nanometer
PCR	polymerase chain reaction
PEG	polyethylene glycol

PEI	polyethyleneimine
pI	isoelectric point
pfu	plaque forming unit
Q	quaternary ammonium
RNA	ribonucleic acid
SDS	sodium dodecyl sulfate
TFF	tangential flow filtration
TMAE	trimethylaminoethyl
VP	virus particles
VSV	vesicular stomatitis virus

REFERENCES

1. Thomas CE, Ehrhardt A, and Kay MA. Progress and problems with the use of viral vectors for gene therapy. *Nat. Rev. Genet.* 2003, 4:346–358.
2. Robbins PD, Tahara H, and Ghivizzani SC. Viral vectors for gene therapy. *Trends Biotechnol.* 1998, 16:35–40.
3. http://www.wiley.co.uk/genetherapy/clinical/ (accessed January 5, 2005).
4. Shenk TE. Adenoviridae: The viruses and their replication. In: *Virology.* Kinpe DM and Howley PM, Eds. Lipincott Williams and Wilkins, Philadelphia, pp. 2201, 2265–2300.
5. Morsy MA, Gu M, Motzel S, Zhao J, Lin J, Su Q, Allen H, Franlin L, Parks RJ, and Graham FL. An adenoviral vector deleted for all viral coding sequences results in enhanced safety and extended expression of a leptin transgene. *Proc. Natl. Acad. Sci. USA* 1998, 95:7866–7871.
6. Schiedner G, Morral N, Parks RJ, Wu Y, Koopmans SC, Langston C, Graham FL, Beaudet AL, and Kochanek S. Genomic DNA transfer with a high capacity adenovirus vector results in improved *in vivo* gene expression and decreased toxicity. *Nat. Genet.* 1998, 18:180–183.
7. Xiao X, Li J, and Samulski RJ. Production of high-titer recombinant adeno-associated virus vectors in the absence of helper adenovirus. *J. Virol.* 1998, 72:2224–2232.
8. Grimm D, Kern A, Rittner K, and Kleinschmidt JA. Novel tools for production and purification of recombinant adeno-associated virus vectors. *Hum. Gene Ther.* 1998, 9:2745–2760.
9. Lundstrom K. Latest development in viral vectors for gene therapy. *Trends Biotechnol.* 2003, 21:117–122.
10. Blouin V, Brument N, Toublanc E, Raimbaud I, Moullier P, and Salvetti A. Improving rAAV production and purification: towards the definition of a scaleable process. *J. Gene Med.* 2004, 6:S223–S228.
11. Atkinson M and Christensen J. Process and product development in the manufacturing of molecular therapeutics. *Curr. Opin. Mol. Ther.* 1999, 1:422–429.

12. Reiser J, Harmison G, Kluepfel-Stahl S, Brady RO, Karlsson S, and Schubert M. Transduction of nondividing cells using pseudotyped defective high-titer HIV type 1 particles. *Proc. Natl. Acad. Sci. USA* 1996, 93:15266–15271.
13. VandenDriessche T, Naldini L, Collen D, and Chuah MKL. Oncoretroviral and lentiviral vector-mediated gene therapy. In: *Methods in Enzymology,* Vol. 346, New York: Academic Press, 2002, pp. 573–589.
14. McTaggart S and Al-Rubeai M. Retroviral vectors for human gene delivery. *Biotechnol. Adv.* 2002; 20:1–31.
15. Sambrook J and Russell DW. *Molecular Cloning: A Laboratory Manual.* 2nd ed., Cold Spring Harbor, NY: Cold Spring Harbor Laboratory Press, 2001, pp. 1.1–1.170.
16. Powell K. DNA Vaccines-back in the saddle again. *Nat. Biotechnol.* 2004, 22:799–801.
17. Huyghe BG, Liu X, Sutjipto S, Sugarman BJ, Horn MT, Shepard MH, Scandella CJ, and Shabram P. Purification of a type 5 recombinant adenovirus encoding human p53 by column chromatography. *Hum. Gene Ther.* 1995, 6:1402–1416.
18. Klyushnichenko V, Bernier A, Kamen A, and Harmsen E. Improved high-performance liquid chromatographic method in the analysis of adenovirus particles. *J. Chromatogr. B* 2001, 755:27–36.
19. Wanlin L, Tuma R, Thomas GJ, and Bamford DH. Purification of viruses and macromolecular assemblies for the structural investigatios using a novel ion exchange method. *Virology* 1994, 201:1–7.
20. Hewish D and Shukla D. Purification of barley yellow dwarf virus by gel filtration on Sephacryl S-1000 Superfine. *J. Virol. Meth.* 1983, 7:223–228.
21. Hjorth R and Moreno-Lopez J. Purification of bovine papilloma virus by gel filtration on Sephacryl S-1000 Superfine. *J. Virol. Meth.* 1982, 5:151–158.
22. Goerke AR, To BCS, Lee AL, Sagar SL, and Konz JO. Development of a novel adenovirus purification process utilizing selective precipitation of cellular DNA. *Biotechnol. Bioeng.* 2005, 83:45–52.
23. Green AP, Huang JJ, Scott MO, Kiersted TD, Beaupre I, Gao G-P, and Wilson JM. A New scalable method for the purification of recombinant adenovirus vectors. *Hum. Gene Ther.* 2002, 13:1921–1934.
24. Kamen A and Henry O. Development and optimization of an adenovirus production process. *J. Gene Med.* 2004, 6:S184–S192.
25. Arcand N, Bernier A, and Transfiguracion J. Adenovirus type 5 (Ad5) chromatographic purification process at 20L scale. *Bioprocess. J.* 2003, 2:72–75.
26. Xie Q, Hare J, Turnigan J, and Chapman MS. Large-scale production, purification and crystallization of wild-type adeno-associated Virus-2. *J. Virol. Meth.* 2004, 122:17–27.
27. Auricchio A, O'Connor E, Hildinger M, and Wilson JM. A single-step affinity column for purification of serotype-5 based adeno-associated viral vectors. *Mol. Ther.* 2001, 4:372–374.
28. Clark RK, Liu X, McGrath JP, and Johnson PR. Highly purified recombinant adeno-associated virus vectors are biologically active and free of detectable helper and wild-type viruses. *Hum. Gene Ther.* 1999, 10:1031–1039.

29. Zolotukhin S, Byrne BJ, Mason E, Zolotukhin I, Potter M, Chesnuut K, Summerford C, Samulski RJ, and Muzycka N. Recombinant adeno-associated virus purification using novel methods improves infectious titer and yield. *Gene Ther.* 1999, 6:973–985.
30. Synder RO and Flotte TR. Production of clinical-grade recombinant adeno-associated virus vectors. *Curr. Opin. Biotechnol.* 2002, 13:418–423.
31. Gao G, Qu G, Burnham MS, Huang J, Chirmule N, Joshi B, Yu, Q-C, Marsh JA, Conceicao CM, and Wilson JM. Purification of recombinant adeno-associated virus vectors by column chromatography and its performance *in vivo*. *Hum. Gene Ther.* 2000, 11:2079–2091.
32. Kaludov N, Handelman B, and Chiorini J. Scalable purification of adeno-associated virus type 2, 4, or 5 using ion-exchange chromatography. *Hum. Gene Ther.* 2002, 13:1235–1243.
33. Brument N, Morenweiser R, Veronique B, Toublanc E, Raimbaud I, Cherel Y, Folliot S, Gaden F, Boulanger P, Kroner-Lux G, Moullier P, Rolling F, and Salvetti. A versatile and scalable two-step ion-exchange chromatography process for the purification of recombinant adeno-associated virus Serotypes-2 and -5. *Mol. Ther.* 2002, 6:678–686.
34. Smith RH, Ding C, and Kotin RM. Serum-free production and column purification of adeno-associated virus type 5. *J. Virol. Meth.* 2003, 114:115–124.
35. Baekelandt T, Eggermont K, Michiels M, Nuttin B, and Debyser Z. Optimized lentiviral vector production and purification procedure prevents immune response after transduction of mouse brain. *Gene Ther.* 2003, 10:1933–1940.
36. Yamada K, MacCarty DM, Madden VJ, and Walsh CE. Lentivirus vector purification using anion exchange HPLC leads to improved gene transfer. *Biotechniques* 2003, 34:1074–1078.
37. Scherr M, Battmer K, Eder M, Schule S, Hohenberg H, Gasner A, Grez M, and Blomer U. Efficient gene transfer into the CNS by lentiviral vectors purified by anion exchange chromatography. *Gene Ther.* 2002, 9:1708–1714.
38. Schauber, CA, Tuerk MJ, Pacheco CD, Escarpe PA, and Veres, G. Lentiviral vectors pseudotyped with baculovirus gp64 efficiently transduce mouse cells in vivo and show tropism restriction against hematopoietic cell types *in vitro*. *Gene Ther.* 2004, 11:266–275.
39. Richieri SP, Bartholomew R, Aloia R, Savary J, Gore R, Holt J, Ferre F, Musil R, Tian HR, Trauger R, Lowry P, Jensen F, Carlo DJ, Maigetter RZ, and Prior CP. Characterization of highly purified, inactivated HIV-1 particles iolated by anion exchange chromatography. *Vaccine* 1998, 16:119–129.
40. Slepushkin V, Chang N, Cohen R, Gan Y, Jiang B, Deausen E, Berlinger Dbinder G, Andre K, Humeau L, and Dropolic B. Large-scale purification of a lentiviral vector by size exclusion chromatography or mustang Q ion exchange capsule. *Bioprocessing J.* 2003, 2:89–95.
41. Scheelf M. Issues of large-scale plasmid DNA manufacturing. In: Rohm HJ and Reed G, Eds. *Recombinant Proteins, Monoclonal Antibodies and Therapeutic Genes.* New York: Wiley-VCH, 1999, pp. 443–469.

42. Birnboim HC and Doly J. A rapid alkaline extraction procedure for screening recombinant plasmid DNA. *Nucleic Acids Res.* 1979, 7:1513–1523.
43. Zhang S, Krivosheyeva A, and Nochumson S. Large-scale capture and partial purification of plasmid DNA using anion exchange membrane capsules. *Biotechnol. Appl. Biochem.* 2003, 37:245–249.
44. Eon-Duval A and Burke G. Purification of pharmaceutical-grade plasmid DNA by anion exchange chromatography in an RNase-free process. *J. Chromatogr. B* 2004, 804:327–335.
45. Prazeres DMF, Schluep T, and Cooney C. Preparative purification of supercoiled plasmid DNA using anion exchange chromatography. *J. Chromatogr. A* 1998, 806:31–45.
46. Levy MS, Collins IJ, Tsai JT, Shamlou PA, Ward JM, and Dunnill P. Removal of contaminant nucleic acids by nitrocellulose filtration during pharmaceutical-grade plasmid DNA processing. *J. Biotechnol.* 2000, 76:197–205.
47. Johnson TR and Han J. Large-scale isolation of plasmid DNA and purification of λ phage DNA using hydroxyapatite chromatography. *Anal. Biochem.* 1983, 132:20–25.
48. Giovannini R and Freitag R. Comparison of different types of ceramic hydroxyapatite for the chromatographic separation of plasmid DNA and a recombinant anti-rhesus D antibody. *Bioseparation* 2001, 9:359–368.
49. Sagar SL, Watson MP, and Lee AL. Case study: capacity challenges in chromatography-based purification of plasmid DNA. In: Rathore AS and Velayudhan A, Eds. *Scale-Up and Optimization in Preparative Chromatography.* New York: Marcel Dekker, 2003, pp. 251–272.
50. Horn N, Meek JA, Budahazi G, and Marquet M. Cancer gene therapy using plasmid DNA: purification of DNA for human clinical trials. *Hum. Gene Ther.* 1995, 6:565–573.
51. Lander RJ, Winters MA, Meacle FJ, Buckland BC, and Lee AL. Fractional precipitation of plasmid DNA from lysate by CTAB. *Biotechnol. Bioeng.* 2002, 79:776–784.
52. Eon-Duval A, MacDuff RH, Fisher CA, Harris MJ, and Brook C. Removal of RNA impurities by tangential flow filtration in an RNase-free plasmid DNA purification process. *Anal. Biochem.* 2003, 316:66–73.
53. Kahn DW, Butler MD, Cohen DL, Gordon L, Kahn JW, and Winkler ME. Purification of plasmid DNA by tangential flow filtration. *Biotechnol Bioeng* 2000, 69:101–105.
54. Diogo MM, Queiroz JA, and Prazeres DMF. Chromatography of plasmid DNA. *J. Chromatogr. A* 2005, 1069:3–22.
55. Thommes J and Kula MR. Membrane chromatography — an integrative concept in the downstream processing of proteins. *Biotechnol Prog* 1995, 11:357–367.
56. Charcosset C. Purification of proteins by membrane chromatography. *J. Chem. Technol. Biotechnol.* 1998, 71, 95–110.
57. Ghose R. Protein separation using membrane chromatography: opportunities and challenges. *J. Chromatogr. A* 2002, 952:13–27.
58. Klein E. Affinity membranes a 10-year review. *J. Membr. Sci.* 2000, 179:1–27.

59. Ljunglof A, Bergvall P, Bhikhabhai R, and Hjorth R. Direct visualization of plasmid DNA in individual chromatography adsorbent particles by confocal scanning laser microscopy. *J. Chromatogr. A* 1999, 844: 129–135.
60. Enders HN, Johnson JA, Ross CA, Welp JK, and Etzel MR. Evaluation of an ion exchange membrane for the purification of plasmid DNA. *Biotechnol. Appl. Biochem.* 2003, 37:259–266.
61. Roper DK and Lightfoot EN. Estimating plate heights in stacked-membrane chromatography by flow reversal. *J. Chromatogr. A* 1995, 702:69–80.
62. Teeters MA, Conrardy SE, Thomas BL, Root TW, and Lightfoot EN. Adsorptive membrane chromatography for purification of plasmid DNA. *J. Chromatogr. A* 2003, 989:165–173.
63. Grunwald AG and Shields MS. Plasmid purification using membrane-based anion-exchange chromatography. *Anal. Biochem.* 2001, 296:138–141.
64. Bencina M, Podgornik A, and Strancar A. Characterization of methacrylate monoliths for purification of DNA molecules. *J. Sep. Sci.* 2004, 27:801–810.
65. Sweeney JA and Hennessey JP. Evaluation of accuracy and precision of adenovirus absorptivity at 260 nm under conditions of complete DNA disruption. *J. Virol.* 2002, 295:284–288.
66. Weggeman M. Development of an Ad35-based Malaria Candidate Vaccine. 2nd Annual Viral Vectors and Vaccines Conference, the Williamsburg Bioprocessing Foundation, WilBio–Europe, Amsterdam, May 25–27, 2005.
67. Konz JO, Lee AL, Lewis JA, and Sagar SL. Development of a purification process for adenovirus: controlling virus aggregation to improve the clearance of host cell DNA. *Biotechnol. Prog.* 2005, 21:466-472.
68. Davidoff AM, Ng CYC, Sleep S, Gray J, Azam S, Zhao Y, McIntosh JH, Karimipoor M, and Nathwani AC. Purification of recombinant adeno-associated virus type 8 vectors by ion-exchange chromatography generates clinical grade vector stock. *J. Virol. Meth.* 2005, 121:209–215.
69. Bataille D, Bortolussi L, Rifki M, Danos O, and Merten OW. Evaluation of Mustang Ion-Exchange Capsules for the Purification of rAAV-2/1 Produced by the Baculovirus/Insect Cell System. 8th Annual Baculovirus and Insect Cell Culture Conference, the Williamsburg Bioprocessing Foundation, Savannah, GA, February 21–24, 2005.
70. Marino MP, Kutner RH, Lajmi AR, Nochumson S, and Reiser J. Development of scalable purification protocols for lentiviral vectors. *Mol. Ther.* 2003, 7:S178.
71. Manilla P, Rebello T, Afable C, Lu X, Slepushkin V, Humeau LM, Schonely K, Ni Y, Binder GK, Levine BL, MacGregor RR, June CH, and Dropulic B. Regulatory considerations for novel gene therapy products: a review of the process leading to the first clinical lentiviral vector. *Hum. Gene Ther.* 2005, 16:17–25.
72. Petsch D and Anspach FB. Endotoxin removal from protein solutions. *J. Biotechnol.* 2000, 76:97–119.

Index

C

D

E

F

G

J

K

L

M

N

O

P

Q

R

S

T

U

V

W